VLSI 91

Proceedings of the IFIP TC10/WG 10.5 International Conference on
Very Large Scale Integration
Edinburgh, Scotland, 20-22 August, 1991

Edited by

A. HALAAS
University of Trondheim
Trondheim, Norway

P. B. DENYER
University of Edinburgh
Edinburgh, Scotland

1992

NORTH-HOLLAND
AMSTERDAM • LONDON • NEW YORK • TOKYO

ELSEVIER SCIENCE PUBLISHERS B.V.
Sara Burgerhartstraat 25
P.O. Box 211, 1000 AE Amsterdam, The Netherlands

Distributors for the United States and Canada:
ELSEVIER SCIENCE PUBLISHING COMPANY INC.
655 Avenue of the Americas
New York, N.Y. 10010, U.S.A.

ISBN: 0 444 89019 X
ISSN: 0926-5473

Printed in The Netherlands

Dedicated to
the Memory of
Sidney Michaelson

PREFACE

Ten years ago, IFIP (International Federation of Information Processing) asked Professor Sidney Michaelson to establish a working group on "VLSI". The aim of this group should be to initiate conferences and workshops all over the world to promote the development of VLSI. This has been achieved with great success and the biannual series of conferences VLSI 81 through to VLSI 89 has been recognised for its high quality. VLSI 91 comes full circle, returning the conference to Edinburgh, the venue for the original VLSI 81.

In 1981 Sidney said that "The major problem in VLSI is really the control of complexity. The hardest part is the control of autonomous yet interacting processes. We do not yet have satisfactory techniques for handling that sort of thing, but I think the techniques we need to develop are independent of whether you are programming or designing a chip".

This statement is still valid. Even if a decade has passed where the focus in the VLSI arena has been on tools of all kinds to assist our imperfect minds and hands, we certainly still have a long way to go before we have become Masters of Complexity.

The papers chosen for this conference by the hard working International Programme Committee reflect the continuing interest in improving design tools and the wide range of engineering concerns which surround the effective exploitation of VLSI.

Professor Michaelson was the initiator of VLSI 91, a conference that for us also would be an opportunity to show him our deep respect in his year of retirement. It was a sad moment during our preparations for VLSI 91 in February that we heard that Sidney had died. His firm voice of justice, and his particular care for the young and less privileged will for many of us be strongly linked to his memory.

These proceedings take a few steps further in our common understanding.

Professor Arne Halaas,
Programme Chairman

ORGANIZATION

Programme Chairman

A. Halaas
University of Trondheim

Programme Committee

G. Birtwistle
University of Calgary

P. B. Denyer (Deputy Chairman)
University of Edinburgh

M. Fourman
University of Edinburgh

C. E. Goutis
University of Patras

J. P. Gray
Algotronix

P. Michel
Siemens

S. Murai
Mitsubishi

T. Nishimukai
Hitachi

G. Saucier
INPG
Grenoble

C. H. Sequin
University of Berkeley

J. Staunstrup
DTH

T. Yanagawa
NEC Corporation

General Chairman

I. Barron
Division Ltd

Organising Committee

Chairman and Treasurer
D. J. Rees
University of Edinburgh

Tutorials
M. Fourman
University of Edinburgh

Proceedings
P. B. Denyer
University of Edinburgh

Exhibitions Coordinator
R. A. McKenzie
University of Edinburgh

Demonstrations Coordinator
D. Rogers
University of Edinburgh

Publicity
A. D. Milne
Wolfson Microelectronics Ltd

Sponsorship
J. P. Gray
Algotronix Ltd

Conference Secretariat

CEP Consultants Ltd
26-28 Albany Street
Edinburgh
EH1 3QH

CONTENTS

Session 8a Routing

Session 8b VLSI Arrays

Keynote Paper

Session 9 Circuit Design 1

Session 10 Circuit Design 2

Session 11 Logic Synthesis and Timing Optimisation

Session 12 Fault Tolerant Arrays

VLSI 91
A. Halaas and P.B. Denyer (Eds.)
Elsevier Science Publishers B.V. (North-Holland)

A Regularly Structured 54-bit Modified-Wallace-Tree Multiplier

T. Sato, M. Nakajima*, T. Sukemura*, G. Goto

Fujitsu Laboratories Ltd., 10-1 Morinosato-Wakamiya, Atsugi 243-01, JAPAN
*Fujitsu Limited, 1015 Kamikodanaka Nakahara-ku, Kawasaki 211, JAPAN

Abstract

This paper describes a regularized structure of a Wallace tree multiplier. Because of the regular structure, the tree multiplier can be designed more easily than a conventional one while retaining the high speed capability of the Wallace tree multiplier. We applied this method to a 54-by-54-bit multiplier. The total pattern area was 3.36 X 3.85 mm, achieving a high density of 6,400 transistors/mm^2 implemented with 0.8-μm triple-metal-layer CMOS technology. The multiplication time of the device is 15 ns, as fast as a previously reported 32-bit multiplier.

Introduction

Recently, high-speed arithmetic operation circuits have been required for high performance computer systems. In many cases, the performance of the system depends on the multiplication time. In order to decrease the latency of the multiplier, parallel multipliers are strongly desired.

To date, two type of parallel multipliers, "array" multipliers and "tree" multipliers, are widely used. The array multipliers are very regular, easy to layout, but suffer from long latency and propagation delay times[1]. On the other hand, the tree multipliers[2] have very high speeds, because they add as many partial products in parallel as possible. However, until now, tree structures have been thought to be too irregular to layout easily[3].

Our goal is to break down "tree" multipliers into recurring structures. By introducing regularity in the Wallace tree multipliers, they become far easier to layout, more compact, and, hence, faster than conventional irregular tree multipliers.

Discussion of Wallace tree

Wallace tree is well known to be effective in constructing fast parallel multipliers at the sacrifice of circuit regularity. The irregular structure of the conventional Wallace-tree multiplier is shown in Fig. 1 for a 54-by-54-bit parallel multiplier which utilizes Booth's 2nd algorithm[5]. In this case, the maximum number of partial products to be added per bit is 28, including one for correcting sign extension terms. A regular structure of modified Wallace-tree multiplier proposed by the authors is shown in Fig. 2. To introduce

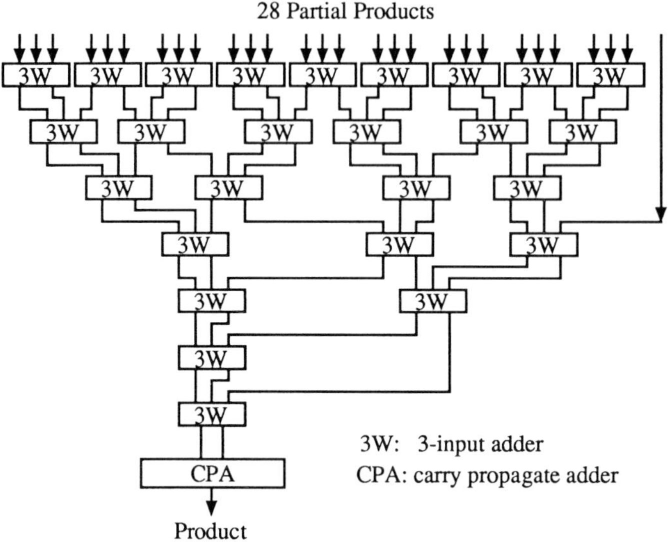

Fig. 1 Conventional structure of Wallace-tree multiplier

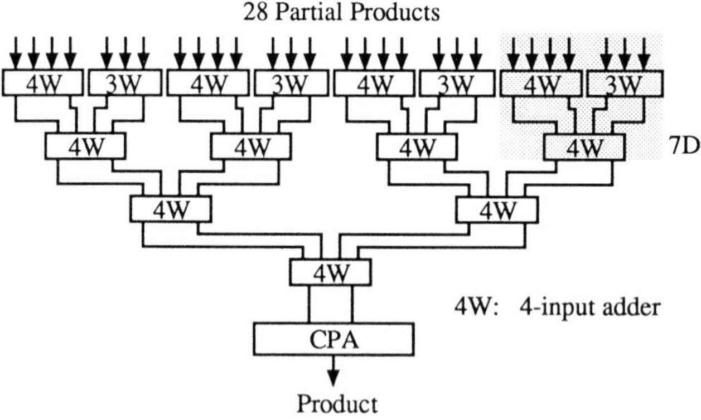

Fig. 2 Regular structure of modified Wallace-tree multiplier

regularity in the construction of the conventional Wallace tree architecture, we have combined together two type of compressors, a 3W(3-2 compressor) and a 4W(4-2 compressor). In this way 28 partial products can be processed with the same 4 circuit groups(the gray area 7D), each of which processes 7 partial products. By using the 7D block as a unit, the multiplier as a whole can be designed hierarchically. Simplifying of the Wallace tree into such repeatable blocks, makes it possible to design a tree multiplier as easily as a conventional array-type multiplier.

Architecture

Figure 3 shows a block diagram of the regular tree multiplier. It consists of four 7D rows, three 4W rows, Booth decoders, wire shifters, and a final carry propagation adder(CPA).

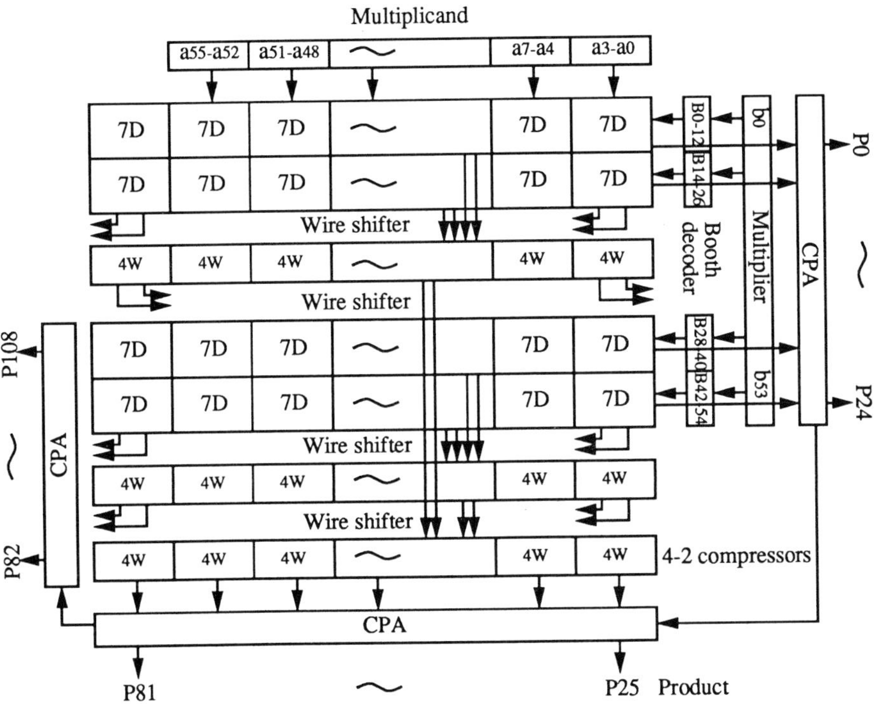

Fig. 3 Block diagram of regular tree multiplier

Almost all area of the die is occupied by the 7D array. A block diagram of the 7D unit is shown in Fig. 4. It produces seven partial products and sums them up. Using Booth's 2nd algorithm, each partial product is processed by a Booth selector(P). Addition is done by the 4-2 compressors(4W) which are shown in Fig. 5 and the full adders(3W). The P blocks are located near 3W

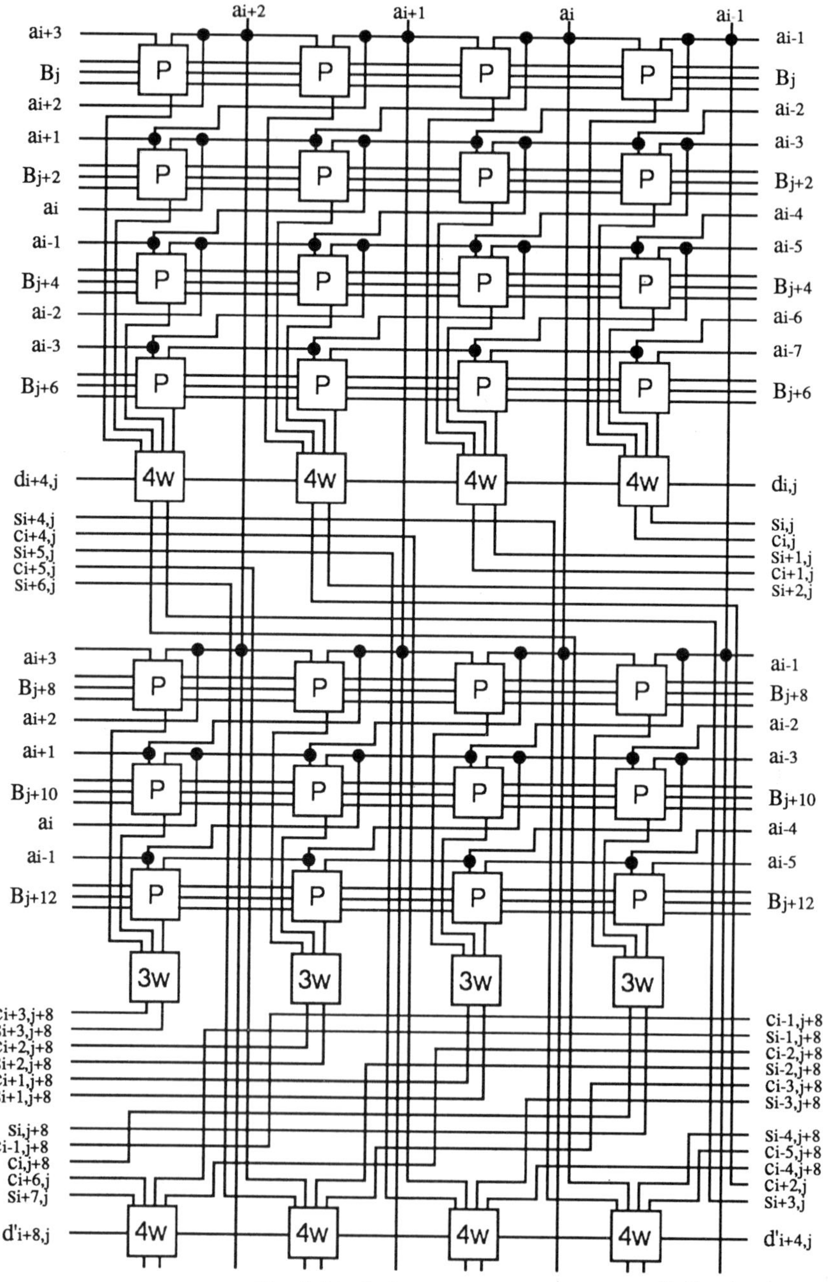

Fig. 4 Block diagram of the unit 7D P: Booth selector

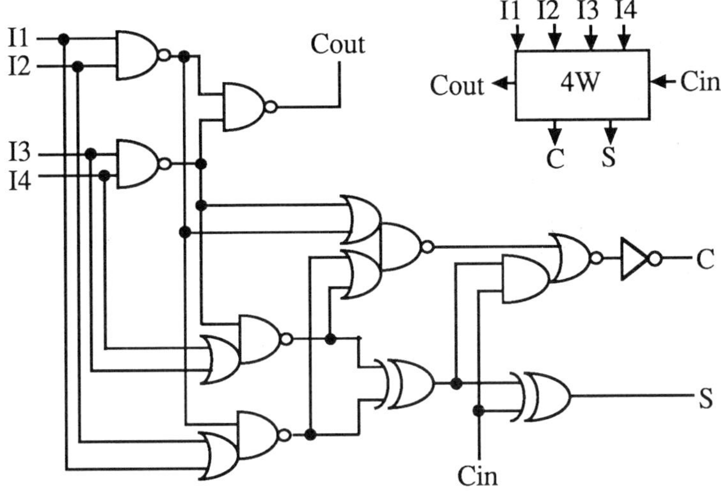

Fig. 5 The 4-2 compressor

and 4W blocks to shorten the critical paths. If the Booth selectors and adders are separated as is the case for the conventional Wallace tree design, the multiplier will need a wide area to pass bus lines through to the trees, and so the delay times of these long critical paths will destroy any high speed capability of the Wallace tree. Figure 4 shows that not only a logical block but also a layout block of 7D can be iteratively used to construct a Modified Wallace tree with minimum design effort. The layout process of a 7D row consists of the following two steps:

i) layout of the 7D macro block including signal lines

ii) place the blocks in order

Therefore the whole layout is quite regular. The flow of signal lines for order matching is also important in the design 7D. In conventional designs, all carry signals($c_{x,x}$) are shifted one place toward the higher bit position (left in Fig. 4) and all sum signals($s_{x,x}$) remain fixed. So the whole multiplier would have a parallelogram shape and waste a lot of area. In Fig. 4, the carry and sum signals are shifted in both directions to make the multiplier a rectangle, thus eliminating wasted area. Therefore, a holding process[6] is not needed and the critical path has effectively been shortened.

The wire shifters are areas for wiring in which carry and sum signals, produced by 4W and 7D, are shifted to adjust ordering in the bit position. In each of the four shift layers, all signals are shifted in a same pattern. Therefore, the shifter areas are also laid out regularly.

The 4W is a 4-2 compressor, which can sum up four partial products concurrently. The compressor has two merits, small die area, and a short propagation delay time. If a full adder(3W) were adopted instead of the 4W, the 3W multiplier would suffer from an increase in vertical signal lines. In our physical design, the width of the multiplier is limited by area needed for vertical signal lines. So a 3W multiplier would become wider than our 4W multiplier. Therefore, 4W is the better choice for the regular tree multiplier than 3W. The latter advantage is discussed in reference [4].

The whole tree is integrated in an area of 3.31 X 3.44 mm with 75,600 transistors. Circuit simulation of the critical path has been done. The estimated propagation delay time of this 54-bit tree is 6.3 ns.

The final carry propagation adder is the new transmission gate adder[7]. It is manufactured on an area of 3.36 X 0.41 mm with 6,900 transistors. Its estimated propagation delay time is 8.5 ns.

The block diagram of the 54-by-54-bit multiplier chip is shown in Fig. 6. Multiplication time can be examined by shifting the phase difference between two clocks, clk0 and clk1.

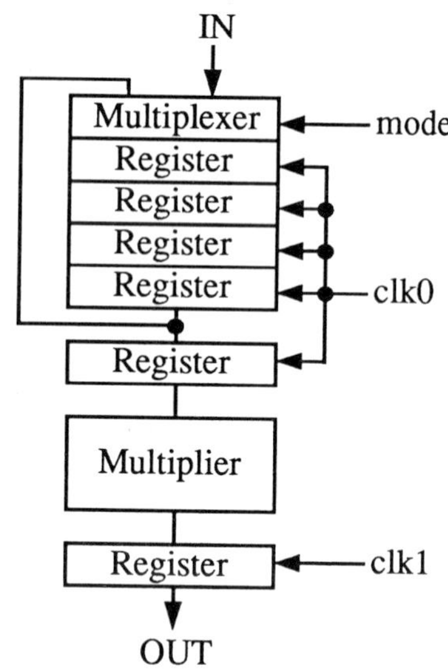

Fig. 6 Block diagram of the chip

Fabrication

The chip has been fabricated using 0.8-µm triple-metal-layer CMOS technology. The design rules are listed in table 1 and a photograph of the multiplier is shown in Fig. 7. The 3.36 X 3.85 mm die was fabricated with 82,547 transistors, this produces a transistor density of 6,400 transistors/mm^2. The regularity of the structure can be clearly seen in the chip layout. The manual layout stage could be dramatically reduced to three weeks as compared with three months when using the conventional design. Most of the design time was devoted to designing blocks such as 7D and 4W.

Table 1 Design rules

Gate length	0.8µm
1st metal pitch	2.5µm
1st metal width	1.0µm
2nd metal pitch	3.1µm
2nd metal width	1.3µm
3rd metal pitch	5.0µm
3rd metal width	2.2µm

Comparison with other approach

Many parallel multipliers have been reported up to date. Some of them are summarized in table 2. Our multiplier is as fast as a 32-bit multiplier[4] based on almost the same process technology and achieves the same density as the same bit-width multiplier based on more advanced 0.5-µm process

Fig. 7 Photograph of the multiplier

Table 2 Comparison of multiplier

Construction of multiplier	Area (mm^2)	Density (Trs/mm^2)	Performance (ns)	Process technology	Reference
32 X 32 (Toshiba)	7.3	3800	15	0.8μm CMOS	[4]
56b MLT-ACM (LSI Logic)	22.1	4100	30	1μm CMOS	[6]
54 X 54 (Toshiba)	12.6	6500	10	0.5μm pseudo-CMOS double metal	[8]
54 X 54 (Fujitsu)	12.9	6400	15	0.8μm CMOS triple metal	VLSI 91

technology[8]. The Toshiba 54 X 54 multiplier is faster than ours, but it uses a p-channel load circuit in the final carry propagation adder and suffers from a through current. Because of the multiplier's high regularity, the design was implemented faster and integrated more densely.

Conclusion

A tree multiplier can be successfully designed from regular, recurring blocks. The structure is quite regular and is easy to layout. As this architecture is applicable to a multiplier with arbitrary bit-width, a tree multiplier generator based on this architecture is strongly desired. With such a system in hand, we will be able to design a long bit multiplier without long design time or slow latency.

Acknowledgement

The authors wish to thank K. Shirasawa, T. Nagasawa, H. Okada, M. Sakate, K. Kuroiwa, K. Hatta, K. Fujita and S. Mori for their continuous support and encouragement.

Reference

[1] Y. Owaki, et al. "A 7.4ns CMOS 16 x 16 Multiplier", 1987 ISSCC Dig. Tech Papers, pp52-53.
[2] Wallace, C. S., "A Suggestion for a Fast Multiplier", IEEE Trans Electronic Computers, Vol, EC-13, Feb. 1964, pp14-17.
[3] Pang, K. F., "Generation Of High Speed CMOS Multiplier-Accumulators", IEEE International Conference on Computer Design, October 1988.
[4] M. Nagamatsu, et al., "A 15ns 32 X 32-bit CMOS Multiplier With an Improved Parallel Structure", 1989 CICC Dig. Tech. Papers, 10.3.

[5] Booth, A. D., “A Signed Binary Multiplication Technique”, Qt. J. Mech. Appl. Math., Vol. 4, Part2, 1951.
[6] Chip C. Stearns, et al., “Yet Another Multiplier Architecture”, 1990 CICC 24.6.
[7] T. Sato, et al, “An 8.5-ns 112-bit Transmission Gate Adder with a Conflict-Free Bypass Circuit”, 1991 VLSI Circuits Symposium, Dig. Tech. Papers, 11.4.
[8] J. Mori, et al, “A 10-ns 54 X 54-b Parallel Structured Full Array Multiplier with 0.5-μm CMOS Technology”, IEEE J. Solid-State Circuits, Vol. 26, No.4, April 1991.

VLSI 91
A. Halaas and P.B. Denyer (Eds.)
Elsevier Science Publishers B.V. (North-Holland)

OCAPI: A prototype for high precision arithmetic

A. GUYOT and Y. KUSUMAPUTRI

TIM3 IMAG
46, Avenue Félix Viallet, 38031 Grenoble Cedex, France

Abstract
This paper first recalls the algortihms for on-line basic arithmetic operations and then presents the prototype (scaled down) of an integrated mathematical processor performing basic operations on either unlimited or very large (2048 bits) integers in an on-line fashion.

1. INTRODUCTION

There are many useful mathematical algorithms that are not easily run on computers, either because they are unstable (very sensitive to rounding errors, most of them dealing with eigenvalues) or because they use very large integers, e.g. cryptography[1] or Gröbner bases[2] .

Software like Bignum[3], Macsyma, Reduce[2], Mathematica, all were developed to work in arbitrary precision (but the first one oriented to computer algebra) by serializing computation on numbers that exceed the dynamic size of the computer word. Recently implementations on parallel computer try to recoup unavoidable serialization by doing many operations in parallel [4].

Nevertheless, high precision arithmetic still needs one or two orders of magnitude improvement in the cost-performance trade-off to be fully accepted. Whatever the coding, high precision numbers would demand too many pins to be transmitted in parallel across chips, so they have to be transmitted serially. Then, on-line algorithms allow an overlap of computation with transmission. In this context, the prototype of a mathematical integrated circuit presented herein has the following characteristics :

1. it performs the basic operations: addition, subtraction, multiplication, division, square-root on either unlimited or very large numbers, depending on the operation.
2. it is linearly scalable to variable precision.
3. it can be pipelined digit by digit with others circuit instances to improve the throughput.

The Avizienis's scheme [5] for carry propagation free operation does not extend to radix 2. So higher radices have been the main focus of interest until recently [6]. In the signed binary digit (SBD for short), each digit is coded on a pair of bits as following :

-1 is coded (1,0),
0 is coded either (0,0) or (1,1) and
+1 is coded (0,1).

Addition of numbers in the so called SBD representation bears a close resemblence with the "carry save addition" and is sometimes named "borrow save addition" [7]. Inside the circuit, digits are transmitted serially, one per clock cycle. While addition, subtraction and multiplication can be performed on-line starting from either the most or the least significant digit, square-root and division can only work in most significant digit first fashion [8].

2. ON-LINE ADDITION

All operations rely on serial addition of sequences of digits or numbers.

2.1. On-line digit addition

The behavior of an MSD first on-line adder can be described by a counter filled digit-by-digit by the inputs A and B and drained digit-by-digit by the output S, a_i, b_i and $s_i \in \{-1,0,1\}$. Since the weight is divided by 2 at each cycle, to conserve the counter value the recursion equation is: counter := 2∗counter + A + B - S∗2 ^{k} ; where k is the latency of the operator (i.e. if the weight of the inputs is 2^i, then the output weight is 2^{i+k} or the delay is k cycles).

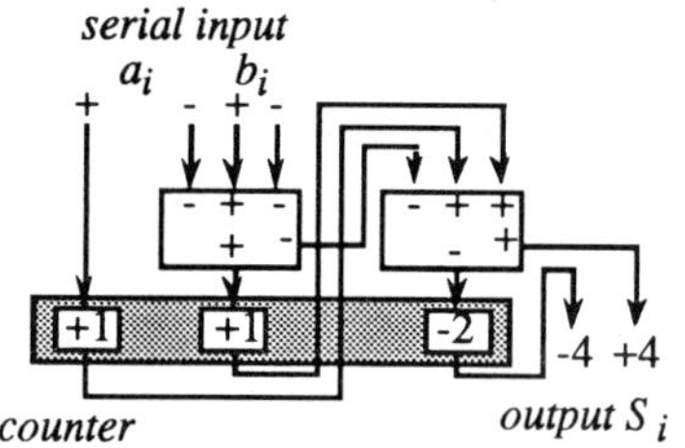


```
Procedure Addserie (ai, bi, si : SBD);
var counter: integer;
 begin;
  counter := 2*counter + ai + bi ;
  if counter > 2 then si := 1            {overflow}
   else if counter < -2 then si := -1 {underflow}
   else si := 0;
  counter := counter - 4*si ;
 end.
```

Figure 1a. On-line digit adder

The counter must not diverge (stay within fixed margins) whatever the sequence of values of a_i and b_i. It is easy to verify that the minimum latency is 2, as demonstrated in [8] (for latencies 0 and 1, sustained maximum values of a_i and b_i makes the counter diverge). The recursion combined with the maximum values of a_i and b_i gives the value of the margin $M = \pm 2$. The pseudo-Pascal procedure translates easily into hardware (fig.1a), if the numeration base of the counter is carefully chosen (here: +1,+1,-2). It is also easy to build an on-line adder with 3 inputs and a latency of 3, that is needed later on in the on-line multiplier.

2.2. On-line numbers addition

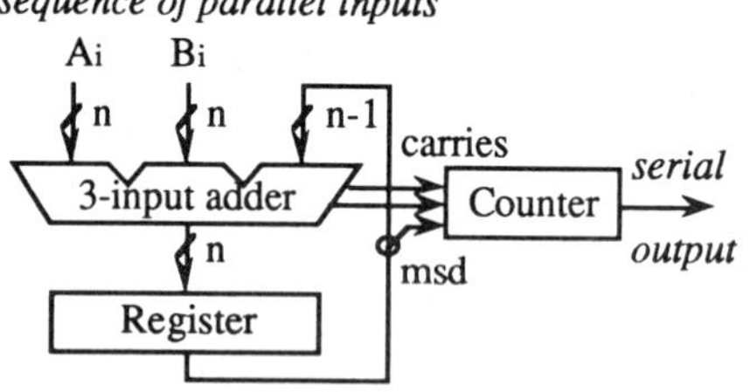

An algorithm similar to the one used for addition of numbers digit by digit can be used to add a sequence A_i of numbers, for i = 1,2,... or to add several sequences A_i, B_i provided that all terms of this sequence(s) are bounded by 2^{k-i}, for some value k. To have a clock speed independent of the size of the numbers, a 3-input carry-propagation-free adder is used [10].

Figure 1b. On-line number adder

3. MULTIPLIER KERNEL

By adding different kinds of counters to a multiplier Kernel, an on-line multiplier or a square-rooter or a divider can be built. This multiplier Kernel follows the on-line paradigm for multiplication developed by Ercegovac & Trivedi [9]. Let A and B be the normalized inputs transmitted serially, starting from the most significant signed binary digit. We have :

$$A = \sum_{i=0}^{n} a_i\, 2^{-i} ; \quad B = \sum_{i=0}^{n} b_i\, 2^{-i} .$$

Let us note

$$A_j = \sum_{i=0}^{j} a_i\, 2^{-i} \text{ and } B_j = \sum_{i=0}^{j} b_i\, 2^{-i} ,$$

this leads to the recursion: $A_j = A_{j-1} + a_j\, 2^{-j}$.

So, to build A_j from A_{j-1} in a register we need to store the incoming digit a_j in the jth position. In our realization (fig. 4) this position is pointed to by a token running in a shift register. Let us define the scaled partial product $P_j = A_j{*}B_j{*}2^j$. By applying the recursion on A_j and B_j , it becomes:

$P_j = 2{*}A_{j-1}{*}B_{j-1}{*}2^{j-1} + A_j{*}b_j + B_{j-1}{*}a_j$,

where $A_j{*}b_j$ is simply obtained by multiplying every digit of A_j by the digit b_j (⊗ in fig. 2c & fig. 4).

As noted in the sub-section 2.2 above, we can output serially the sequence of sums: $P_j = 2{*}P_{j-1} + A_j{*}b_j + B_{j-1}{*}a_j$, since subsequent recursion steps dealing with less significant digits will have no effect on the already computed most significant digits.

4. OPERATORS' BUILDING BLOCKS

Each basic operator combines a multiplier Kernel and a specific counter :

4.1. On-line Multiplier

P = A * B

The on-line multiplier is made of the multiplier Kernel and a counter, as shown in fig.2a. This counter is simply a serial adder that converts three pairs of bits coming from the multiplier Kernel into one output pair. This counter outputs serially the product P with 3 clock-cycles on-line delay with regard to the serial inputs.

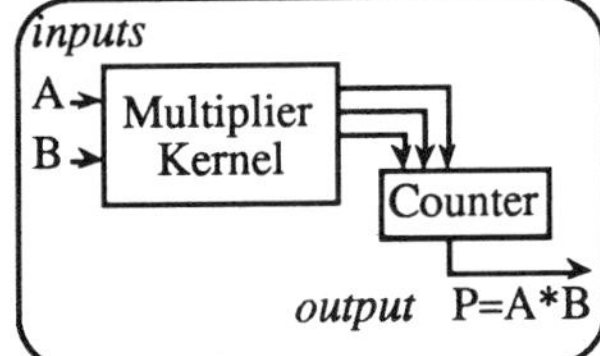

Figure 2a. Multiplier

4.2. On-line Square-Rooter

R = √A

The counter here needs 4 bits in 2's complement (fig. 2b). The serial output $r_i \in \{-1,0,1\}$ is computed as follows : **if** counter ≠ 0 **then** r_i = sign(counter) **else** $r_i = 0$. Note that the square-rooter is sensitive to the parity of the operand first bit's weight. If this parity is not odd then the result has to be multiplied by $\sqrt{2}$. An odd-weighted first digit can be obtained by recoding.

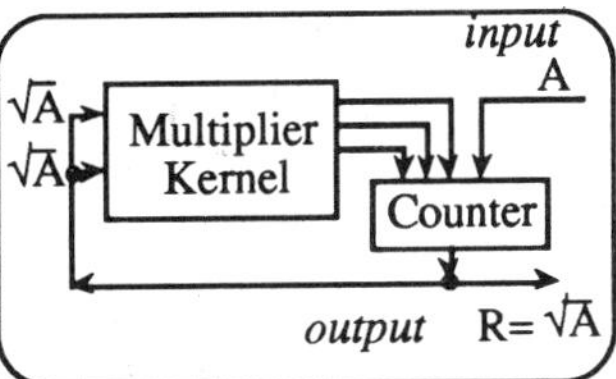

Figure 2b. Square-Rooter

4.3. On-line Divider Q = A/B

While the principle of the on-line division is rather simple, the implementation is more tricky (fig. 2c). As a matter of fact, the counter here is a low precision estimate of the partial remainder, and to make it converge the firsts digits of divisor B have to be greater than a certain value, say ∂. The division will proceed in 2 steps. First, feed the SBDs of B into EB until EB $\geq \partial$. Second, feed the SBDs of the dividend A into the counter and the next digits of B into the multiplier Kernel. At each cycle, the counter receives 3 SBDs from the multiplier Kernel and the next SBD from the dividend A.

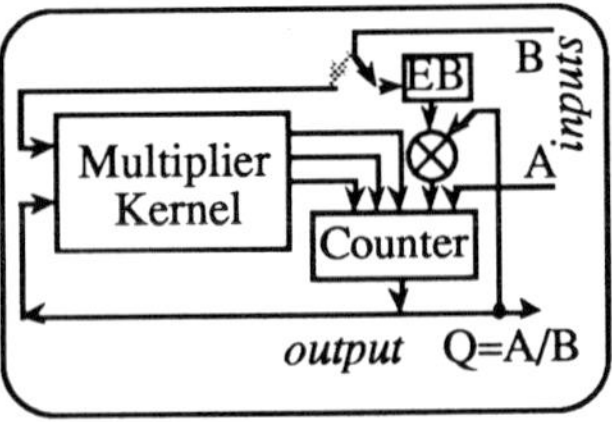

Figure 2c. Divider

Let p be the sum of these three SBDs from the multiplier Kernel and one SBD from A, so : $-4 \leq p \leq +4$. The recursion equation becomes:

$$counter := 2*counter - EB*q_i + p \quad (1)$$

and the stability condition independent of the actual value EB is :

$$-\partial + |p| \leq counter \leq \partial - |p| \quad \forall p. \quad (2)$$

To determine the value of each quotient digit $q_i \in \{-1,0,1\}$, the range of counter is divided into three regions by a threshold s^+ and s^-. So :

if $-s^- < counter < +s^+$ **then** $q_i = 0$ (3)

if $-\partial + |p| \leq counter \leq -s^-$ **then** $q_i = -1$ (4)

if $+s^+ \leq counter \leq \partial - |p|$ **then** $q_i = +1$ (5)

By giving one worst case value 4 to $|p|$ and applying the recursion (1), equations (2) and (5) we get the threshold value: $|s| \geq \pm 4$. These results combined with equations (2) and (3) give the minimum absolute value $|\partial| \geq 14$ of the first digits EB of B and 10 for the counter.

5. CHIP DESIGN

A prototype circuit performing either the on-line multiplication or the on-line division has been designed. It was fabricated by the French brokerage service CMP in October 1990[11].

5.1. Chip Organization

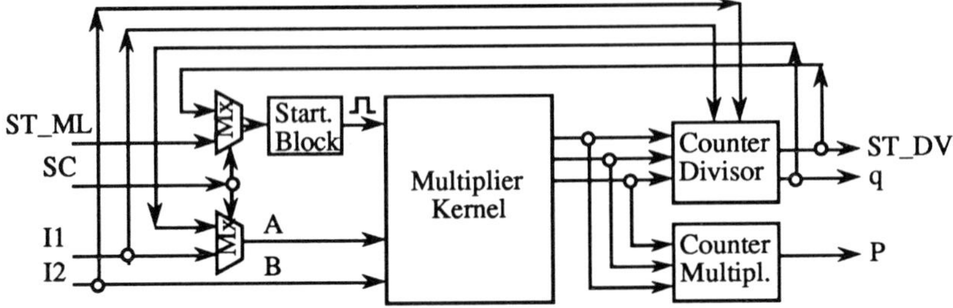

Figure 3. Chip organization

In figure 3, two different counters are connected to a multiplier Kernel whose inputs are selected by the control pin SC according to the operation mode. To activate the multiplier Kernel, the token in the shift register is fired by the input ST_ML in the multiplication mode and by ST_DV in the division mode. In the division mode ST_DV also indicates that EB is ready.

5.2. Design of the multiplier Kernel

The multiplier Kernel is assembled from m cells connected serially. The jth cell contains the jth position of the shift register for the token (top of fig. 4), the latches for a_j and b_j, two digit multipliers (⊗ in fig. 4) and a 3-input carry-propagation free adder CPFA. B_{j-1} is obtained by delaying B_j (right of fig. 4 : register D0). The most significant digit starts at the first cell and the last significant digit ends at the mth cell.

Each cell contains 216 transistors layed out in full custom to achieve a high density (4.5 Kt/mm^2). A 600 decimal digit Kernel (or m = 2048 signed binary digits) would occupy about 1 cm^2.

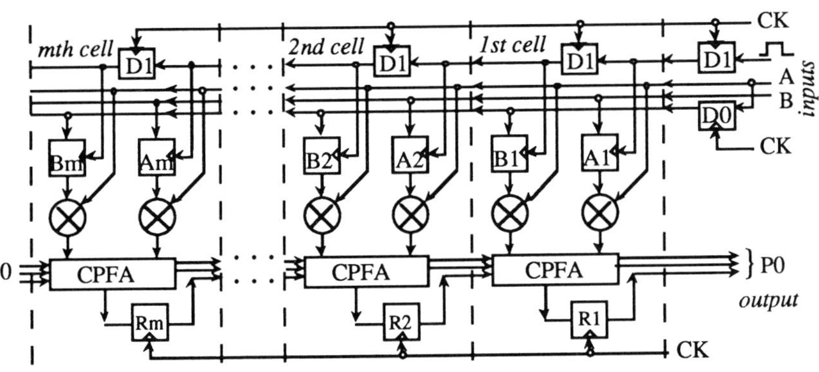

Figure 4. Structure of the multiplier Kernel

5.3. Prototype chip

In the prototype chip fabricated, the multiplier Kernel consists of 64 regular cells. These regular cells are coiled up in four rows to fill up the area of the chip (see fig. 5).

This prototype chip contains two counters, one for the multiplication function and the other for the division. The placement of these counters is arranged such that they fill the empty rectangular surface.

The regular cell is a full custom design, the others blocks are made with standard cells from the ES2 library.

The technology used is CMOS 2μ double metal. The long connections are in metal-2 and the power and ground routing are in metal-1.

The dimension of the chip is about 11mm^2, with about 16,000 transistors in the inner core. To realize a chip with thousands of regular cells, a technology less than 2μ becomes necessary.

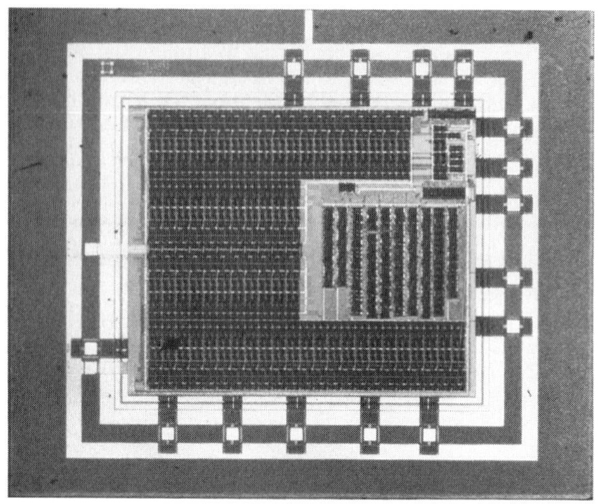

Figure 5. The prototype chip

6. FUTURE WORK

This VLSI circuit under design is to be placed close to a microprocessor and plugged into the memory bus, in order to significantly reduce the time required for high precision software computation. Due to the limited word size of the memory and pad limitaion of the chip, long numbers have to be transmitted serially in time O(n).

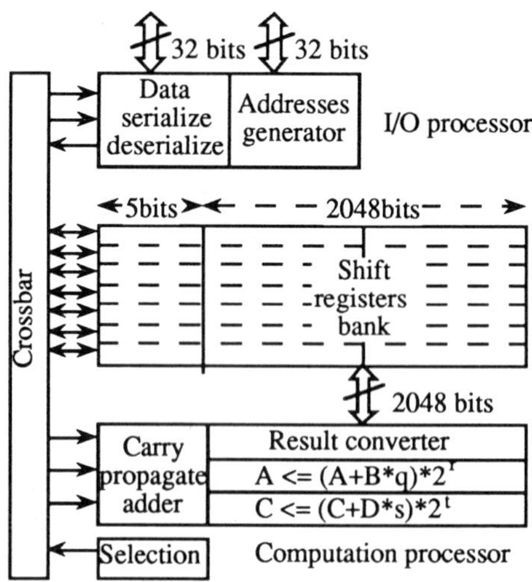

Figure 6. Coprocessor architecture

The goals of the architecture are :

1. To overlap computation and communication by on-line computation plus pipelining of load in, execution and store back [13].
2. To minimize external bus occupancy by keeping intermediate results or local variables in an on-chip register bank.
3. To benefit from a very large operator to avoid multiple precision operations. Internal redundant notation keeps the carry propagation path short [5,10]. The operator must take advantage of the parallel transmission of internal variables with the register bank and of the serial transmission of the external variables.
4. To offer add, subtract, multiply, divide, GCD, extended GCD, square root [13] and also provide support for rational and matrix operations.
5. To efficiently use the silicon, so the datapath and the operator are configurable to perform several operations in parallel on short operands and/or on operands transmitted in parallel.

The I/O processor acts as a DMA. It spies on the buses for the address and value of the first word of the operand, and then requests the bus to generate the address sequence for the next word [14]. Each register of the bank is accessible in parallel or in serial mode. The most significant bit occupies the right most position. One parallel and six serial accesses can take place simultaneously. A crossbar allows to serially load or store registers while computing on others in the pipeline mode, or to bypass the bank in on-line mode. The crossbar includes a serial adder. For extended GCD, three of the registers are used as a stack to record the sequence of operations.

Two adder/subtractor-shifters, each with two registers for redundant numbers, compose the computation processor. For multiply, divide or square root, they are used together in the on-line mode or independently in the parallel/serial mode to perform two different operations. In figure 6, A,B,C,D represent registers for large numbers, $q,s \in \{-1,0,1\}$ and $r,t \in \{0,1,2\}$. Registers B and D can be loaded serially, from left to right [14]. A pair of registers converts on-the-fly, serial signed binary results into conventional representation (-1 or 0 for the most significant position, 0 or 1 elsewhere) [15].

7. CONCLUSION

We have described a processor for very large integers whose execution time is fully overlapped by transmission, thus achieving the maximum throughput. For division of very large integers the chip is expected to run about 10^4 faster than a T800 [4]. Simulation shows that higher level algorithms like GCD and extended GCD should also be included [4,6,12].

8. REFERENCES

1 A. Vandemeulebroecke, E. Vanzieleghem, T. Denayer, C. Trullemans and P. Jespers, "A new carry free division algorithm and application to a single chip RSA processor", Proc. ESSCIRSC'89, Vienna, 1989.

2 H. Melenk et al., "On Gröbner bases computation on supercomputers using Reduce", FB Mathematik und informatik der Ferunversität Hagen, 1988.

3 J.L. White, "Reconfigurable retagetable bignums: A case study in efficient portable Lisp", Proc. ACM Conf. Lisp & functional programing, 1986.

4 J.L. Roch, "The PAC System, general presentation", Proc. CAP90. E. Kaltofeln ed., North Holland, 1990.

5 A. Avizienis, "Signed-digit number representation for fast parallel arithmetic", IRE Transaction on Electronic Computer, vol. EC-10, September 1961.

6 T.M. Carter, "Cascade: hardware for high/variable precision arithmetic", Proc.IEEE Symposium on Computer Arithmetic, Santa Monica, September 1989.

7 C.Y. Chow and J.E. Robertson, "Logical design of a redundant binary adder", Proc.IEEE Symposium on Computer Arithmetic, Como, May 1987.

8 J. Duprat, Y. Herreros and J.M. Muller, "Some results about on-line computation of functions", Proc.9th IEEE Symposium on Computer Arithmetic, Santa Monica, September 1989.

9 M.D. Ercegovac, K.S. Trivedi, "On-line algorithms for division & multiplication", IEEE Transactions on Computers. vol. C-26 no.7, July 1977.

10 A.Guyot, Y. Hereros and J.M. Muller, "Janus, an on line multiplier-divider for large numbers ", Proc. 9th IEEE Symposium on Computer Arithmetic, Santa Monica, September 1989.

11 H. Delori, A. Guyot and J.F. Paillotin, "French MPC activity report", IMAG report 1990.

12 A. Guyot, "OCAPI: Architecture of a VLSI coprocessor for the GCD and extended GCD of large numbers", Proc. 10th IEEE Symposium on Computer Arithmetic, Grenoble, 1991.

13 V. Oklobdzija and M. Ercegovac, " An on-line square root algorithme", IEEE Transactions on Computers, vol. C31-1, 1982.

14 M. Cosnard, A. Guyot, B. Hochet, J.M. Muller, H. Ouaouicha, Ph. Paul and E. Zysman, "The FELIN arithmetic coprocessor", Proc. 8th IEEE Symposium on Computer Arithmetic, Como, 1987.

15 M.D. Ercegovac and T. Lang, "On-the-fly conversion of redundant into conventional representation", IEEE Transactions on Computers, vol. C36-7, 1987.

VLSI 91
A. Halaas and P.B. Denyer (Eds.)
Elsevier Science Publishers B.V. (North-Holland)
1992 IFIP.

Design of a Highly Pipelined 2nd Order IIR Filter Chip

O.C. Mc Nally, J.V. Mc Canny, R.F. Woods

*IFI Institute of Advanced Microelectronics,
Dept. of Electrical and Electronic Eng.,
The Queen's University of Belfast
Ashby Building, Stranmillis Rd.,
Belfast BT9 5AH, Northern Ireland*

ABSTRACT

The design of a high performance bit parallel second order IIR filter chip is described. The chip in question is highly pipelined, uses most significant bit first arithmetic and consists mainly of arrays of simple carry save adders. It has been designed in 1.5 um double level metal CMOS technology, accepts 12 bit input data and coefficient values and can operate at up to 40 megasamples per second. The highly regular nature of the architecture has been exploited for test pattern generation. It is shown how small, but important modifications to the basic architecture, can significantly improve testing. As a result, 100% fault coverage can be achieved using less than 1000 test vectors.

1.INTRODUCTION

Until recently it was the view that fine grain pipelining could not be employed to increase the sampling rate of recursive digital filters such as IIR filters. The effect of introducing m levels of pipelining into the feedback loop of such a circuit introduces an m cycle latency, therefore negating any potential speed-up. In a recent series of papers the authors and their co-workers [1,2,3] have reported important advances in this field. In particular, they have shown how fully pipelined bit parallel IIR filters can be designed in which the latency is only two clock cycles and this is independent of wordlength. This is achieved through the use of most significant digit (msd) first arithmetic. The circuit latency can then be reduced to one clock cycle by using one level of the "scattered look-ahead" technique described by Parhi [4].

The purpose of this paper is to describe a new demonstrator chip which has been designed to exploit the results of this research. The chip in question has been designed using VLSI Technology's 1.5 micron double level metal CMOS standard cell system and implements a 2nd order IIR filter section with a full coefficient range. The chip is highly pipelined, uses most significant bit first arithmetic and consists mainly of arrays of simple carry save adders. It has been designed to operate every cycle and to accept 12 bit input data and coefficient values. It also contains on board circuitry necessary for the detection and correction of overflow. This work significantly advances a previous chip design undertaken in this laboratory [5]. This implemented only a first order IIR filter section and did not incorporate many of the improvements in circuit efficiency which have been developed since and which are included in the chip presented in this paper.

The paper is structured as follows. Section 2 gives an overview of the architecture on which the chip design is based and discusses issues such as overflow detection and correction in a system in which a carry-save redundant number system is used. Section 3 considers testing and describes how methods developed by Marnane and Moore [6] for other regular arrays can be extended to the more demanding problem of testing msd first IIR filter arrays. Details of the full chip design will then be presented in section 4.

2. SECOND ORDER SECTION IIR FILTER

The architecture used as the basis of the chip design has been described in detail previously [1,2]. Only an overview is given here. The circuit in question implements a full second order IIR filter function which can be expressed as

$$y_n = u_n - b_1.y_{n-1} - b_2.y_{n-2} \qquad (1)$$

where

$$u_n = c_0(x_n + a_1.x_{n-1} + a_2.x_{n-2}) \qquad (2)$$

The value x_n represents a sampled input data stream and y_n is the filtered output. The values a_1, a_2, b_1 and b_2 represent the filter coefficients. For the circuit considered the coefficients have their full range ie. $|b_1|<2$, and $|b_2|<1$.

It has been shown in previous papers [1-2] that there are major problems in applying fine grain pipelining to a recursive function such as equation (1) if conventional lsb first arrays are employed. For this reason a large number of cycles are required (typically of the order of the input wordlength) before the bits required in the subsequent computation (ie the most significant bits) are available to be fed back. It has however been demonstrated in references [1] and [2], that a fully pipelined bit parallel circuit for implementing equations (1) and (2) can be designed with a latency of only two cycles. This is achieved by using most significant digit (msd) architectures to implement the recursive part of the computation. This can be reduced to a single cycle by iterating the transfer function so that the output y_n depends on y_{n-2} and y_{n-4} rather than y_{n-1} and y_{n-2} [4]. A major attraction of combining these two methods is that only one iteration is required and the number of iterations is wordlength independent. This contrasts with least significant bit (lsb) first systems where the number of iterations and hence the amount of hardware depends on the wordlength. Expansion of equations (1) and (2) in this manner results in equation (3).

$$y_n = u_n - b_1.u_{n-1} + b_2.u_{n-2} + (b_1^2 - 2b_2).y_{n-2} - b_2^2.y_{n-4} \qquad (3)$$

Assuming the poles to be complex conjugate, it then follows that $b_1 = -2r.\cos\beta$ and $b_2 = r^2$ and equation (3) becomes

$$y_n = u_n + 2r\cos\beta.u_{n-1} + r^2.u_{n-2} + 2r^2\cos 2\beta.y_{n-2} - r^4.y_{n-4} \qquad (4)$$

The coefficients $2r\cos\beta$ and $2r^2\cos 2\beta$ lie between -2 and +2, while r^2 and r^4 are in the range -1 and +1.The gain c, lies between 0 and +1, and the a_i coefficients between -2 and +2.

An architecture for implementing the computation in equation (4) is shown schematically in figure 1. This consists of four multiplier/accumulator blocks for computing the non-recursive part of equation (4). The delays in dashed boxes are specific to a design with 12 bit coefficients and 12 bit x inputs. Details of the multiplier/accumulator blocks used in the recursive part are shown at the subcell level in figure 2. This architecture consists mainly of carry save adder cells in which the data is organised most significant digit first. This array is pipelined every third row, as indicated by the solid black dots. This circuit exploits the fact that carry save is a redundant number system, with the cells along the edge of the array performing the parallel addition of sum and carry data streams which are generated in the bulk of the array. Figure 2 also includes special edge cells OVA, OVB and OVC which are used for overflow and saturation. A more detailed explanation of this array operation is given in reference [2].

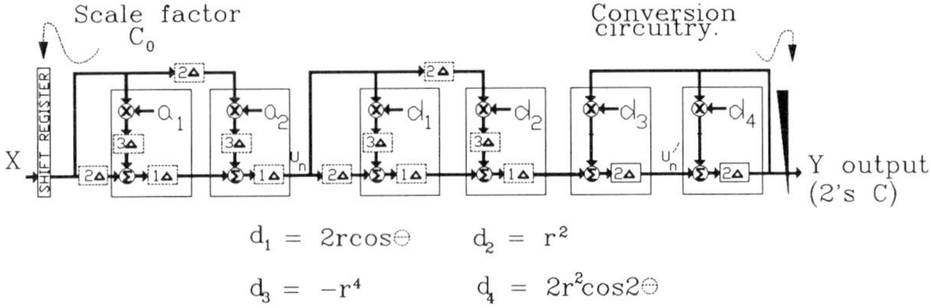

Figure 1 Second Order Section IIR Filter with Scattered Lookahead

The msb array in figure 2 can also be used for the computation of the non-recursive terms. However, overall circuit design is simpler if lsb first arrays are used [2]. The chip which has been designed therefore uses msd first arrays for the recursive computation and lsb first arrays for the non-recursive part. The conversion array, shown at the output of figure 1, takes a 12 (in general n) digit input in signed binary number (SBNR) representation and produces a 13 (in general n+1) bit conventional 2's complement output [1].

The throughput rate of the circuit in figure 1 determines the speed of operation in the filter and is given by $th_r = 1/(t_l+t_r+2t_a+t_{acc})$, where t_l, t_r, t_a and t_{acc} are the delays through a latch, a row (t_r=one fulladder plus multiply cell delay), a fulladder and an accumulator.

A detailed description of the complete IIR filter was written in the ELLA hardware description language and has been simulated right down to the bit level. Analogue waveforms have been generated, sampled and the resulting digital results used in the simulation. In this way it was possible to check the functionality of the filter and to show that it could meet an arbitrary specification. The DSP package, Hypersignal, from Hypercyption, was also used to design filter coefficients for various filter specifications. This software also produced ideal outputs for the specification used. One such example, obtained in this way is given in figure 3(a). The same coefficient values were used in the ELLA simulation, and the spectral analysis graph in figure 3(b) obtained from it. This shows good agreement with the ideal result. One important point to note in using this filter is that a maximal signal to noise (SNR) ratio can be obtained, because scaling may be reduced to allow larger signals, at the expense of saturating the output. Scaling provides a tradeoff between quantization (or roundoff) noise and saturation [7].

3. TESTING OF THE IIR FILTER ARRAY

The test approach which has been adopted is based on the work of Marnane and Moore[6]. This exploits the regularity of arrays to improve test. It is unique in that different edge cells and the effect of time can be incorporated into test models obtained. This scheme has been extended to cope with the component arrays required in the IIR filter in figure 1. The results of this work are summarised below, while further details are given in reference [3].

The strategy employed initially focused on a single processing element (P.E.). One can either apply the stuck-at, CMOS switch level, or functional fault models at the gate or transistor level of the PE. The functional fault model was chosen for its greater generality. In the case of a PE with four inputs a total of sixteen test vectors are needed. The next stage in the process involved the determination of vectors which would propagate these tests to other PE's along the main directions within the array. This was done using a test vector dependence graph (TVDG), which is derived from the truth tables for each of the PE's. A major consideration, was to consider how the effect of faults

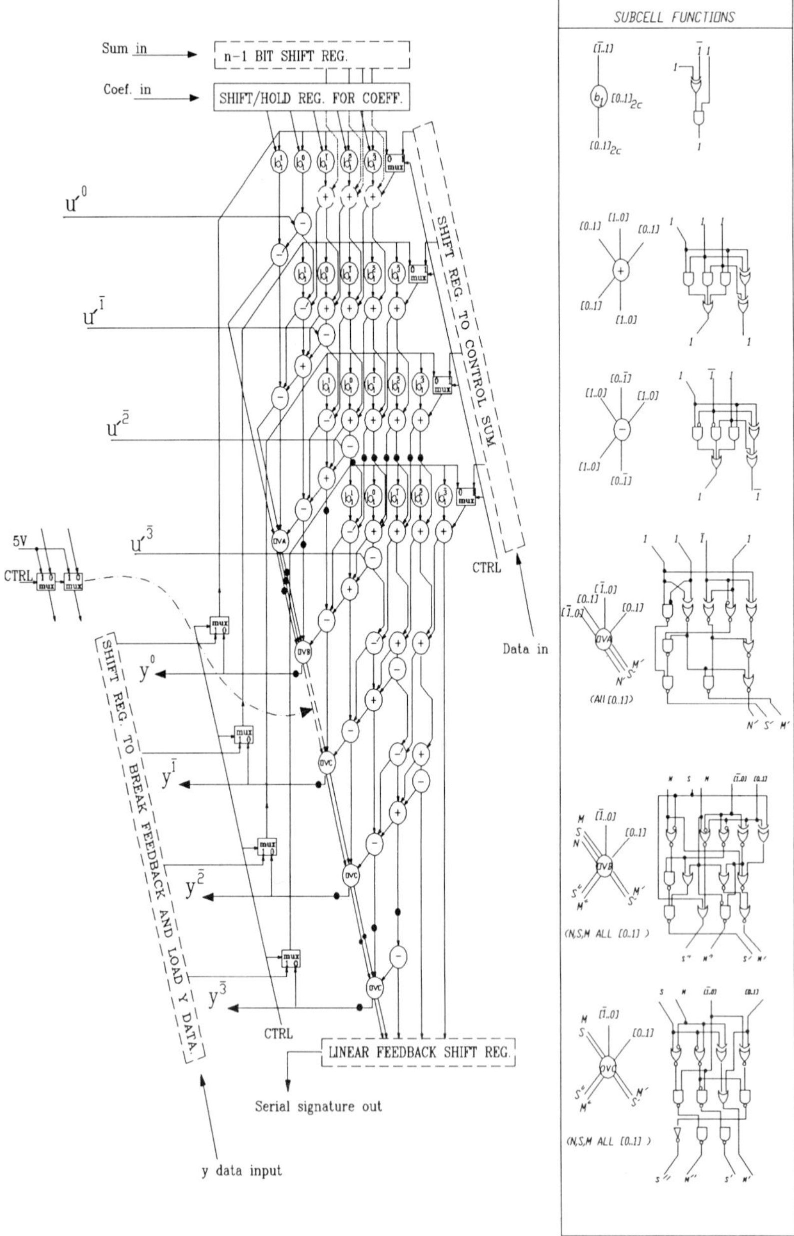

Figure 2 MSD first architecture with test circuitry included

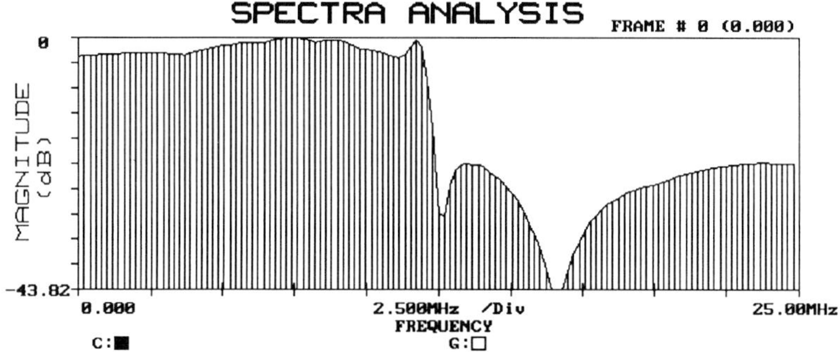

Figure 3a Output from Hypersignal

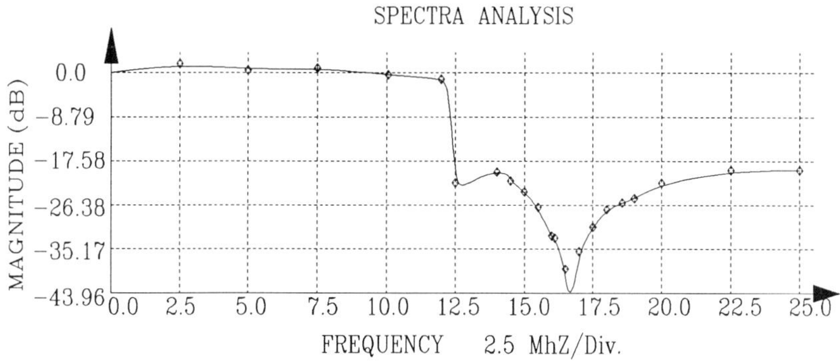

Figure 3b Simulated output from ELLA model

FILTER SPECIFICATION

DESIGN:	ELLIPTIC
SAMPLING FREQUENCY (Hz):	50000000
CENTER FREQUENCY (Hz):	6000000
BANDWIDTH (Hz):	12000000
TRANSITION BANDWIDTH1 (Hz):	00
TRANSITION BANDWIDTH2 (Hz):	400000
STOPBAND ATTENUATION (dB):	20
FIR LENGTH/IIR ORDER:	4
RIPPLE (dB)/STEIGLITZ ERR:	3
FILTER TYPE (LP/HP/BP/BS):	LOW PASS

generated within the array could be propagated to the array boundary. This was done using fault effect propagation graphs, which were again derived from the PE truth tables.

An important issue which had to be addressed was the effect of different boundary cells. These were modelled as boundary restrictions [3,6]. These cells can significantly influence the size and nature of the test sets chosen. Moreover, it was found that simple modifications can be made to the basic architecture which dramatically reduce the number of test vectors needed.

With the approach adopted two different test sets can be generated, namely cyclical and linear test sets [3,6]. The former are preferable as they consist of matrices of test sets which repeat in both the x and y directions and have the important property that they are independent of the array size. With linear test sets, test vectors are applied to one row per test set. Therefore test time increases linearly with array size.

The overall system (figure 1) consists of two recursive msd first multiplier arrays, four non-recursive lsb first multiplier arrays and circuitry to convert the IIR filter outputs from redundant radix two format into a two's complement one.

One of the msd arrays is shown in figure 2. The main core of this circuitry consists of conventional carry-save multiply/ full-add cells, with multiply/ half-add cells along the top and right edges. The accumulator, which consists of adders and overflow cells, is shown along the left-hand edge. This array has a number of boundary restrictions due to these edge cells. It has been shown in [3] that relaxing these edge restrictions produces better fault coverage and times. This is the reason for the extra shift registers and multiplexers in figure 2.

The strategy adopted in testing the lsb first arrays is similar to that used for the msd arrays. These circuits also comprise of an array of carry-save adders but are simpler in that they do not require circuitry along the left edge as in the msd case. As with the msd arrays, the generation of test patterns is considerably simplified by using half adders in the top row of the circuit.

The circuitry required for data conversion has been described in detail previously and basically consists of a triangular array of simple cells. Two possible test approaches are possible. Either the test vectors are generated in the last recursive block and fed into the conversion array in parallel, or an additional two shift registers, each of 12 units, are used to produce the input y digits for test. The second requires a slightly longer test time but allows the conversion array to be isolated from the other arrays during test. It has therefore been used for this reason.

The most straightforward way to test the overall IIR filter system is to start with the components on the right hand side of the circuit and work backwards in a right to left manner ie. the output conversion circuitry is tested first, then the first msd array with coefficient d_4, the msd array with d_3 and so on. Details of the number of test vectors and the number of clock cycles required in a system with 12 bit coefficients and 12 bit data are as follows. In the case of the msd first arrays 100% fault coverage is achieved using 4 cyclic and 4 linear test sets. For these a total of 864 clock cycles are required, of which 216 are required for testing, while the rest are necessary for loading the test vectors. In the case of the lsb arrays, the complete array can be tested using a total of 19 cyclical tests plus one linear test set. This requires 165 clock cycles, 76 cycles to load the shift registers and 89 cycles to perform the actual tests. The data input loading from the shift registers to the conversion array requires 130 clock cycles, with a further 88 clock cycles for test.

4. DETAILS OF IIR FILTER CHIP

A detailed layout of the full IIR filter chip is shown in figure 4. The floorplan of the final chip follows from the architecture in figure 1. Input signals enter the chip in a bit parallel fashion at the top right side, as shown, and pass through the scaling shift register. These are then clocked through the first two lsb arrays, which are associated with coefficients a_1 and a_2. The output from the second is used in the third and fourth lsb arrays, with coefficients d_1 and d_2. Results output from the fourth lsb array are input directly to the msd first arrays. These occupy the bottom half of the chip. The output conversion circuit is located in the bottom right hand corner and this produces a two's complement result.

In all the arrays data movement occurs in the rising edge of the main clock CLKM1. This clock is distributed as a tree network with local buffering. Each buffer drives up to nine latches. This network has been designed so that clock skew is minimised. The chip has been designed so that the coefficient values are programmable. These are contained in holding registers which sit at the top of each of the multiplier/ add arrays. They are loaded serially to reduce the number of input pads required. The outputs from each holding register are connected to a set of multiplexers which control their input to each of the multiplier/ add arrays. During normal circuit operation the coefficient values are connected directly under the control of signal CTRLAB. However the inputs to the multiplier/add arrays can alternatively all be set to zero. This is important for both testing and chip initialisation.

4.1 Least significant bit first arrays

Each array used in the non-recursive section of the filter perform a simple multiply and addition, $u_{out} = a.x_{in} + u_{in}$, which proceeds least significant bit first. A detailed layout of the arrays used is shown in figure 4.

The shift /hold register for the coefficient is at the top of the layout. An additional shift register to load 'sum bits' is also required at the top of the array. This register is used to simplify test, and reduce test times. The input x (or multiplier) data enters the array in parallel along the right hand edge. A column of latches are necessary along this edge to ensure the additive input, is delayed by two clock cycles before entering the array. (see fig. 1) A linear feedback shift register (LFSR) is also included in this column. It collects sum bits from the array, and produces a unique serial signature output, again for test purposes.

The main core of the array is devoted to multiplication of the x data by the coefficient. Successive bits of the input x are multiplied by the complete coefficient word in successive rows, starting with the least significant bit of x in the top row. The core employs conventional carry save techniques with pipelining after the fourth, eighth and twelfth rows. This level of pipelining has been chosen to produce a circuit whose speed is compatible with the msd first arrays used.

The accumulating products and additive input are added using a variable block adder VBA [8] which is placed at the bottom of each LSB array. A bit parallel two's complement result is obtained. The average LSB array occupies an area of 0.19 x 0.26 cm.

4.2 Most significant digit first arrays

The most significant digit first arrays used in the recursive part of the filter also employ carry-save principles, and therefore consist mainly of simple gated full adders. The accumulator circuit required is at the left side of the layout in MSD2 and the right side in MSD1. The additive input, u_1, from the fourth lsb array is a two's complement number and is added into the top row of MSD1. The carry/ sum SBNR output ($u^{\cdot}$) from MSD1 is added to the accumulator of MSD2 in a bit parallel format, with the most significant digit being input at the top. In this case the coefficients are d_3 & d_4 respectively.

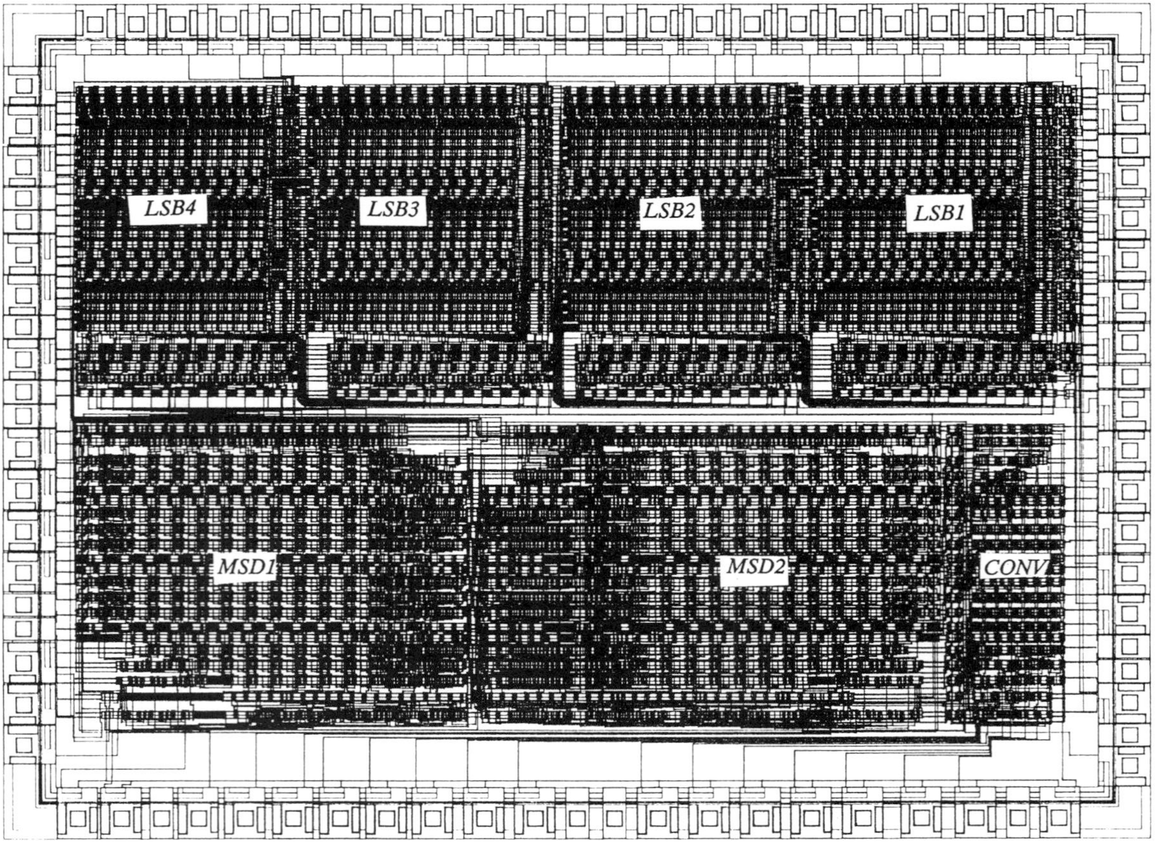

Figure 4 Second Order section IIR filter chip layout

The coefficient loading is the same as described above and the hardware required is at the top of each array. Two shift registers are positioned between the middle of the two MSD arrays. These provide test inputs in place of the normal feedback y digits. The feedback lines are broken through the use of a set of multiplexers under the control of CTRLYSM.

The msd first arrays both produce output values (y) in the form of radix 2 signed binary numbers (SBNR). These are obtained in groups of three from the accumulator at the left hand side of MSD2. The first group which are output from the top left hand edge are y^0 y^{-1} and y^{-2}, the next group of three are obtained from the cells below and so on. A total of 12 output digits are produced in this way. The bottom parts of both layouts each include a linear feedback shift register which is also used for testing the chip.

4.3 Conversion array

The conversion array is used to transform the 12 digit y output into a 13 bit two's complement word. This array uses the 'on the fly' conversion algorithm of Ercegovac and Lang [9]. This performs the conversion at the same time as the data is deskewed. This array with test hardware occupies an area of 0.09 x 0.24 cm. It accepts y digits in groups of three from the array MSD2. These are in a clock skewed format starting with the most significant digits. To facilitate test, the conversion array is partitioned from the other chip hardware by the introduction of shift registers plus multiplexers. A control signal CLTRLCV determines whether the test or normal input signals are applied to the circuit.

5. DISCUSSION OF RESULTS

The IIR filter which has been designed contains a total of 60,000 transistors and offers a sampling rate of up to 40 megasamples per second. It is contained in an 84 pin package and has a maximum power dissipation of 1W. Internal core power supplies and external pad power buses have been separately supplied, thereby reducing switching noise problems. The chip occupies an area of 0.88 cm by 0.64 cm and can be exhaustively tested in just 2,500 clock cycles using less than 1000 test vectors.

It is useful to compare the chip which has been designed with that described by Parhi and Hatamian [10]. This chip implements a fourth order section IIR filter and consists of two second order sections cascaded together. It has 10 bit coefficients, operates on 10 bit data and produces 15 bit output. The architecture of the chip is based on a 'scattered look ahead' technique and it uses fine grain pipelining. The chip also employs conventional least significant bit first techniques and consists of 17 multiply/accumulate units. It is a full custom design implemented in 0.9um double level metal CMOS technology and contains 78,000 transistors. They also quote a sampling rate of 100 Mhz for the chip.

A direct one to one comparison between the chip and that presented here is difficult because of the differences in filter order, the technology (used 0.9um CMOS as opposed to 1.5 um CMOS) and the method of design (ie full custom as opposed to using standard cells). However, a number of important points can be made.
(1) The number of multiply/add blocks required to implement a fourth order using our approach is 12 compared to 17 required by Parhi and Hatamian [10]. In terms of hardware our method is much more efficient for a given filter specification.
(2) The propagation delay between stages in Parhi and Hatamian's chip is of the order of five full adder delays. This is comparable to the chip described in this paper and one can speculate that sampling rates in the region of 100 Megasamples per second are possible if a 0.9 um CMOS technology were used to implement the design presented here.

(3) Perhaps the most important advantage of the approach used here is a constant sampling rate for all wordlengths. In particular, the number of multiply blocks required is not dependent on the wordlengths of either the data or the coefficients. This is not the case in a system which uses least significant bit first arithmetic. For example, if the wordlength is 16 bits and one required a system in which each multiplier/accumulator block had 4 levels of pipelining then a total of 10 multiplier arrays are required if lsb first only arrays were used as opposed to 6 multiplier/accumulate arrays if msb first arrays were used. The attraction of using msb first techniques therefore becomes apparent when one considers a broad range of possible system specifications.

6. CONCLUSIONS

In this paper we have described the design of a novel high performance second order IIR filter chip. The chip can operate at up to 40 megasamples per second, has 12 bit inputs and outputs and contains only 60,000 transistors. The recursive computations have been implemented using most significant bit first arithmetic, implemented in the main using simple carry save adders. The design undertaken has conclusively demonstrated the benefits of using most significant bit arithmetic in that the hardware overhead involved is only marginally greater than required for conventional least significant bit first arithmetic. This means that a broad range of recursive DSP chips can be designed in which fine grained pipelining can be used to achieve high throughput rates. One must therefore conclude that the most significant bit first arithmetic must have a major role to play in the design of a wide range of future DSP chips.

Acknowledgement : Financial support for O.C. Mc Nally from the Department of Education for N Ireland (DENI) and STC (Monkstown) Ltd is gratefully acknowledged. Funding from the UK Science and Research Council to support in the chip design described is also acknowledged.

REFERENCES

[1] S.C. Knowles, J.G. Mc Whirter, R.F. Woods and J.V. Mc Canny "Bit-level systolic architectures for high performance IIR filtering", *Journal of VLSI Signal Processing*, Feb 24 1989, pp9- 24.

[2] O C Mc Nally, J V Mc Canny and R F Woods, "Optimised bit level architectures for IIR filtering", *IEEE International Conference on Computer Design: VLSI in Computers & Processors*, Massachusetts, September 1990.

[3] O.C Mc Nally, W.P. Marnane and J.V. Mc Canny, "Design and test of a bit parallel 2nd order IIR filter structure", *Proc. IEEE Int. Conf. on Acoustics, Speech, and Signal Processing*, Toronto, Ontario, May 14-17, 1991.

[4] K.K. Parhi and D.G. Messerschmitt, "Pipelined VLSI Recursive Filter Architectures using Scattered Look-ahead and Decomposition ", *Proc. IEEE Int. Conf. on Acoustics, Speech, and Signal Processing*, New York, April 1988, pp. 2120-2123.

[5] R.F. Woods, J.V. Mc Canny, S.C. Knowles and O.C. Mc Nally, "A high performance IIR digital filter chip", *Proc. IEEE Int. Conf. on Circuits and Systems*, New Orleans, May 1990

[6] W.P. Marnane and W.P. Moore, "A Computational Approach to Testing Regular Arrays", in "*Systolic Array Processors*", eds. J.V. Mc Canny, J.G. Mc Whirter, E.Swartzlander, Prentice-Hall, 1989, pp 577-586.

[7] L.B. Jackson, "*Digital filters and signal processing*", Second edition, Kluwer Academic Publishers, 1989, pp310-315

[8] V.G. Oklobdzija, "Simple and efficient CMOS circuit for fast VLSI adder realization", Proc. *IEEE Int. Conf. on Circuits and Systems*, 1988

[9] M.D. Ercegovac and T. Lang, "On-the-Fly Conversion of Redundant into Conventional Representations," *IEEE Trans on Computers*, vol. C-37, no. 7, July 1987, pp. 895-897

[10] K.K. Parhi and M Hatamian,"A High Sampling Rate Recursive Digital Filter Chip", *Proc. of the 3rd-VLSI Signal Processing Workshop*, IEEE Press, Monterey, USA, Nov. 1988

VLSI 91
A. Halaas and P.B. Denyer (Eds.)
Elsevier Science Publishers B.V. (North-Holland)

Design of a Fully Parallel Viterbi Decoder

Jens Sparsø[a], Steen Pedersen[a] and Erik Paaske[b]

[a]Department of Computer Science, Building 344, Technical University of Denmark, DK-2800 Lyngby, Denmark. (E-mail: jsp@id.dth.dk)

[b]Institute of Circuit Theory and Telecommunication, Building 343, Technical University of Denmark, DK-2800 Lyngby, Denmark.

Abstract

In this paper we describe the design of a full custom VLSI implementation of a fully parallel Viterbi decoder for the $R = 1/2$, $K = 7$ code standardized by CCSDS, [1]. Although it is a university design using simplified design rules etc., the area and speed of the design compare very favorably with state of the art commercial components. This has been achieved (1) by the development of a new regular and area efficient floor plan for the *Add-Compare-Select module* (a parallel computing structure based on a shuffle-exchange type interconnection network), (2) by a simple and fast bit-slice implementation of the Add-Compare-Select processing elements, and (3) by the development of a new algorithm for the *Path Storage and Selection* module. The paper presents these new developments and reports on the performance of the chip. It contains approx. 100.000 transistors and the area including pad-cells is 64 mm^2. The Add-Compare-Select module and the Path Storage and Selection module operate at speeds up to 30 MHz and 21 MHz respectively under worst case conditions ($V_{DD} = 4.75\ V$ and $T_A = 70\ ^OC$).

1 Introduction

Convolutional coding with Viterbi decoding is widely accepted as an efficient method to improve the bit error rate on digital communication channels with low to moderate signal to noise ratios. In particular the *code rate* $R = 1/2$, *constraint length* $K = 7$ code recommended by CCSDS [1], has been used for a number of applications, and extended use of this code is expected in the future.

In this paper we consider a fully parallel implementation of the Viterbi decoding algorithm which can process one set of input signals and produce one decoded bit per clock cycle. This is essential at high information bit rates (i.e. above 5-10 Mbit/sec). The computational complexity of such a decoder is considerable. It involves the implementation of 64 identical processing elements connected by a shuffle-exchange type network of 128 5-7 bit busses, and an active memory structure. The two blocks are commonly denoted the *Add-Compare-Select* (ACS) module and the *Path Storage and Selection* (PSS) module respectively.

With the purpose of studying the implementation of these structures and gaining experience with VLSI systems design in general, we have designed such a chip. During the 3 year project period a series of 5 prototype chips have been designed and fabricated. In his paper we describe the final design that combines the ACS and PSS modules in one chip. The design is based on a new floor plan for the ACS module [2] and a new algorithm for the PSS module [3].

The paper is organized as follows. Section 2 introduces the Viterbi algorithm, and sections 3 and 4 briefly describe the above mentioned new developments and relate them to existing commercial designs [4], [5] as well as more general results and discussions [6], [7], [8]. Finally section 5 describes our design philosophy and some key results.

2 The Viterbi Decoding Algorithm

The Viterbi algorithm is a dynamic programming algorithm that finds the optimal path (the path with the minimum weight) through a lattice or trellis diagram. Figure 1(b) shows an example of such a lattice relating to the simple $R = 1/2$, $K = 3$ code that we will use for explanation. A possible encoder for this code is shown in figure 1(a). The code rate, R, denotes the number of encoded bits generated by the encoder for each input bit, and the constraint length, K, denotes the number of bits in the input sequence from which the encoded bits are computed.

The nodes in a column of the lattice correspond to the possible states of the encoder at each time interval, and the edges in a section of the lattice correspond to the possible state transitions. Thus each possible sequence of input data corresponds to a path through the lattice, and the problem of decoding can be formulated as finding the maximum likelihood path through the lattice, given the received and distorted sequence of data. For the $K = 7$ decoder a section of the lattice consists of $2^{K-1} = 64$ nodes and 128 edges.

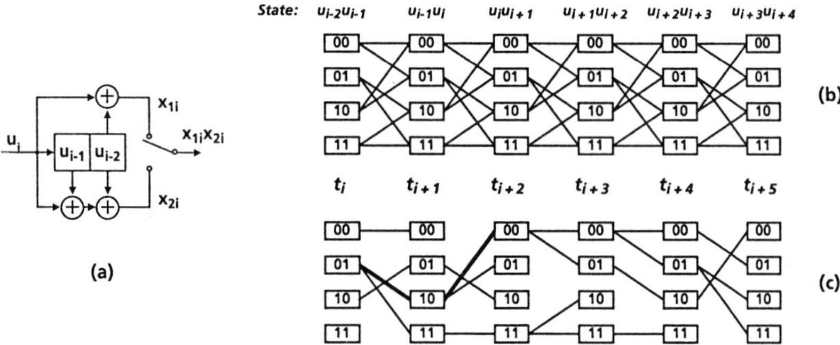

Figure 1: (a) Simple $R = 1/2$, $K = 3$ convolutional encoder. (b) A section of the corresponding state transition lattice. (c) Example of a reduced lattice.

The Viterbi algorithm operates in three steps normally implemented in three circuit modules:

1. **The input module** converts the received sequence of data into a sequence of weights that are associated with the edges of the lattice. These input weights are called *Branch Metrics* (BM).

2. **The ACS module** maintains a set of accumulated weights, one for each state in a column of the lattice. These accumulated weights are called *path metrics* (PM). For the two paths entering a state, the accumulated weight is calculated by adding the weight associated with the state transition and the PM of the preceding state. The two sums are compared, the smaller of the two is stored as the new PM of the state, and the decision (that represents the most likely path into the state) is output. In this way the ACS module generates a 64 bit decision vector in each clock cycle,

and the sequence of decision vectors represents a reduced lattice with only one path entering each node, figure 1(c).

3. **The PSS module** selects from the so-called surviving paths in the reduced lattice, the maximum likelihood path from which the transmitted sequence of data can be reproduced. The surviving paths merge into a single path if one goes back far enough through the lattice, as indicated by the boldfaced edges in figure 1(c). The *decoding depth, L,* denotes the number of trace back steps needed to reach this globally surviving path (with a sufficiently high probability), and L is a characteristic of the implementation. The PSS module provides storage for at least the L most recent decision vectors, and using one of the algorithms described in section 4 it outputs the decoded data sequence.

The input module can be implemented using a simple ROM-based table lookup procedure, and in the project we have focused our attention on the ACS and PSS modules.

3 The Add-Compare-Select Module

Implementation of the ACS module in a fully parallel decoder involves the realization of a structure similar to a section of the lattice, i.e. an array of 64 processing elements (ACS elements) connected by a shuffle-exchange type network with 128 busses. The width of these busses corresponds to the number of bits used to represent the PM's, and typical values are 5, 6 or 7 bits. In addition a number of global signals have to be distributed to all processing elements (branch metrics, clock signals and power). The major design issue is to minimize the area required to implement the interconnection network by careful floor planning.

3.1 Previously Reported Approaches

Several authors have proposed locally connected structures. Gulak and Kailath discuss some of these in [9]. In most of the proposed structures, localized communication is paid for by reduced utilization of the processing elements. The potential of these structures is that they bridge the gap between the obvious serial solution, with only one processing element, and the fully parallel structure. As we are aiming for a fully parallel implementation, these structures are not relevant.

Another approach is based on the observation that the necessary interconnections can be provided by a shuffle-exchange connected structure [6]. Layout of shuffle-exchange structures have been studied by others, and proven optimal layout structures, based on the so-called Thomson grid model, have been derived [7], [10]. The Thomson grid model assumes that the area of the processing elements can be neglected in comparison with the area required by the interconnections. As described in [2] this assumption does not hold in the case of the Viterbi decoder and such layout organizations does not yield sufficiently dense implementations.

Finally we mention that no information is available on the topological organization of the PE's in the Qualcomm chip [5], and that the NTT chip [4] uses a brute force approach with 3 metallization layers.

3.2 The Two-Column Ring Topology

In our design studies we aimed at two goals: (1) a regular floor plan with a simple distribution of global signals, and (2) optimization of the interconnection network.

It is a well known practice to implement the ACS elements in pairs, as they share the same PM-inputs two by two, figure 2(a)-(b) [11]. This reduces the number of "global"

PM-busses by a factor of two. Notice, that the remaining interconnections correspond to the well known perfect shuffle, and the topology presented below therefore applies to shuffle-exchange structures as well (by forming "exchange pairs" of processing elements).

To further optimize, we arranged the double ACS elements (DACS elements) in such a way that most communication is between neighbours. The DACS elements can be laid out forming one directed cycle, in which one of the two outgoing PM-busses of a DACS element connects directly to the neighbour DACS element. By organizing the ring as two columns, routing of the remaining PM-busses - one quarter of the original number - can be done in the channel between the two columns. Figure 2(d) illustrates this for the simple eight node example.

The key to the ring organization was found in a body of mathematical literature normally not related to floor planning [12], [13]. The Viterbi decoder and the shuffle-exchange interconnection networks are examples of deBruijn graphs. A deBruijn graph is a directed and connected graph in which the number of vertices is a power of two, and in which every vertex has exactly two ingoing and two outgoing edges. For this type of graph it is always possible to find a Hamilton cycle - i.e. a path which goes through all of the vertices once and only once. Figure 2(c) shows the graph for the 8-node example. In this simple case there is only one Hamilton cycle (0,1,3,2,0), and it is easy to find. The general case is more complicated and requires computer programs. Figure 5(a) in section 5 shows the resulting floor plan for the Viterbi decoder with 64 ACS processing elements, and further details can be found in [2].

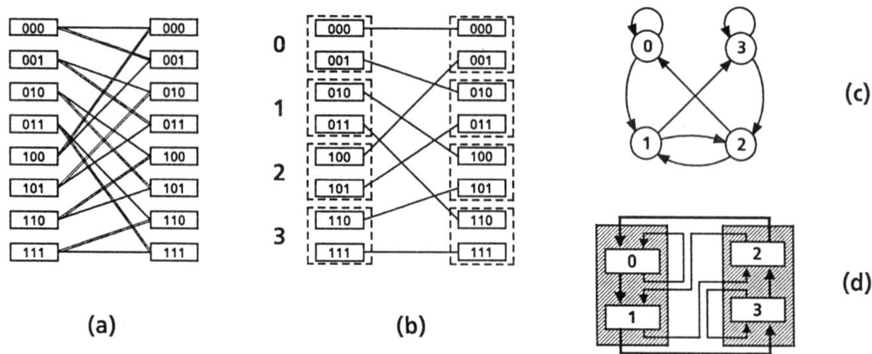

Figure 2: The basic steps leading to the two-column ring topology: (a)-(b) organization in pairs, (c) identification of Hamilton cycles, here (0,1,3,2,0), and (d) the resulting two-column floor plan.

The two-column ring topology has the advantage that routing of the remaining PM-busses has been reduced to a channel routing problem that is well understood. Furthermore, the structure is very regular and the global signals can easily be routed vertically through the four columns of ACS elements.

3.3 Circuit Design in the ACS Processing Elements

We use 7 bits to represent the PM's and the layout is organized as 7 identical bit-slices. The circuit design is straightforward and with only one exception we use static circuits. The comparator is implemented as a subtractor and the adders and the comparator use ripple carry propagation [14, figures 8.2, 8.4, 8.5]. This is a simple *and* fast solution, because the comparator starts calculating as soon as the least significant bits of the sums are calculated. Because of this overlap of carry propagation the total delay for an add-compare operation equals the delay in only 9 circuit stages (e.g. 7 carry stages in the adder, the most significant sum stage in the adder and the most significant carry stage in the comparator). In our opinion the use of carry look-ahead techniques as reported in [4] does not pay because it focuses on the adders in isolation instead of the combined add-compare operation. Figure 3 shows the logic diagram and floor plan of an ACS processing element.

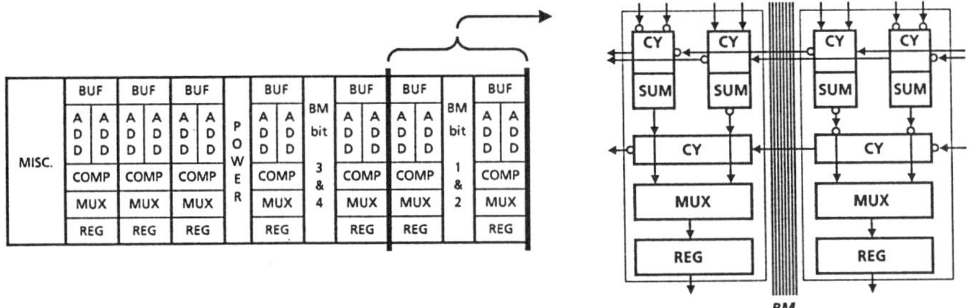

Figure 3: Floor plan and logic diagram of an ACS element.

4 The Path Storage and Selection Module

There are traditionally two different approaches for realizing the PSS module [8], [11]. These are named the register exchange algorithm (REA) and the trace back algorithm (TBA). The REA with a decoding depth $L = 30$ was used by NTT [4], while Qualcomm [5], as far as we know, used the TBA with $L = 80$. In the following we briefly describe the REA, the TBA, and the new combined Register Exchange and *Block* Trace Back Algorithm, which we denote the Combined Block Algorithm (CBA). Further details can be found in [3].

4.1 Trace Back Algorithm

The TBA stores the decision vectors in a memory, traces back L steps in order to reach the surviving path, and then traces further back, generating a sequence of decoded bits (in reverse order). During the trace back process, new decision vectors arrive and they must be stored for later processing. Consequently a straightforward implementation requires 3 blocks of RAM (each capable of storing L decision vectors). More elaborate structures with a smaller storage overhead are possible by speeding up the trace back process (cf. [8]). This can be done using more parallelism i.e. by using more but smaller RAM blocks or operating the trace back circuitry at a higher speed enabling more than one trace back step per received decision vector. Notice finally, that all variants of the TBA only process one bit per trace back step.

The advantages of the TBA are that the required storage can be implemented using dense RAM-structures, and that it avoids complex interconnect. The disadvantages are the storage overhead and the more complex operations, resulting in a smaller speed potential (as compared to the REA). We have estimated the area for an L=60 realization in a 2 μm CMOS process to be approx. 45 mm^2 [3].

4.2 Register Exchange Algorithm

The REA does not store the sequence of the decision vectors. Instead the decision vectors are processed as they are generated by the ACS module. This is facilitated by implementing the path memory as a set of lattice connected shift registers, figure 4. The length of the shift registers are L (the decoding depth), and in this example there are 4 shift registers. The sequence of data in a shift register represents the path leading to the state at the given time. As the paths merge into one the oldest bits in the shift registers will be identical and the decoded data sequence can be tapped from any of the outputs.

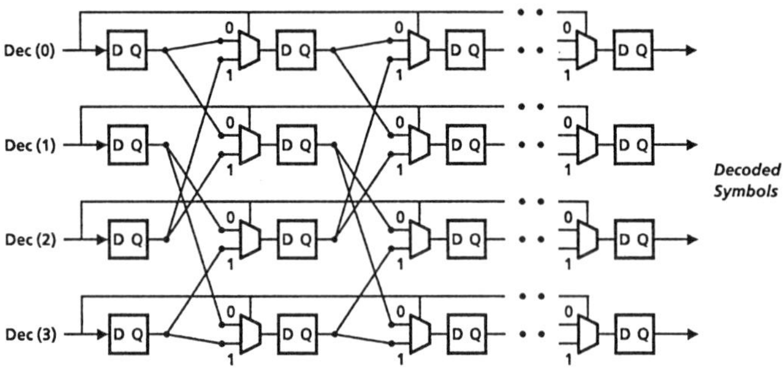

Figure 4: Register exchange circuit block (for the simple $K = 3$ code).

The advantages of the REA structure are regularity and simplicity, which makes it suitable for VLSI implementation. Furthermore, the simplicity results in a good speed potential. The disadvantages are the many instances of the interconnection network and that the storage has to be implemented as discrete flip flops. Altogether this results in a relatively large and inefficiently used area. We have estimated the area for an L=60 realization in a 2 μm CMOS process to be approx. 75 mm^2 [3].

4.3 The Combined Block Algorithm (CBA)

As mentioned above, the two traditionally used approaches for realization of the PSS-module both have serious drawbacks when it comes to VLSI implementation. Therefore, we have studied how to change the structure into a new algorithm where we combine the small number of memory locations of the REA with the compactness of realizing them as RAM cells as in the case of the TBA.

This is achieved by using the REA to produce small segments of length l of the surviving paths, and then performing a trace back based on segments instead of bits. This speeds up the trace back process by a factor of l and thus reduces the storage overhead significantly. A block diagram sketching the principle of the CBA is found in figure 5(a) in section 5. A detailed description is found in [3]. It should be noted that for a given decoding depth, L, many different implementations are possible depending on the choice of l.

The advantage of the CBA is that the storage overhead is moderate and that most of the storage can be implemented in a dense RAM structure, resulting in a very small area. The disadvantage is the complexity of designing both a register exchange structure and a trace back structure. The speed potential of the CBA is similar to the TBA.

The area will depend on the block length l and it is possible to minimize the area with proper selection of l. For a decoding depth $L \simeq 60$ a minimum area of 25 mm^2 is obtained for $l = 5$ assuming a 2 μm CMOS process. In our design we used the more convenient $l = 8$, resulting in an area of approx. 28 mm^2 [3]. Table 1 summarizes the area of the different PSS-structures. Obviously, the absolute figures to a certain degree reflect our proficiency in layout, but the relative figures are fairly unbiased and they clearly show the small area requirement of the CBA.

Algorithm	Area	Normalized
Register Exchange Algorithm (REA)	75 mm^2	2.7
Trace Back Algorithm (TBA)	45 mm^2	1.6
Combined Block Algorithm (CBA)	28 mm^2	1

Table 1: Area of different implementations of the PSS module (2 μm CMOS).

5 Implementation

The combined ACS-PSS chip is the last in a series of five prototype chips designed and fabricated during the 1988-90 project period, table 2. The first two chips contained a pair of ACS processing elements and the purpose of these chips was to verify the design of the processing element and the installation of the MAGIC design system (from Berkeley) that we have used for layout. The third chip contained the ACS module and the fourth chip contained the PSS module. All chips have been fabricated via NORCHIP (the Scandinavian CMOS IC prototype implementation service).

Viterbi decoder chip series	Chip 1 DACS element	Chip 2 DACS element	Chip 3 ACS module	Chip 4 PSS module	Chip 5 ACS+PSS module
Submission date	Apr. 88	Oct. 88	May 89	Aug. 89	Jun. 90
Technology	AMS 3 μm	ES2 2 μm	AMS 2 μm	AMS 2 μm	AMS 1.2 μm
Transistors	1,700	1,700	50,000	50,000	100,000
Size (incl. pads)	-	-	6×9 mm^2	5×9 mm^2	8×8 mm^2
Speed (worst case)	10 MHz	26 MHz	19 MHz	18 MHz	30/21 MHz

Table 2: Chips designed and fabricated during the project.

5.1 Design Philosophy

All chips have worked on first silicon and at the predicted speed. We attribute this to an overall design strategy of minimizing all potential risks, and a permanent awareness that the project goal was a working chip. Also, the expertise acquired from the first rather simple designs significantly contributed to the success of the following more complex chips. Below is a description of the most important points:

Functional modelling has been essential throughout the project, and much effort has been put into the creation of a detailed model of the entire transmission system including the encoder and a noisy transmission channel. This *system model* was used in several ways both in chip specification and in layout verification. First, the influence of several implementation parameters (e.g. number range and rescaling strategy) on the system performance (coding gain) was investigated. Next, different algorithms for the PSS module were evaluated. The model was written in Pascal and for the parts of the decoder that have been implemented it was refined to be bit-true at the clock-cycle level. These sections of the refined model constituted the *chip specification.* Finally - and very important - the refined model was used to produce test vector sets at chip, module and block level. We could simply transmit data through the model and record test vectors at the interfaces.

Simple and safe circuit design. We have used simple and safe circuits: static CMOS logic, static registers and 6-transistor static RAM-cells. The design is purely synchronous and uses a 2-phase non-overlapping clock system.

Timing estimates: We relied primarily on manual RC-calculations supplemented by the detailed timing measurements from chip 1 and 2. In addition we used the SPICE circuit simulator to produce accurate *relative* timing figures (e.g. sizing of transistors and balancing the clock distribution), and accurate analog waveforms (e.g. charge sharing problems in the RAM-cell design). These SPICE simulations was based on simplified equivalent circuit diagrams set up manually.

Design for testability: Test has been facilitated by the use of scan path's and partitioning. The 64 bit wide interface between the ACS and PSS modules can be accessed via I/O-pads allowing the ACS and PSS modules to be tested separately. Furthermore the register exchange block in the PSS module can be bypassed allowing direct write into the RAM block. In addition all four clock signals (ϕ_1, $\overline{\phi_1}$, ϕ_2 and $\overline{\phi_2}$) can be applied from external sources, and this (indirectly) enabled us to examine for example the read and write time of the RAM in the PSS-module.

5.2 Results

Figure 5 shows a floor plan and a microphotograph of the combined ACS-PSS chip. The chip was fabricated in a double metal 1.2 μm CMOS technology. The layout was based on a set of simplified design rules for the 2 μm process and the masks were derived by scaling down this layout by a factor of 0.8. Therefore, the area and speed potential of the 1.2 μm technology is not fully utilized.

In relation to the previous chips, a number of circuit level enhancements have been made. The most important improvement is that the transistors in the carry chains in the ACS processing elements have been sized (in the previous chips all transistors were identical). The ACS module now operates at 30 MHz and the PSS module at 21 MHz, table 2. As the operations performed by the PSS module are less complex than the operations in the ACS module this is clearly unsatisfactory. The timing bottleneck in the PSS module has been identified to be the read cycle in the RAM block. This is due to the rather simple RAM cell realization (without sense amplifiers). The speed of the PSS module could therefore be enhanced by using a more elaborate design (for example, from a commercial RAM compiler).

(a)

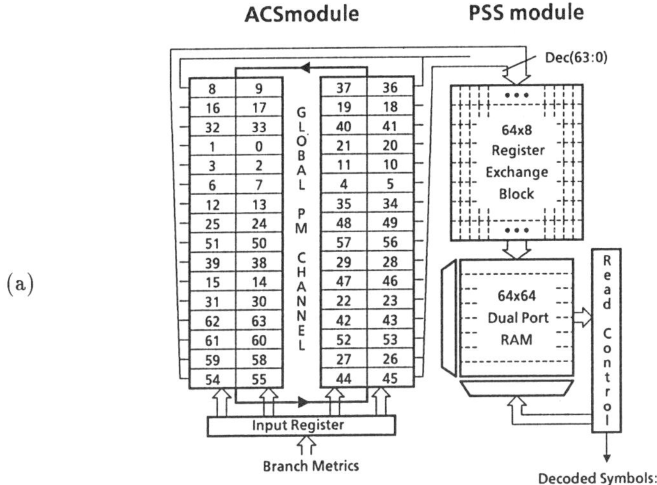

(b)

Figure 5: (a) Floor plan showing some details of the ACS module and the PSS module, (b) microphotograph showing the produced chip.

6 Conclusion

We have designed a high speed Viterbi decoder for the R = 1/2 and K = 7 code, which is standardized by CCSDS, using a full custom CMOS technology.

The proposed design utilizes a topologically optimized realization of the ACS module which highly reduces the area used for interconnecting the 64 processing elements. Furthermore, a new area effective algorithm has been developed for the PSS module.

A series of chips has been produced using these techniques, and the possibility of designing a Viterbi decoder for speeds above 30 Mbit/sec utilizing today's technology has been demonstrated.

Acknowledgements

Part of this work was supported by the Danish Technical Research Council under grant no. STVF/FTU 5.17.5.6.27.

References

[1] Consultative Committee for Space Data Systems, *Recommendation for Space Data System Standards, Telemetry Channel Coding*, CCSDS 101.0-B-2 Blue Book, January 1987.

[2] J. Sparsø, H. N. Jørgensen, E. Paaske, S. Pedersen and T. Rübner–Petersen, "An Area-Efficient Topology for VLSI Implementation of Viterbi Decoders and other Shuffle-Exchange Type Structures," *IEEE Journal of Solid State Circuits,* vol. 26, no. 2, pp. 90-97, Feb. 1991.

[3] E. Paaske, S. Pedersen, J. Sparsø, "An Area-Efficient Path Memory Structure for VLSI Implementation of High Speed Viterbi Decoders," Department of Computer Science, Technical University of Denmark, Report no. ID-TR:1990-76, Sept. 1990. (Submitted to *INTEGRATION, the VLSI Journal*).

[4] T. Ishitani, K. Tansho, N. Miyahara, S. Kubota, and S. Kato, "A Scarce State Transition Viterbi Decoder for Bit Error Correction," *IEEE Journal of Solid State Circuits,* vol. 22, pp. 575-582, August 1987.

[5] *Q-1401 Single-Chip Viterbi Decoder,* Qualcomm Incorporated, 10555 Sorrento Valley Road, San Diego, CA 92121.

[6] P. G. Gulak, E. Shwedyk, "VLSI Structures for Viterbi Receivers: Part 1 General Theory and Applications," *IEEE Journal on Selected Areas in Communications*, vol. 4, no. 1, pp. 142-154, January 1986.

[7] F. T. Leighton, G. L. Miller, "Optimal Layouts for Small Shuffle-Exchange Graphs," *Proc. of VLSI 81,* pp. 289-299, Ed. J. P. Gray, Academic Press 1981.

[8] D.J. Coggins, D.J. Skellern, R.A. Keaney, J.J. Nicolas, "A Comparison of Path Memory Techniques for VLSI Viterbi Decoders," *Proceedings of VLSI 89,* Münich, Germany, August 1989, pp.379-388, Ed. G. Musgrave and U. Lauther, Elsevier Science Publishers B.V., 1981.

[9] P. G. Gulak, T. Kailath, "Locally Connected VLSI Architectures for the Viterbi Algorithm," *IEEE Journal on Selected Areas in Communications,* vol. 6, no. 3, pp. 527-537, April 1988.

[10] D. Kleitman, F. T. Leighton, M. Leapy and G. L. Miller, "New Layouts for the Shuffle-Exchange Graph," in *Proc. 13th Annual ACM Symposium on the Theory of Computing.*, pp. 278-292, May 1981.

[11] G. C. Clark, Jr. and J. B. Chain, *Error-Correcting Coding for Digital Communications,* Plenum Press, 1981.

[12] S. W. Golomb, *Shift Register Sequences,* Holden-Day, San Francisco, 1967.

[13] H. Fredricksen, "A survey of full length nonlinear shift register cycle algorithms," *SIAM Review,* vol. 24, no. 2, pp.195-221, April 1982.

[14] N. Weste and K. Eshraghian, *Principles of CMOS VLSI Design – A Systems Perspective,* Addison-Wesley, 1985.

VLSI 91
A. Halaas and P.B. Denyer (Eds.)
Elsevier Science Publishers B.V. (North-Holland)

Pipelined BIT-Serial SYNthesis of Digital Filtering Algorithms

R. Nagalla and L.E. Turner

Department of Electrical and Computer Engineering,
University of Calgary, Calgary, Alberta, Canada, T2N 1N4

Abstract

This paper describes a technique for the synthesis of pipelined bit-serial digital signal processing (DSP) systems. The technique takes into account requirements on throughput (sample rate), system word length and design size. The program BITSYN evaluates different resource sharing strategies (multiplexing) and generates a minimum gate design, a high sample rate design, and a range of designs between these limits. BITSYN accepts as input a behavioral description in the form of a signal flow graph (SFG) and generates output in the FIRST[1] language.

1. INTRODUCTION

Architectural synthesis is the transformation of an abstract behavioral (algorithmic level) specification of a digital system into a register transfer level (RTL) structure that realizes the specified behavior, while satisfying a set of goals and constraints. The behavioral description gives the functional mappings from inputs to outputs and can be specified in the form of a signal flow graph (SFG) or a netlist. The RTL structure is a set of components such as arithmetic units, registers, multiplexers, their interconnections and the hardware required to control data transfers between them. These structures are represented using a hardware description language such as the input description language for the FIRST[1] silicon compiler. One objective of the synthesis process is to find a structure that best meets the constraints of clock rate, area or size and system word length (swl) while achieving the goals such as maximizing the sample rate and minimizing the size.

The steps involved in the design automation process of a digital system from algorithm to implementation are shown in figure 1. Hardware compilation is the process of converting RTL structural information into different integrated technology targets such as custom integrated circuit masks, or a data file which defines the configuration of a gate array device.

Synthesizers such as SEHWA[2], HAL[3], MAHA[4], S(P)LICER[5], SPAID[6] have been developed for bit-parallel systems. This paper proposes a methodology for the architectural synthesis of pipelined bit-serial systems, implemented in a design tool called BITSYN (Bit-serial Synthesis). BITSYN evaluates different resource sharing (multiplexing) strategies and generates multiplexed bit-serial digital signal processing (DSP) systems.

Pipelined bit-serial systems[1] are built from a set of functional operators. Each operator performs a function bit by bit, on one or more input words. Each operator will have

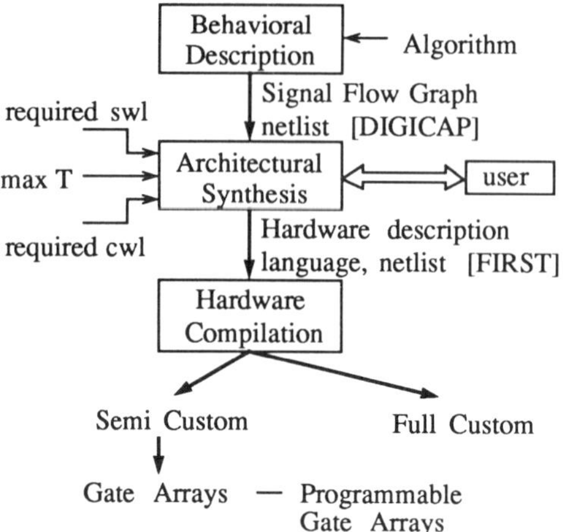

Figure 1. Design process of a digital signal processing system.

one or more control inputs generated by a control generator module. These control inputs indicate the arrival of a new input word or words and are used to initiate (or terminate) the operation for each set of data words.

In many cases, pipelined bit-serial systems offer implementations which are efficient in area-time measure[1]. The hardware required for bit-serial primitives is typically much smaller than bit-parallel primitives at the expense of the data throughput rate. The bit-serial operators are pipelined at the bit-level so that their internal components are used at the best possible throughput rate. In serial systems, the bit sequential transmission on single wires leads to other inherent advantages such as efficient communication within and between chips, and simpler routability in an integrated circuit.

2. BIT SERIAL ARCHITECTURAL SYNTHESIS

This paper discusses an approach to the synthesis of pipelined bit-serial digital architectures that execute DSP algorithms. The signal processing algorithm (SPA) accepts a new data set every sample period T, where T is the reciprocal of the throughput rate. Given an input sample and the current state, the SPA produces a new state and an output sample. In general, a SPA can be executed in bit-parallel, digit-serial[7] or bit-serial implementations. This type of classification is based on the number of signal bits which are used simultaneously in a functional operator. An additional classification is the number of functional operators in an implementation. In an operator-parallel implementation, each SPA operation is executed by a separate unit, resulting in the fastest data throughput rate and a large hardware cost. In an operator-serial or multiplexed implementation where similar operations are executed by fewer functional units than the number of operations, a slower data throughput rate and smaller architecture results. The program BITSYN is used to evaluate a range of multiplexed pipelined bit-serial implementations as well as a full operator-parallel implementation.

The automatic design of the data path from a functional description is performed in four different phases : translation, scheduling, allocation, and binding. In the translation step, the algorithmic description is translated into a data flow graph (DFG). Each node in a DFG represents an operation to be performed by a functional operator. In bit-parallel systems, the scheduling operation is the process of partitioning the DFG into a number of control steps (clock cycles) and assigning each operator to a control step. The number of control steps are then minimized within the limits of the hardware resources. Operations which occur in different control steps are mutually exclusive and can share functional operators. Allocation in bit-parallel systems is the assignment of operations to hardware, that is, the actual mapping of the operations to specific functional operators. The registers and data transfer components such as multiplexers and buses are also allocated to realize the partitioned DFG. The final step in the design process is binding the functional operators such as multipliers, adders etc. to their pre-designed gate-level structures.

The different properties of bit-serial operators must be known before discussing a synthesis strategy for bit-serial systems. The bit-serial multiplier is typically the largest element in size and has the largest latency. The hardware requirements and the latency of a multiplier are a direct function of the coefficient word length (cwl), not the swl. There is typically an order of magnitude or more difference between the size of all other operators such as adders, subtracters, multiplexers, registers and the bit-serial multiplier. For example, in a particular FIRST[1] implementation, a bit-serial multiplier with a cwl of 4 bits requires 260 gates whereas an adder requires only 30 gates. Thus in bit-serial systems, hardware can be significantly reduced by reducing the number of multiplier operators.

The system latency is equal to the sum of latencies of all the operators on a critical path and is dependent on the system implementation. For a critical path, the time at which each operator output is available is the same as the time at which each output is needed. In recursive systems the system word length (swl) can not be less than the system latency. Thus the implemented structure imposes a minimum bound on the swl. The swl is also limited by the required swl constraint specified by the user. The required swl is determined from an analysis of the non-linear effects of finite precision arithmetic[8]. If the system latency is the same as the required swl, then no extra delay elements (one bit registers) are needed and the operators acts as delay stages between the state inputs (current state) and the state outputs (next state). If system latency is less than the required swl, then additional delay elements must be introduced.

In bit-serial systems, the allocation step is replaced by resource sharing and register allocation procedures. In the resource sharing process, the objective is to reduce the number of multiplier operators while meeting the sample rate and swl constraints. The multiplier operators are shared using 2:1 multiplexers in a number of levels. Scheduling must take the latencies of each of the operators into account and calculate a time schedule for all operators. In bit-serial systems, two designs with the same number of multiplexing levels can have different sample rates due to different system word lengths, whereas in bit-parallel systems, two designs with the same control steps will have the same sample rate.

BITSYN accepts as input a compact, high level, signal flow graph (SFG) description which is similar to the SFG description used in DIGICAP[8]. The SFG description is in the form of branches connecting a set of nodes. A typical SFG description for a 2nd order wave digital filter[9] is shown in figure 2. BITSYN generates designs using the high level description language supported by the FIRST silicon compiler[1].

2.1. Resource sharing

The SFG information given in figure 2 is converted to data flow graphs[DFG] or precedence graphs as shown in figure 3. These graphs are built using a tree structure in which the leaf nodes represent the data inputs and current states (state inputs) and the parent nodes at the top of the trees represent the outputs and next states (state outputs). The DFG defines the order in which nodes must be evaluated for correct operation. A node in the graph denotes an arithmetic function and it can be evaluated if and only if all of its child node values have been previously computed. For each of the outputs and the state outputs, separate DFGs are constructed. A node in these graphs has always two child

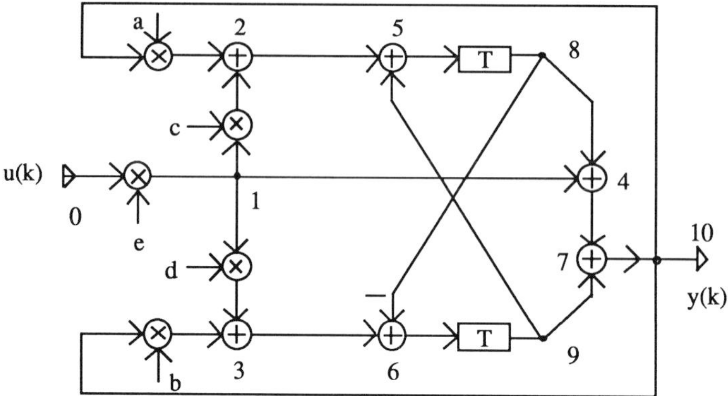

Figure 2. Second-order wave digital filter signal flow graph

nodes, but it can be a child to any number of nodes. A node output may be required at different levels of the same or different graphs.

An operator-parallel implementation can be achieved by replacing every node in the DFG's with the corresponding bit-serial operator (multiplier, adder etc.). Thus the DFG without any modification provides the data path structure with full operator-parallelism for pipelined bit-serial systems. Resources are shared to realize multiplexed architectures .

Whenever a resource is shared for two or more operations, multiplexers and demultiplexers are needed to channel the inputs to and outputs from the resource and registers may be needed to store the intermediate outputs. Individual adders are not shared because the gate cost of a single adder is not much more than the gate cost of a multiplexer. Thus the main objective is to share multiplier operators, as well as other operators (adders,

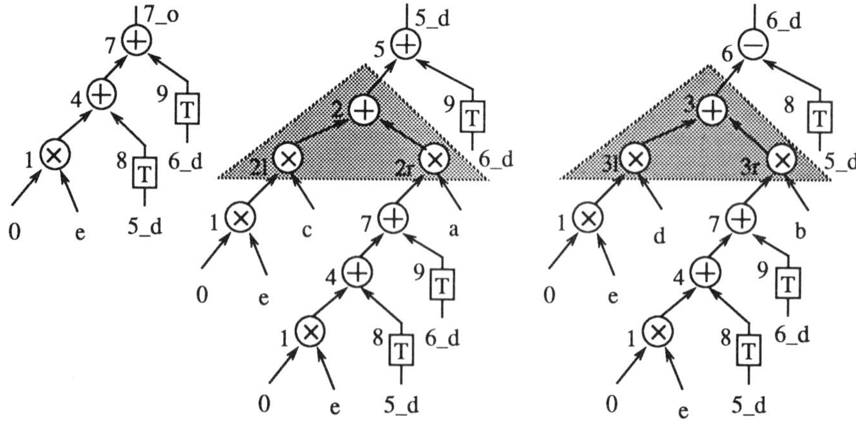

Figure 3. 2nd order wave digital filter : precedence graphs - no multiplexing

subtracters, etc.) which can be grouped with the multipliers. If two nodes are connected to the multiplier nodes through an equal number of similar nodes, then they can be shared along with the multipliers. The algorithm processes the precedence graphs, and modifies them as the resource sharing process continues. The different steps involved in the resource sharing process are explained with the help of the DFG's of the second order wave digital filter shown in figure 3.

Consider a pair of multiplier operations or nodes in the DFG. Examine all the child and parent nodes of both multipliers and find the node pairs which are alike in their operations. The node pairs may include other multiplier pairs. This is done by checking recursively using depth first and breadth first searches. For the multipliers (21,31) in figure 3 the pairs of nodes that can be shared along with (21,31) are (2,3) and (2r, 3r). After finding the two sets of nodes S1 = {21,2,2r} and S2 = {31,3,3r} which can be shared, the set which will remain in the graph and share other set's operations is selected. If a node of S1 is a child of any node of S2 then the set S1 must be computed before S2 to ensure proper data flow. The set S1 performs the operations of the other set during the OFF period of the multiplexing control signal and the set S2 becomes redundant. In this example, either S1 or S2 can perform each others operations.

The resultant resource shared DFGs are shown in figure 4. The nodes with an inverted 'Y' inside the two level multiplexers and are named 'xn..' (n = 1, in the example) where n denotes the level of multiplexing. The multiplexers are controlled by control signals 'cn'. Figure 5 illustrates the relation between the system clock, swl and different control signals. The nodes identified with 'Y' are conceptual demultiplexers. These nodes are useful in scheduling and in register allocation. The nodes named 'Ln..' indicate that the child of this node is valid during the ON period of the control signal Cn. Similarly the

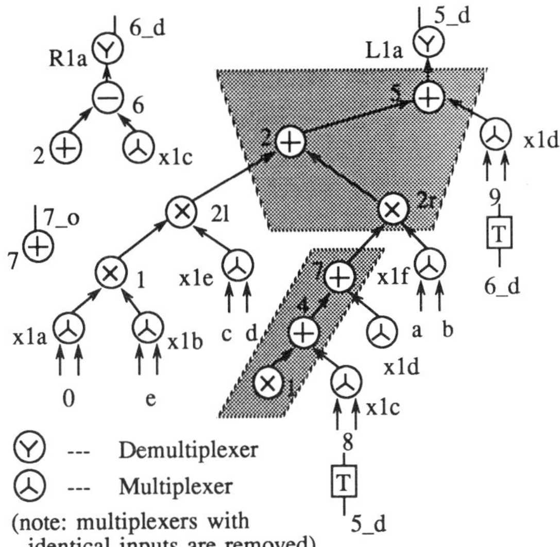

Figure 4. 2nd order wave digital filter : precedence graphs - multiplexing level = 1

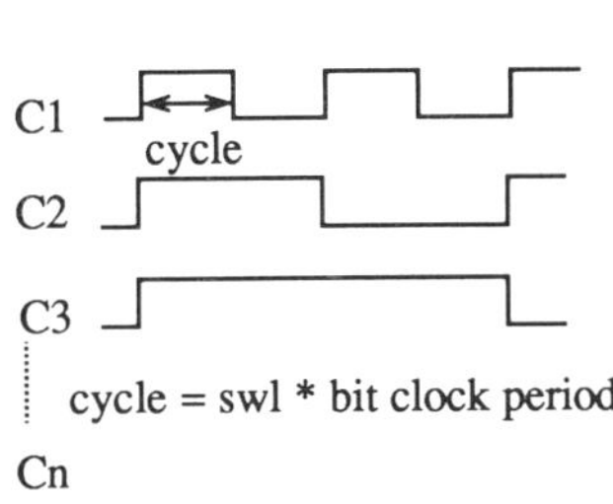

Figure 5. Control signals for multiplexed architectures

nodes with names 'Rn..' are valid during the OFF period of the control signal Cn. Thus after the first level of multiplexing, the number of multiplier operators is reduced from five to three. The resource sharing process continues with further levels of multiplexing to reach a minimum number of multipliers providing that the specified constraints (swl, cwl, sample rate) are not violated. The sets S1 = {2r,2,5} and S2 = {1,4,7} can be shared in the second level of multiplexing. Since all the nodes in S2 are children to the set S1, S2 must be computed before S1. Different structures are possible by starting with different multiplier pairs instead of (21,31). If two more sets of nodes (S3,S4) which are alike in their pattern exists, and are exclusive or concurrent in their operations with sets (S1,S2) then S3 and S4 can be multiplexed along with (S1,S2).

2.2. Scheduling

Scheduling is the process of determining the time at which each operation occurs. The timing information found through this process is used to calculate the maximum system latency which is the lower bound on the swl in recursive structures. Three different times, 'Tin', the time at which all node input data is required, 'Tout', the earliest time at which all node output data is available and 'Tneeded', the time at which node output data is needed, are computed for each node in the DFG. If a node output is connected to many other nodes, there may be more than one 'Tneeded' time for that node. These times are calculated in two phases. All the inputs at the bottom of the precedence graphs are initially assigned to have a 'Tout' at time zero. The time 'Tout' of all the demultiplexer nodes is also set to zero. In phase 1, the times are computed by traversing from bottom to top of the graphs such that for each operator :

Tout = Tin + latency of the operator.
Tneeded = Tout.

In phase 2, the times are updated by traversing from the top to the bottom of the graphs. Each node may have more than one 'Tneeded' time if that node is a child to more than one node. The times are modified as

Tout = minimum[Tneeded1,Tneeded2, ...].
Tin = Tout - latency of the operator.

The second phase of calculations ensures that all the nodes at the bottom of the graphs have appropriate 'Tin' values instead of the assumed zero time. In this fashion the system latency is calculated correctly as the maximum of the difference between 'Tout' of the next state node and 'Tin' of current state node.

2.3. Register allocation

In a pipelined bit-serial system, registers or delay elements are used to synchronize the arrival of data when, due to the different number and latency of operators in different data paths, there is a mismatch in the timing of the input signals at any operator. If a node has different 'Tout' and 'Tneeded' times then the number of one bit registers needed is equal to the difference of the two times. When resources are shared, the results computed in the first half period of the multiplexing cycle have to be stored or delayed until the other results are computed. The conceptual demultiplexers are used for this purpose. The nodes with names 'Ln..' are replaced by an appropriate number of 1 bit registers where as the nodes with names 'Rn..' are ignored because the input signals to these nodes are valid during the second half period of the multiplexing control signal and need no delay elements. In the case of the 'n' th level demultiplexer, the number of delay elements required is equal to 2^{n-1} * swl. All the inputs and state inputs are updated once in each sample period and when resources are shared these inputs may be used throughout the period as input values are circulated using circulating buffers. If an input has a 'Tin' value which is different from zero, then that input can be tapped at an appropriate delay from the circulated buffer without using extra delay elements.

2.4. Algorithm

The proposed BITSYN algorithm is based on an exhaustive search of the possible designs where the constraints such as minimum sample rate and maximum gate count are used to limit the range of the exhaustive search. The different steps involved in the BITSYN algorithm are summarized as follows :

1. Build the precedence graphs or DFG's. (next state and output calculations)
2. Prepare a list of all the possible pairs of multiplier nodes. Set the level of multiplexing to 1.
3. For each multiplier node pair {
   - 3a. Perform the resource sharing operation. That is, find two groups of nodes which create similar patterns around each multiplier and introduce multiplexers and conceptual demultiplexers into the DFG. Remove the nodes whose operations are shared, from the DFG's.
   - 3b. Schedule the resultant DFG's and allocate the registers. The design structure is then analyzed to find the number of equivalent gates, swl and normalized sample frequency. The number of equivalent gates is equal to the sum of the gates for each and every operator needed to implement the design in a gate array process.
     swl = maximum of {system latency, required swl}
     normalized sample frequency = $1/(2^n * \text{swl})$ where 'n' is the level of multiplexing.
   - 3c. Store the design. The multiplier node whose operation is shared does not exist after multiplexing. The node pairs containing such a multiplier node can be removed from the list of multiplier node pairs. The modified list is used in the further resource sharing process.
   - 3d. If at any level of multiplexing, the overhead costs are more than multiplier savings and no further reduction in the system latency is possible, the resource sharing process stops at that node. If a design does not satisfy the constraints such as the minimum sample rate and the maximum gate count, no further resource sharing process is necessary.
   - 3e. If the list of multiplier node pairs is not empty then increment the level of multiplexing, go to step 3 and repeat the recursive procedure.

   }

3. BIT-SERIAL DESIGNS

BITSYN generates pipelined bit-serial designs in a three dimensional space. The three dimensions are normalized sample frequency, gate count and the system latency. The normalized sample frequency is defined as the ratio of sample frequency to the system bit clock frequency. The system latency is the lower bound on the swl in recursive structures. The individual designs are not saved in FIRST netlist format. Instead to save memory, all the required transformations, that is the multiplier pairs and their order of multiplexing, are saved. The FIRST netlist of a selected design can be generated by performing the transformations when required. If two solutions have the same normalized sample frequency and system latency, then only the solution with the fewest gates is saved. If they have the same sample frequency and different system latencies then both solutions are stored. If the swl or cwl specifications are changed then the designs have to be recomputed to determine a new sample frequency and gate count. BITSYN thus allows the designer to chose different designs with a variety of different hardware costs, sample rates and system wordlengths. Different example designs generated using BITSYN will now be described. In all examples, the gate count (hardware cost) for each design has been extracted from actual working gate array implementations of the FIRST primitives. In tables 1 and 2 the times required to evaluate each set of filter designs is given in cpu seconds for the BITSYN 'C' language program running on a Sun SPARC 1+ CPU.

3.1. Second order wave digital filter (Wave-2)

A variety of designs generated using BITSYN for the second order wave digital filter[9] (signal flow graph in figure 2) have been tabulated in table 1. The cwl has been set to 12 bits (signed) and various values for the swl have been selected. For designs marked with #, a user specified swl of 32 bits has been selected (which is greater than the minimum swl). Note that there is a significant reduction in the gate count for designs where two or fewer multipliers are used.

3.2. Second order direct form digital filter (Direct-2)

In table 1 the design of a second order biquadratic direct form digital filter[10] has been evaluated using BITSYN. This filter has a 12 bit (signed) cwl and the swl is determined by the critical path latency. Note that this filter can be used to implement the same biquadratic transfer function as the Wave-2 digital filter. For level 0 multiplexing, this direct form design requires fewer gates and has a slightly higher sample frequency as compared to the wave filter design (due to the lower critical path latency). For multiplexing level three (one multiplier) the wave filter design requires somewhat fewer gates.

Table 1
Second order biquadratic filters designed using BITSYN (cwl = 12 bits)

Structure	Design No	Level of multip-lexing	No. of multi-pliers	System latency (bits)	Gate count	Normalized sample frequency	Sample frequency (KHz) *
Wave-2	1 m	0	5	25	4632 m	0.04000 m	1600.0 m
[4.5 @]	1 #				4977 #	0.03125 #	1250.0 #
	2 m	1	3	26	3485 m	0.01923 m	769.2 m
	2 #				4060 #	0.01562 #	625.0 #
	3 m	2	2	24	2715 m	0.01042 m	416.7 m
	3 #				3185 #	0.00781 #	312.5 #
	4 m	3	1	25	2444 m	0.00500 m	200.0 m
	4 #				2924 #	0.00391 #	156.3 #
Direct-2	1 m	0	5	23	4352 m	0.04348 m	1739.2 m
[19.5 @]	2 m	1	3	24	3425 m	0.02083 m	833.3 m
	3 m	2	2	24	2960 m	0.01042 m	416.7 m
	4 m	3	1	24	2611 m	0.00521 m	208.3 m
	5 m	3	1	25	2531 m	0.00500 m	200.0 m

m swl = system latency = minimum swl   # swl = 32 bits
@ cpu time in seconds on Sparc 1+ CPU   * bit clock frequency = 40 MHz

Table 2
Fifth order low pass elliptic filters designed using BITSYN (cwl = 12 bits)

Structure	Design No	Level of multip-lexing	No. of multi-pliers	System latency (bits)	Gate count	Normalized sample frequency	Sample frequency (KHz) *
LDI-5	1	0	9	68	12043	0.01471	588.2
[3228.5 @]	2	1	7	49	9600	0.01020	408.2
	3	2	5	26	6810	0.00962	384.6
	4	2	5	27	6586	0.00926	370.4
	5	3	4	26	6399	0.00481	192.3
	6	3	3	27	5869	0.00463	185.2
	7	4	2	29	6412	0.00216	86.2
	8	5	1	29	8816	0.00108	43.1
Wave-5	1	0	8	71	11563	0.01408	563.4
[810.6 @]	2	1	6	66	11546	0.00758	303.0
	3	2	5	45	9650	0.00556	222.2
	4	2	4	46	9333	0.00543	217.4
	5	3	3	25	7526	0.00500	200.0
	6	3	3	28	6358	0.00446	178.6
	7	4	2	29	6486	0.00216	86.2
	8	5	1	29	7110	0.00108	43.1

swl = system latency = minimum swl for all designs
@ cpu time in seconds on Sparc 1+ CPU * bit clock frequency = 40 MHz

3.3. Fifth order bilinear LDI filter (LDI-5)

The BITSYN program has been used to analyze designs of a fifth order Bilinear LDI (elliptic) low pass digital filter[11]. Designs generated for this low sensitivity ladder filter structure using BITSYN are given in Table 2. Note that in this particular filter architecture, the implementation which requires the fewest gates is the one in which three multipliers are used. For designs with fewer than three multipliers, the extra cost of registers required to delay signals is larger than the savings in the gate count due to the sharing of multiplier operators.

3.4. Fifth order wave digital filter (Wave-5)

The BITSYN program has been used to evaluate designs for a fifth order wave digital low pass (elliptical) filter[12] which has been studied extensively in high level synthesis[6][13]. In table 2 the design alternatives for this filter are provided. This filter implements the same class of transfer functions as the 5th order LDI filter. Note that the smallest design is larger than that of the LDI but has similar swl and sample rate specifications.

4. CONCLUSIONS

In this paper we have presented a technique for the synthesis of pipelined bit-serial DSP systems and described a program BITSYN for this synthesis. A variety of recursive digital filter implementations have been evaluated and contrasted using BITSYN. The design space for this synthesis includes the size (gate count), sample rate and minimum system latency (critical path). BITSYN relieves the designer of the involved and tedious process of scheduling complicated recursive digital filtering algorithms for pipelined bit-serial implementations. The potential applications for small size bit-serial designs which meet the required sample rate, swl and cwl specifications will increase as the clock speed of semiconductor technologies increases. The large swl inherent in these bit-serial designs may be used advantageously to overcome the effects of finite precision arithmetic in practical designs.

The simple multiplexing strategy used (figure 5) can in some cases contribute to an increase in the design size at higher levels of multiplexing. More complex control structures will be considered to overcome this limitation. In addition, strategies to multiplex groups of add, subtract and shift units will also be investigated.

Acknowledgements

This work is supported by the Natural Science and Engineering Research Council of Canada and the Alberta Microelectronics Centre.

5. REFERENCES

1 P. Denyer and D. Renshaw, "VLSI signal processing : A bit-serial approach", Reading, MA: Addison-Wesley, 1985.

2 N. Park and A. Parker, "SEHWA : A software package for synthesis of pipelines from behavioral specifications", IEEE Trans. on Computer-Aided Design, vol. 7, pp. 356-370, Mar 1988.

3 P. Paulin, J. Knight, and E. Gyrczyc, "HAL: A multi-paradigm approach to automate data path synthesis", in Proc. 23rd Design Automation Conf., pp. 263-270, 1986.

4 A. Parker, J. Pizarro and M. Mlinar, "MAHA : A program for data path synthesis", in 23rd Design Automation Conf., pp. 461-466, 1986.

5 B. Pangrie and D. Gajski, "State synthesis and connectivity binding for microarchitecture compilation", in Proc. ICCAD86, Nov 1986.

6 B. S. Haroun and M. I. Elmasry, "Architectural synthesis for DSP silicon compilers", IEEE Trans. on Computer-Aided Design, vol. 8, pp. 431-447, April 1989.

7 R. Hartley and P. Corbett, "Digit-serial processing techniques", IEEE Trans. on Circuits and Systems, vol. 37, June 1990.

8 L. E. Turner, D. A. Graham and P. B. Denyer, "The analysis and implementation of digital filters using a special purpose CAD tool", IEEE Trans. on Education, vol. 32, pp. 287-297, Aug 1989.

9 K. Meerkotter and W. Wegener, "A new second-order digital filter without parasitic oscillations", Arch. Elektronik & Ubertragungstech., 29, pp 312-314, 1975.

10 A. Antoniou, "Digital Filters: analysis and design", Reading, McGraw-Hill, 1979, pp. 77.

11 L. E. Turner and B. K. Ramesh, "Low sensitivity digital LDI ladder filters with elliptic magnitude response", IEEE Trans. on Circuits and Systems, vol. CAS-33, pp. 697-706, July 1986.

12 P. Dewilde, E. Deprettere and R. Nouta, "Parallel and pipelined VLSI implementations of signal processing algorithms", in VLSI and Modern Signal Processing, S. Y. Kung, H. J. Whitehouse, T. Kailath, Ed. Englewood Cliffs, NJ: Prentice-Hall, 1985, ch. 15, pp. 257-276.

13 R. Potasman, J. Lis, A. Nicolau and D. Gajski, "Percolation based synthesis", in 27th ACM/IEEE Design Automation Conf. pp. 444-449, 1990.

VLSI 91
A. Halaas and P.B. Denyer (Eds.)
Elsevier Science Publishers B.V. (North-Holland)

Symbolic Model Checking with Partitioned Transition Relations

J. R. Burch E. M. Clarke D. E. Long
School of Computer Science
Carnegie Mellon University

Abstract

We significantly reduce the complexity of BDD-based symbolic verification by using *partitioned transition relations* to represent state transition graphs. This method can be applied to both synchronous and asynchronous circuits. The times necessary to verify a synchronous pipeline and an asynchronous stack are both bounded by a low polynomial in the size of the circuit. We were able to handle stacks with over 10^{50} reachable states and pipelines with over 10^{120} reachable states.

1 Introduction

Although methods for verifying sequential circuits by searching their state transition graphs have been investigated for many years, it is only recently that such methods have begun to seem practical. Before, the largest circuits that could be verified had about 10^6 states. Now it is easy to check circuits that have many orders of magnitude more states [3, 5, 6, 7]. The reason for the dramatic increase is the use of special data structures such as binary decision diagrams (BDDs) [2] for encoding the state transition graphs of such systems.

In this paper, we show how to process state transition graphs more efficiently than in our previous work [5, 6]. Our new approach involves using multiple BDDs, which are implicitly conjuncted or disjuncted, to represent the graphs. We call this kind of representation a *partitioned transition relation*. The BDDs that make up the partitioned transition relation are derived in a natural way from the structure of the circuit being verified. We illustrate the power of the technique by verifying an asynchronous stack [10] and a synchronous pipeline circuit [5]. Using a partitioned transition relation, we were able to verify a stack 32 bits wide and 2 cells deep. For comparison, we were unable to verify a stack only 1 bit wide and 1 cell deep when using a single BDD to represent the transition relation because the transition relation required more than 350,000 BDD nodes. For a

This research was sponsored in part by the Defense Advanced Research Projects Agency (DOD), ARPA Order No. 4976, and in part by the National Science Foundation under contract numbers CCR-8722633 and MIP-8858807.

pipeline with 4 registers, each 32 bits wide, the partitioned transition relation required less than 2,500 BDD nodes, while using a single BDD required nearly 340,000 nodes, a savings of nearly a factor of 140. On a Sun 4, the verification time improved from approximately 14,000 seconds (projected) to 995 seconds, a factor of about 14. We were also able to handle example pipelines with over 10^{120} reachable states.

There are several other methods that use BDDs in the verification of sequential circuits. Bryant and Seger [3] use a symbolic switch-level simulator to check pre- and post-conditions specified in a restricted form of temporal logic. The logic allows boolean conjunction and the next time modality (**X**). Coudert, Berthet, and Madre describe a system for showing equivalence between deterministic finite automata [7]. Their system performs a symbolic breadth-first search of the state space reachable by the product of the two automata. None of these methods can easily handle nondeterministic systems. With transition relations, it is very natural to model examples like the cache coherency protocol for the Encore Gigamax, which McMillan has recently investigated [11]. A major feature of the Gigamax architecture is an asynchronous, and hence nondeterministic, interconnection network. The use of abstraction to hide certain details of the cache replacement policy also gives rise to nondeterminism in this example.

2 Symbolic verification

Given a circuit, let V be its set of boolean state variables. We identify a boolean formula over V with the set of valuations which make the formula true. A valuation of the variables corresponds in a natural way to a state of the circuit; hence the formula may be thought of as representing a set of circuit states. The BDD for the formula is in practice a concise representation for this set of states. In the remainder of the paper, we will denote sets of states using S and T. We denote the BDD representing the set S by $S(V)$, where V is the set of variables that the BDD depends on. In addition to representing sets of states of a circuit, we must represent the transitions that the circuit can make. To do this, we use a second set of variables V'. A valuation for the variables in V and V' can be viewed as designating a pair of states in the circuit, and we can represent sets of pairs using BDDs as above. We will refer to sets of pairs of states as transition relations. If N is a transition relation, then we write $N(V, V')$ to denote the BDD that represents it.

There are many finite state verification methods that can make effective use of this representation [5, 7]. For our purposes, the important property of these algorithms is that the basic step is performing computations of the following form:

$$S'(V') = \exists_{v \in V} \left[S(V) \wedge N(V, V') \right].$$

(The notation above indicates a series of nested existential quantifications, one for each variable in V.) This expression, called a relational product, gives the set of states S' reachable in one step from the set of states S in a circuit with transition relation N. It is crucial to be able to do this computation efficiently. A special algorithm is typically used to do this operation in one pass over the BDDs $S(V)$ and $N(V, V')$. By using such an algorithm, it is possible to avoid building the BDD for $S(V) \wedge N(V, V')$, which would often be impractically large. Unfortunately, the BDD $N(V, V')$ itself is often very big. Up to this point, being forced to construct this BDD has been the major stumbling block in trying to verify complex circuits. In the following sections, we describe how to overcome this problem by using a partitioned transition relation to represent N.

3 Deriving transition relations

The first step in verifying a circuit is to derive its transition relation. Our goal is to reflect the structure of the circuit in the structure of the transition relation, so that the transition relation can be stored and manipulated more efficiently.

For a synchronous circuit with n state variables, we let $V = \{v_0, \ldots, v_{n-1}\}$ and $V' = \{v'_0, \ldots, v'_{n-1}\}$. For each state variable v_i, there is a piece of combinational logic which determines how it is updated. Let f_i be the function computed by this logic. Then the value of v_i in the next state is given by

$$v'_i = f_i(V).$$

These equations are used to define the relations

$$N_i(V, V') = (v'_i \Leftrightarrow f_i(V)).$$

In a legal transition of the circuit, each N_i must be true; hence the transition relation for the circuit is

$$N(V, V') = N_0(V, V') \wedge \cdots \wedge N_{n-1}(V, V').$$

Thus, the transition relation for a synchronous circuit can be expressed as a conjunction of relations.

In practice, each N_i can often by represented by a small BDD (typically fewer than 100 nodes). However, the size of the BDD representing the entire transition relation may grow as the product of the sizes of the individual parts, and thus may be prohibitively large. In the past, this has been the major limitation of symbolic model checking. For our new method, we instead represent the transition relation by a list of the parts, which are implicitly conjuncted. We call this representation a conjunctive partitioned transition relation.

Asynchronous circuits can be modeled with a conjunctive partitioned transition relation, like synchronous circuits, and can also be represented by a disjunctive partioned transition relation. To simplify the description of how these forms of transition relation are computed, we assume that all the components of the circuit have exactly one output, and have no internal state variables. It is straightforward to generalize the method to handle cases where this assumption does not hold.

In asynchronous circuits, there can be an arbitrary delay between when a transition is enabled and when it actually occurs. We can model this by allowing each component to nondeterministically choose whether to transition its output, resulting in a conjunctive partitioned relation with n parts, all of the form

$$R_i(V, V') = N_i(V, V') \vee (v'_i \Leftrightarrow v_i),$$

where $N_i(V, V') = (v'_i \Leftrightarrow f_i(V))$ gives the quiescent value of the output of each component. For some components, such as C-elements and flip-flops, the function $f_i(V)$ may depend on the current value of the output of the component, as well as the inputs.

The above model for asynchronous circuits allows wires to transition concurrently. We can also use an interleaving model, which allows only one wire to transition at a time. This idea can be used to construct a disjunctive partitioned transition relation, as follows. First, apply distributivity to the conjunction of the R_i, giving a disjunction of 2^n terms. Each of these terms corresponds to the simultaneous transitioning of some subset of the n wires in the circuit. Second, keep only those terms that correspond to exactly one wire transitioning. This results in a disjunction of the form

$$N(V, V') = Q_0(V, V') \vee \cdots \vee Q_{n-1}(V, V')$$

where

$$Q_i(V, V') = (v'_0 \Leftrightarrow v_0) \wedge \cdots \wedge (v'_{i-1} \Leftrightarrow v_{i-1}) \wedge N_i(V, V') \wedge (v'_{i+1} \Leftrightarrow v_{i+1}) \wedge \cdots \wedge (v'_{n-1} \Leftrightarrow v_{n-1}).$$

We represent the full transition relation as a list of the $Q_i(V, V')$, which are implicitly disjuncted.

4 Computing relational products

As noted earlier, computing relational products is a fundamental operation in many symbolic verification methods. This section describes how relational products can be computed using the representations described in the previous section. These techniques significantly increase the size of circuits that can be verified compared to previous methods.

For a disjunctive partitioned transition relation, the relational product computed is of the form

$$S'(V') = \exists_{v \in V} \big[S(V) \wedge (N_0(V, V') \vee \cdots \vee N_{n-1}(V, V'))\big].$$

This relational product can be computed without ever constructing the BDD for the full transition relation by rewriting $S'(V')$,

$$S'(V') = \exists_{v \in V} \big[S(V) \wedge N_0(V, V')\big] \vee \cdots \vee \exists_{v \in V} \big[S(V) \wedge N_{n-1}(V, V')\big].$$

Thus, we are able to reduce the problem of computing $S'(V')$ to one of computing a series of relational products involving relatively small BDDs. This technique was used previously for verifying asynchronous circuits [5]. Much larger asynchronous circuits could be verified using this method than with a monolithic transition relation.

For a conjunctive partitioned transition relation, the relational product computed is of the form

$$S'(V') = \exists_{v \in V} \big[S(V) \wedge (N_0(V, V') \wedge \cdots \wedge N_{n-1}(V, V'))\big]. \tag{1}$$

The main difficulty in computing $S'(V')$ without building the conjunction is that conjunction does not distribute over existential quantification. The method given below overcomes this difficulty.

Our new technique is based on two observations. First, circuits exhibit locality, so many of the $N_i(V, V')$ will depend on only a small number of the variables in V and V'. Second, although conjunction does not distribute over existential quantification, subformulas can be moved out of the scope of existential quantification if they do not depend on any of the variables being quantified. We will take advantage of these observations by conjuncting the $N_i(V, V')$ with $S(V)$ one at a time and quantifying out each variable v when none of the remaining $N_i(V, V')$ depend on v. More formally, the user must choose a permutation ρ of $\{0, \ldots, n-1\}$. This permutation determines the order in which the $N_i(V, V')$ are conjuncted. For each i, let D_i be the set of variables in V that $N_i(V, V')$ depends on. Also, let

$$E_i = D_{\rho(i)} - \bigcup_{k=i+1}^{n-1} D_{\rho(k)}.$$

Thus, E_i is the set of variables contained in $D_{\rho(i)}$ that are not contained in $D_{\rho(k)}$ for any k larger than i. The E_i are pairwise disjoint and their union is equal to V. The relational product in equation 1 can be computed as

$$\begin{aligned} S_1(V, V') &= \exists_{v \in E_0} \left[S(V) \wedge N_{\rho(0)}(V, V') \right] \\ S_2(V, V') &= \exists_{v \in E_1} \left[S_1(V, V') \wedge N_{\rho(1)}(V, V') \right] \\ &\vdots \\ S'(V') &= \exists_{v \in E_{n-1}} \left[S_{n-1}(V, V') \wedge N_{\rho(n-1)}(V, V') \right]. \end{aligned}$$

The ordering ρ has a significant impact on how early in the computation state variables can be quantified out. This affects the size of the BDDs constructed and the efficiency of the verification procedure. Thus, it is important to choose ρ carefully, just as with the BDD variable ordering. In practice, we have found it fairly easy to come up with orderings which give good results.

In the previous section, we described how a circuit could be represented by a set of $N_i(V, V')$, each depending on exactly one variable in V'. While this is almost always more efficient than constructing the full transition relation, it may not be the best choice. As long as the BDDs do not get too large, it is better to combine several of the $N_i(V, V')$ into one BDD by forming their disjunction or conjunction.

5 Verifying asynchronous circuits

Asynchronous circuits can be verified in two steps. First, compute the set of states the circuit, composed with an environment, can reach from a given set of initial states. Then check that no hazard can occur in any of the reachable states. Finding the reachable states is the most computationally expensive of these two steps. In practice, checking for hazards is usually done as the reachable states are computed. This is similar to Dill's [9] method for verifying safety properties of asynchronous circuits.

The set of reachable states is found by computing the least fixed point S of

$$S(V') = S_0(V') \vee \exists_{v \in V} \left[S(V) \wedge N(V, V') \right],$$

where S_0 is the initial set of states and N is the transition relation of the circuit. We use frontier set simplification to speed up the computation of this fixed point [5, 7]

There are significant differences in the complexity of doing reachability analysis using conjunctive and disjunctive partitioned transition relations. Consider two uncoupled systems M' and M'' with disjoint sets of state variables V' and V''. Let M be the composition of these two systems. This is an unrealistic example, but it helps illustrate what happens when computing the reachable states of loosely coupled systems. The BDD $S(V)$ representing the set of reachable states of M is equal to $S'(V') \wedge S''(V'')$, where $S'(V')$ ($S''(V'')$) is the BDD for the reachable states of M' (M''), and $V = V' \cup V''$. An efficient way to order the BDD variables of the combined system in this case is to have all the variables of one component (say M') before any of the variables in the other component. Then the number of BDD nodes in $S(V)$ is equal to the sum of the nodes in $S'(V')$ and $S''(V'')$, independent of whether conjunctive or disjunctive partitioning is used. However, the sizes of the BDDs representing the intermediate state sets are potentially different for the two methods.

Let $S_i(V)$, $S_i'(V')$ and $S_i''(V'')$ be the BDDs representing the states reachable in i steps by M, M' and M'', respectively, using non-interleaved semantics. Similarly, let $T_i(V)$, $T_i'(V')$ and $T_i''(V'')$ be the BDDs representing the states reachable in i steps by M, M' and M'', respectively, using interleaved semantics. In the conjunctive case, $S_i(V) = S_i'(V') \wedge S_i''(V'')$, so the size of each $S_i(V)$ is equal to the sum of the sizes of $S_i'(V')$ and $S_i''(V'')$, just as for the set of reachable states. However, for the disjunctive case,

$$T_i(V) = \bigvee_{k=0}^{i} T_k'(V') \wedge T_{i-k}''(V'').$$

Thus, interleaving semantics introduces an artificial correlation between the local states of M' and M'' in the $T_i(V)$. The $T_i(V)$ are generally much larger than the $S_i(V)$, since each $T_i(V)$ must contain $T_k''(V)$ for all $k \leq i$. Because of this effect, reachability analysis with disjunctive partitioning is less efficient than with conjunctive partitioning.

We can make disjunctive partitioning more efficient by modifying the breadth first search used for reachability analysis. To search the reachable states of M, first compute states reachable by transitions of wires in M'. Then compute the states reachable from that set by transitioning on wires in M''. This is equal to the global reachable state set, since M' and M'' are uncoupled. Separately computing local fixed points for the two parts of the system in this way removes the artificial correlation described above. In general, for a circuit C divided into loosely coupled subcircuits C_j, we compute the reachable states of C by repeatedly computing local fixed points for each C_j until a global fixed point is reached. This idea can be extended to a hierarchy with any number of levels.

6 An asynchronous stack

In this section, we compare conjunctive and disjunctive partitioned transition relations for verifying asynchronous circuits by considering an asynchronous lazy stack due to Martin [10]. To determine the asymptotic performance of the various methods discussed earlier, we performed a reachability analysis for stacks with varying depth d and word width w. This is sufficient to determine the asymptotic complexity of verification, even though we did not check for hazards. Hazard checking increases the times by a constant factor.

The stack consists of an array of d cells, each cell consisting of a control part, a data part and a completion tree. The data part of each cell consists of w storage elements. The completion trees signal when all the storage elements in a cell have completed the current data transfer.

The verification system that we use is written in a combination of C and LISP. The BDD package is written in C and is roughly comparable in performance to the package described by Brace, Rudell and Bryant [1].

We studied how verification time varied with w for four different methods:

1. Disjunctive partitioning using standard breadth first search. We combined the transition relations for the gates making up each individual control part, each of the individual storage elements, and each completion tree.

2. Disjunctive partitioning using modified breadth first search and the same partitioning of the transition relation as above. At the top level, the hierarchy used for local fixed point computation consisted of the environment and each cell as a unit. Each cell was broken into the control part, the completion tree and the data part. The data part was further subdivided into $\lceil \lg(w) \rceil$ levels, each of two parts.

3. Conjunctive partitioning using the same partitioning of the transition relation as above. We used the following ordering ρ of the parts of transition relation: the environment at the top of the stack; the control part and data parts of each cell, ordered from the top of the stack to the bottom; the completion trees, also ordered from the top of the stack to the bottom; and the environment at the bottom of the stack.

4. Conjunctive partitioning using the same partitioning as above, but with the control and data parts within each cell combined into one BDD. The ρ used above is modified in the obvious way.

In all cases, we used an initial state set in which each cell could be full or empty and the data in each cell was arbitrary. Using a more restricted set of initial states, such as having all cells initially empty, can increase the verification time by as much as a factor of d.

A graph of the search times versus stack width for the various methods is shown in figure 1. We found that disjunctive partitioning with breadth first search were feasible only for small examples. Disjunctive partitioning with modified breadth first search and conjunctive partitioning were all much more efficient. Search times using methods 2 and 3 grew slightly faster than quadratically. Method 4 gave a growth rate of roughly $w^{1.5}$. Using this method, we were able to find the reachable states of a 32 bit wide, depth 2 stack in under an hour of CPU time on a Sun 4. This circuit had over 989 boolean state variables and 10^{50} reachable states.

The BDDs in the transition relation are all of constant size, except for those representing the completion trees. These BDDs are growing as $w^{\lg 3}$, but for the values of w

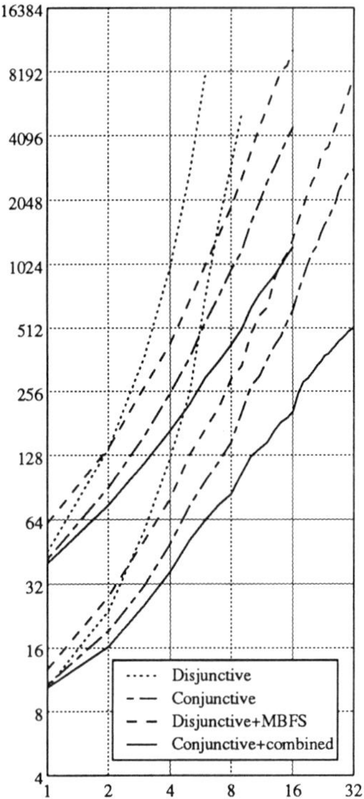

Figure 1: Search times in seconds for stacks of various widths, with $d = 1, 2$

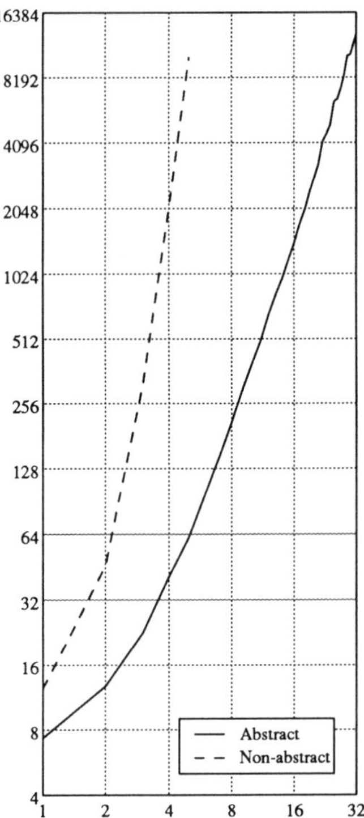

Figure 2: Search times in seconds for stacks of various depths, with $w = 1$

we considered, they are still quite small. For larger w, it might be necessary to split the completion trees into more than one BDD.

We also explored how the search time varied with the depth of the stack, using conjunctive partitioning. The number of steps needed to compute the reachable states grows quadratically in d. The states which require the largest number of steps to reach are states in which internal signals within the stack control are not stable. Thus, we were able to avoid the quadratic search depth by replacing the control part of each cell by an abstract model having only external signals. We separately verified that the abstract model correctly describes the external behavior of the control part. With this abstraction, the number of steps needed to find all reachable states is linear in d, and the search time is cubic in d (see figure 2).

Although this kind of abstraction can greatly improve the efficiency of verifiers that explicitly enumerate states, it is usually not nearly as helpful when used with symbolic verifiers. For example, the search times for stacks of depth one improve only about 20 percent when the abstract model of the control part is used. The effect that abstraction has on the search depth, as described above, is an exception to this rule.

7 A synchronous pipeline

We also considered the verification of a synchronous pipeline circuit. This circuit, described in an earlier paper [5], performs three-address arithmetic and logical operations on operands stored in a register file. We experimented with a number of versions of the pipeline with varying numbers of registers, register widths, numbers of pipe stages, and numbers of operations. The verification times grow as low polynomials in all dimensions. We also ran several more realistic examples. The largest of these was a pipeline with 8 registers, each 32 bits wide, 2 pipe registers, and one operation. This example had 406 state variables resulting in more than 10^{120} reachable states, and the verification took 4 hours and 20 minutes of CPU time on a Sun 4. Details of the verification of the pipeline can be found in [4].

8 Discussion and future research

Using partitioned transition relations significantly improves the efficiency of symbolic verification. We verified a stack with over 950 state variables and more than 10^{50} reachable states and a pipeline with more than 400 state variables and over 10^{120} reachable states. We also studied the asymptotic performance of our verification methods. This kind of asymptotic analysis is an important way to compare different techniques.

For deterministic systems, a transition function vector can be used to represent how a circuit transitions from one state to another. In this method, a separate BDD is used for each state holding node of the system. This BDD represents the function computed by the combinational logic driving the associated node. Coudert *et al.* [7, 8] describe a number of algorithms for manipulating transition functions. They note that the monolithic transition relation can require many more BDD nodes than the corresponding transition function vector [8]. However, they report that computations with transition relations are faster than those using transition functions. Partitioned transition relations provide the speed of transition relations and the memory efficiency of transition functions.

Touati *et al.* [12] proposed another method for representing transition relations as implicit conjunctions. They use the *constrain* operator of Coudert *et al.* [7] to eliminate the state set $S(V)$ in equation 1. Then they compute the resulting conjunction as a balanced binary tree, quantifying out each variable in V when all the BDDs depending on that variable have been combined. We believe that this method is inferior to the one proposed here because the constrain operator may introduce dependencies on any of the variables in $S(V)$. This makes it impossible to compute in advance a schedule for quantifying out the variables in V, which in turn reduces the practicality of caching results between relational product computations. In addition, if $S(V)$ depends on most of the variables in V, it may not be possible to quantify out many variables before performing the final conjunction. They also suggest having one transition relation per state variable. In our experience, it is often better to combine parts of the transition relations to reduce overhead; this idea is also applicable to their method. We implemented their method and tested it on some of the examples in section 7. For a pipeline with four 8 bit registers, one pipe register and one operation, our method was more than five times faster. In addition, for some of the relational product computations, the intermediate BDDs using their method were more than an order of magnitude larger than the final result.

References

[1] K. S. Brace, R. L. Rudell, and R. E. Bryant. Efficient implementation of a BDD package. In *27th ACM/IEEE Design Automation Conference*, 1990.

[2] R. E. Bryant. Graph-based algorithms for boolean function manipulation. *IEEE Trans. Comput.*, C-35(8), 1986.

[3] R. E. Bryant and C.-J. Seger. Formal verification of digital circuits using symbolic ternary system models. In R. Kurshan and E. M. Clarke, editors, *Workshop on Computer-Aided Verification*. DIMACS, 1990.

[4] J. R. Burch, E. M. Clarke, and D. E. Long. Representing circuits more efficiently in symbolic model checking. In *28th ACM/IEEE Design Automation Conference*, 1991.

[5] J. R. Burch, E. M. Clarke, K. L. McMillan, and D. L. Dill. Sequential circuit verification using symbolic model checking. In *27th ACM/IEEE Design Automation Conference*, 1990.

[6] J. R. Burch, E. M. Clarke, K. L. McMillan, D. L. Dill, and J. Hwang. Symbolic model checking: 10^{20} states and beyond. In *LICS*, 1990.

[7] O. Coudert, C. Berthet, and J. C. Madre. Verification of synchronous sequential machines based on symbolic execution. In J. Sifakis, editor, *Automatic Verification Methods for Finite State Systems*, volume 407 of *LNCS*. Springer-Verlag, 1989.

[8] O. Coudert, J. C. Madre, and C. Berthet. Verifying temporal properties of sequential machines without building their state diagrams. In R. Kurshan and E. M. Clarke, editors, *Workshop on Computer-Aided Verification*. DIMACS, 1990.

[9] D. L. Dill. *Trace Theory for Automatic Hierarchical Verification of Speed-Independent Circuits*. ACM Distinguished Dissertations. MIT Press, 1989.

[10] A. J. Martin. A synthesis method for self-timed VLSI circuits. In *Proceedings: IEEE International Conference on Computer Design*, 1987.

[11] K. L. McMillan and J. Schwalbe. Formal verification of the Gigamax cache consistency protocol. In *International Symposium on Shared Memory Multiprocessing*, 1991.

[12] H. J. Touati, H. Savoj, B. Lin, R. K. Brayton, and A. Sangiovanni-Vincentelli. Implicit state enumeration of finite state machines using BDD's. In *ICCAD*, 1990.

VLSI 91
A. Halaas and P.B. Denyer (Eds.)
Elsevier Science Publishers B.V. (North-Holland)

Integration of Formal Methods with System Design

Eleanor M. Mayger and Michael P. Fourman [1]
Abstract Hardware Ltd, Uxbridge, UB8 3PH, U.K.

Abstract
This paper presents the application of formal synthesis techniques to derive a hardware implementation of an error-correcting algorithm. We show how an executable high level specification is refined, using the LAMBDA/DIALOG design assistant, to a register-transfer level netlist suitable for compilation to silicon using established logic synthesis techniques. LAMBDA[2] uses computer-assisted formal reasoning to provide formally verified logical correctness for interactive design. It allows user control of architectural decisions by relating behaviour and structure, allowing the user to specify bindings between abstract behaviour and structure. Lambda supports both data refinement and hierarchical design. User-interaction is *via* a graphical *schematic capture* style interface.

1 Related Work

In response to technology push, there is increasing automation of the design *process*. Logic Synthesis is now a reality ([Brayton *et al.*, 1988]) — silicon structures are generated from a specification of function. Behavioural Synthesis is increasingly successful at automating state assignment, scheduling and allocation [Camposano and Tabet, 1989]. However, there are few tools supporting system-level design. A programme for developing a design assistant based on Computer-Assisted Formal Reasoning was proposed in [Fourman *et al.*, 1988]. Here we report on the LAMBDA/DIALOG design assistant, which implements and extends these ideas, and its application to a significantly complex case study.

We are concerned to integrate user-interaction with automated synthesis *tactics*. Similar concerns motivate [Dutt *et al.*, 1990], however, we are also concerned to use a representation that supports high-level specification (including features such as user-defined datatypes and recursively defined functions), and refinement of both data and timing. Our approach is also related to that of [Hanna *et al.*, 1989] where it is proposed to develop a logic of design; instead, we use higher-order logic to represent design.

Other formal tools have been applied to hardware *verification*. In particular, HOL [Gordon, 1988] and the Boyer-Moore theorem prover [Boyer and Moore, 1988] have been used to verify the VIPER microprocessor [Cohn, 1988] and the FM8501[Hunt, 1987]. However, these tools tend to be suited more to academics than engineers and have been used for *post design* verification; they are not integrated with the design process.

---

[1]The case study described in this paper was supported under U.K. D.T.I. I.E.D. Project Number 1292 "Formal System Design".

[2]Logic And Mathematics Behind Design Automation

LAMBDA has drawn on many ideas first used in other systems. Its hardware model is similar to HOL, in that components are described as relations between signals (functions). HOL uses an approach introduced in LCF[Gordon *et al.*, 1979], which is suited to ad-hoc post design verification, but does not easily model design refinement. Instead, like ISABELLE [Paulson, 1987], LAMBDA uses higher-order unification. It uses novel techniques to provide close control of this unification, allowing interactive synthesis as well as verification. It represents a *partial* design as a logically derived *rule*, relating the partial design, the original specification and any outstanding tasks. When there are no outstanding tasks, the rule expresses the fact that the design implements the specification. There is no need for post-design work on formal verification, because the design process has already generated the *proof* that the implementation is correct.

2 Introduction

We use a case study– the synthesis of a hardware implementation of an error-correcting algorithm– to illustrate techniques that provide machine support for high-level design. We show how an executable high-level specification is refined, using the LAMBDA/DIALOG design assistant, to a register-transfer level netlist suitable for compilation to silicon using established logic synthesis techniques.

Although any step in the synthesis (e.g. scheduling see [Fourman and Mayger, 1989]) can, in principle, be automated, the toolset also allows interaction. This leaves the designer free to use his own skill and judgement during the course of the design. LAMBDA will record his decisions and ensure that information and constraints, (involving e.g. timing), are maintained and propagated correctly throughout the whole design process[Mendler, 1990].

Like most Theorem Provers, LAMBDA can make use of equational reasoning to rewrite design requirements into an equivalent, but simpler form. LAMBDA also supports tactical and inductive proof. In addition, it uses unification, which permits synthesis; the addition of new components in order to achieve a design. LAMBDA was designed to be used by engineers rather than logicians, so the user interface is an important element of the toolset. A window interface, called DIALOG, allows the user to interact with the Theorem Prover using a schematic editor.

3 Case Study

3.1 Problem Description

This example (we call it *erco*) concerns the hardware implementation of an error correcting algorithm for a compact disc player. [3] The algorithm concerns an 8 bit (i.e. 256 element) finite field `GF`, represented by a polynomial of degree 7 over the booleans, with addition `++` given by **xor**ing the elements of two polynomials together, pairwise. Multiplication `**` is determined by the characteristic polynomial of the field (i.e. an equation giving x^8 in terms of 1 to x^7).

The system takes as inputs polynomials of degree 31 over `GF` (given by arrays of length 32), and calculates 4 *syndroms* for each one. It attempts to make all these 4 values zero by modifying the polynomial at at most two locations. It outputs the modified polynomial together with an error flag, which is set to true iff the modification cannot be performed.

An algorithm specifying how this is to be achieved is written in the functional language, ML. This algorithm was supplied to us by Philips, and their implementation of it is of the order of 10,000 gates, and 6 man-months design time.

[3] We would like to thank Philips, in particular Ton Kalker, for providing the specification.

3.2 Design stages

We describe several distinct phases of the design process in this paper. We have

- **Specification.** We discuss the use of a functional language to give an abstract specification of the algorithm.
- **Properties of the Specification** It is useful to investigate properties of the specification so as to check that it is sensible. We show how formal proof can be a useful complement to simulation, for this purpose.
- **DIALOG Set-up.** Different design areas require different knowledge, and different intermediate modules. We show how DIALOG is customised to a particular design area.
- **Synthesis.** Here, we discuss some of the steps in the interactive synthesis of an implementation. We give examples of refinement, partitioning and constraint propagation. LAMBDA and DIALOG are used to produce a "correct by construction" design and proof.

We spent about six man-months working on this example. Most (95 %) of this time was spent on writing (re-usable) libraries to extend the toolset. If we were to undertake a similar example now the time required should be of the order of one or two weeks to produce a design to gate-level. The top level, together with most of the second level, of a hierarchic implementation was produced. The final description was in terms of a net-list involving components such as `GF` adders, and some timing information. About 3000 blocks/gates were produced. The reason for the discrepancy between this number and the one reached by Philips is partly that our implementation was incomplete, we only did the 'interesting bits' of the design; and partly that the gates used were at a slightly higher level of description. (e.g. we refer to components such as `COMP#(w1,w2)`, which tests to see whether two `GF`s are equal, without expanding it out as an 8 bit comparator). In addition, the implementations embody different optimisations.

3.3 Specification

LAMBDA represents a design state as a logical *rule*. Because the system only allows you to create valid rules, it also constrains you to produce correct designs. The system creates new rules/designs by transforming old ones. Thus at the start of any design/proof, we need a logically valid rule. Sometimes this will be a previously derived design state, at other times it will be the truism –*If I can implement this specification then I can implement this specification.*
As the design progresses, this will be transformed [4] to – *If I have these components and achieve these additional tasks then I will have implemented this specification*, and finally – *If I have this complete implementation and have these timing conditions then I have implemented this specification.* Here the timing conditions will typically be information about how and when inputs and outputs are supplied.
To create the initial rule above, we push the goal – *I can implement the specification.*
Thus at the start of the design cycle we must produce a formal specification of the tasks to be achieved. Usually this involves the conversion of a rather imprecise (and sometimes even contradictory) set of requirements in English[5] into a formal (unambiguous) statement of the goals to be attained by an implementation. In general, the writing of a formal specification is a difficult task, since it requires the author to decide exactly what the desired properties

[4]see [Finn *et al.*, 1989] for details of how rules are transformed.
[5]or German, Japanese, ...

of a system should be. An informal (i.e. English) specification can be interpreted in more than one way, and can easily be re-interpreted to give the *moving goalpost* nightmare of system development. No one (whether using formal methods or not) can avoid the problems that will occur if a system is incorrectly specified. The advantage of using a proof system to generate an implementation is that logical errors are restricted to errors in the specification, rather than being allowed to proliferate between each stage of the design. Our use of unification allows specifications to include parameters (e.g. latency) which are instantiated during the course of the design.

A specification language needs to be able to describe both high level system behaviour and low-level net-list. When performing top-down design, the implementation at one level is the next stage's specification, so we need a very flexible language. We base our logical specification language upon the programming language ML [Milner *et al.*, 1990], together with various logical connectives, such as `forall` (which expresses, for example, that forall times, some property is true). ML is a strongly typed functional language. It has the following desirable features from the point of a specification (and programming) language:

- **strong typing** All type errors are caught at compilation time.
- **polymorphism** Each legal expression in ML has a type. The types are allowed to be as general as possible. Thus we can define a 1 unit delay as

  ```
  val DELAY1#(i,o) = forall t. o (t + 1) === i t;
  ```

  Here the signals `i` and `o` are regarded as functions from time (represented by natural numbers 0,1,2,...) to some set of values. These values might be booleans true/false (i.e. hi/lo), 8 bit integers, or some (as yet undefined) representation of natural numbers. Thus one single DELAY definition can be used to represent delays on all types of expression.
- **functional language** Functions are first-class objects, which can be passed in as arguments to other functions. This gives you far greater flexibility of description than a procedural language, because relationships can be described without having to specify how they are to be implemented. The main reason for using a language such as ML is that it allows you to describe all levels of the problem (from high-level system behaviour to low-level net-list or HDL) in *one* uniform language. It also allows you to integrate the description of both software and hardware.

Using ML as our specification language has the advantage that functional specifications can be read into an ML compiler to create ML functions, and hence simulated. The LAMBDA toolset is also implemented in and accessed from ML.

Specifications may vary as to how many constraints they put upon the implementation. In this particular example, we used a specification supplied by Philips, which was written in terms of a number of ML functions. One such function, `syndrom` is shown below.

```
fun syndrom []     a = GFzero
  | syndrom (h::t) a = h ++ a ** syndrom t a;
```

Here `syndrom` acts on a list of `GFs`. It is defined recursively, giving an 8 bit zero if the list is empty, and acting on a list with first element `h` and remainder `t` to give `h` added to a multipled by the `syndrom` applied to `t`. When used in the top-level specification function `decode1`, which relates the input polynomials to the output polynomials and error flag, it is applied to the 32 element list which represents the input polynomial as a list of GFs.

The complete erco specification consisted of 220 lines of ML (including comments). We did not include any interface requirements on how the polynomials were to be supplied (serially, in parallel, ...), which allowed greater freedom to the implementation. (Such requirements could have been included, if desired.) The specification also did not specify the width of numbers passed in and out of internal functions. Thus we have a function `ems`, which is used to modify a polynomial to make its `syndroms` zero, and return the modified polynomial and the number of modifications. The specification states that it returns a natural number: the designer must discover how to represent that number.

When the user enters a specification into the LAMBDA toolset, the system interprets that description and converts it to a selection of logical rules. These rules can then be used during the design process to define the logical meaning of the specification.

3.4 Properties of the Specification

One application for LAMBDA is to check/prove various properties of the specification. This is useful for two reasons. Firstly, any such properties can then be assumed during the implementation. Secondly, it is important to try to check that the specification is sensible. (If we start with an incorrect specification, we will certainly end up with a bad implementation.) Thus in this particular specification a linear squaring function was defined, and one desirable property to check is that

```
square x = x ** x
```

There are two ways to do this. Firstly, there is simulation, a practical possibility in this example because the problem is combinatoric. There are 256 possible cases. The ML compiler can be used as a simulator, by reading the functions defined in the specification direct into the compiler. It is then simple to write a test function that tests whether the two functions are equal for a particular **GF**, and then to map this function over all the possible cases. This is a good check, because if it failed there would be no point in attempting to *prove* that the functions are equivalent.

We could also formally prove a *rule* which states that the two functions are the same. Here we do not simulate, but leave the value of each bit as a variable, and use the definitions of square and sum to rewrite the problem into 8 subgoals of the form

`x1 = (x2 &&`[6] `x4) xor x3 xor (x4 && x2) xor (x1 xor x3)`

which can be solved if one uses the fact that `xor` is associative and commutative and the facts that

`x xor x = false`   `x xor false = x`

However, this is obviously not the sort of manipulation one wishes to perform by hand, so we wrote a *tactic* to automatically reorder the `xor`s and `&&`s so as to eliminate redundant terms.

It is worth noting that results which have been 'proved' by simulation could be assumed without further need for proof. We are free to declare any result (whether true or false) to be an axiom (true rule) of the system. But one should only do this if one is convinced (e.g. by simulation) that the result is true, because assuming an incorrect fact would invalidate future proofs. Such assumptions are dangerous because one might, for instance, make a typing mistake when entering the result, or one might have false confidence in the simulator.

[6] where && is boolean conjunction

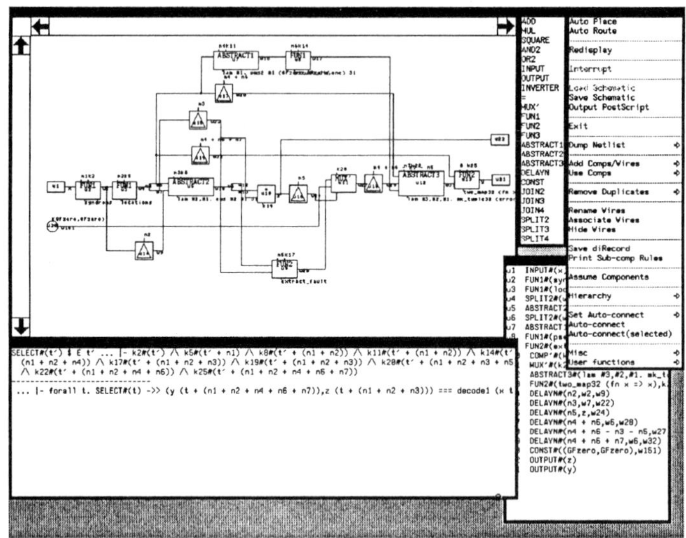

Figure 1: Top level synthesis of decode1

3.5 DIALOG Set-Up

DIALOG provides an engineer-friendly interface to LAMBDA. DIALOG allows the *structural* parts of the design state, namely any components, to be represented as a schematic. Figure 1 shows DIALOG in use. One window, the schematic window, shows the components being used in the implementation. Below it a text window shows the non-structural part of the rule which represents the current design state. To its right is a component menu consisting of all the components known to DIALOG; this is customisable. To the right of the component menu the main menu allows standard options such as 'Autoplace', together with LAMBDA-specific operations that drive the Theorem Prover; it includes a 'User Menu' which consists of operations supplied by the user. Finally, a window prints out a netlist of all the components appearing in the schematic window. It is worth mentioning the 'Dump Netlist' option in the main menu. LAMBDA has been designed to deal with high-level system design, and therefore needs to interface to the user's chosen favourite lay-out packages in order to get down to silicon. There are now a number of useful tools to automate low-level design. It is straight-forward to automatically translate LAMBDA's net-list into the format required by the chosen silicon compiler (e.g. VHDL, HILO, ...)

Before entering DIALOG, the user must specify the components to appear in the component menu, together with various user operations (e.g. scheduling) for the main menu. This allows us to introduce suitable libraries of subcomponents to the synthesis process. Each component in the menu must be supplied with the following information

- component symbol, list of terminal types
- a *tactic* which determines how the component transforms the design state

- an additional menu (possibly empty) which gives any other ways in which the component can be used.
- zero-delay information to allow DIALOG to check for zero-delay loops and an optional user-defined design check to impose additional constraints on the use of the component.

Some components, such as the AND gate and inverter, are already available in the system library. Others, such as ADD and MUL, (for performing addition and multiplication on the 8 bit field), were defined before starting the synthesis. The main purpose of the design exercise was to drive and define library and tool development for the system. Thus we also developed and refined generic (re-usable) libraries to implement functional abstractions during the course of the synthesis.
Although in this example the DIALOG menus were created by the same person who was then using it to perform the design process, in general this need not be the case. One DIALOG set-up can be used for a whole class of problems, and so a user can operate the system using a predefined set up, without needing to understand the details of that set-up.

3.6 Top Down Implementation

The erco specification is large enough to require hierarchic design. We adopted top-down partitioning (although it should be noted that LAMBDA is equally amenable to bottom up design.) Thus the first stage of the design consisted of the production of a top-level implementation for the function `decode1`, whose definition includes reference to functions `syndrom`, `ems` and `locations`.
To implement it, the user starts by pushing the goal
The output signals y, z at some times $t + m_1$, $t + m_2$ *are* `decode1` *applied to the input signal at the earlier time t.*
LAMBDA then creates the rule,
If the output signals at times $t + m_1$,$t + m_2$ *are* `decode1` *applied to the input signal at time t, then the output signals at times* $t + m_1$,$t + m_2$ *are* `decode1` *applied to the input signal at time t.*
This is the starting point for the synthesis. The user now goes into DIALOG, and sees the schematic corresponding to the above statement (one input and two outputs). Then, by using the components available in the component menu, the 'if' part of the rule representing the design state is gradually simplified/eliminated. A possible implementation is shown in Figure 1. This contains various `FUN` and `ABSTRACT`[7] components, together with a multiplexer and some delays. The definition of a FUN1 is shown below.

```
val FUN1#(f,SELECT,n,x,z) =
  forall t. SELECT#(t) ->> z (t + n) = f (x t)
```

`FUN1#(f,SELECT,n,x,z)` is defined as having an output `z` which is given by the function `f` applied to the input `x` `n` time units previously, provided that the `SELECT` parameter was then true. Thus the component can be used to represent components with arbitrary functionality (e.g. f is `syndrom` or `locations`), and arbitrary latency. The `SELECT` parameter allows the component to be un-pipelined, since it specifies when the calculation is to be started.
In this design, such components are very useful, because one would not wish to define separate components for every function which is used in the specification.

[7] similar to FUNs

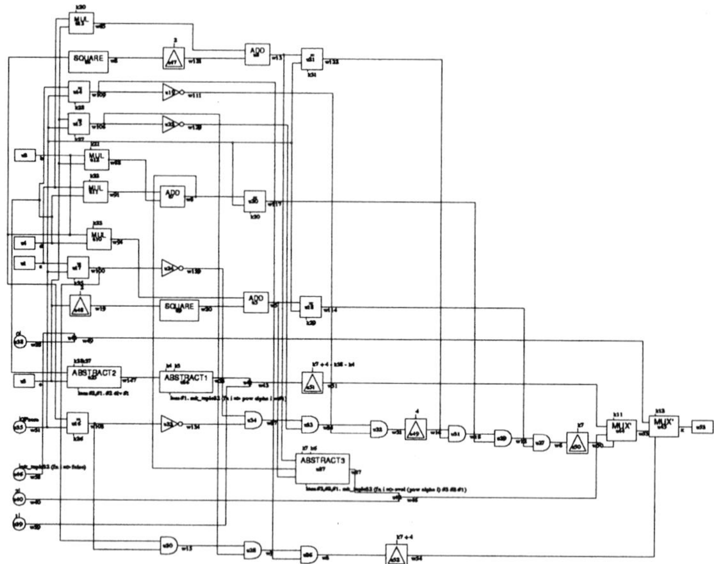

Figure 2: Second level: synthesis of locations

At this first level of design, the latencies of the individual functional blocks are as yet unknown. This results in delays such as $n_1 + n_2 - n_3 - n_4$, where the n_i are some of these latencies. Now the necessity that delays should be non-negative places a constraint on the lower level design, namely that $n_1 + n_2 \geq n_3 + n_4$. The overall latencies m_1 and m_2 are in terms of these latencies, (e.g. m_1 is $n_1 + n_2 + n_4 + n_6 + n_7$.) The form of the SELECT parameters is also unspecified – with the proviso that if the SELECT parameter for `decode1` is true at time `t`, the SELECT parameters for the other functional blocks must be true at the related times when the corresponding functions are calculated, (i.e. LAMBDA has introduced conditions for the control signals for the lower level design).

3.7 Lower level synthesis

Each functional block in the first level implementation is used as a specification for further (lower level) implementations. DIALOG supports hierarchical top-down development by allowing the user to start from a component used at one level of the implementation, generate a sub-implementation for it, and then return to the top level and, if required, replace the component by its implementation. Thus Figure 2 shows an implementation for the `locations` function, which contains some low level gates, and some parts that still need to be synthesised at a lower level.

One feature common to many of the blocks designed was that they involved iteration over the 32 elements of the GF polynomial. The function `syndrom` is defined by structural recursion over a list, and is applied to a 32 element list. Although one could expand out its definition and then synthesise it, it is extremely tedious and prohibitively expensive to do so, especially since the structure of the resulting block is essentially 32 blocks of the same element.

It was therefore decided to develop a general construct CHAIN to allow functional blocks to be chained together, so that syndrom above could be realised by creating a block with inputs h, carry_in and output carry_out, with

```
carry_out = h ++ a ** carry_in
```

and then applying the construct CHAIN to join 32 of these blocks together, feeding the carry_out of one block into the carry_in of the next.
The construct decided on is shown (pictorially) in Figure 3. Here the number n is equal to the length of the lists t_{in} and t_{out}. The reason for passing this parameter in is that it allows each block in the chain to depend on its position in the chain, thus allowing us to chain arbitrary blocks together.

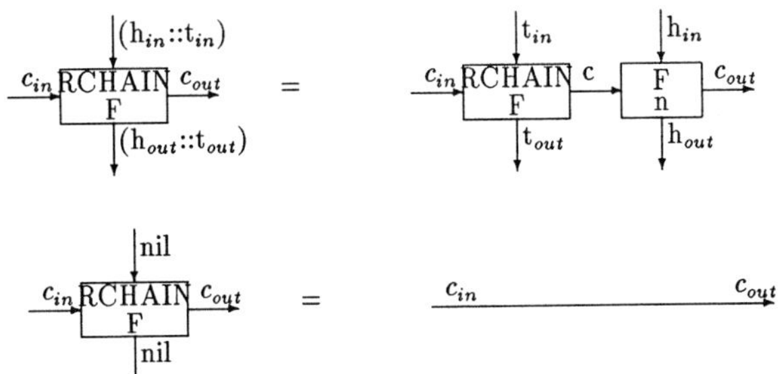

Figure 3: RCHAIN on lists

In this example, we can use a rule for RCHAIN to show that syndrom can be implemented with

```
F#(N,x,cin,y,cout) =
  forall t. SELECT#(t) ->>
    cout (t + (n * N + n)) = x t ++ a ** cin (t + n * N)
```

where n is a latency which will be calculated during the synthesis of F, and SELECT is a select parameter for the syndrom. F was then synthesised, using an adder(of latency 1), multiplier(latency 3) and a delay (see Figure 4), giving a value for n of 4. Note that the expression F#(N,x,cin,y,cout) does not give any value to the output y, thus meaning that this wire can be omitted on the corresponding component.
Finally, the RCHAIN implementing the 32 element syndrom was expanded out into 32 blocks of F.

3.8 Abstraction refinement

We have already mentioned that the highest level specification refers to ems, which returns a natural number together with a GF polynomial. The highest level implementation uses a functional block for ems, which is specified in terms of this natural number, not a finite bit representation. At some level of the implementation, this conversion must be made. An analysis of the definition of ems, which iterates over the 32 elements of a GF polynomial, suggests that 5 bits might be sufficient. In fact, we can do much better. The number produced by ems is compared with the output of the locations function, which can be proved to be 0,1 or 2. This

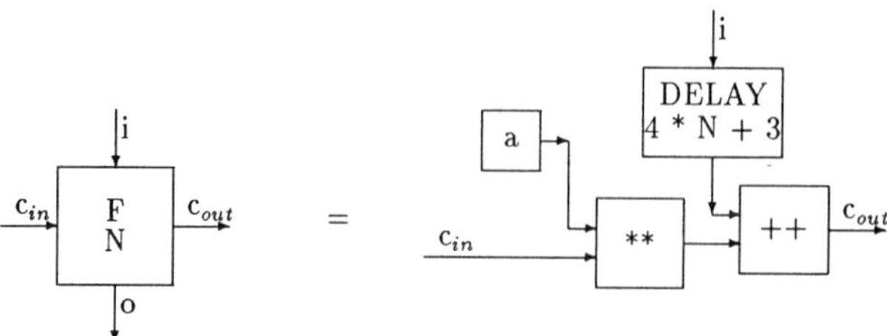

Figure 4: Synthesis of link F for syndrom

allows us to *prove* that a 2 bit representation is sufficient, because we only need to know whether `ems` produces 0,1,2, or > 2. Thus we replaced the `ems` block by a block which outputs the two bit number obtained by applying `natRep` to the number of errors, where `natRep` converts 0,1,2, < *any number greater than* 2 > to 4 distinct opcodes.

4 Conclusions

We applied the LAMBDA toolset to design an implementation for an industrial example, the error correcting code for a CD player. The case study allowed us to take a high-level functional description and implement it in terms of lower level functional blocks and net-list. The use of formal logic to describe both the problem and the design process guaranteed the logical correctness of any implementation so produced. The 'proof' is integrated with the 'design' and so there is no need for post design verification.
Comparisons of design times are difficult in this example, because most of the time spent using LAMBDA was involved in creating generic libraries which would be re-usable in later examples. Once these libraries had been created, the (interactive) run to create the top level `decode1` implementation shown in Figure 1 was about 45 minutes.
Constraints for timing were produced as a by-product of the synthesis. Implementation dependent details, such as the exact form of the `SELECT` parameters required to signal the start of calculation in an unpipelined component, can be propagated down through the levels of design until their exact form can be chosen. The methodology also permits postponement of decisions about opcode assignment and length until these assignments can be optimised. The design also automatically records such decisions. We controlled problems of scale by making use of hierarchy and by using the `RCHAIN` construct to avoid the expansion of recursive and iterative functions. The use of a functional specification allowed us to simulate parts of the specification so as to check it for consistency.

5 Future Work

The core of the system, in terms of libraries which permit reasoning about abstractions such as time (natural numbers), is now in place. The component libraries are still very basic and need to be extended. We anticipate that much of this extension will be done by the users as they decide on the components they need.
The main thrust of the toolset's development is now example driven. We are engaged in a

number of collaborative projects which will lead down to silicon. For example, we are improving our representations of busses to allow multiply driven busses, and now have a more elegant procedure to allow abstraction definition.

References

[Boyer and Moore, 1988] R.S. Boyer and J.S. Moore. *A Computational Logic Handbook.* Academic Press, Boston, 1988.

[Brayton *et al.*, 1988] R.K. Brayton, R. Camposano, G. DeMicheli, R.H.J.M. Otten, and J.T.J. van Eijndhoven. The Yorktown silicon compiler system. In D. Gajski, editor, *Silicon Compilation.* Addison-Wesley, 1988.

[Camposano and Tabet, 1989] R. Camposano and R.M. Tabet. Design representation for the synthesis of behavioural VHDL models. In *Proceedings CHDL'89*, Washington DC, June 1989.

[Cohn, 1988] Avra J. Cohn. A proof of correctness of the Viper microprocessor: The first level. In G. Birtwhistle and P.A. Subrahmanyam, editors, *VLSI Specification, Verification and Synthesis.* Kluwer Academic Publishers, 1988.

[Dutt *et al.*, 1990] N. D. Dutt, T. Hadley, and D. D. Gajski. An intermediate representation for behavioral synthesis. In *Proceedings of 27th ACM/IEEE Design Automation Conference.* IEEE, 1990.

[Finn *et al.*, 1989] S. Finn, M. Fourman, M. Francis, and R. Harris. Formal system design—interactive synthesis based on computer-assisted formal reasoning. In Dr. Luc Claesen, editor, *IMEC-IFIP International Workshop on Applied Formal Methods for Correct VLSI Design, Volume 1*, pages 97–110. Elsevier Science Publishers, B.V. North-Holland, Amsterdam, 1989.

[Fourman and Mayger, 1989] M.P. Fourman and Eleanor M. Mayger. Formally based systems design—Interactive hardware scheduling. In *VLSI '89.* North-Holland, 1989.

[Fourman *et al.*, 1988] M.P. Fourman, W.J. Palmer, and R.M. Zimmer. Proof and synthesis. In *Proceedings ICCD '88*, Rye Brook, NY, 1988.

[Gordon *et al.*, 1979] Michael J. Gordon, Robin Milner, and Christopher P. Wadsworth. A mechanised logic of computation. In *Edinburgh LCF: Lecture Notes in Computer Science—Volume 78.* Springer-Verlag, 1979.

[Gordon, 1988] Mike Gordon. HOL: A proof generating system for higher-order logic. In G. Birtwhistle and P.A. Subrahmanyam, editors, *VLSI Specification, Verification and Synthesis.* Kluwer Academic Publishers, 1988.

[Hanna *et al.*, 1989] F.K. Hanna, M. Longley, and N. Daeche. Formal synthesis of digital systems. In Dr. Luc Claesen, editor, *IMEC-IFIP International Workshop on Applied Formal Methods for Correct VLSI Design, Volume 2*, pages 532–548. Elsevier Science Publishers, B.V. North-Holland, Amsterdam, 1989.

[Hunt, 1987] W.A. Hunt. The mechanical verification of a microprocessor design. In *HDL Descriptions to Guaranteed Correct Circuit Designs*, pages 89–132. North-Holland, 1987.

[Mendler, 1990] M. Mendler. Constrained proofs: A logic for dealing with behavioural constraints in formal hardware verification. In *Workshop on Designing Correct Circuits, Oxford.* Springer-Verlag, 1990.

[Milner *et al.*, 1990] R. Milner, M. Tofte, and R. Harper. *The Definition of ML.* The MIT Press, Cambridge, Massachusetts, 1990.

[Paulson, 1987] L. Paulson. Natural deduction proof as higher-order resolution. *Logic Programming*, 3:237–258, 1987.

VLSI 91
A. Halaas and P.B. Denyer (Eds.)
Elsevier Science Publishers B.V. (North-Holland)

Deriving bit-serial circuits in Ruby

Geraint Jones[a] and Mary Sheeran[b]

[a]Programming Research Group, Oxford University Computing Laboratory, 11 Keble Road, Oxford OX1 3QD, England; Electronic mail: *gj@comlab.oxford.ac.uk*

[b]Department of Computing Science, University of Glasgow, Glasgow G12 8QQ, Scotland; Electronic mail: *ms@dcs.glasgow.ac.uk*

Abstract

The action of bit-serial arithmetic circuits is often explained in purely pictorial terms. In contrast, this paper describes an attempt to deal with the systematic development of bit-serial arithmetic circuits within a mathematical framework which we have previously used to develop parallel circuits. A well-known bit-serial adder is formally shown to implement the specification of an adder, without any recourse to detailed arguments about snapshots or specific arguments about sequences of inputs.

1 DESIGN BY CALCULATION

Ruby [1] is a language of relations and functions that supports a style of development by calculation. A circuit is represented by a relation between the signals at its terminals, and forms of circuit are captured by functions which operate on those relations.

For example, if the second terminal of a two-terminal component R is connected to the first of a similar component S, they behave as $R;S$, defined by $x\ (R\,;S)\ y \Leftrightarrow \exists z.\, x\ R\ z\ \&\ z\ S\ y$. At another extreme, the parallel composition of two circuits with no connection between them is described by $[R,S]$ where $\langle x_1,x_2\rangle\ [R,S]\ \langle y_1,y_2\rangle \Leftrightarrow x_1\ R\ y_1\ \&\ x_2\ S\ y_2$. The converse of a relation, defined by $x\ R^{-1}\ y \Leftrightarrow y\ R\ x$, represents a circuit which is like R but with its left and right connections swapped.

The composition operators can easily be shown to satisfy many laws such as $[P,Q]\,;[R,S] = [P;R, Q;S]$, which is true for all P, Q, R, and S. Similarly, $[R,S]^{-1} = [R^{-1},S^{-1}]$ and $(R;S)^{-1} = S^{-1};R^{-1}$. These and many similar laws can be used for what is essentially algebraic manipulation of expressions describing circuits, in the course of which there is little need to consider the details of what the circuit does.

Ruby relations can be used to describe more or less abstract views of a circuit, and in particular it is possible to describe within the one framework the process of refining what might be thought of as a specification into something that might be an implementation. For example, if *add* is the relation between a pair of natural numbers and the natural number which is their sum, and if *bin* is the relation between a tuple of bits and the natural number of which they are a binary least-significant bit first representation, then $[bin, bin]\,; add\,; bin^{-1}$ describes a 'circuit' which relates two such tuples of bits to a similar tuple of bits representing their sum.

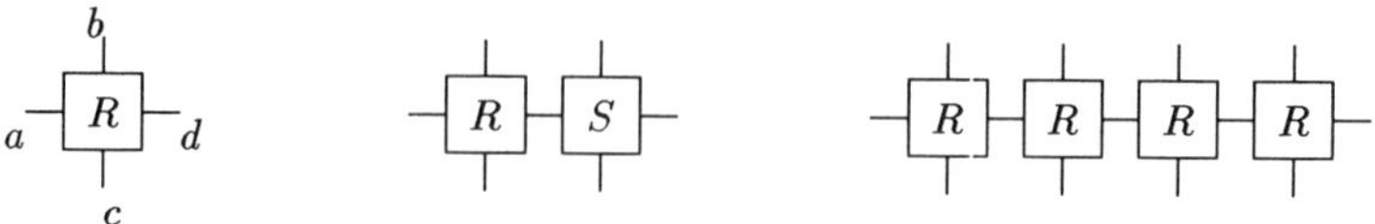

Figure 1: $\langle a, b\rangle\ R\ \langle c, d\rangle$ and layouts suggested by $R \leftrightarrow S$, and $\mathsf{row}\ R$

2 NOTATION USED IN THIS PAPER

We will need a number of specific circuit constructors in the calculations of this paper. These are chosen from a large collection of common forms discussed in more detail in, for example, reference [1]. One reason for having such a great number is that there is a rich collection of equations relating them, which makes calculation easier.

Many of the forms are abbreviations, which capture common patterns, for example we write $R \setminus S$, read 'R conjugated by S', for $S^{-1} ; R ; S$. We write $\mathsf{fst}\ R$ for the parallel composition of R and the identity relation, and $\mathsf{snd}\ S$ for the identity relation in parallel with S. It follows that $\mathsf{fst}\ R ; \mathsf{snd}\ S = [R, S] = \mathsf{snd}\ S ; \mathsf{fst}\ R$ and so on. Distinguish between fst and snd and the particular pieces of wiring represented by the projection relations π_1, π_2, which are defined by $\langle x, y\rangle\ \pi_1\ x$ and $\langle x, y\rangle\ \pi_2\ y$.

We write R^n for the repeated composition of n copies of R, so that $R^1 = R$, and $R^{n+1} = R^n ; R$. Repeated parallel composition is captured by $\mathsf{map}\ R$, a relation between equal-length lists, relating two lists if corresponding elements of the lists are related by R.

$$x\ (\mathsf{map}\ R)\ y \quad \Leftrightarrow \quad \#x = \#y\ \&\ \forall i.\, x_i\ R\ y_i$$

Those relations which are equivalence relations, that is T for which $T = T^{-1} = T^2$, serve the rôle of types in our calculations. A constant is just a type with only one equivalence class: that is a type k for which $k = k ; any ; k$ where any is the universal type, the relation for which $x\ any\ y$ for any x and y. A type T for which $x\ T\ y$ only if $x = y$, is just a data constraint. In particular, for any number n we write n for the identity relation on lists of that length, so for example 2 is the type of all pairs. Constraints capture observations such as $C ; R = C ; S$, that two circuits, R and S, have the same behaviour so long as the signals in their domain meet some constraint, C.

We will need to deal specially with some relations between pairs and pairs. If R relates pairs to pairs, we will often want to divide its domain signals between the left and top, and the range signals between the bottom and right, as shown in figure 1. For such relations we define $R \leftrightarrow S$, read 'R beside S', by

$$\langle a, \langle b, c\rangle\rangle\ (R \leftrightarrow S)\ \langle\langle x, y\rangle, z\rangle \quad \Leftrightarrow \quad \exists p.\, \langle a, b\rangle\ R\ \langle x, p\rangle\ \&\ \langle p, c\rangle\ S\ \langle y, z\rangle$$

which suggests the layout also shown in figure 1. Complementary to this is $R \updownarrow S$, read 'R below S' and defined by $R \updownarrow S = (R^{-1} \leftrightarrow S^{-1})^{-1}$. The generalisation of beside-composition to lists is a row of components, defined by

$$\begin{aligned}&\langle a, bs\rangle\ \mathsf{row}\ R\ \langle ys, z\rangle \\ &\quad \Leftrightarrow \quad \#bs = \#ys\ \&\ (\exists ps.\, a = ps_0\ \&\ \forall i.\, \langle ps_i, bs_i\rangle\ R\ \langle ys_i, ps_{i+1}\rangle\ \&\ ps_{\#bs} = z)\end{aligned}$$

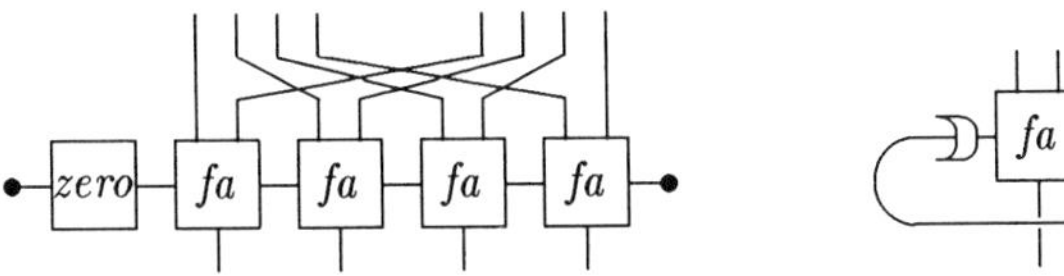

Figure 2: an instance of the bit-parallel adder $zip\ ;\ \pi_2^{-1}\ ;\ \mathsf{fst}\ zero\ ;\ \mathsf{row}\ fa\ ;\ \pi_1$, and the bit-serial adder $\mu(\mathsf{fst}\ \mathcal{D}\ ;\ fa)$ which we derive from it

and an instance of its layout is also shown in figure 1.

We will also need an interleaving relation, *zip*, which relates a pair of equal length lists to a list of the same length each element of which is a pair, defined by

$$\langle x, y\rangle\ zip\ z \quad \Leftrightarrow \quad \#x = \#y = \#z\ \ \&\ \ \forall i.\ \langle x_i, y_i\rangle = z_i$$

We can use this to represent the interleaving of two buses of wires.

3 A BIT-PARALLEL ADDER

We have previously [2] shown several calculations which start from specifications consisting of representation relations and abstract arithmetic operations, and leading to expressions which describe bit-parallel circuits. Here we will just quote the result of such a calculation.

Using the abstract adder, *add*, which relates two natural numbers to their sum, and the representation relation *bin*, which relates a tuple of bits to the number which it represents, we would like to specify that an adder circuit should implement the relation $[bin, bin]\ ;\ add\ ;\ bin^{-1}$. This relates a pair of tuples of bits to the tuple of bits which represents the sum of the numbers represented by the pair.

This specification is not readily implementable as a circuit because its interface is not of a fixed size: a circuit would have to have a particular width of input and of output. Let N be the identity on N-tuples, then $A = [N\,;bin, N\,;bin]\,;add\,;(N\,;bin)^{-1}$ is a circuit which takes lists of bits of width N and produces an output of width N.

Since only numbers less than 2^N are represented by $N\ ;\ bin$, this relation only adds pairs of inputs which represent numbers that have a sum less than 2^N. It is therefore not a total function from all pairs of N-lists, but all our circuits eventually implement total functions – they produce *some* output for every input.

Let sum_{2^N} be the identity relation on pairs of number which sum to less than 2^N, then $sum'_{2^N} = sum_{2^N} \setminus [N\,;bin, N\,;bin]^{-1}$ is the constraint which we need, being the identity on the N-bit representations of pairs which sum to less than 2^N. Certainly, $A = sum'_{2^N}\ ;\ A$, but we can also refine $sum'_{2^N}\ ;\ A$ into an implementation

$$A \quad = \quad sum'_{2^N}\ ;\ zip\ ;\ \pi_2^{-1}\ ;\ \mathsf{fst}\ zero\ ;\ \mathsf{row}\ fa\ ;\ \pi_1\ ;\ N \tag{1}$$

where fa is a full-adder. The term $zip\ ;\ \pi_2^{-1}\ ;\ \mathsf{fst}\ zero\ ;\ \mathsf{row}\ fa\ ;\ \pi_1\ ;\ N$ is the circuit itself, which is a total function on all pairs of N-bit representations, and has a fixed size. Its form suggests the layout shown in figure 2.

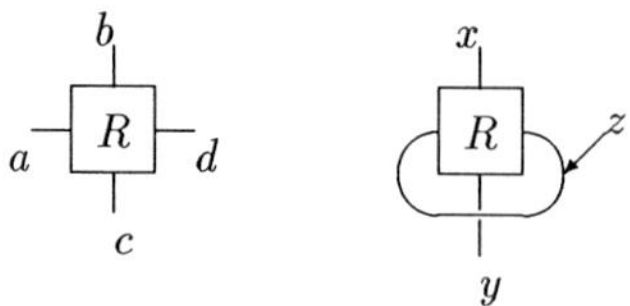

Figure 3: Layouts suggested by $\langle a, b\rangle\, R\, \langle c, d\rangle$ and $x\,(\mu\, R)\, y$, defined to be $\exists z.\, \langle z, x\rangle\, R\, \langle y, z\rangle$

4 TIME

We argue [3] that Ruby supports simple reasoning about sequential circuits. Suppose R describes a 'stateless' (combinational) circuit, and that it is provided with a succession of (potentially different) inputs at a succession of times. If the circuit is allowed sufficient time to settle into a stable state before being clocked again, that state is described by the relation R. That is to say, the whole 'clocked' or 'sequential' behaviour of the circuit is described by $\mathsf{map}\, R$, seen as a relation between two time-sequences.

We can give a similar account of each of the operators which appear in Ruby expressions, giving them a new interpretation which combines the sequential behaviours of the component expressions to yield the sequential behaviour of the whole. For example, the parallel composition of R and S is a relation between two time-sequences of pairs: at any time, the pair in the domain is related to the pair in the range at that time if their first components would have been related by the R at that time, and their second by S. That is to say, the sequences related by the clocked interpretation of the parallel composition of two circuits are obtained by zipping together those related by the components.

It can be proved, by induction over the language, that any expression E in Ruby in which each of the components is stateless is also itself stateless, that is that the sequential interpretation of E is $\mathsf{map}\, E$. In order to be able to talk about circuits which are not stateless we introduce the primitive circuit $\mathcal{D}$, read 'delay', defined by $x\, \mathcal{D}\, y \Leftrightarrow \forall t.\, x_t = y_{t+1}$. That is to say, the signal in the range of $\mathcal{D}$ at any time is the same as the signal in its domain at the previous time. It might be implemented by a latch, register, or delay element taking the domain signal as input, and returning the range signal as output. Similarly $\mathcal{D}^{-1}$, the 'anti-delay', predicts its domain signal in the range. Although it might seem unrealistic, it could be implemented by a latch operating in the other direction.

Although many of the Ruby operators have complicated interpretations as operations on sequential behaviours, reference [3] shows that any well-typed equation which is true in the simpler combinational interpretation of Ruby remains true in the new sequential interpretation. This is the case even if the variables in the equation are allowed to range over all (Ruby-definable) sequential circuits, even ones that are not stateless. This means that we can often prove things about sequential circuits just by the much simpler expedient of showing that the corresponding thing is true of combinational circuits.

Of course, only equations which make sense in the combinational interpretation are amenable to this approach. There are a few things which we need to know and that are particular to sequential circuits, and for these we have to resort to checking their validity directly in the sequential interpretation. For example, if k is a constant circuit, which

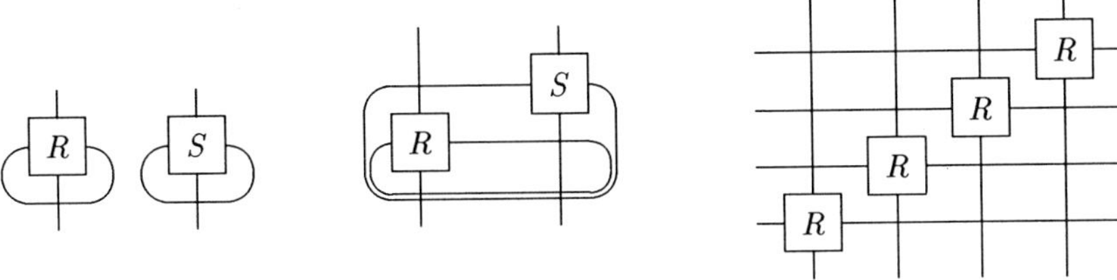

Figure 4: $[\mu\, R, \mu\, S]$ compared to $\mu([R, S] \setminus zip^{-1})$, and an instance of $\mathsf{map}\, R \setminus zip^{-1}$

holds its connections at the same value at all times, then

$$k \;=\; k\,;\mathcal{D} \tag{2}$$

which must be an equation about sequential circuits, since $\mathcal{D}$ has no other interpretation.

The sequential interpretation of equation 1 describes a combinational circuit which implements a word-serial, bit-parallel adder; but it has a long ripple-propagation path along the carry chain, and elsewhere [3, 4] we have pipelined such circuits, using Leiserson's retiming regime [5]. From the pipelining calculation we would derive requirements on the data-skew of the input, and consequences for the data-skew of the output.

5 FEEDBACK

Any bit-serial design must have an internal state which contains the carry forward from one bit's calculation, and a feedback mechanism which delivers it to the next bit's calculation. We can deal with the state using $\mathcal{D}$, and we introduce a new construction to deal with feedback,

$$x\,(\mu\, R)\, y \;\Leftrightarrow\; \exists z.\, \langle z, x\rangle\, R\, \langle y, z\rangle$$

the definition of which suggests the layout shown in figure 3.

The feedback constructor has a rich collection of simple equations which relate it to other parts of the language, for example

$$A\,;\mu\, R\,;B \;=\; \mu(\mathsf{snd}\, A\,;R\,;\mathsf{fst}\, B) \tag{3}$$

$$\mu(\mathsf{fst}\, P\,;R) \;=\; \mu(R\,;\mathsf{snd}\, P) \tag{4}$$

which explain, respectively, the interface of $\mu\, R$, and how the representation of the 'state' on the feedback wire may be changed. Notice that the truth of equation 4 does not depend on any properties of P: it need not be a function, nor the inverse of a function, nor onto, nor total. This one equation, therefore, captures many different kinds of 'change of state' transformations.

An unused feedback loop can always be eliminated

$$\mu(\mathsf{fst}\, any\,;R\,;\mathsf{snd}\, any) \;=\; \pi_2^{-1}\,;R\,;\pi_1 \tag{5}$$

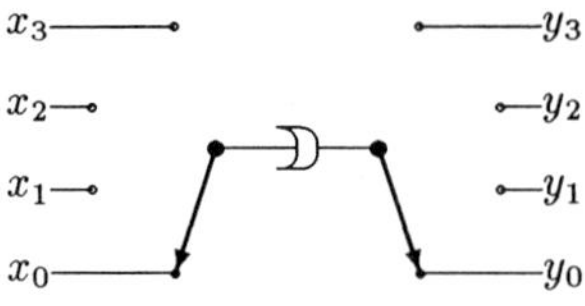

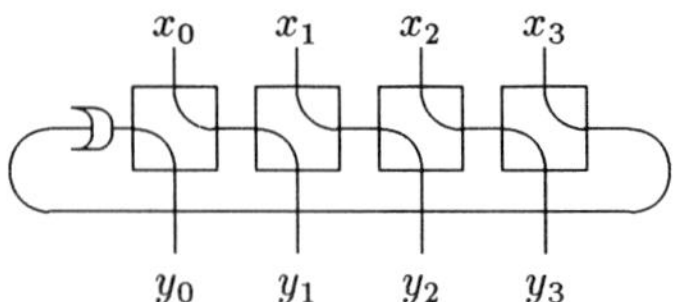

Figure 5: $\mathcal{D} \setminus \underline{bundle}$ and the corresponding $\mu(\mathsf{fst}\,\mathcal{D} \,;\, \mathsf{row}\,2)$

and feedback loops can be composed in series and parallel, $\mu\, R \,;\, \mu\, S \;=\; \mu(S \updownarrow R)$ and $[\mu\, R, \mu\, S] = \mu([R, S] \setminus zip^{-1})$. The latter equation is illustrated in figure 4, and generalises to arbitrarily wide parallel compositions, because $\mathsf{map}\,\mu\, R \;=\; \mu(\mathsf{map}\, R \setminus zip^{-1})$. The kernel of the right-hand side suggests the layout shown at the right of figure 4, and $\mu(\mathsf{map}\, R \setminus zip^{-1})$ would be obtained by joining corresponding wires on the left and right.

6 SLOWING

So far, all the circuits have been operating at the same rate. We can also describe circuits which have regions running on clocks with different frequencies [4]. An N-slow circuit [5] is one in which every instance of $\mathcal{D}$ has been replaced by $\mathcal{D}^N$. The N is a constant in the whole of this paper, but we could deal with different values of N by decorating a couple of the constructors with the scaling factor to which they relate.

In order to explain slowing in Ruby, we will introduce a new primitive which describes the demultiplexing of a sequence of values into N separate signals.

$$x \;\underline{bundle}\; y \quad \Leftrightarrow \quad x_{Nt+i} = y_{t,i}$$

Each y_i is a sequence delivered, or sampled, at $1/N$th the rate of x. We underline $\underline{bundle}$ as a reminder that – like $\mathcal{D}$ – it has only a sequential interpretation. That means, for example, that in equations like

$$[\underline{bundle}, \underline{bundle}] \,;\, zip \quad = \quad 2 \,;\, \underline{bundle} \tag{6}$$

the only possible reading is the sequential one.

It can be proved, by an induction over all expressions in the language, that an N-slow version of any Ruby-definable circuit R behaves like N independent copies of R with the inputs being demultiplexed and the outputs being interleaved.

$$\mathsf{slow}\, R \quad = \quad (\mathsf{map}\, R) \setminus \underline{bundle}^{-1} \tag{7}$$

This shows that although $\mathsf{slow}\, R$ was defined by a manipulation of the text of expressions, it depends only on the meaning of the expression, so that if $R = S$ then $\mathsf{slow}\, R = \mathsf{slow}\, S$. Since slowing was defined by a substitution, it is immediate that it distributes through all the operators, so $\mathsf{slow}(R\,;S) = \mathsf{slow}\, R\,; \mathsf{slow}\, S$, and $\mathsf{slow}[R, S] = [\mathsf{slow}\, R, \mathsf{slow}\, S]$, and so on. Moreover, $\mathsf{slow}\, R = R$ for any stateless R, and $\mathsf{slow}\,\mathcal{D} = \mathcal{D}^N$.

Since $\mathcal{D}$ is itself Ruby-definable, $\mathcal{D}^N = \mathsf{slow}\,\mathcal{D} = (\mathsf{map}\,\mathcal{D}) \setminus \underline{bundle}^{-1}$ by equation 7. A similar result, which we will need in this paper, describes a way of implementing isolated

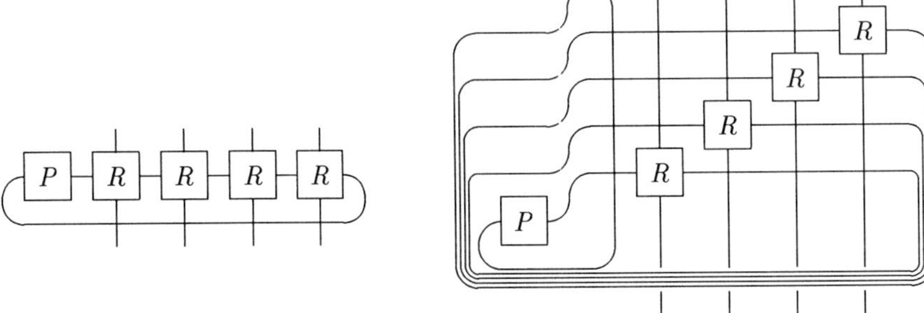

Figure 6: instances of $\mu(\mathsf{fst}\, P\,;\, \mathsf{row}\, R)$ and $\mu(\mathsf{fst}\, \mu(\mathsf{fst}\, P\,;\, \mathsf{row}\, 2)\,;\, (\mathsf{map}\, R \setminus zip^{-1}))$

single delays – ones which are not a part of a $\mathcal{D}^N$ – in circuits that are otherwise slow. Consider $\mathcal{D} \setminus \underline{bundle}$, which can be read as a delay sandwiched between a multiplexer and a demultiplexer, as illustrated in figure 5. It is the relation given by

$$x\ (\mathcal{D} \setminus \underline{bundle})\ y \quad \Leftrightarrow \quad \begin{cases} x_{t,i} = y_{t,i+i}, & \text{if } 0 \leq i < N-1 \\ x_{t,N-1} = y_{t+1,0} & \end{cases}$$

and from this it can be proved that

$$\mathcal{D} \setminus \underline{bundle} \quad = \quad (\mu(\mathsf{fst}\, \mathcal{D}\,;\, \mathsf{row}\, 2)) \setminus N \tag{8}$$

The construction row 2 on the right-hand side is a common idiom: the 2 is just two wires in parallel, and a row of them is a shift, which is made into a cyclic shift by the μ, and the signal that is carried around is delayed, as illustrated in figure 5.

7 SLOW COMPONENTS WITH BOTH STATE AND FEEDBACK

A bit-serial arithmetic circuit will be implemented by a bit-operation clocked once per bit, and slowed to the speed at which N-bit words are presented. So we will be interested in folding word-level circuits, ones with row R in them, to get forms like $\mu(\ldots \mathsf{slow}\, R \ldots)$ which implement the same operation with bit-level components R.

A general result about feedback and row, which will prove useful in this is that

$$\mu(\mathsf{fst}\, P\,;\, \mathsf{row}\, R) \quad = \quad \mu(\mathsf{fst}\, \mu(\mathsf{fst}\, P\,;\, \mathsf{row}\, 2)\,;\, (\mathsf{map}\, R \setminus zip^{-1})) \tag{9}$$

The right-hand side suggests the tangled layout shown in figure 6. Notice that this result, which we quote here without proof, is one which can be checked in the simple combinational interpretation of the language, and then promoted to the sequential interpretation in which it would otherwise be much more complicated to prove.

Armed with this result, and earlier observations, we proceed to expand a candidate bit-level circuit to see what word-level operation it would implement

$$
\begin{aligned}
& \mu(\mathsf{fst}\,\mathcal{D}\,;\,\mathsf{slow}\,R) \\
= & \{\text{ equation 7 and } \mu\, R = \mu(R \setminus 2)\,\} \\
& \mu(\mathsf{fst}\,\mathcal{D}\,;\,\mathsf{map}\,R \setminus (2\,;\,\underline{bundle})^{-1}) \\
= & \{\text{ equation 6 }\} \\
& \mu([\mathcal{D}\,;\,\underline{bundle}, \underline{bundle}]\,;\,(\mathsf{map}\,R \setminus zip^{-1})\,;\,[\underline{bundle}^{-1}, \underline{bundle}^{-1}]) \\
= & \{\text{ equation 4 }\} \\
& \mu([\mathcal{D} \setminus \underline{bundle}, \underline{bundle}]\,;\,(\mathsf{map}\,R \setminus zip^{-1})\,;\,\mathsf{fst}\,\underline{bundle}^{-1}) \\
= & \{\text{ equation 8 }\} \\
& \mu([\mu(\mathsf{fst}\,\mathcal{D}\,;\,\mathsf{row}\,2) \setminus N, \underline{bundle}]\,;\,(\mathsf{map}\,R \setminus zip^{-1})\,;\,\mathsf{fst}\,\underline{bundle}^{-1}) \\
= & \{\text{ equation 4 }\} \\
& \mu([\mu(\mathsf{fst}\,\mathcal{D}\,;\,\mathsf{row}\,2)\,;\,N, \underline{bundle}]\,;\,(\mathsf{map}\,R \setminus zip^{-1})\,;\,[\underline{bundle}^{-1}, N]) \\
= & \{\text{ since } zip \text{ relates only pairs of lists of equal length }\} \\
& \mu([\mu(\mathsf{fst}\,\mathcal{D}\,;\,\mathsf{row}\,2), \underline{bundle}]\,;\,(\mathsf{map}\,R \setminus zip^{-1})\,;\,\mathsf{fst}\,\underline{bundle}^{-1}) \\
= & \{\text{ equation 3 }\} \\
& (\mu(\mathsf{fst}\,\mu(\mathsf{fst}\,\mathcal{D}\,;\,\mathsf{row}\,2)\,;\,(\mathsf{map}\,R \setminus zip^{-1}))) \setminus \underline{bundle}^{-1} \\
= & \{\text{ equation 9 }\} \\
& (\mu(\mathsf{fst}\,\mathcal{D}\,;\,\mathsf{row}\,R)) \setminus \underline{bundle}^{-1}
\end{aligned}
$$

and then conjugating on both sides with $\underline{bundle}$, since $N = \underline{bundle}^{-1}\,;\,\underline{bundle}$ we obtain

$$
(\mu(\mathsf{fst}\,\mathcal{D}\,;\,\mathsf{slow}\,R)) \setminus \underline{bundle} \quad = \quad (\mu(\mathsf{fst}\,\mathcal{D}\,;\,\mathsf{row}\,R)) \setminus N \tag{10}
$$

which we will use to replace the word-level operation, involving $\mathsf{row}\,R$, on the right-hand side by the bit-serial operation, $\mu(\mathsf{fst}\,\mathcal{D}\,;\,\mathsf{slow}\,R)$, on the left.

8 LIMITING THE CARRY-AROUND

It remains only to show that word-level addition can be implemented by a circuit with feedback, of the form $\mu(\mathsf{fst}\,\mathcal{D};\mathsf{row}\,R)$. The intuition behind this step of the development is that such a circuit will do the right thing – provided only that the input is so constrained that the state being fed back can be guaranteed always to be just that constant *zero* which is required in equation 1. Let b be any constraint strong enough that

$$
\mathsf{snd}\,b\,;\,\mathsf{row}\,R \quad = \quad \mathsf{snd}\,b\,;\,\mathsf{row}\,R\,;\,\mathsf{snd}\,k \tag{11}
$$

for some constant k. Then

$$
\begin{aligned}
& b\,;\,\pi_2^{-1}\,;\,\mathsf{fst}\,k\,;\,\mathsf{row}\,R\,;\,\pi_1 \\
= & \{\text{ equation 5 and equation 3 }\} \\
& \mu([any\,;\,k, b]\,;\,\mathsf{row}\,R\,;\,\mathsf{snd}\,any) \\
= & \{\text{ equation 11 }\} \\
& \mu([any\,;\,k, b]\,;\,\mathsf{row}\,R\,;\,\mathsf{snd}(k\,;\,any))
\end{aligned}
$$

$= \{\text{ equation } 4\}$

$\quad \mu([k\,;\,any\,;\,any\,;\,k, b]\,;\,\mathsf{row}\,R)$

$= \{\text{ since } any \text{ is a type, } k \text{ is a constant}\}$

$\quad \mu([k, b]\,;\,\mathsf{row}\,R)$

$= \{\text{ equation } 2\}$

$\quad \mu([k\,;\,\mathcal{D}, b]\,;\,\mathsf{row}\,R)$

$= \{\text{ equation } 4\}$

$\quad \mu([\mathcal{D}, b]\,;\,\mathsf{row}\,R\,;\,\mathsf{snd}\,k)$

$= \{\text{ equation } 11\}$

$\quad \mu([\mathcal{D}, b]\,;\,\mathsf{row}\,R)$

$= \{\text{ equation } 3\}$

$$b\,;\,\mu(\mathsf{fst}\,\mathcal{D}\,;\,\mathsf{row}\,R) \qquad (12)$$

which has just the form required to apply equation 10.

9 ASSEMBLING THE RESULTS

To complete the development of the adder we need a constraint on the input to meet the hypothesis in equation 11 with $k = zero$. We will take $b = sum'_{2^N-1} \setminus zip$, although with a bit more work we could have got a slightly better result.

$sum'_{2^N-1}\,;\,A \quad = \{\text{ equation 1, since } sum'_x\,;\,sum'_y = sum'_{\min(x,y)}\}$

$\quad sum'_{2^N-1}\,;\,zip\,;\,\pi_2^{-1}\,;\,\mathsf{fst}\,zero\,;\,\mathsf{row}\,fa\,;\,\pi_1\,;\,N$

$= \{\text{ since } sum'_x = zip\,;\,zip^{-1}\,;\,sum'_x\}$

$\quad zip\,;\,b\,;\,\pi_2^{-1}\,;\,\mathsf{snd}\,zero\,;\,\mathsf{row}\,fa\,;\,\pi_1\,;\,N$

$\quad$ where $b = sum'_{2^N-1} \setminus zip$

$= \{\text{ equation 12, since } \mathsf{snd}\,b\,;\,\mathsf{row}\,fa = \mathsf{snd}\,b\,;\,\mathsf{row}\,fa\,;\,\mathsf{snd}\,zero\}$

$\quad zip\,;\,b\,;\,\mu(\mathsf{fst}\,\mathcal{D}\,;\,\mathsf{row}\,fa)\,;\,N$

$= \{\text{ equation 10, since } b = b\,;\,N\}$

$\quad sum'_{2^N-1}\,;\,zip\,;\,(\mu(\mathsf{fst}\,\mathcal{D}\,;\,\mathsf{slow}\,fa)) \setminus \underline{bundle}$

$= \{\text{ since } fa \text{ is stateless, so } fa = \mathsf{slow}\,fa\}$

$\quad sum'_{2^N-1}\,;\,zip\,;\,(\mu(\mathsf{fst}\,\mathcal{D}\,;\,fa)) \setminus \underline{bundle}$

$= \{\text{ equation 3, using } fa = \mathsf{snd}\,2\,;\,fa\}$

$\quad sum'_{2^N-1}\,;\,zip\,;\,\underline{bundle}^{-1}\,;\,2\,;\,\mu(\mathsf{fst}\,\mathcal{D}\,;\,fa)\,;\,\underline{bundle}$

$= \{\text{ equation 6, and } sum'_x = sum'_x\,;\,zip\,;\,zip^{-1}\}$

$\quad sum'_{2^N-1}\,;\,[\underline{bundle}^{-1}, \underline{bundle}^{-1}]\,;\,\mu(\mathsf{fst}\,\mathcal{D}\,;\,fa)\,;\,\underline{bundle}$

and composing on both sides with $[\underline{bundle}, \underline{bundle}]$ on the left and $\underline{bundle}^{-1}$ on the right

$sum''_{2^N-1}\,;\,[\underline{bundle}\,;\,bin, \underline{bundle}\,;\,bin]\,;\,add\,;\,(\underline{bundle}\,;\,bin)^{-1}$

$\quad = \quad sum''_{2^N-1}\,;\,\mu(\mathsf{fst}\,\mathcal{D}\,;\,fa)$

$\quad$ where $sum''_x = sum_x \setminus [(\underline{bundle}\,;\,bin)^{-1}, (\underline{bundle}\,;\,bin)^{-1}]$

The term $\mu(\mathsf{fst}\,\mathcal{D}\,;fa)$ is the actual circuit, illustrated in figure 2, consisting of a stateless full-adder with a single one-bit latch on a feedback loop which takes the carry signal. The left-hand side of the equation shows that all three numbers are represented by $\underline{bundle};bin$, that is as N-bit binary numbers given bit-serially with least-significant bit first. Finally, the sum''_{2^N-1} is a constraint which describes the way that input must be supplied: correct operation is guaranteed if the input represents pairs which each sum to less than 2^N-1.

10 CONCLUSION

We have shown how we can deal with the systematic development of a bit-serial adder in the Ruby notation which had previously been used only to develop parallel circuits. The development of the adder itself makes no recourse to arguments about when inputs are provided nor when states are recorded.

Although almost all of the results quoted in this paper are used in the development of the adder, few of them are specific to this circuit. The bulk of the results could equally appear in many other developments. Indeed of all the equations, only those in which *add* or *fa* appear involve facts particular to the adder.

The material in this paper about time and delays was originally developed to deal with pipelining of bit-parallel circuits; that dealing with feedback was for calculating with state-machines; the work on $\underline{bundle}$ and slowing was originally for treating circuits in which only a proportion of the components are active on each cycle. By combining techniques from these three areas we have been able to deal with bit-serial circuits. There is scope for taking this work further, for example although $\underline{bundle}\;;bin$ describes only least-significant bit first representations, we can perform similar calculations with $\underline{bundle}$; *reverse* ; *bin*, to deal with most-significant bit first representations.

This gives us some confidence that we could go on to develop a general theory of serialisation, making it possible to reason precisely about the balance between the time and space efficiency of circuits which we can know with great confidence to be otherwise identical in behaviour.

REFERENCES

1 G. Jones and M. Sheeran, *Circuit design in Ruby*, in [6]. 13–70.

2 G. Jones and M. Sheeran, *Relations and refinement in circuit design*, in [7]. 133–152.

3 G. Jones and M. Sheeran, *Timeless truths about sequential circuits*, in [8]. 245–259.

4 M. Sheeran, *Retiming and slowdown in Ruby*, in [9]. 289–308.

5 C. E. Leiserson and J. B. Saxe, *Retiming Synchronous Circuitry*, Tech. Report 13, Digital Systems Research Center, Palo Alto, California, 1986.

6 Jørgen Staunstrup (ed.), *Formal methods for VLSI design*, North-Holland, 1990.

7 C. C. Morgan, J. C. P. Woodcock (eds.), *3rd Refinement Workshop*, (Proc BCS FACS workshop, Hursley, January 1990) Springer 1991.

8 S. K. Tewksbury, B. W. Dickinson and S. C. Schwartz (eds.), *Concurrent computations: algorithms, architecture and technology*, Plenum Press, New York, 1988.

9 G. J. Milne (ed.), *The fusion of hardware design and verification*, North-Holland, 1988.

VLSI 91
A. Halaas and P.B. Denyer (Eds.)
Elsevier Science Publishers B.V. (North-Holland)

Structuring Hardware Proofs: First steps towards Automation in a Higher-Order Environment

Klaus Schneider, Ramayya Kumar and Thomas Kropf

University of Karlsruhe, Institute of Computer Design and Fault Tolerance, P.O. Box 6980, 7500 Karlsruhe, Germany

Abstract

Most proofs of hardware in an higher-order logic environment follow a definite pattern. This observation is used to give a methodology for hardware proofs in order to isolate the situations where the designer's creativity is required, and to automate the remaining tedious proof tasks. The interactive HOL theorem prover has been extended by generalized hardware specific tactics for simplifying proofs and an automatic theorem prover, called FAUST, for proving the simplified subgoals.

1. INTRODUCTION

Although formal verification of hardware has been the focus of extensive research in the recent past [1-3], it has not yet been embedded within the toolkit of normal circuit designers. The main reasons for this are twofold — the existing automatic approaches can handle only a limited class of circuits, and the powerful interactive approaches can be driven only by logicians. In this paper we present a method for structuring hardware proofs within an interactive environment, thus isolating the creative and the routine steps so that effective assistance for the creative steps and an automation of the latter is made possible. Our approach therefore bridges the gap between such verification techniques and a normal circuit designer.

The automatic approaches which are mostly based on variations of propositional logic, concentrate on proving tautologies [4], or verifying the correctness of finite state machines [5, 6]. However, when complete systems or data paths are to be verified, their space/time requirements increase tremendously. Furthermore, the incapability of the inherent propositional logic to easily express the concepts of hierarchy and modularity limit their use to the verification of small parts of hardware systems.

The interactive approaches on the other hand, are mostly based on higher order logic [7, 8]. This logic suffers from the basic drawback that it is incomplete and therefore does not lend itself to automation. However, its expressive power is a great asset for the compact specification and verification of generic hardware [9-11]. Higher order logic can be used to naturally specify the input and output signals of the hardware devices as functions of time, and recursively specify generalized n-bit regular structures. It also allows the hierarchical and modular verification of systems, which is a primary requirement in tackling complex systems.

HOL is a widely used proof assistant for interactive hardware verification in a higher-order logic environment [12, 13]. It is based on natural deduction and uses a small number of axioms and inference rules for deriving theorems. It also allows the use of user-defined procedures or tactics for combining the inference rules and theorems, thus resulting in a mechanization of small parts of proofs. However, verifying hardware in HOL requires a great deal of knowledge in mathematical logic and proof techniques [14].

Our experiences in interactive hardware verification with HOL and LAMBDA [15] and a thorough study of reports have shown that there exists a definite structure in hardware proofs, irrespective of the kind of circuit to be verified, when specifications are to be verified against register-transfer level implementations. On closer examination, we have observed that a large portion of this proof process can be automated and the creative parts, which need interactions, can be well isolated. This automation can greatly relieve a normal circuit designer from proving the correctness of tedious but easily provable formulae, such as

"∀ f . ∃ l . (l t ↔ f t) ∧ P (l t) ↔ P (f t)".

In order to automate large portions of proofs we have used HOL as a platform for implementing an automatic prover called FAUST (First-order Automation using Unification within a Sequent calculus Technique), which is based on a modified form of sequent calculus called restricted sequent calculus ($\mathcal{RSEQ}$). $\mathcal{RSEQ}$ lends itself to an efficient implementation and furthermore the resultant proofs are readable and understandable, since it closely reflects the semantics of the various logical connectives.

The paper is organized as follows – in Section 2, we investigate the structure of hardware proofs in a higher order environment. Section 3 describes the theory behind the restricted sequent calculus ($\mathcal{RSEQ}$). Experimental results are reported in section 4 and section 5 concludes the paper.

2. STRUCTURE OF HARDWARE PROOFS

In the following we assume formal descriptions of a circuit specification S at an abstraction level i and an implementation I at the next lower level i+1, both given in higher-order logic as described e.g. in [9].

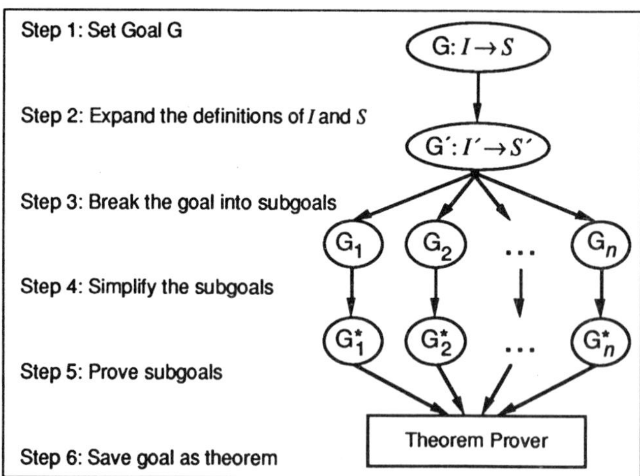

Figure 1. Structure of a hardware proof

Given the formal descriptions of the specification S and the implementation I, the goal of hardware verification is to show the behavioural equivalence of the two descriptions, i.e. $I \leftrightarrow S$. If the specification is incomplete, e.g. only certain safety or liveness properties are to be verified, such an equivalence cannot be shown and instead the implication of the specification from the implementation has to be proven, i.e. $I \rightarrow S$.

In many cases it is possible to structure the steps in hardware verification as depicted in Figure 1, where internal lines (step 4) are connections between different modules of the implementation which are not primary inputs or outputs.

In the rest of this section, these six steps are elaborated using the parity circuit as an example [12]. An informal specification of the synchronous even parity circuit is as follows:

Initially the output (out) is set to "T" (true). At every n+1th clock, the output is T iff there have been an even number of T's on the input line - in.

Step 1: Define the specification and implementation and set the goal

The informal specification is converted to a formal one by choosing functions and abstract datatypes which closely reflect the notions in the informal specification, so that the validation of the formal specification is easy to perform. The use of hardware description languages and automatic conversions to formal specifications allow the use of usual specification techniques [16].

A possible formal specification of the informally specified "even parity circuit" is as follows:

$\forall$ in, out. PARITY_SPEC(in, out) := $\forall$ t . ((out 0 $\leftrightarrow$ T) $\wedge$ (out (suc t) $\leftrightarrow$ EVEN (in,out))

where the predicate EVEN is defined as

$\forall$ in, out. EVEN(in, out) := $\forall$ t . (in (suc t) $\leftrightarrow$ $\neg$ out t)

The predicate EVEN, encodes the informal specification — at all time instants, EVEN is true, iff "in t+1" is equivalent to the complement of "out t".

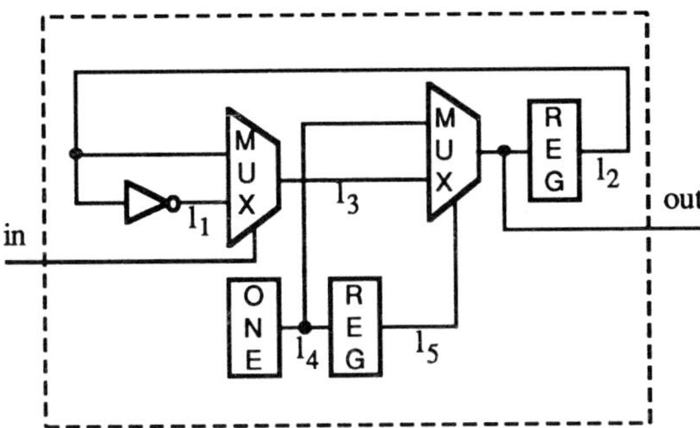

Figure 2. Parity Implementation

The implementation on the other hand, can be automatically derived from netlists as a conjunction of predicates, each of which correspond to the specification of some previously verified component. A black-box treatment of the implementation yields universally quantified input/output lines and existentially quantified internal lines [9]. The formal description of the parity circuit implementation given in Figure 2 is as follows :

$$\forall \text{ in, out . PARITY\_IMP(in,out)} := \exists\, l_1, l_2, l_3, l_4, l_5.\ \forall\, t.$$
$$\begin{array}{l} \text{NOT\_SPEC}\ (l_2\ t, l_1\ t) \wedge \\ \text{MUX\_SPEC}\ (\text{in } t, l_1\ t, l_2\ t, l_3\ t) \wedge \\ \text{REG\_SPEC}\ (\text{out}, l_2) \wedge \\ \text{ONE\_SPEC}\ (l_4\ t) \wedge \\ \text{REG\_SPEC}\ (l_4, l_5) \wedge \\ \text{MUX\_SPEC}\ (l_5\ t, l_3\ t, l_4\ t, \text{out } t) \end{array}$$

The goal to be proved for the parity example is as follows:

$$\forall \text{ in, out. PARITY\_IMP(in, out)} \leftrightarrow \text{PARITY\_SPEC(in, out)}$$

Step 2: Expand the definitions of *I* and *S*

Both specification and implementation usually rely on the usage of predefined predicates and modules to keep the descriptions hierarchical, modular and understandable. To perform the proof task, these definitions have to be expanded. The complex datatypes are also refined, to allow rewriting, e.g. a natural number is refined to $\text{suc}^n\ 0$, . Minor user interaction is required here, in order to control the granularity of expansions.

Table 1
Formal specifications of the library components

Component	Definition
NOT_SPEC (in,out)	$\forall$ in,out. (out $\leftrightarrow \neg$ in)
ONE_SPEC (out)	$\forall$ out. (out $\leftrightarrow$ T)
MUX_SPEC (sel,in_1,in_2,out)	$\forall$ sel, in_1, in_2, out. (out $\leftrightarrow$ ((sel $\rightarrow \text{in}_1$) $\wedge$ ($\neg$sel $\rightarrow \text{in}_2$)))
REG_SPEC (in,out)	$\forall$ in, out. ($\forall$ t. ((out 0 $\leftrightarrow$ F) $\wedge$ (out (suc t) $\leftrightarrow$ in t)))

Applying this step on the parity example using the definition of the predicate EVEN and the components as specified in Table 1, generates the following formula.

$$\begin{array}{l} \forall \text{ in, out}. \\ \quad \exists\, l_1, l_2, l_3, l_4, l_5.\ \forall\, t.\ (l_1\ t \leftrightarrow \neg\ l_2\ t) \wedge \\ \qquad (l_3\ t \leftrightarrow ((\text{in } t \rightarrow l_1\ t) \wedge (\neg \text{ in } t \rightarrow l_2\ t))) \wedge \\ \qquad (\forall\, t_1.\ (l_2\ 0 \leftrightarrow F) \wedge \\ \qquad\qquad (l_2\ (\text{suc } t_1) \leftrightarrow \text{out } t_1)) \wedge \\ \qquad (l_4\ t \leftrightarrow T) \wedge \\ \qquad (\forall\, t_2.\ (l_5\ 0 \leftrightarrow F) \wedge \\ \qquad\qquad (l_5\ (\text{suc } t_2) \leftrightarrow l_4\ t_2)) \wedge \\ \qquad (\text{out } t \leftrightarrow ((l_5\ t \rightarrow l_3\ t) \wedge (\neg\ l_5\ t \rightarrow l_4\ t))) \\ \leftrightarrow \\ \forall\, t.\ ((\text{out } 0 \leftrightarrow T) \wedge \\ \qquad (\text{out (suc } t) \leftrightarrow (\text{in (suc } t) \leftrightarrow \neg \text{ out } t)) \end{array}$$

Step 3: Break the goal into subgoals

This is the creative step, where the user has to use his knowledge in breaking up the goal into subgoals, apply proof strategies like induction and use the lemmas needed. Many design specific heuristics can be built in and design decisions be incorporated, to aid the user in this task [17]. However, due to the very nature of the problem, automating this step is impossible in many cases. In the simple parity example this step is superfluous and can be skipped.

Step 4: Simplify the subgoals

Having broken up the original goal into subgoals, the subgoals can then be *automatically* simplified. This step is also called the unwind step and consists of three main operations, which are executed consecutively, namely

- (i) conversion to a specialized prenex normal form,
- (ii) elimination of internal lines, and
- (iii) simplification of formulae which contain "true" and "false".

Prenex normal form is an equivalent form of the original formula, containing quantifiers at the beginning of the formula [18]. This operation consists of applying certain quantifier rewrite rules on the formulae in the subgoals, so as to generate a prenex normal form, which is a prerequisite for the application of steps (ii) and (iii). A complete list of the rewrite rules needed for transforming hardware specific formulae is given in [19]. We list only those rules needed for the parity example in this paper.

<u>Rule 1</u>: $(\forall\, t.\, (P \wedge Q))\ '\ (\forall\, t.\, P) \wedge (\forall\, t.\, Q)$

<u>Rule 2</u>: $(\forall\, t.\, P)\ '\ P$ if t is not free† in P

<u>Rule 3</u>: $(\forall\, t.\, P)\ '\ (\forall\, z\ [P]_t^z{}^{\dagger\dagger})$ if z is not free in P

Repeated application of these rules on the parity example results in the following description

$$
\begin{array}{l}
\forall\, in, out\,.\\
\quad \exists\, l_1, l_2, l_3, l_4, l_5.\ \forall\, t\,.\ (l_1\, t \leftrightarrow \neg\, l_2\, t) \wedge\\
\qquad\qquad (l_3\, t \leftrightarrow ((in\, t \rightarrow l_1\, t) \wedge (\neg\, in\, t \rightarrow l_2\, t))) \wedge\\
\qquad\qquad (l_2\, 0 \leftrightarrow F) \wedge\\
\qquad\qquad (l_2\, (suc\, t) \leftrightarrow out\, t) \wedge\\
\qquad\qquad (l_4\, t \leftrightarrow T) \wedge\\
\qquad\qquad (l_5\, 0 \leftrightarrow F) \wedge\\
\qquad\qquad (l_5\, (suc\, t) \leftrightarrow l_4\, t) \wedge\\
\qquad\qquad (out\, t \leftrightarrow ((l_5\, t \rightarrow l_3\, t) \wedge (\neg\, l_5\, t \rightarrow l_4\, t)))\\
\quad \leftrightarrow\\
\quad \forall\, t\,.\ ((out\, 0 \leftrightarrow T) \wedge\\
\qquad (out\, (suc\, t) \leftrightarrow (in\, (suc\, t) \leftrightarrow \neg\, out\, t))
\end{array}
$$

The elimination of internal lines also comprises the application of some rewrite rules, so that internal lines are eliminated. The two rules that are needed for the parity example are:

<u>Rule 4</u>: $\forall\, P, Q.\ (\exists\, l.\ \forall\, t.\ (l\, t \leftrightarrow Q\, t) \wedge P\, (l\, t))\ '\ (\forall\, t.\ P\, (Q\, t))$

<u>Rule 5</u>: $\forall\, P, Q.\ (\exists\, l.\ \forall\, t.\ (l\, 0 \leftrightarrow a) \wedge (l\, (suc\, t) \leftrightarrow Q\, t) \wedge P_1\, t \wedge P_2\, (l\, t))$

'

$(\forall\, t.\ P_1\, t \wedge P_2\ \ 0 \wedge P_2\, (suc\, t))$

Rule 4 intuitively states, that those internal lines which are outputs of combinational components "Q" can be directly eliminated by replacing them by their definitions. In the example, after minor syntactical changes, i.e. by reducing the scope of existentially quantified

---

† A free variable is a variable which is not bound by any quantifier (for details see [18])

†† $[P]_t^z$ denotes the substitution of all occurences of t in P, by z

variables to their actual appearance, an application of rule 4 on the internal lines l_1, l_3 and l_4 results in the following formula:

$$
\begin{aligned}
&\forall\, in, out\,.\\
&\quad \exists\, l_2, l_5.\ \forall\, t\,.\ (l_2\, 0 \leftrightarrow F) \wedge\\
&\qquad\qquad (l_2\, (suc\ t) \leftrightarrow out\ t) \wedge\\
&\qquad\qquad (l_5\, 0 \leftrightarrow F) \wedge\\
&\qquad\qquad (l_5\, (suc\ t) \leftrightarrow T) \wedge\\
&\qquad\qquad (out\ t \leftrightarrow ((l_5\, t \rightarrow ((in\ t \rightarrow \neg\, l_2\, t) \wedge (\neg\, in\ t \rightarrow l_2\, t))) \wedge (\neg\, l_5\, t \rightarrow T)))\\
&\quad \leftrightarrow\\
&\quad \forall\, t\,.\ ((out\ 0 \leftrightarrow T) \wedge\\
&\qquad (out\ (suc\ t) \leftrightarrow (in\ (suc\ t) \leftrightarrow \neg\, out\ t))
\end{aligned}
$$

Rule 5 analogously leads to the elimination of those internal lines which correspond to the outputs of sequential components. However, since sequential components may be used for storing internal states, complete elimination of the internal lines is not possible in all cases. Applying this rule using l_5 as P_1 and out as P_2 for eliminating l_2, yields the following formula:

$$
\begin{aligned}
&\forall\, in, out\,.\\
&\quad \exists\, l_5.\ \forall\, t\,.\ (l_5\, 0 \leftrightarrow F) \wedge\\
&\qquad\qquad (l_5\, (suc\ t) \leftrightarrow T) \wedge\\
&\qquad\qquad (out\ 0 \leftrightarrow ((l_5\, 0 \rightarrow ((in\ 0 \rightarrow \neg\, F) \wedge (\neg\, in\ 0 \rightarrow F))) \wedge (\neg\, l_5\, 0 \rightarrow T)) \wedge\\
&\qquad\qquad (out\ (suc\ t) \leftrightarrow ((l_5\, (suc\ t) \rightarrow ((in\ (suc\ t) \rightarrow \neg\, out\ t) \wedge\\
&\qquad\qquad\qquad\qquad (\neg\, in\ (suc\ t) \rightarrow out\ t))) \wedge\\
&\qquad\qquad\qquad (\neg\, l_5\, (suc\ t) \rightarrow T)))\\
&\quad \leftrightarrow\\
&\quad \forall\, t\,.\ ((out\ 0 \leftrightarrow T) \wedge\\
&\qquad (out\ (suc\ t) \leftrightarrow (in\ (suc\ t) \leftrightarrow \neg\, out\ t))
\end{aligned}
$$

Rule 4 can now be applied again to eliminate the internal line l_5.

The constants "T" and "F" appearing in the formulae due to the initializations of the sequential components, can be eliminated by using the usual logical simplification rules, yielding:

$$
\begin{aligned}
&\forall\, in, out\,.\\
&\quad \forall\, t\,.\ ((out\ 0 \leftrightarrow T) \wedge\\
&\qquad (out\ (suc\ t) \leftrightarrow ((in\ (suc\ t) \rightarrow \neg\, out\ t) \wedge (\neg\, in\ (suc\ t) \rightarrow out\ t))))\\
&\leftrightarrow\\
&\quad \forall\, t\,.\ ((out\ 0 \leftrightarrow T) \wedge\\
&\qquad (out\ (suc\ t) \leftrightarrow (in\ (suc\ t) \leftrightarrow \neg\, out\ t)))
\end{aligned}
$$

Step 5: Automatic proof of the simplified subgoals

Having reduced the subgoals into a simple form, an automated theorem prover can then be used to prove each subgoal separately. The next section is essentially devoted to a brief description of such a prover called FAUST. In many cases, FAUST is capable of proving these subgoals without further aid. However, to speed up the proof process and to contain the space and runtime requirements, it may be necessary to guide the proof process by proving some lemmas. An analysis of the subgoals could lead to an automatic suggestion of the lemmas to be proved before the automatic prover is started.

Step 6: Update library for future use

The specification and the correctness theorems generated for the design can then be stored within a library and recalled later while designing components which use the currently verified component. A hierarchy can thus be achieved in future proofs as the specification and the correctness theorems are sufficient and the current component need not be broken down to its implementation anymore.

In this section, we have shown that although hardware verification remains to be interactive, most of the drudgery can be automated. The extent to which hardware verification has been automated up to now can be compared by referring e.g. to the tutorial paper on HOL [12].

3. THE THEORY BEHIND FAUST

Although higher order logic has been used in the last section to specify and verify hardware, it is apparent, that only few constructs are necessary, which exceed the expressiveness of first order logic. However we have observed that, it is possible to automate proofs of statements in this restricted higher order logic, by means of a calculus which grounds on first order techniques.

A theorem prover which is capable of proving subgoals according to the proof structure given in section 2, must be able to prove higher-order formulae. Moreover, to gain confidence in the resultant proofs, they have to be readable. This motivated us to implement the prover FAUST, based on a modification of the well-known, proof-tree based, sequent calculus $\mathcal{SEQ}$ [18,20], called "Restricted Sequent Calculus ($\mathcal{RSEQ}$)". The modifications were instigated by the need for an efficient implementation of the γ rules (universal/existential quantifier elimination). In the following we briefly present the underlying ideas of $\mathcal{RSEQ}$. A detailed presentation may be found in [19, 22].

The α-, β- and δ- rules in $\mathcal{SEQ}$ are deterministic and hence not critical for the construction of the proof tree. On the other hand, the γ-rules are nondeterministic as an arbitrary term may be chosen for substituting a bound variable. When the formula to be proved is large, often γ-rules have to be applied during the early phases of the proof tree construction. However, the "right" choice of the substitution heavily influences the length and complexity of the further proof. Hence, the concept of a *metavariable* is introduced so as to postpone the decision for an appropriate substitution term, until enough decision information is available. The process is performed as follows. If one arrives at a point where a γ-rule has to be applied, a search for an appropriate term is not undertaken. Instead, a new metavariable is introduced and the construction of the proof tree proceeds further. When the construction process is ripened to an extent that a suitable term can be found, then the term is substituted for the metavariable. The set of variables is split up into two disjoint sets of "normal" variables $\mathcal{V}$ and metavariables $\mathcal{V}_M$.

The introduction of the metavariable poses problems during the δ-rule application, as the constant substituted must be new, i.e. it does not occur in any of the nodes of the proof tree constructed so far. Since the terms corresponding to the existing metavariables are still unknown, a *restriction* is placed on the metavariables, so that the terms to be substituted for the metavariable do not contain the constants introduced by the δ-rules. To this effect a set called the *forbidden set* (fs_m), $fs_m \subseteq \mathcal{V}$ is defined, for each used metavariable $m \in \mathcal{V}_M$, containing all those constants introduced by δ-rule applications after the creation of the metavariable *m*. An alternative approach which uses skolemization instead of restrictions is given in [18].

The sequents that appear in $\mathcal{RSEQ}$ are called *restricted sequents* and have the form $\Gamma \vdash \Delta \parallel \mathcal{R}$, where $\Gamma, \Delta \subseteq \mathcal{F}$ (the set of formulae) and $\mathcal{R} \subseteq \mathcal{V}_M \times 2^{\mathcal{V}}$. The restriction $\mathcal{R}$ of each sequent belonging to $\mathcal{RSEQ}$ consists of list of pairs (m, fs_m).

The terms for substitution of the metavariables are found via *meta-unification*. This is a modification of the normal Robinson's unification algorithm [23] in such a manner, that only metavariables are considered as substitutable sub-terms. However, only those substitutions σ are allowed which satisfy the property :

$\forall \tau \in fs_m$. τ does not occur in $\sigma(m)$ for all $(m, fs_m) \in \mathcal{R}$

The proof tree construction otherwise basically proceeds in a manner similar to the normal sequent calculus techniques. The introduction of the metavariables and restrictions requires a modified set of γ and δ-rules. For a detailed description of the complete rule-set and the proof tree construction, the reader is referred to [19,22].

FAUST, which is based on $\mathcal{RSEQ}$ has been implemented within the HOL theorem proving environment and is written in ML. Our initial idea was to implement FAUST by using the tactics provided within HOL, so that the reliability of the proofs obtained is high. However, this slowed down the speed of automatic proving so drastically, that we felt the need for implementing a stand-alone prover capable of interacting with HOL. This interaction has been achieved by introducing the proofs completed by FAUST as theorems, using the "mk_thm" (make theorem) function in HOL. Since this puts us on thin ice, FAUST also generates a HOL tactic, which can then be used to validate the automatic proofs *automatically* within a normal HOL session.

4. EXPERIMENTAL RESULTS

FAUST was first tested for its correctness by using the propositional and first-order formulae taken from [24, 25]. The runtimes of some complex formulae are given in Table 2 and were measured on SUN 4/65 using the ML-package included in the usual HOL-system. The problem called Andrew's challenge (P_{34} in Table 2) was solved by generating 86 subgoals as compared to 1600 subgoals generated by resolution provers. Additionally, we have observed that specialized HOL tactics can be developed for difficult problems such as Uruquart's problems, which was then solved in linear time.

Table 2
Runtimes of Benchmark-Formulae [25]

Formula	Time (Seconds)
P_{12}	0.2
P_{21}	0.4
P_{25}	0.4
P_{26}	0.9
P_{28}	0.7
P_{34}	17.6

Having gained confidence about the correctness of our prover we have looked at some combinational circuits which also required a matter of seconds. At present we have proved the correctness of only small sequential circuits such as parity, serial adder, flipflops, and minmax [26]. The parity, serial adder and flipflop examples did not require any interaction and were proved in a few seconds. Since the minmax example needs interactive steps, we can only qualitatively compare it with a normal HOL run. There, each step is interactive, whereas using our approach we had to interact with HOL only while proving the correctness of a comparator module.

5. CONCLUSIONS AND FUTURE WORK

In this paper it has been shown that most hardware proofs can be performed by following the sequence of steps given in section 2. The creative steps involved in proving the correctness are few in number and most of the other steps can be automated. Furthermore we have elucidated that, although one needs higher order for specifying hardware, it is a restricted form which can be handled by first-order proving techniques. For this purpose, a modified form of sequent calculus has been proposed which allows an efficient implementation.

Currently, we are improving the efficiency of our prover further. We are also working on embedding our approach within a commercial design framework [27], so that verification proceeds hand in hand with design.

6. REFERENCES

1 G. Birtwistle, P.A. Subrahmanyam (Eds.): Current Trends in Hardware Verification and Automated Theorem Proving; Springer Verlag, 1988.

2 P. Camurati, P. Prinetto: Formal Verification of Hardware Correctness: Introduction and Survey of Current Research; IEEE Computer, July 1988, pp. 8-19.

3 V. Stavridou, H. Barringer, D.A. Edwards: Formal Specification and Verification of Hardware: A Comparative Case Study; Proc. 25th Design Automation Conference (DAC 88), 1988, pp. 197-204.

4 E. Cerny, C. Mauras: Tautology Checking Using Cross-Controllability and Cross-Observability Relations; Proc. International Conference on Computer-Aided Design (ICCAD 90), 1990, pp. 34-37.

5 O. Coudert, C. Berthet, J.C. Madre: Verification of Synchronous Sequential Machines Based on Symbolic Execution; Proc. Workshop on Automatic Verification Methods for Finite State Systems, Grenoble, June 1989.

6 J.R. Burch, E.M. Clarke, K.L. McMillan, D.L. Dill, L.J. Hwang: Symbolic Model Checking: 10^20 States and Beyond; Proc. 5th Annual Symposium on Logic in Computer Science, 1990.

7 A. Camilleri, M. J. C. Gordon, T. Melham: Hardware Verification using Higher-Order Logic; Borrione (Ed.), Proc. IFIP Workshop on "From H.D.L. Descriptions to Guaranteed Correct Circuit Design", Grenoble 1986, North-Holland, pp.43-67.

8 S. Finn, M. Fourman, M. Francis, B. Harris: Formal System Design - Interactive Synthesis based on Computer Assisted Formal Reasoning; Proc. Intl. Workshop on Applied Formal Methods for Correct VLSI Design, Leuven, November 1989.

9 F.K. Hanna, N. Daeche: Specification and Verification of Digital Systems Using Higher-Order Predicate Logic; IEE Proc. Pt. E, Vol. 133, No. 3, September 1986, pp. 242-254.

10 M. J. C. Gordon: Why High-Order Logic is a good Formalism for Specifying and Verifying Hardware; Milne/Subrahmanyam (Eds.), Formal Aspects of VLSI Design, Proc. Edinburgh Workshop on VLSI 1985, North-Holland 1986, pp. 153-178.

11 J. Joyce: More Reasons Why Higher-Order Logic is a Good Formalism for Specifiying and Verifying Hardware; Proc. International Workshop on Formal Methods in VLSI Design, Miami, January 1991.

12 M. Gordon: A Proof Generating System for Higher-Order Logic; VLSI Specification, Verification and Synthesis, Eds. Birwistle G. and Subrahmanyam P.A., Kluwer, 1988.

13 Proceedings of the Third HOL Users Meeting; Aarhus University, October 1990.

14 P. Loewenstein: Experiences Using a Theorem Prover for Hardware Verification; Proc. International Workshop on Formal Methods in VLSI Design, Miami, January 1991.

15 Abstract Hardware Limited: LAMBDA - Logic and Mathematics behind Design Automation; User and Reference Manuals, Version 3.1, 1990.
16 R. Boulton, M. Gordon, J. Herbert, J. van Tassel: The HOL Verification of ELLA Designs; Proc. International Workshop on Formal Methods in VLSI Design, Miami, January 1991.
17 S. Kalvala, M. Archer, K. Levitt: A Methodology for Integrating Hardware Design and Verification; Proc. International Workshop on Formal Methods in VLSI Design, Miami, January 1991.
18 M. Fitting: First-Order Logic and Automated Theorem Proving; Springer Verlag, 1990.
19 K. Schneider: Ein Sequenzenkalkül für die Hardware-Verifikation in HOL; Diploma Thesis, Institute of Computer Design and Fault-Tolerance, University of Karlsruhe, 1991.
20 J.H. Gallier: Logic for Computer Science: Foundations of Automatic Theorem Proving; Harper & Row Computer Science and Technology Series No. 5, Harper & Row Publishers,New York, 1986.
21 L.C. Paulson: Natural Deduction as higher-order Resolution; Journal of Logic Programming, Vol. 3, 1986, pp. 237-258.
22 K. Schneider, R. Kumar, T. Kropf: Automating most parts of hardware proofs in HOL; Proc. Workshop on Computer Aided Verification, Aalborg, July 1991.
23 J.A. Robinson: A Machine-oriented logic based on the resolution principle; Journal of the ACM, Vol.12, pp.23-41, 1965.
24 D. Kalish, R. Montague: Logic: Techniques of Formal Reasoning; World, Harcourt & Brace, 1964.
25 F.J. Pelletier: Seventy-Five Problems for Testing Automatic Theorem Provers; Journal of Automated Reasoning, Vol.2, pp.191-216, 1986.
26 L. Claesen: Preface; Proc. Intl. Workshop on Applied Formal Methods for Correct VLSI Design, Leuven, November 1989.
27 Cadence Design Systems Inc.: User Manuals; July 1989.

VLSI 91
A. Halaas and P.B. Denyer (Eds.)
Elsevier Science Publishers B.V. (North-Holland)

DOMINO: Deterministic Placement Improvement with Hill-Climbing Capabilities

KONRAD DOLL, FRANK M. JOHANNES, AND GEORG SIGL
Institute of Electronic Design Automation, Department of Electrical Engineering,
Technical University of Munich, D-8000 Munich 2, Germany

This paper addresses the problem of cell placement for ASICs. Problem sizes encountered today require economical use of computation time. On the other hand, to produce near-optimum placements the algorithm applied must not get trapped far from the global optimum. The algorithmic situation is often simplified by assuming that deterministic algorithms cannot escape local minima, while randomized algorithms have the ability to reach the global optimum of a combinatorial problem. For the placement problem we demonstrate that, as a third possibility, a deterministic algorithm can be tailored to escape local optima. Our deterministic algorithm DOMINO produces up to 10% better results than TimberWolfSC 5.4, one of the best randomized algorithms, requiring 5–10 times less computation time.

1 Introduction

Currently there are two challenging demands in automated layout synthesis of Application Specific Integrated Circuits (ASICs). Primarily, the *quality* of the final layout should be improved to maximize the circuit's performance and to minimize chip area. On the other hand, layout design tools must be able to treat circuits with *complexities* of tens of thousands of cells. Either goal can only be approached at additional computing cost. For circuits of moderate complexities, methods are known that satisfy both demands. But, particularly for very large high-performance circuits, either excessive computation times have to be accepted or layout quality must be compromised.

To support the layout design of very large circuits, we focused our research on placement algorithms, since during placement the most crucial design decisions have to be made. To facilitate the placement and the routing tasks and to allow for using the large cell libraries available, we adopt the popular column-oriented layout style with cells of equal widths and differing heights. This style is widely used for standard-cell circuits and for conventional gate-arrays. It has also been successfully applied to the sea-of-gates layout style. For solving the placement problem, algorithms using constructive and iterative improvement techniques have been proposed (see [Prea88,Leng90], e. g.).

Constructive algorithms fall into two major subclasses, the partitioning algorithms, e. g. [Laut79,SuKe88], and the so-called analytical algorithms, e. g. [Otte82,Blan85,ShDu85]. Recently, high-quality solutions were obtained with algorithms combining both strategies [TsKH88,KlSJ88,KSJA91,SiDJ91]. Algorithms of this class are usually fast, produce good results because of their global view of the problem, but are restricted in the choice of objectives and in general will not yield the global optimum of the placement problem.

Iterative placement improvement algorithms aim at improving existing solutions, especially placements obtained with constructive algorithms. Typically, in one iterative step they select a small and local subproblem to be solved by exact or heuristic methods.

These algorithms also divide into two subclasses that apply deterministic and random techniques.

Deterministic placement improvement methods offer a rather free choice of objectives, but they can get trapped in a local optimum. In this paper we propose a novel deterministic strategy that has the ability of escaping local optima and that requires considerably less computation time than randomized algorithms.

Iterative improvement methods based on randomized algorithms never reject better solutions, but also accept intermediate placements of inferior quality with low probabilities. Thus, they have the ability to escape local optima and to approach the global optimum arbitrarily close, if sufficient computation time is provided. Since this is not always practicable particularly for large circuits, layout quality must be compromised. There are two basic randomized methods, i. e. evolution-based algorithms [KlBa90,CoPa86] and Simulated Annealing (SA) (see [KiGV82,Rome89,SeLe87], e. g.). For functions of continuous variables the SA method was merged with a local search technique and computation time reductions of two orders of magnitude were reported [Marc88].

For algorithms applying the SA principle it has been proved that they will provide a solution arbitrarily close to the global optimum, if given enough time. The semi-custom placement and routing package TimberWolfSC [SeLe87] is a recognized application that combines elaborate heuristics with the SA principle. In this paper we use TimberWolfSC Version 5.4[1] as a reference for evaluating our placement algorithm.

Randomized iterative improvement strategies for column-oriented layout styles must cope with overlapping cells and empty space between them. Typically, in an iteration step one cell is randomly moved to a new location, two cells are randomly exchanged, or a set of badly placed cells is simultaneously removed from their current locations using a stochastic method and placed in improved positions. Because cells are allowed to have different heights, these moves will in general produce overlapping cells as well as empty space between cells. TimberWolfSC solves this problem by adding to the objective function a penalty term to penalize cell overlaps. The weighting factor balancing the terms strongly impacts the quality of the final configuration and has to be determined by experiments. The Simulated Evolution procedure [KlBa90] terminates each iterative placement step by a cell realignment that removes all overlaps and unused spaces.

Although in each iteration a *local* subproblem is solved, resolution of overlaps and free space can yield *global* destruction of the placement by chain reactions and can slow down or even destroy process convergence. Thus, the method of overlap treatment is crucial to the iterative placement process. We avoid global cell realignment and penalty terms as well. Instead, we used a simple but effective strategy to maintain overlap-free and compact placements throughout the iterative process. This strategy consists of constructing placements like growing crystals and has been applied in compaction with success [ShSS86].

In Section 2 we present an outline of our novel deterministic placement improvement process. A more detailed description follows in Section 3. The ideas are implemented in the placement program DOMINO. The results given in Section 4 show that this deterministic algorithm has the ability of escaping local minima and that especially for very large circuits placements of quality better than TimberWolfSC 5.4 can be obtained requiring significantly less computation time. The concluding Section contains some remarks and generalizations.

---

[1]We would like to thank Prof. C. Sechen of Yale University for making TimberWolfSC 5.4 available to us.

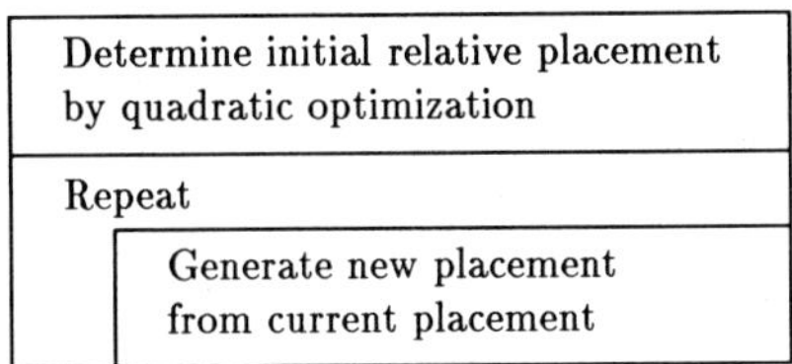

Figure 1: Top-level structure of the procedure

2 Outline of the Procedure

For a high-quality final placement, a global view of the problem is definitely necessary at least during the initial stages of the algorithm. For example, SA methods use a high temperature (a high probability for long-distance moves) for this purpose and Simulated Evolution applies a hierarchical evolution-based process to find a globally optimized placement. We prefer analytical algorithms like the GORDIAN procedure [KlSJ88,KSJA91, SiDJ91], which is composed of alternating global optimization and partitioning steps. It was developed for circuits of tens of thousands of cells. The main advantage of GORDIAN is the simultaneous treatment of all cells during all levels of partitioning. The so formulated problem can be solved efficiently by applying quadratic programming techniques. By abstracting from their real geometries cells are treated as points to be optimally arranged in the plane with respect to an objective function modeling wiring lengths. The resulting placement will in general contain overlapping cells. It is often called a point, global, or relative placement, since in a global view it reflects the optimal cell adjacencies. We use it as initial placement for an iterative improvement process as shown in Figure 1.

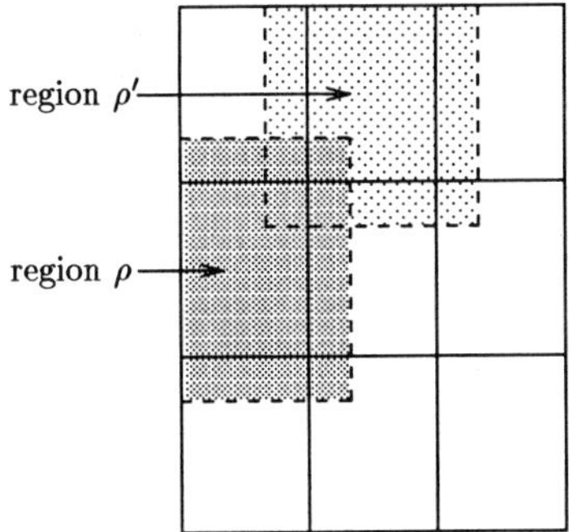

Figure 2: Dividing the layout area into regions

The iterative process generates a sequence of intermediate placements. During each generation step the current placement is used for guiding the construction of a new placement. Thus, the proposed method can be viewed as a constructive algorithm embedded in an iterative improvement procedure. After each generation, a placement free of interspersed spaces and overlapping cells is available. The process finishes when after several generations no significant improvement is achieved.

Each generation problem is geometrically divided into similar local subproblems. A subproblem consists of a subset of cells to be rearranged. For that purpose the layout area is represented by an array of overlapping regions, as shown in Figure 2. To each region ρ all cells $\mathcal{M}_\rho$ currently placed inside of it are assigned. In the special case of the initial generation, when we are given a relative placement with overlaps, cells are assigned to regions according to their relative positions by recursively bipartitioning the set of cells with alternating horizontal and vertical cuts. Thus, even if the relative placement algorithm has produced a clustered placement, the cells are spread out over all regions.

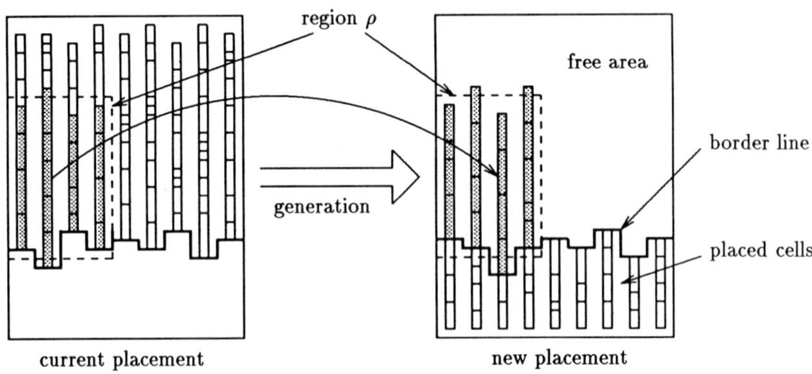

Figure 3: Generation of a new placement from the current placement

Suppose we are generating a new placement from the current placement shown in Figure 3 by meandering through the regions from bottom to top. We have already produced a partial new placement without overlaps and unused spaces in the lower part of the layout area. This area is separated from the free area above it by a border line. Region ρ is selected next and all cells associated with it are rearranged above the border line between the lateral borders of region ρ. By applying a transportation algorithm units of cell area are transported to locations according to a cost function approximating wire lengths.

This procedure is repeated for all regions with the border line migrating from bottom to top until a complete new placement has been generated, which is then treated as the current placement for the next generation. During successive generations, different orderings of regions are used. As adjacent regions overlap, cells can move from one region to another during the iterative process.

For solving the optimization problem to determine new cell positions in a region, various objectives can be chosen. We have selected an approved objective to be minimized, i. e. the sum of the half perimeters of the minimum rectangles enclosing the pins of each net. Moreover, we decided to use the transportation algorithm to obtain subproblem solutions efficiently. Since exactly calculated half perimeters cannot be used as transportation costs (see Section 3), we choose to neglect some pins during the calculation if necessary. Thus, in some cases we may obtain a suboptimal result for the current region with respect to our objective (half perimeters). Since we cannot easily predict how this somewhat inaccurate model will affect this objective, there is some uncertainty that will allow us to escape local optima with a low probability.

The technique of using models for solving engineering problems, such as cell placement, is common to most solution methods. The result, although based on a model abstracting from reality, is regarded as the solution of the real problem. The defects and inaccuracies caused by the abstraction process are usually qualified as minor side effects not impairing the quality of the solution. But, as far as we know, these effects have never been viewed as a vehicle to escape local minima.

But what about convergence of this iterative process? The net model has to be constructed in such a way that it is accurate enough in a global sense. The amount of inaccuracy, on the one hand, must be small enough to effect convergence, and on the other hand it must be high enough to allow for climbing out of local minima. The better the initial global placement is the more accurate the model can be. The choice has to be validated experimentally and is justified by excellent placement results (see Section 4).

3 Placement Improvement

To improve the current placement in a region ρ, the cells (or modules) $m_\mu \in \mathcal{M}_\rho$ have to be assigned new positions above the border line without introducing overlaps or free space. As already mentioned, we apply a deterministic method of low polynomial complexity to use computation time as economically as possible. To account for the different cell heights we divide each cell $m_\mu \in \mathcal{M}_\rho$ into s_μ subcells of equal sizes. Thus, the area of a cell is modeled by a suitable number of subcell area units and the interconnections of a cell with its neighbors are transferred to each of its subcells. Typically, we choose the subcell size equal to the greatest common divisor of the cell sizes. By formulating our problem as a transportation problem, we can assign all subcells to locations simultaneously and without overlaps.

The cost function used with the transportation problem is based on estimated lengths of all nets connected to cells in region ρ. For the solution of the transportation problem to be the exact solution of the placement problem, the cost values of different pairs (subcell, location) must be independent of each other. Wire length estimations are independent for all nets connecting one cell in region ρ with cells not in region ρ. But, for nets connecting two or more cells inside the same region wire lengths can only be estimated from the current placement and therefore the transportation result needs not be optimal with respect to our objective (half perimeters). As already mentioned this gives us the ability of escaping local minima.

Linear assignment, a special case of transportation, has already been used for placement improvement [Aker81]. To make cost values independent, in [Stei61] only unconnected cells are selected in one iterative step, or in [KlBa90] nets connecting cells to be placed simultaneously are neglected.

Each placement improvement step consists of four major operations, to be described in more detail in the following subsections.

3.1 Providing Locations

We first have to determine a set of candidate locations $\mathcal{L}_\rho$, where subcells can be positioned by the transportation algorithm. We provide locations for all subcells above the border line between the lateral borders of region ρ. Columns are filled with locations to form a straight line on top as shown in Figure 4.

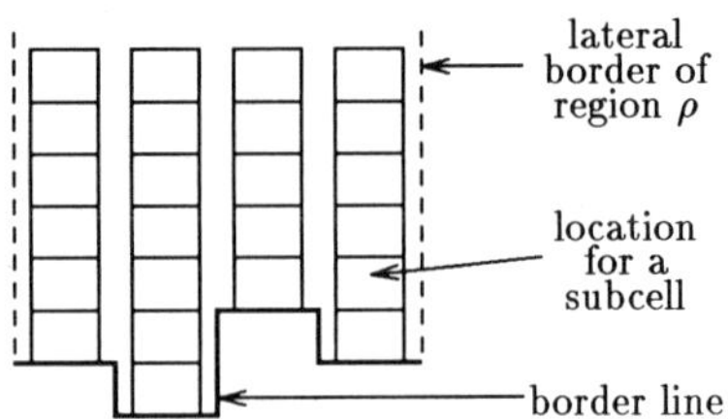

Figure 4: Providing locations

3.2 Determining Transportation Costs

The cost $c_{\mu\nu}$ to assign a subcell of cell m_μ to a location l_ν is calculated by

$$c_{\mu\nu} = \sum_{\sigma \in \mathcal{N}_\mu} L_\sigma \tag{1}$$

where $\mathcal{N}_\mu$ denotes the set of nets connected to cell m_μ and L_σ is the estimated length of net σ.

Let $\mathcal{P}_\sigma$ denote the set of all cells interconnected by net σ, and let $\Gamma(\mathcal{P}_\sigma)$ be the half perimeter of the smallest rectangle enclosing the pins of net σ. If only one cell of $\mathcal{P}_\sigma$ is in $\mathcal{M}_\rho$, then $L_\sigma = \Gamma(\mathcal{P}_\sigma)$, because L_σ does not depend on the positions of other cells in region ρ.

For a net connecting more than one cell in region ρ, we introduce a net model enabling the independent placement of all cells in region ρ. For that purpose we define the set $\mathcal{P}_\sigma^o = \mathcal{P}_\sigma \setminus \mathcal{M}_\rho$ containing all cells in $\mathcal{P}_\sigma$ outside region ρ. If $\mathcal{P}_\sigma^o \neq \{\}$, then for calculating $c_{\mu\nu}$

$$L_\sigma = \Gamma(\mathcal{P}_\sigma^o \cup \{m_\mu\}). \tag{2}$$

If $\mathcal{P}_\sigma^o = \{\}$, net σ connects only cells inside region ρ. To prevent such nets from moving too far from their old places we introduce a virtual cell m_ϕ, to which all cells in $\mathcal{P}_\sigma$ are connected and which is positioned in the center of gravity of the cells in $\mathcal{P}_\sigma$. We set $\mathcal{P}_\sigma^o = \{m_\phi\}$ in equation 2 to calculate L_σ.

3.3 Transportation

The transportation problem can be formulated as a minimal-cost maximal flow problem [Leng90] on a network (directed graph) as in Figure 5. This network consists of a source node S supplying subcells (units of cell area), a destination node D, a set of $|\mathcal{M}_\rho|$ cell nodes m_μ, and a set of $|\mathcal{L}_\rho|$ location nodes l_ν. Since each location can hold at most one subcell, all capacities of arcs leading from location nodes to node D are set to one. The capacities of arcs between node S and cell nodes are s_μ, such that cell m_μ can supply at most s_μ subcells. The cost of assigning a subcell of cell m_μ to location l_ν is $c_{\mu\nu}$ from Subsection 3.2. By using the flow augmentation method (see [Leng90], e. g.) the transportation algorithm can efficiently assign subcells to locations at minimum total transportation cost.

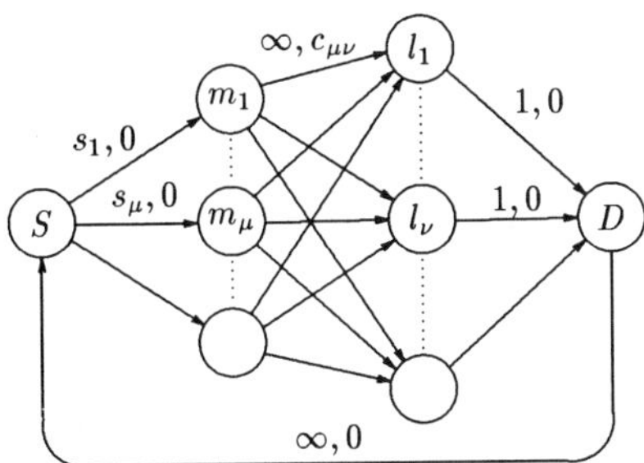

Figure 5: Transportation network (arcs labeled with capacity, cost)

3.4 Moving the Cells

We determine the new positions of the cells $m_\mu \in \mathcal{M}_\rho$ from the locations computed for the subcells. In experiments we observed, that almost all subcells of a cell lie side by side after transportation. Therefore, we place a cell in the column holding most of its subcells. For each cell, the center of gravity of its subcells determines the position of the cell within a column. We stack the cells one over another in this order, thereby avoiding any overlaps and unused spaces. Since the cells can move only inside the current region, this operation does not impair global convergence, but provides extra chances of leaving local optima.

3.5 Choice of Parameters

In our approach two parameters must be adjusted:

The first parameter is the size of a region. If it is small, we consider only few cells simultaneously and therefore require more generations to reach a good result. For large regions with many cells, we found the computation times of the transportation program to increase with $|\mathcal{M}_\rho|^{2.5}$ on the average. At the same time, the probability of getting trapped in a local minimum was reduced. As a compromise we chose the regions to contain 20 to 40 cells.

The second parameter is the size of the subcells. By choosing the highest common divisor of the cell sizes we obtain an exact model of the cell areas. To save computation time we increase the size of the subcells at the beginning of the iterative process. Since with larger subcells the capacities s_μ represent the cell areas only inaccurately, the probability to escape local minima is increased during the early iterations.

4 Experimental Results

The DOMINO results were compared with the placement results of TimberWolfSC 5.4 and GORDIANL on the basis of several standardcell benchmark examples from [WLS90] and one very large example *AVQ*. Computations were carried out on a DECstation 5000/200. The quadratic optimization procedure for the initial placement and the iterative placement procedure are written in C and FORTRAN 77, respectively.

The characteristics of the circuits are shown in Table 1. In Table 2, wire lengths after placement, layout areas after final routing, and computation times for placement are

	primary1	*primary2*	*biomed*	*industry3*	*AVQ*
cells	752	2 907	6 417	15 032	21 046
pads	81	107	97	332	68
nets	904	3 029	5 742	21 924	21 316
pins	2 941	11 229	21 040	68 290	80 322
columns	18	29	37	38	60

Table 1: Characteristics of the examples

		prim1	*prim2*	*biomed*	*industry3*	*AVQ*
TWSC 5.4	half perimeter [*mm*]	1 027	3 937	2 268	51 320	6 012
	min. span. tree [*mm*]	1 157	4 688	3 467	55 010	6 716
	chip core area [mm^2]	22.35	89.08	52.60	n.a.	n.a.
	cpu time [*sec*]	766	5 249	14 434	56 112	97 566
DOMINO	half perimeter [*mm*]	1 057	3 989	1 908	46 290	5 204
	min. span. tree [*mm*]	1 192	4 705	2 964	49 440	5 792
	chip core area [mm^2]	21.97	89.40	47.60	n.a.	n.a.
	cpu time [*sec*]	99	931	1 767	8 643	15090
GORDIANL	half perimeter [*mm*]	1 083	4 283	2 301	50 080	6 032
	min. span. tree [*mm*]	1 218	4 999	3 337	53 460	6 644
	chip core area [mm^2]	22.49	93.34	48.51	n.a.	n.a.
	cpu time [*sec*]	75	617	960	6 403	11011

Table 2: Results (DOMINO cpu times include initial placement)

compiled. Wire length is measured by the sum of the half perimeters and by the sum of the Manhattan minimum spanning trees of all nets. The circuits were routed with the layout system VENUS [HoMS87]. As the minimum layout areas of the *primary* benchmark circuits are given by their pads, we have reduced the pads in size, routed the circuit, and are listing the internal core area without pads in Table 2.

The results show that DOMINO produces placements comparable in quality to TimberWolfSC 5.4 for the smaller circuits *primary1* and *primary2*. With the large circuits *biomed*, *industry3*, and *AVQ* DOMINO results are about 10–15% better than TimberWolf and GORDIANL results with respect to estimated wire length. This significant reduction also yields a smaller layout area after final routing of *biomed*. It is important to notice that these excellent results can be obtained with 5–10 times less computation time than TimberWolfSC 5.4, where the additional computation time needed for the itarative procedure after determining an initial relative placement with GORDIANL is moderate. This shows that particularly for large circuits TimberWolfSC has to compromise layout quality, because it is too expensive to provide sufficient computation time to approach the global optimum futher. Even if we spend more computation time, for circuit *industry3* TimberWolfSC only yields an estimated wire length (half perimeter) of 47 110 *mm* within 215 000 seconds. This result is only as good as DOMINO's although TimberWolfSC needs 23 times more computation time.

The ability of escaping local minima is illustrated in Figure 6 for the first 13 generations. It shows the estimated wire length after each rearrangement of the cells in a region. The dots mark the wire length after a generation. In the first generation, when a placement without overlapping cells is constructed from a relative placement, wire length increases.

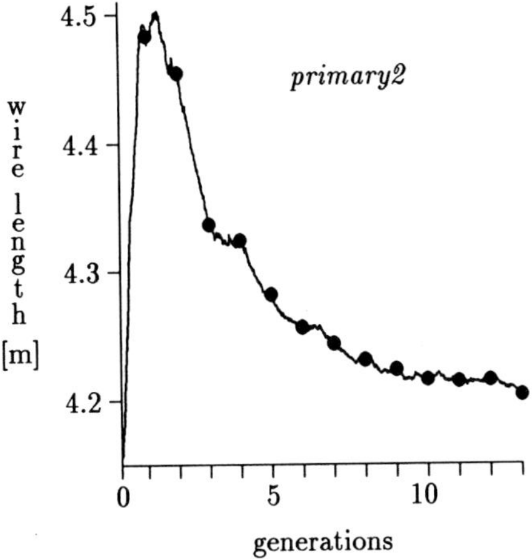

Figure 6: Wire length versus iteration count

For the following generations wire lengths generally decrease. Intermediate increases demonstrate the hill-climbing facilities of the method.

5 Conclusions

We have taken a somewhat naive view to cell placement for column-based layout styles that resulted in a deterministic algorithm with the ability to escape local minima by exploiting defects and inaccuracies of models abstracting from reality. A simple and efficient strategy to maintain an overlap-free placement throughout the iterative process was introduced. Excellent results have been obtained by our preliminary version of DOMINO with respect to placement quality and computation time.

The method presented in this paper for column-based layout styles can be generalized for macro-cell designs by modifying the procedure for avoiding overlaps and unused spaces. In addition, by reanimating the ideas proposed in [Loos79] placement could be combined with routing to obtain even better final layouts.

Finally, we would like to point out that the ideas applied in this paper to a special placement problem are of more general nature. Since most engineering problems are known to be intractable in polynomial time, models abstracting from reality are often used to allow a formal and efficient treatment of the given problem. The solution obtained on the basis of the model is then regarded as a (partial) solution of the real problem. The defects and inaccuracies of the model are mostly regarded as minor effects that can be neglected. The important point is, that this fact also has the positive aspect of providing chances to escape local minima.

6 Acknowledgement

The authors would like to thank Professor K. Antreich for his continuous advice and support. This work was partially supported by the German National Science Foundation (Deutsche Forschungsgemeinschaft) under Grant An 125/7.

References

[Aker81] S. B. Akers, "On the Use of the Linear Assignment Algorithm in Module Placement," *Proc. 18th DAC*, pp. 137–144, 1981.

[Blan85] J. P. Blanks, "Near-optimal placement using a quadratic objective function," *Proc. 22nd DAC*, pp. 609–615, 1985.

[CoPa86] J. P. Cohoon and W. D. Paris, "Genetic placement," *ICCAD*, pp. 374–377, 1986.

[HoMS87] E. Hörbst, C. Müller-Schloer, and H. Schwärtzel, *Design of VLSI Circuits Based on VENUS.* Springer-Verlag, 1987.

[KiGV82] S. Kirkpatrick, J. C. D. Gelatt, and M. P. Vecchi, "Optimization by Simulated Annealing," *Science*, pp. 671–680, 1983.

[KlSJ88] J. M. Kleinhans, G. Sigl, F. M. Johannes, "Gordian: A new global optimization / rectangle dissection method for cell placement," *Proc. ICCAD*, pp. 506–509, 1988.

[KSJA91] J. M. Kleinhans, G. Sigl, F. M. Johannes, and K. J. Antreich, "Gordian: VLSI placement by quadratic programming and slicing optimization," *IEEE Trans. CAD*, pp. 356–365, 1991.

[KlBa90] R. M. Kling and P. Banerjee, "Optimization by simulated evolution with application to standard cell placement," *Proc. 27th DAC*, pp. 20–25, 1990.

[Laut79] U. Lauther, "A min-cut placement algorithm for general cell assemblies based on a graph representation," *Proc. 19th DAC*, pp. 1–10, 1979.

[Leng90] T. Lengauer, *Combinatorial Algorithms for Integrated Circuit Layout.* J. Wiley 1990

[Loos79] K.J. Loosemore, "Automatic layout for integrated circuits," *ISCAS*, pp. 665–668, 1979.

[Marc88] M. Marchesi, "A new class of optimization algorithms for circuit design and modelling," *Proc. ISCAS*, pp. 1691–1695, 1988.

[Otte82] R. H. J. M. Otten, "Eigensolutions in top-down layout design," *Proc. ISCAS*, pp. 1017–1020, 1982.

[Prea88] B. Preas and M. Lorenzetti, *Physical Design Automation of VLSI Systems.* Benjamin/Cummings, 1988.

[Rome89] F. I. Romeo, *Simulated Annealing: Theory and applications to layout problems.* Memorandum UCB/ERL M89/29, 1989.

[SeLe87] C. Sechen and K. W. Lee, "An improved simulated annealing algorithm for row-based placement," *Proc. ICCAD*, pp. 478–481, 1987.

[ShDu85] L. Sha and R. W. Dutton, "An analytical algorithm for placement of arbitrarily sized rectangular blocks," *Proc. 22nd DAC*, pp. 602–608, 1985.

[ShSS86] H. Shin, A. L Sangiovanni-Vincentelli, C. H. Sequin, "Two-Dimensional Compaction by Zone-Refining," *Proc. 23rd DAC*, pp. 115–122, 1986.

[SiDJ91] G. Sigl, K. Doll, F. M. Johannes, "Analytical placement: A linear or a quadratic objective function?," *Proc. 28th DAC*, 1991.

[Stei61] L. Steinberg, "The backboard wiring problem: A placement algorithm," *SIAM Review*, pp. 37–50, 1961.

[SuKe88] P. R. Suaris and G. Kedem, "An algorithm for quadrisection and its application to standard cell placement," *IEEE Trans. CAS-35*, pp. 294–303, 1988.

[TsKH88] R.-S. Tsay, E. S. Kuh, and C.-P. Hsu, "Proud: A fast sea-of-gates placement algorithm," *Proc. 25th DAC*, pp. 318–323, 1988.

[WLS90] International Workshop on Layout Synthesis, MCNC, North Carolina, 1990.

VLSI 91
A. Halaas and P.B. Denyer (Eds.)
Elsevier Science Publishers B.V. (North-Holland)

A Flow-oriented Approach to the Placement of Boolean Networks

Stefan Mayrhofer, Massoud Pedram *, Ulrich Lauther
Siemens AG - Corporate Research and Development, Munich
* Dept. of EECS, University of California, Berkeley

Abstract

The consideration of physical design issues during logic synthesis requires the placement of Boolean networks. Since Boolean networks are described by a directed acyclic graph (DAG) it is possible to move from a netlist-oriented view of the placement problem to a flow-oriented view. Based on the locations of the I/O pads we determine in a first step a preferred direction for every edge of the DAG. We then try to find a placement of the Boolean network such that the actual directions of all edges follow their preferred directions as much as possible. By considering the signal flow we obtained improved timing behaviour of circuits at a moderate increase in the total wire length.

1 Introduction

Automatic placement for VLSI chip design has been extensively investigated in the past. Most of the approaches proposed so far are based on one of the following optimization techniques: mincut partitioning [1], [2], simulated annealing [3], or quadratic optimization [4], [5]. While placement algorithms based on mincut bi-partitioning are very fast, the quality of the final solution may vary drastically. Simulated annealing gives excellent solutions at the expense of very high computation times. More recently, a placement algorithm based on quadratic optimization has been proposed where the uniform distribution of the modules over the placement area is achieved by alternating global optimization and bi-partitioning steps over several levels of hierarchy. From the module coordinates calculated during the global optimization steps, initial bi-partitions are derived which are then improved using the Fiduccia-Mattheyses heuristic [6]. In this paper we introduce an extension of this approach which has been applied to the problem of the placement of Boolean networks. This problem arises in layout-oriented synthesis of combinational logic where the objective is to incorporate physical design considerations into the logic minimization, decomposition, and technology mapping. The idea is to generate a placement of a Boolean network (i.e. a DAG representation of logic equations) and then to backannotate the placement information into the logic design process [7]. It is very important that the placement solution reflects not only the global connectivity of the Boolean network but also produces the shortest directed path between any pair of primary input and primary output. Recent work shows that by considering layout issues during synthesis circuits with smaller area and higher performance can be achieved. A more detailed description of this approach is given in [8].

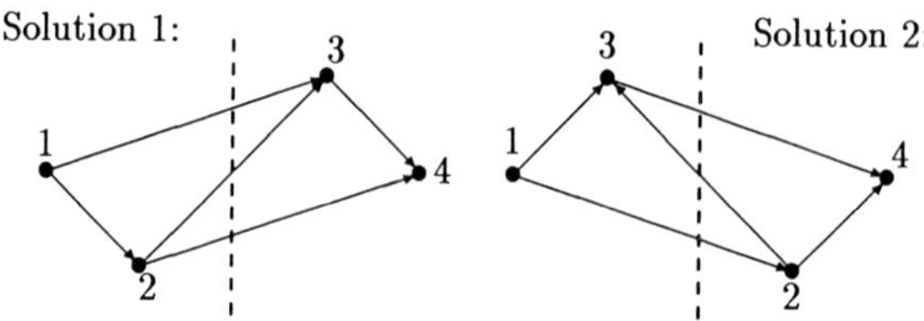

Figure 1: Comparison of two bi-partitionings of a DAG

In order to capture the structure of the Boolean network more effectively during the placement, we move from a netlist-oriented view of the placement problem to a flow-oriented view. Particularly in the bi-partitioning steps there are degrees of freedom which have not been exploited so far. The following simple example illustrates this point. Fig. 1 shows a very small Boolean network together with two bi-partitioning solutions which are equivalent in terms of the number of edges of the DAG crossing the cutline. (Note that modules 1 and 4 have already been assigned to the left and the right regions respectively .) While for the first partitioning solution, every path from module 1 to module 4 crosses the cutline at most once, in the second partitioning solution the path connecting the modules { 1,2,3,4 } crosses the cutline three times. Clearly, the first solution is - from a timing point of view - more desirable. Therefore, during the bi-partitioning step, not only the number of edges connecting modules in both partitions but also the edge directions should be considered. This requirement motivates the development of a flow-oriented bi-partitioning heuristic.

The paper is organized as follows. After giving the basic terminology in the next section, we show how a preferred direction for the edges of a DAG can be determined. We then explain the new objective function for the bi-partitioning steps which an important part of the proposed placement approach. This objective function can be combined very effectively with the well known concepts of the Fiduccia-Mattheyses heuristic [9]. Finally, we discuss first results achieved with the new approach.

2 Basic definitions

Given a set M of modules and a set N of nets, a netlist specifies all modules M_ν connected by the net $\nu \in N$. For Boolean networks, we distinguish between the source μ_ν^+ and the sinks $\mu \in M_\nu^-$ of every net $\nu \in N$ (see Figure 2).

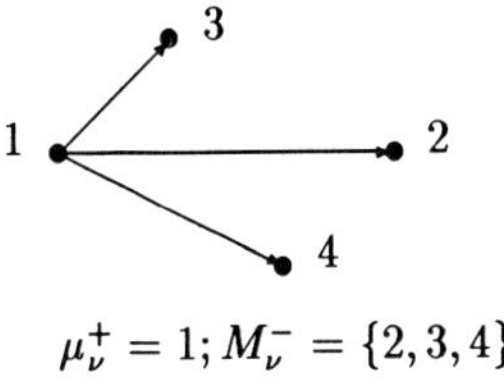

$$\mu_\nu^+ = 1; M_\nu^- = \{2, 3, 4\}$$

Figure 2: The source-sink net model

This leads to a transformation of the netlist description into a directed acyclic graph $D = (M, E)$. The set of nodes in the DAG is identical to the set of modules. The set of edges is derived from the source-sink netmodel in the following way: for every net $\nu \in N$ the set of edges E_ν is defined as $E_\nu = \{(\mu_\nu^+, \mu) \mid \mu \in M_\nu^-\}$ and the set of edges E of the DAG is given by $E = \cup_{\nu \in N} E_\nu$. A sequence $(\mu_i, \mu_{i+1}), (\mu_{i+1}, \mu_{i+2}), \cdots, (\mu_{j-1}, \mu_j)$ of edges $e \in E$ defines a directed path in the DAG from module μ_i to module μ_j : $p(\mu_i, \mu_j)$.

Throughout this paper we distinguish between the set of primary inputs I, the set of primary outputs O, and the set of internal modules G. A Boolean network can be partitioned into a set of logic cones where each logic cone corresponds to a primary output and all its transitive fanin nodes. More formally, for every primary output $o \in O$, the corresponding cone C_o is described by the set C_o^M of modules and the set of edges C_o^E of the cone as follows:

$$\begin{aligned} C_o^M &= \{\mu \in I \cup G \mid \exists p(\mu, o)\} \\ C_o^E &= \{(\mu_i, \mu_j) \in E \mid \mu_i \in C_o^M \wedge \mu_j \in C_o^M\} \\ C_o^S &= \{\mu \in I \mid \mu \in C_o^M\} \end{aligned}$$

The last equation characterizes the primary input support of a cone, i.e. all modules in the set $\mu \in C_o^M$ which are primary inputs. Based on the definitions given above, we derive in the following section an algorithm to determine the preferred direction φ_e^p for all edges $e \in E$. The goal of the flow-oriented placement approach is to place the modules in such a way that the edges follow their preferred direction as much as possible.

3 Cone analysis

For the small example given in section 1, it is obvious that the preferred direction for the edges in the DAG is from left to right. However, we have to give a general procedure for assigning a preferred direction to each edge of the Boolean network. As will be shown in this section this can be done by analysing the cones for all primary outputs in the Boolean network. The procedure **cone_analysis** described in this section is based on the assumption that the locations (x_μ, y_μ) of all primary inputs and primary outputs $\mu \in I \cup O$ are prespecified and do not change during the calculation of the placement. In order to calculate the preferred directions for all edges $e \in E$ we determine in the first step for every primary output $o \in O$ the sets C_o^M, C_o^E, and C_o^S characterizing the corresponding cone by executing a depth first search starting at $o \in O$. Since the positions of modules $\mu \in C_o^S$ in the support of a cone are prespecified, we can calculate the center of gravity for a support C_o^S as

$$x_o^S = \frac{1}{\mid C_o^S \mid} \sum_{\mu \in C_o^S} x_\mu \qquad y_o^S = \frac{1}{\mid C_o^S \mid} \sum_{\mu \in C_o^S} y_\mu \quad .$$

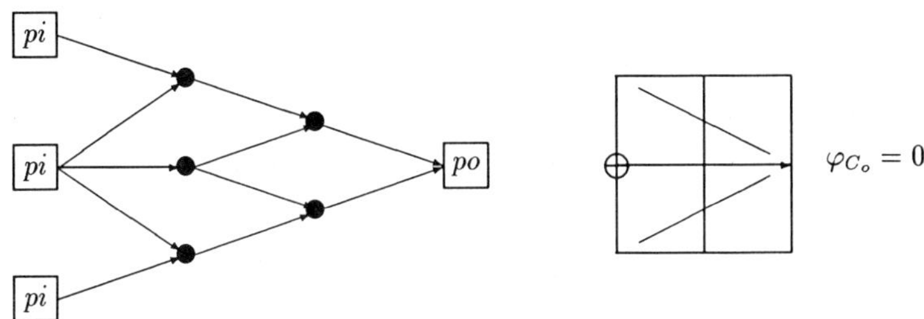

Figure 3: A simple example for a cone C_o

In the next step, we determine the direction φ_{C_o} of the cone which is defined as the orientation of the axis from the center of gravity of C_o^S to the primary output $o \in O$. All edges $e \in C_o^E$ of the cone are then assigned this direction as preferred direction φ_e^p (see Figure 3). However, in general an edge may lie in several cones with different directions. If this is the case, we derive the preferred direction φ_e^p as the mean value of the directions of all cones in $C_e = \cup_{o \in O}\{C_o \mid e \in C_o^E\}$. Therefore the I/O pads should be placed in a way that the directions of different cones don't vary too much. The cone analysis can now be summerized in the following way:

procedure cone_analysis()

input: $D = (M, E), \ \ (x_\mu, y_\mu) \ \forall \mu \in (I \cup O)$
output: $\varphi_e^p \ \forall e \in E$
complexity: $O(\mid O \mid \cdot \mid E \mid)$

1.) **forall** ($e \in E$) $\varphi_e^p = 0$
2.) **forall** ($o \in O$)
  2.1) determine C_o^E and C_o^S by executing a depth first search starting at $o \in O$
  2.2) calculate x_o^S, y_o^S, and φ_{C_o}
  2.3) **forall** ($e \in C_o^E$) $\varphi_e^p = \varphi_e^p + \varphi_{C_o}$
3.) **forall** ($e \in E$) $\varphi_e^p = \varphi_e^p / \mid C_e \mid$

Due to the fact that the locations of the primary inputs and the primary outputs are prespecified the procedure **cone_analysis** needs to be executed only once before we calculate the placement. Clearly the initialization step 1.) and the calculation of the mean value in step 3.) can be done in $O(\mid E \mid)$. The complexity of the procedure is determined by step 2.) where we do for each of the $\mid O \mid$ primary outputs a cone analysis based on depth first search which takes $O(\mid E \mid)$ steps [10].

4 Bi-partitioning of directed acyclic graphs

Based on the results of the cone analysis, we formulate the cost function for the bi-partitioning of a DAG. This function is used to evaluate a bi-partitioning of a set M of modules into two disjoint subsets A and B with $M = A \cup B$. For the Fiduccia-Mattheyses heuristic, the cost of a net $\nu \in N$ is derived from the assignment of the modules $\mu \in M_\nu$ to the subsets A and B. The cost is one if $M_\nu \cap A \neq \emptyset \wedge M_\nu \cap B \neq \emptyset$ and zero otherwise [9]. In this iterative improvement heuristic a given bi-partitioning is modified by moving modules between the subsets A and B such that the total cost for all nets decreases.

However, since we are operating on a DAG and not on a netlist, we have to examine all edges $e \in E_\nu$ of a given net. Fig. 4 illustrates that the direction φ_e of an edge depends on the direction of the cutline and on the assignment of the net source to either subset A or B. For a vertical (horizontal) cut, we assume the sets A and B to be ordered from left to right (bottom to top). We set $\varphi_e = 0$ ($\varphi_e = \frac{\pi}{2}$) if $\mu_\nu^+ \in A$ and $\varphi_e = \pi$ ($\varphi_e = \frac{3\pi}{2}$) if $\mu_\nu^+ \in B$.

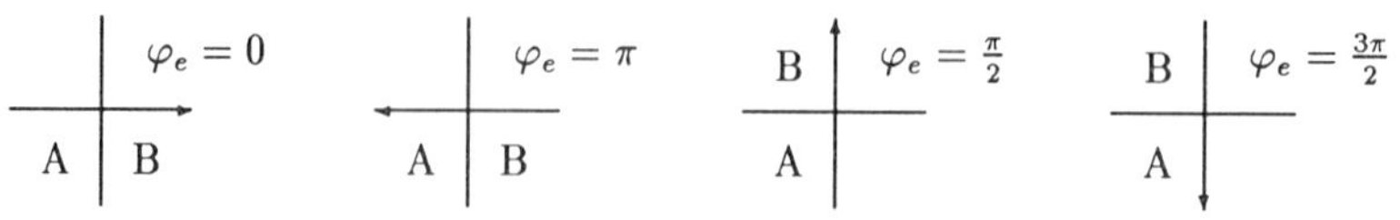

$\varphi_e = 0$ A B $\varphi_e = \pi$ A B   B $\varphi_e = \frac{\pi}{2}$ A   B $\varphi_e = \frac{3\pi}{2}$ A

Figure 4: The directions of an edge crossing a cutline

In order to compare the actual direction of every edge against its preferred direction, we capture for every net the flow of all edges $e \in E_\nu$ in a vector $\alpha_\nu = \{aa_\nu, ab_\nu, ba_\nu, bb_\nu\}$.

Consider for example the first case shown in Fig. 5. The source is assigned to subset A while the sinks are assigned to subsets A and B. Here, the number aa_ν of edges in the subset A is 2 and the number ab_ν of edges crossing the cutline from left to right is 2. The second case in Fig. 5 gives an example where we have $ba_\nu = 3$ edges that cross the cutline from right to left and $bb_\nu = 1$ edge in subset B. Since the source of a net is either assigned to subset A or subset B, two components of this vector are always zero.

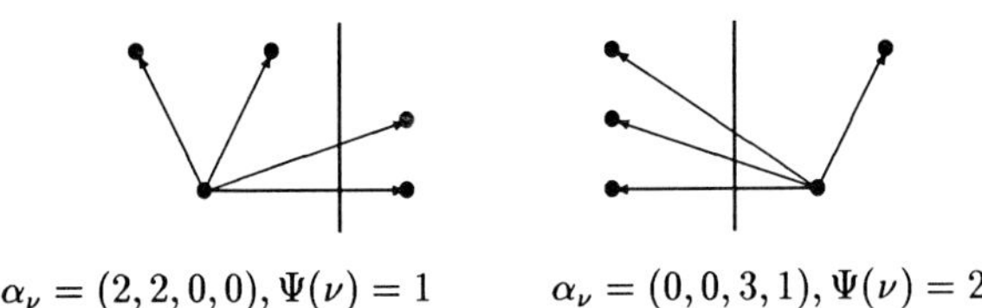

$\alpha_\nu = (2,2,0,0), \Psi(\nu) = 1$ $\qquad$ $\alpha_\nu = (0,0,3,1), \Psi(\nu) = 2$

Figure 5: Examples for the vector α_ν, $\varphi_e^d = 0 \;\; \forall \;\; e \in E_\nu$

Based on this description of the flow of edges we now formulate the cost function for the bi-partitioning of a DAG. The basic idea is to penalize the deviation of the direction of an edge from its preferred direction by an additive term in the cost function:

$$\Psi(\nu) = \begin{cases} 0 \;, & \text{if } ab_\nu = 0 \;\wedge\; ba_\nu = 0 \\ 1 \;, & \text{if } ab_\nu > 0 \;\vee\; ba_\nu > 0 \text{ and } \forall_{(\mu_\nu^+,\mu)\in E_\nu} \;\; | \varphi_e - \varphi_e^p | < \frac{\pi}{2} \\ 2 \;, & \text{if } ab_\nu > 0 \;\vee\; ba_\nu > 0 \text{ and } \exists_{(\mu_\nu^+,\mu)\in E_\nu} \;\; | \varphi_e - \varphi_e^p | \geq \frac{\pi}{2}. \end{cases}$$

As long as no edge $e \in E_\nu$ of a net $\nu \in N$ is cut we set the cost of this net to zero. If one or more edges cross the cutline the cost is one if all edges follow their preferred directions and two if there exist one or more edges that violate this condition. This formulation takes into account that for multipin nets the evaluation of a partitioning solution in terms of the actual number of edges crossing the cutline may be overly pessimistic [11]. It also can be easily incorporated into the Fiduccia-Mattheyses heuristic. The only modification necessary is the update of the vectors α_ν after the move of a module for all nets connected to this module. All possible cases are discussed in Table 1. Every update operation can be executed in constant time so that the complexity of one pass in the bi-partitioning heuristic remains linear in the total number of edges of a DAG.

Table 1: Update of α_ν after the move of a module $\mu \in M_\nu$

module μ moved from	$\mu = \mu_\nu^+$	$\mu \in M_\nu^- \wedge \mu_\nu^+ \in A$	$\mu \in M_\nu^- \wedge \mu_\nu^+ \in B$
A $\rightarrow$ B	$bb_\nu = ab_\nu$ $ba_\nu = aa_\nu$ $aa_\nu = ab_\nu = 0$	$aa_\nu = aa_\nu - 1$ $ab_\nu = ab_\nu + 1$	$ba_\nu = ba_\nu - 1$ $bb_\nu = bb_\nu + 1$
B $\rightarrow$ A	$aa_\nu = ba_\nu$ $ab_\nu = bb_\nu$ $bb_\nu = ba_\nu = 0$	$ab_\nu = ab_\nu - 1$ $aa_\nu = aa_\nu + 1$	$ba_\nu = ba_\nu + 1$ $bb_\nu = bb_\nu - 1$

5 Experimental results

The proposed bi-partitioning algorithm has been implemented in C language and is now integrated in the GORDIAN placement package [6]. We applied the proposed method to several MCNC logic benchmarks [12] which were optimized by *misII* [13]. Table 2 shows characteristics of these benchmark circuits:

Table 2: Examples

	circuit 1	circuit 2	circuit 3
modules	552	1636	1657
nets	562	1658	1717

We compared the standard Fiduccia-Mattheyses heuristic and the flow-oriented approach proposed in this paper. The cpu-times for the calculation of the three placement solutions were 14.9 sec., 60 sec. and 66 sec. on a DEC3100 Workstation for both bi-partitioning approaches. In order to evaluate how good the structure of a Boolean network is captured we calculated the maximal delays from all primary inputs to all primary outputs after final placement by using a standard block-oriented delay calculation technique such as in [14]. We assume that for every module the internal delay is given. For every net the delay was estimated using the output resistance of the source module driving the net and the value of the lumped capacitance derived from the net bounding box. Here we assume that the values for horizontal and vertical capacitance per unit length of interconnect are given. The wiring length after final placement was estimated as the sum of the length of minimal rectilinear spanning trees of all nets. As shown in Table 3 the flow-oriented approach to the bi-partioning problem led to considerable improvements in the timing behaviour of the circuit after final placement. The comparison in Table 4 shows

Table 3: Delay in timing units

	circuit 1	circuit 2	circuit 3
Fiduccia	$1.014 \cdot 10^3$	$1.51 \cdot 10^4$	$3.23 \cdot 10^4$
Ψ	$0.881 \cdot 10^3$	$1.24 \cdot 10^4$	$3.01 \cdot 10^4$
improvement	15 %	22 %	7 %

that the improved timing behaviour is achieved through a very moderate increase in the wiring length. Experiments also indicate that the I/O pad placement is of major

Table 4: Wire length

	circuit 1	circuit 2	circuit 3
Fiduccia	$9.26 \cdot 10^4$	$4.29 \cdot 10^5$	$4.92 \cdot 10^5$
Ψ	$9.37 \cdot 10^4$	$4.59 \cdot 10^5$	$5.01 \cdot 10^5$
increase	2 %	7 %	2 %

importance for the flow-oriented approach. A pad placement which does not take into account the structure of the Boolean network may lead to poor results with respect to timing improvements.

6 Conclusions and future work

In this paper we have shown that a flow-oriented approach to the placement of Boolean networks is able to capture the underlying logic structure to a large extent. Based on the cone analysis of Boolean networks, we are able to formulate a new cost function for the bi-partitioning of directed acyclic graphs. A possible extension would be to perform the cone analysis on every level of the hierarchy, thus obtaining more accurate calculation of the preferred direction of an edge. Currently, we are extending our new approach to the placement of sequential circuits.

7 Acknowledgements

The authors would like to acknowledge Prof. E.S. Kuh of University of California at Berkeley for valuable discussions and suggestions.

References

[1] U. Lauther, "A min-cut placement algorithm for general cell assemblies based on a graph representation", *ACM/IEEE Proc. 16th Design Automation Conference* , pp. 1-10, 1979.

[2] S. Mayrhofer and U. Lauther, "Congestion driven placement using a new multi-partitioning heuristic", *IEEE International Conference on Computer-Aided Design ICCAD-90*, pp. 332-335, 1990.

[3] C. Sechen, "VLSI Placement and global Routing using Simulated Annealing", *Kluver Academic Publishers*, 1988.

[4] K.J. Antreich, F.M. Johannes, and F.H. Kirsch, "A new approach for solving the placement problem using force models", *IEEE Int. Symp. on Circuits and Systems, Proc. ISCAS*, pp. 481-486, 1982.

[5] C.-K. Cheng and E.S. Kuh, "Module placement based on resistive network optimization", *IEEE Trans. on CAD*, vol. CAD-3, pp. 218-225, July 1984.

[6] J.M. Kleinhans, G.Sigl, F.M. Johannes and K.J. Antreich , "GORDIAN: VLSI placement by quadratic programming and slicing optimization", *IEEE Trans. on CAD*, vol. CAD-10, No. 3, pp. 356-365, 1991.

[7] M. Pedram and N. Bhat, "Layout-oriented technology mapping", to appear in the *ACM/IEEE Proc. 28th Design Automation Conference* , 1991.

[8] M. Pedram, N. Bhat, K. Chaudhary, S. Mayrhofer and E.S. Kuh, "Layout considerations in combinational logic synthesis", *Proc. MCNC Workshop on Logic Synthesis*, paper 11.1, May 1991.

[9] C. M. Fiduccia and R.M. Mattheyses, "A linear-time heuristic for improving network partitions", *ACM/IEEE Proc. 19th Design Automation Conference* , pp. 175-181, 1982.

[10] A.V. Aho, J.E. Hopcroft, and J.D. Ullman, "The design and analysis of computer algorithms", *Addison-Wesley Publishing Company*, 1974.

[11] D.G. Schweikert, B.W. Kernighan, "A proper model for the partitioning of electrical circuits", *Proc. Design Automation workshop*, pp. 56-62, 1972.

[12] "Logic synthesis and optimization benchmarks - user guide", *Microelectronics Research Center of North Carolina*, 1988.

[13] R.K. Brayton, R. Rudell, A.L. Sangiovanni-Vincentelli and A. Wang, "MIS: Multiple-level interactive logic optimization system", *IEEE Trans. on CAD*, vol. CAD-6, No. 6, pp. 1062-1081, 1987.

[14] R.B. Hitchcock,Sr., G.L. Smith, and D.D. Cheng, "Timing analysis of computer hardware", *IBM J. Res. Develop., Vol. 26, No. 1, January 1982* , pp. 100 - 105.

VLSI 91
A. Halaas and P.B. Denyer (Eds.)
Elsevier Science Publishers B.V. (North-Holland)

Bounds on Net Delays for Physical Design of Fast Circuits

Habib Youssef[a], Rung-Bin Lin[b], and Eugene Shragowitz[b]

[a] Computer Engineering Department, King Fahd University, Box 1734, Dhahran 31261, Saudi Arabia

[b] Computer Science Department, University of Minnesota, 4-192 EE/CSci Building, 200 Union Street S.E., Minneapolis, Minnesota 55455

Abstract

In the past the dominant approach to solving timing problems in layout was based on sorting logical paths according to their criticality and assigning of different weights to nets. Timing-driven layout procedures used these weights to bias layout process. Undesirable outcome of this was that some noncritical paths became critical after layout. An alternative to the weight-based approach is development of length bounds on all nets. In this paper, we will discuss various formulations of this problem and suggest a new algorithm for its solution.

1. Introduction

With further increase in the switching speed of logic devices and in the size of VLSI chips, propagation delay on interconnections becomes an important factor in circuit performance. In general, physical layout is performed iteratively with the assistance of timing verification tools. Pre-layout timing analysis [1-4] with the estimated propagation delay on nets is often sufficient to pinpoint the part of the circuit with timing problems. At this point, if the predicted circuit performance is worser than the desired one, the circuit is redesigned to shorten the logic paths or the faster circuits are used to increase the speed. After layout, wire delay information is extracted from the circuits and timing verification is repeated to assure that the circuit can operate within the projected clock cycle. If the desired timing performance has not been achieved, physical layout is iterated until the required timing performance is met or redesigning of the circuit is considered. In the high performance designs the timing performance of a circuit becomes unpredictable prior to physical layout due to increasing portion of interconnection delays in a clock cycle. Therefore, it might be very difficult to obtain a circuit free from timing problems simply by repeating physical layout process without considering the timing factors during layout. Moreover, it is very undesirable to perform this process many times.

*This work is supported in part by the Mentor Graphics Corporation under Grant 9106895.
**Dr. Youssef would like to thank KFUPM for its support in attending the conference.

Net delay is mainly dependent on the length of net, on the output driving capability, and on the capacitance of loads at its fanouts. Among these, the output driving capability and the input capacitance loads are not subject to changes during physical layout. Therefore, net length will determine the variable part of net delay in the final layout. From this point of view, it seems quite reasonable to minimize the total net length during physical layout. Indeed, for the past few years, minimization of total net length was one of the common objectives in physical layout. It was hoped, that by employing this objective the minimal total net delays could be achieved, and it will result in the minimal average net delay. For the circuits with slow logic devices, the variable part of net delays only accounts for a small portion of total path delays. In this situation, physical layout with this objective is quite often able to obtain a circuit free from the timing problems. On the contrary, in the high performance designs, the variable part of net delay is a decisive factor in circuit performance. Minimization of total net length during physical layout may produce excessive long interconnection delays on some paths.

More recently, several layout methodologies have been proposed to produce timing-problem-free circuits for high performance designs. Two major ideas may be identified in these methodologies. The first one is to rank nets according to their criticality, and the second one is to develop constraints on length of nets. Although some successes [5-9] were reported, the first approach has major shortcoming. Noncritical paths may become critical after layout. The second approach - development of constraints on length of nets seems more promising. Several works addressed this issue lately [10-14]. In the following text we will discuss possible formulations of this problem and demonstrate why it is necessary to introduce new algorithm for finding bounds on net length.

2. Problem definition

Let $G=(V,E)$ be a directed graph representing a VLSI circuit. V and E are respectively the sets of vertices and edges of the graph. $V = V_T \cup V_N$, where $V_T = V_{source} \cup V_{sink}$ is the set of vertices which represent sources and sinks. In VLSI circuits, memory elements function as sources or sinks, primary inputs function as sources, and primary outputs function as sinks. V_N is the set of vertices which are not the sources or sinks. The edges of G represent the signal nets which connect the logic cells. It is assumed that the logic cells between sources and sinks are combinational.

A path is defined as an ordered sequence of vertices of the graph G from one of source vertices to one of sink vertices. Let V_π be the set of vertices of path π, and E_π be the set of signal nets within that path. A circuit is free from timing problems if every signal path in the design has larger delay than its earliest required arrival time, $ERAT$, and has smaller delay than its latest required arrival time, $LRAT$. Therefore, each path can be characterized by the following two inequalities:

$$T_\pi \le LRAT_\pi \,, \; and\, T_\pi \ge ERAT_\pi \,, \qquad (1)$$

where T_π is the delay along path π. The path delay of π is expressed by the following formulae,

$$T_\pi = \sum_{v\in V_\pi} SD_v + \sum_{e\in E_\pi} SC_e\ , \tag{2}$$

where SC_e is the net delay of net $e\in E_\pi$, and SD_v be the switching delay of a logic cell v.

Let Π be the set of all possible paths in the circuit graph. For each path $\pi\in\Pi$, we define the following two bounds which are also known as long path slack and short path slack,

$$U_\pi = LRAT_\pi - \sum_{v\in V_\pi} SD_v\ ,\ and\ L_\pi = ERAT_\pi - \sum_{v\in V_\pi} SD_v\ . \tag{3}$$

From (1-3) follows, for the final layout to be free from all timing problems, that the net delays SC_e's along each path π must satisfy the following linear inequalities:

$$\begin{cases} \sum_{e\in E_\pi} SC_e \le U_\pi \\ \sum_{e\in E_\pi} SC_e \ge L_\pi \\ \quad \pi\in\Pi \end{cases} \tag{4}$$

Now the problem for finding the timing constraints for all nets can be defined as follows :

Given a circuit G described above, find the values of net delays SC_e for $e\in E$ such that the system of inequalities (4) holds.

The linear system defined in (4) has in general larger number of inequalities than the number of variables, and is based exclusively on satisfying the set of constraints. However, it can be transformed into an optimization problem by providing an objective function.

3. Mathematical programming formulations

Several possible objective functions can be used with system of inequalities (4).

(a). $F = Min \sum_{\pi\in\Pi} \left[LRAT_\pi - AAT_\pi\right]^2$, where AAT_π is the actual arrival time of path π. This objective function is trying to assign the delays to nets such that the actual delays for all paths will be as close as possible to their latest required arrival times. This formulation can be interpreted as a quadratic programming problem.

(b). $F = Max \sum_{e\in E} SC_e$, where E is the set of all nets. This objective function was first used in [12] to maximize the total timing delays of nets over the circuit.

(c). $F = Min \sum_{\pi\in\Pi} \left[LRAT_\pi - AAT_\pi\right] = Max \sum_{\pi\in\Pi} AAT_\pi$, where AAT_π is defined above. This objective function is the first order formulation of (a). However, to inspect it closely, it is a linear combination of net delays with different coefficients. The coefficient for each net is equal to the number of paths which traverse the corresponding net. It seems quite natural to assign large coefficients to nets which are traversed by large number of paths.

(d). $F = Min\ Max\ SC_e$, this formulation minimizes maximal delay among all nets in the circuit.

All models from this group have some common problems :

(i) The number of constraints in each of formulations (a)-(d) is equal to the number of paths in the circuit. For any real circuit this number is very large and can easily go to tens of thousands. Enumeration of all paths is a difficult problem by itself. This factor makes selection of models (a)-(d) by linear or nonlinear programming impossible.

(ii) A typical feature of the solution by linear programming is the large number of zero elements in the solution. Zero delays on large number of nets do not have a good physical interpretation.

(iii) Models (a)-(d) do not take into consideration the electrical characteristics of circuits other than switching delays. Meanwhile, such parameters of circuits as driving capabilities, etc., should make an impact on length of nets.

In VLSI circuits, long path problems are most difficult to avoid. To large extent, the speed of a circuit is determined by the propagation delay along its longest path. On the contrary, short path problems are easier to correct and they are less likely to happen. Usually, they are prevented by the insertion of extra circuit components to lengthen all suspected short paths. Hence, it is reasonable to deal only with the long path problems, and the linear system (4) is reduced to the long path inequalities. For linear systems of that form, a solution of interest would be to seek for all nets the maximal interconnection delay values which satisfy the long path inequalities of system (4).

4. Specialized methods for solving system (4)

4.1. Zero-slack Algorithm

This algorithm which was introduced in [10-11] generates one of the solutions of system (4) by introducing new variables - component slacks. *The component slack, μ_v, of a component v is defined as the difference between the latest required arrival time and the actual arrival time on that component.* With component slack being positive, an amount of net delay can be added to component delay (i.e., switching delay) without violating the timing constraints on paths. Details of computing the latest required and actual arrival times can be found in [11]. This algorithm has time complexity $O(n(n+m))$, where n and m are the number of nets and number of components, respectively. The basic steps of the zero-slack algorithm are described as follows :

(i) Component slack is computed for each component of the circuit.

(ii) A set of consecutive components with least positive slack is identified along a path. If no such subset is found, then exit.

(iii) The component slack is divided among the components which were identified in (ii). The slack assigned to each component is added to its delay. Go to (i).

The following properties of the zero-slack algorithm can be stated :

(i) The zero-slack algorithm provides one of the solutions for system (4).

(ii) Distribution of slack according to the zero-slack algorithm does not depend on electrical properties of components.

(iii) While at least one component will have a zero slack at each iteration, the number of paths with zero slack at the end of the process is not defined.

4.2. Optimal assignment algorithm

The optimal assignment algorithm [11] was suggested as an alternative to the zero-slack algorithm for solving system (4). It is devised to add an equal delay of SC to every component at each iteration of algorithm. The basic steps of this algorithm are described as follows :

(i) A delay SC is added to each component.

(ii) The actual arrival times from all inputs are propagated through the circuit until the outputs are reached.

(iii) The actual arrival time is compared to the latest required arrival time at each output component to obtain a potential range of SC. Since each output component will produce one range for SC. The exact value of SC is obtained by taking the minimum range from these ranges and then taking the maximum value from this minimum range.

The time complexity for this algorithm is $O(n(n+m)d)$, where n is the number of nets, m is the number of components, and d is the largest number of components on the paths.

This algorithm has the following properties :

(i) This algorithm finds one of the solutions for system (4).

(ii) Path slacks are distributed simultaneously and uniformly to all components.

(iii) The distribution of slacks has no connection to electrical properties.

5. Specialized method for solving system (4) based on electrical characteristics.

5.1. Minimax algorithm

The *Minimax* algorithm was designed to find timing bounds for all nets, which satisfy system (4) and simultaneously are related to the electrical characteristics of cells and their interconnections. One of key ideas of this approach comes from the fact that each net may belong to multiple paths. The propagation delay of this net should not exceed the minimal value among the maximal delays allowed by these paths. The second idea is to distribute slack on the path among constituent nets according to relative weight of their electrical characteristics, e.g., proportionally to the number of fanouts, driving strength, etc. These two ideas result in the following mathematical formulation of the problem :

$$\begin{cases} u_\eta^\pi = U_\pi \times \dfrac{\rho_\eta}{\sum\limits_{\eta \in E_\pi} \rho_\eta} & \text{for } all\ \pi \in \Pi \\ u_\eta^* = \min\limits_{\pi \in \Pi_\eta} u_\eta^\pi & \text{for } all\ \eta \in E \end{cases} \tag{5}$$

Here $\dfrac{\rho_\eta}{\sum\limits_{\eta \in E_\pi} \rho_\eta}$ defines the relative weight of the net η along the path π. Π_η is the set of the paths which traverse net η. u_η^π is the net delay distributed to η according to its relative weight on the path π. It is easy to see that the solution of system (5) simultaneously satisfies system

(4). This follows from $u_\eta^* \le u_\eta^\pi$ for any π. Calculation of the u_η^* requires finding $\min\limits_{\pi\in\Pi_\eta} U_\pi \times \frac{\rho_\pi}{\sum\limits_{\eta\in E_\pi} \rho_\eta}$. In this expression ρ_η (weight of the net) does not depend on the path and therefore this problem reduces to finding $u_\eta^* = \rho_\eta \min\limits_{\pi\in\Pi_\eta} \frac{U_\pi}{\sum\limits_{\eta\in E_\pi} \rho_\eta}$. It was shown in [14] that the problem of finding u_π^* for all $\eta \in E$ is NP-Hard. But special structure of the functions $U_\pi = LRAT_\pi - \sum\limits_{v\in V_\pi} SC_v$ and $P_\pi = \sum\limits_{\eta\in E_\pi} \rho_\eta$ allows to obtain a good approximation algorithm for this problem. U_π is defined on the edges of the graph and is an additive function. $P_\pi = \sum\limits_{\eta\in E_\pi} \rho_\eta$ is also an additive function defined on edges. *Minimax-PERT* algorithm which is an approximated solution of the system (5) exploits this property. This algorithm finds the lower bound on net delays defined by the *Minimax* algorithm :

$$u_\eta^* = \rho_\eta \min_{\pi\in\Pi_\eta} \frac{U_\pi}{\sum\limits_{\eta\in E_\pi} \rho_\eta} \ge \rho_\eta \times \frac{\min\limits_{\pi\in\Pi_\eta} U_\pi(\eta)}{\max\limits_{\pi\in\Pi_\eta} \sum\limits_{\eta\in E_\pi} \rho_\eta}. \tag{6}$$

Here $U_\pi(\eta)$ is the slack on the path π traversing net η. $U_\eta^{min} = \min\limits_{\pi\in\Pi_\eta} U_\pi(\eta)$ corresponds to the path, which has the minimal slack at the output among all the paths traversing net η. $P_\eta^{max} = \max\limits_{\pi\in\Pi_\eta} \sum\limits_{\eta\in E_\pi} \rho_\eta$ corresponds to the path with the largest weight among all the paths traversing the net η. U_η^{min} and P_η^{max} may be found independently by the PERT-like algorithm [15]. The time complexity of this algorithm is $O(n+m)$. From the fact, that $u_\eta^* \le u_\eta^\pi$ for any η follows that after the first iteration of *Minimax-PERT* some positive slacks will remain on the majority of paths. This suggests to use *Minimax-PERT* repeatedly until the slacks on the longest paths traversing each net become sufficiently small. According to our experimental data, the number of iterations required for such distribution is only 4 to 5 for the circuits with thousands of cells. It is useful to point out that computations of P_η^{max} for all η should be done only once and does not need to be repeated. This algorithm asymptotically converges to the solution with zero slacks on the longest paths traversing each net. The basic steps of the *Minimax-PERT* can be stated as follows :

(i) U_η^{min} and P_η^{max} for each net $\eta \in E$ are computed by PERT-like algorithm.

(ii) Each net η is assigned a delay bound $u_\eta = \rho_\eta \times \frac{U_\eta^{min}}{P_\eta^{max}}$.

(iii) Repeat (i) and (ii) until slacks on the longest paths traversing each net are smaller than a threshold value.

The properties of the *Minimax-PERT* algorithm can be stated as follows :

(i) The solution found by this algorithm satisfies the system (4).

(ii) Net delays are distributed with respect to electrical characteristics.

5.2. Experimental Results

This algorithm was implemented in LISP and was used in calculation of timing bounds for a layout system [9] developed at the University of Minnesota. The algorithm was tested on four real designs (Table 1) provided by Control Data Corporation and demonstrated good performance.

Table 1
The four test cases

Chip	Description	N_{io}	N_{cells}	N_{nets}	*Clock period*
AA	micro-code sequencer	202	2268	3505	100 ns
AB	32-bit multiplier	231	3230	4752	100 ns
AD	main memory controller	210	4144	5470	100 ns
ETA	portion of a controller	232	5093	6367	21 ns

Table 2
Repetitive applications of *Minimax-PERT* to the ETA chip

	Initially	*After 1st iteration*	*After 2nd*	*After 3rd*	*After 4th*
Percent of paths with zero slack	0	72.47%	88.44%	91.75%	93.7%
$\bar{U}$	14.149 *ns*	2.896 *ns*	1.841 *ns*	1.417 *ns*	1.179 *ns*

Table 2 gives a summary of the repetitive execution of *Minimax-PERT* for the *ETA* chip. Initially, all paths have positive slacks, and the average slack on the longest paths traversing each net is equal to 14.149 *ns*. After one iteration, the percentage of those paths with zero slack is equal to 72.47%, and the average slack of the remaining paths with positive slacks is equal to 2.896 *ns*. After 4 iterations 93.7% of those paths has zero slack, and the remaining 6.3% has an average slack equal to 1.179 *ns*.

6. Conclusion

Three groups of algorithms for finding timing bounds for nets were described in this paper. The first group contains various mathematical programming formulations. Practical implementation of these models is limited by the size of the problems and also by some undesired properties of solutions found. The second group of algorithms provides the solution to original problem, but these solutions may cause difficulties in implementations, because they are not related to the electrical parameters of circuits. The algorithm from the third group - *Minimax-PERT* solves the problem with respect to the electrical characteristics of circuits. This algorithm was implemented and demonstrated good characteristics in experiments.

7. References

1 Lionel C. Bening, Thomas A. Lane, and James E. Smith, "Developments in logic network path delay analysis", in ACM/IEEE Proc. of the 19th Design Automation Conference, 1982, pp. 605-609.

2 Jacques Benkoski, Erik Vanden Meersch, Luc J. M. Claesen, and Hugo De Man, "Timing Verification Using Statically Sensitizable paths", in IEEE Transaction on CAD, Vol. CAD-9, No. 10, October 1990, pp. 1073-1084.

3 Daniel Brand & Vijay S. Iyengar, "Timing Analysis using Functional Relationships", in Proc. of ICCAD'86, pp. 126-129.

4 Robert B. Hitchcock, Sr., Gordon L. Smith, and David D. Cheng, "Timing Analysis of Computer Hardware", in 1982 IBM J. RES. DEVELOP., Vol. 26, No. 1, pp. 100-116.

5 Michael Burstein and Mary N. Youssef, "Timing Influenced Layout Design", in ACM/IEEE Proc. of the 22nd Design Automation Conference, 1985, pp. 124-130.

6 A. E. Dunlop et al., "Chip layout optimization using critical path weighting", in ACM/IEEE Proc. of the 21st Design Automation Conference, 1984 pp. 133-136.

7 Michael A. B. Jackson and Ernest S. Kuh, "Performance-driven Placement of Cell Based IC's", in ACM/IEEE Proc. of the 26th Design Automation Conference, 1989, pp. 370-375.

8 Suphachai Sutanthavibul and Eugene Shragowitz, "An adaptive Timing-Driven Layout for High Speed VLSI", in ACM/IEEE Proc. of the 27th Design Automation Conference, 1990, pp.90-95 .

9 Suphachai Sutanthavibul, " Algorithms for Performance-Driven Physical Design of VLSI", Computer and Information Sciences, University of Minnesota, PhD Thesis, October 1990.

10 Peter S. Hauge, Ravi Nair, and Ellen J. Yoffa, "Circuit placement for predictable performance", in Proc. of ICCAD'87, pp. 88-91.

11 Ravi Nair, C. Leonard Berman, Peter S. Hauge, and Ellen J. Yoffa, "Generation of Performance constraints for layout", in IEEE Transaction on CAD, Vol. CAD-8, No. 8, August 1989, pp. 860-874.

12 Eugene Shragowitz, Habib Youssef, and Lionel C. Bening, "Predictive Tools in VLSI System Design: Timing Aspects", in Proc. of COMPEURO'88, pp. 48-55.

13 Habib Youssef and Eugene Shragowitz, "Timing Constraints for Correct Performance", in Proc. of ICCAD'90, pp. 24-27.

14 Habib Youssef, "Timing Analysis of Cell Based VLSI Designs", Computer and Information Sciences, University of Minnesota, PhD Thesis, January 1990.

15 T. I. Kirkpatrick and N. R. Clark, "PERT as an Aid to Logic Design", in 1966 IBM J. Res. Develop., Vol. 10, No. 2, pp. 135-141.

VLSI 91
A. Halaas and P.B. Denyer (Eds.)
Elsevier Science Publishers B.V. (North-Holland)

Area Minimization of IC Power/Ground Nets by Topology Optimization

Karl-Heinz Erhard[a] and Frank M. Johannes[b]

[a] Corporate Research and Development
Siemens AG, D-8000 Munich 83, Germany

[b]Institute of Electronic Design Automation, Department of Electrical Engineering
Technical University of Munich, D-8000 Munich 2, Germany

Abstract
This paper presents a new method for the layout of single-layer power/ground nets in VLSI macrocell circuits. Our aim is to minimize the total area of the power/ground nets subject to voltage drop, electromigration and design rule constraints. In extension to former methods, not only all wire widths of the power/ground nets are optimized, but also their topology.

1 Introduction

The design of VLSI macrocell circuits is usually divided into placing the modules on the chip and routing the power/ground (p/g), clock, and signal nets in the channels between the modules.

The p/g nets are usually routed before the signal and clock nets and are laid out completely on the metal layer(s) because of its low resistivity. As p/g topologies either general graphs or trees are used. We suspect that any general p/g graph can be replaced by a set of p/g trees with the same or less total area, satisfying the same reliability constraints (see below) as imposed on the general p/g graph. Additional advantages of trees are that single-layer routing without the use of vias is possible and the design algorithms become simpler and more efficient. A power (ground) tree consists of a set of rectilinear wire segments, where the root of the tree is a power (ground) pad and the leaf nodes are power (ground) pins of the modules of the integrated circuit.

Routing of p/g nets consists of two tasks:
1) design of the interconnect topology and
2) determination of the width of each segment of the p/g topology.

Z. Syed and A. El Gamal [SG82] solve the first problem by applying "traffic rules" to the channels between the modules in order to obtain a planar routing of p/g nets. Other approaches try to minimize the total wire length of the p/g trees [RM81, Mou83].

When solving the second problem, several constraints must be taken into consideration:

Firstly the width of every segment must be greater than a certain value given by fabrication technology. Secondly the current density in each p/g segment must not exceed a certain limit in order to avoid metal migration [D'H71]. Thirdly voltage drops along p/g routes must be restricted to guarantee correct logic operation of the modules [CB86].

The width of each p/g segment must be large enough to satisfy all these constraints. Since the portion of total chip area covered by p/g nets can be large, it is reasonable to minimize the p/g routing area subject to the imposed constraints.

This (nonlinear) optimization problem can be solved by methods of feasible directions [CB85, Cho87, DMS89], penalty function methods [CB88] or augmented Lagrangian methods [CB86, CB87].

For given constraints the minimal value of p/g routing area depends on the choice of the p/g topology. Chowdhury showed in [Cho87], that no correlation exists between minimal routing area and total wire length of p/g topologies. For this reason generating a p/g topology with minimal total wire length (as do [RM81, Mou83]) need not lead to minimal p/g routing area.

The method presented in this paper optimizes both the p/g topology and the widths of the p/g segments with regard to the total p/g routing area. In the next section we give an outline of our approach. More details of the proposed method are described in sections 3 through 5. Results for macrocell circuits are presented in section 6.

2 Outline of the approach

Our method consists of the following two steps:
(1) Optimization of the p/g topology
(2) Optimization of the widths of the p/g segments.
In both steps the objective is to minimize the total p/g routing area.

Step (1) is based on the fact that a characteristic permutation π of all modules can be assigned to any planar p/g topology. This will be described in the next section. We can minimize the total p/g routing area subject to the imposed constraints by varying the parameter π, which results in an optimized p/g topology. To determine the optimal p/g topology it would be necessary to calculate the minimal p/g routing area for every permutation precisely. Since this is too expensive, we make two simplifications:

a) Since certain p/g topologies can be proved to be non-optimal with regard to total routing area, these topologies need not be considered for our optimization. We decribe all *feasible* permutations by certain Hamiltonian paths in a so-called topology graph (see section 4).

b) For every feasible permutation we determine the minimal p/g routing area (subject to the imposed constraints) only approximately: By weighting all edges of the topology graph the sum of the weights of all edges of the corresponding Hamiltonian path is an estimation of this area (see section 5).

In this way we can solve our optimization problem by searching a *minimal* (feasible) Hamiltonian path in the topology graph. For this problem we use an efficient heuristic based on the exchange of edges of the Hamiltonian path [GN72].

Since the obtained permutation π, which minimizes the *approximate* value of the p/g routing area, may differ from the permutation π_{opt}, which is optimal with regard to the *precise* value of the p/g routing area, we apply the following procedure:

We generate a small set Π of permutations which minimize the approximate value of the p/g routing area. Then we perform step (2) of our approach for every $\pi \in \Pi$ and choose the permutation with the smallest *precise* value of the p/g routing area.

In step (2) we optimize the width of every segment of the obtained p/g topologies by applying the method of feasible directions, similar to [DMS89].

3 Characterization of P/G Topologies

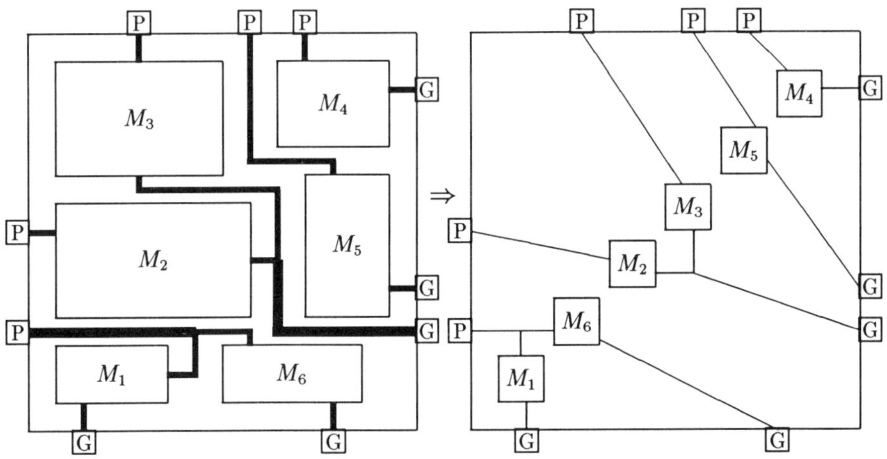

Figure 1: Describing P/G Topologies by Permutations

The basic idea of our method is to assign a permutation π to a planar p/g topology. For illustration consider a p/g topology supplying six modules (see figure 1). We suppose that every power pad is placed on the left or top edge of the chip, and every ground pad on the right or bottom edge. Moreover we assume every module to have one power pin and one ground pin. In case of several power (ground) pins we choose a representative power (ground) pin and connect it locally with the other power (ground) pins. These assumptions apply also to the following sections.

Now assume all modules to be movable and all p/g segments to be rubber bands. After moving every module towards the diagonal from the lower left to the upper right corner of the chip while maintaining planarity, it is easy to read a permutation of the modules on this diagonal. For our example we obtain $\pi = (1, 6, 2, 3, 5, 4)$, if we read from left to right.

4 Feasible Permutations

It is obvious that p/g topologies containing detours are non-optimal with regard to total routing area. We call a permutation *feasible* if it is corresponding to a p/g topology *without* detours.

For the next section it is useful to describe all feasible permutations by Hamiltonian paths in a directed topology graph $T = (V, E)$ with $|V| = m$, where the vertices represent the modules. To define the edges of T, consider any permutation π containing the modules M_i and M_j. To exclude detours, module M_j must not lie outside region R_i^l, if M_j is *left* of module M_i within π (see figure 2). Therefore an edge (j, i) from vertex j to vertex i shall exist if and only if M_j lies within R_i^l (i.e. M_j is allowed to be left of M_i within π). If M_j is *right* of M_i within π, the feasible region for M_j is R_i^r (see figure 2). So an edge (i, j) from vertex i to vertex j shall exist if and only if M_j lies within R_i^r.

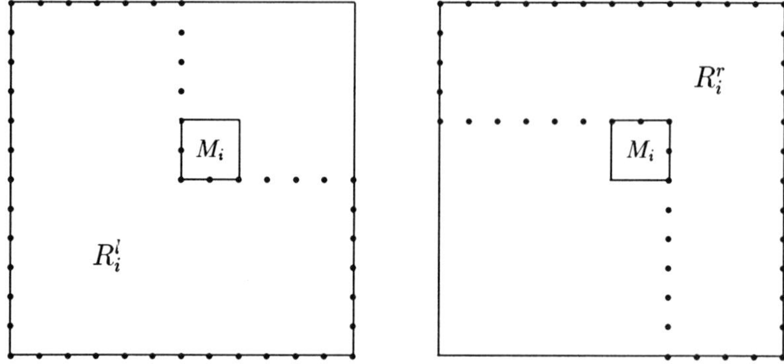

Figure 2: Regions R_i^l and R_i^r

In graph T every Hamiltonian path $H = (v_1, v_2, ..., v_m)$ satisfying the constraint

$$\forall_{k,l \in \{1,2,...,m\}, k<l} : \quad (v_k, v_l) \in E$$

corresponds to a feasible permutation. It should be noted that not every Hamiltonian path in T is feasible (i.e. fulfills this constraint) because the relation corresponding to T is non-transitive.

Figure 3 shows the topology graph T for the example in figure 1. The permutation

$\pi = (1,6,2,3,5,4)$ is feasible. However, the permutation $\pi = (1,3,6,2,5,4)$, which corresponds to the Hamiltonian path with the bold edges, is not feasible since $(3,2) \notin E$.

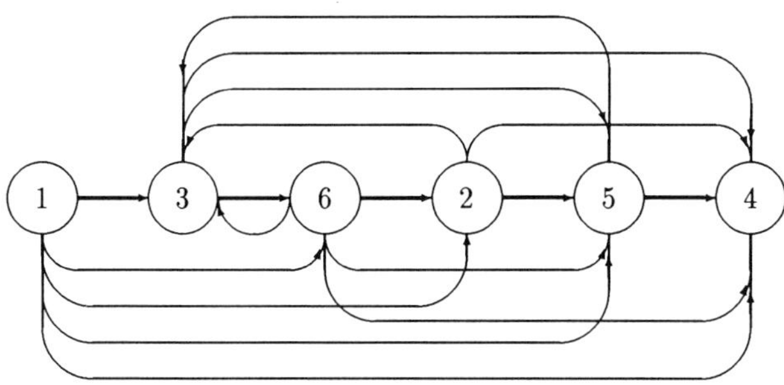

Figure 3: Topology graph T

5 Calculation of P/G Routing Area

Instead of calculating the precise value $A_{min}(\pi)$ of the (constrained) minimal p/g routing area for all feasible permutations π, we use an approximation as mentioned in section 2. For this purpose we weight each edge (i,j) of the topology graph T by the value $a(i,j)$, which is the minimal p/g routing area if *only* the modules M_i and M_j are routed. Thus we neglect the influence of the other modules on the routing area covered by the p/g routes supplying the modules M_i and M_j. Therefore the sum of the weights of all edges of a certain (feasible) Hamiltonian path in T is an estimation of $2 \cdot A_{min}(\pi)$, where π is the corresponding permutation. (The multiplier 2 is due to the fact that each module except the first and last one within π is taken into account twice.)

In order to calculate $a(i,j)$, we suppose the following p/g topology:
As the branching point of the two power routes supplying the power pins of the modules M_i and M_j, we consider the upper left corner of the enclosing rectangle of the two power pins, and connect it to the nearest point on the left or upper border of the chip.

Analogously we consider the lower right corner of the enclosing rectangle of the two ground pins as the branching point of the two ground routes, and connect it to the nearest point on the right or lower border of the chip (see figure 4).

Chowdhury and Breuer have shown in [CB88], that using *separate* power (ground) routes for *all* modules ("star scheme") may result in less p/g area than supplying *all* modules over a common power (ground) trunk ("normal scheme"). To improve the quality of the solution even further we decide for each *module pair* which scheme is better in terms of area requirement. Now let $a_c^p(i,j)$ and $a_s^p(i,j)$ denote the minimal routing area of the power routes in the case of a common and separate power routes, respectively. $a_c^g(i,j)$ and $a_s^g(i,j)$ shall denote the minimal area of the ground routes in case of a common and

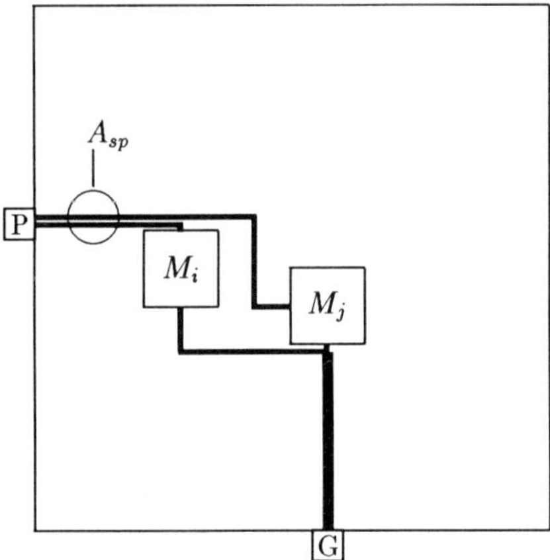

Figure 4: P/G Topology for two modules

separate ground routes, respectively. To calculate these four values we apply the optimization technique already used in step (2) of our approach (cf. section 2). To compute $a_s^p(i,j)$ and $a_s^g(i,j)$ we also consider the spacing area A_{sp} (cf. figure 4) between the separate p/g routes (as it has been done for the star scheme in [CB88]). If $a_c^p(i,j) \leq a_s^p(i,j)$, we use a common else separate power routes. Analogously, if $a_c^g(i,j) \leq a_s^g(i,j)$, we use a common else separate ground routes. Thus,

$a(i,j) := \min(a_c^p(i,j), a_s^p(i,j)) + \min(a_c^g(i,j), a_s^g(i,j))$.

For every edge (i,j) of T, the information about the decision whether to use a common or separate power (ground) routes is saved for the purpose of generating the optimized p/g topology: If the edge (i,j) is contained in the resulting Hamiltonian path, the power (ground) routes supplying the modules M_i and M_j in the optimized p/g topology are either common or separate according to the saved information.

Thus in the optimized p/g topology the decision for common or separate power (ground) routes is taken individually for each module pair. In contrast to this, Chowdhury and Breuer [CB88] have taken the decision globally for all modules.

6 Experimental results

We implemented our method in the C language and applied it to the benchmark macrocell circuits ami33 and ami49 [IWL90] and two randomly generated examples. For placing the modules on the chip we used the placement program GORDIAN [KSJ88, KSJA91].

We compared our approach with the method applying "traffic rules" to the channels between the modules [SG82] in order to generate the p/g topology, combined with step

(2) of our approach (for the purpose of optimizing the widths of all p/g segments). In the table, A_{tr} denotes the obtained minimal p/g routing area subject to the imposed constraints in the case of applying traffic rules, and A denotes the obtained p/g routing area for our approach. Furthermore it contains the cpu times for the execution of our program on a 8 Mips Siemens WS30 workstation running UNIX BSD4.2.

Table
Minimal P/G Routing Area

	ami33	ami49	s32	s64
# modules	33	49	32	64
A_{tr} in μm^2	63 792	3 932 992	49 065	209 266
A in μm^2	60 834	2 984 216	39 268	192 303
cpu time in seconds	14.8	18.6	18.4	39.0

On all circuits we imposed the following constraints:
We chose the values of the maximal allowable voltage drop between a power (ground) pad and a power (ground) pin randomly between $0.3V$ and $0.5V$. Moreover we assumed the upper limit for the current density in each p/g segment to be $1.0mA/\mu m$. The minimum wire width was to be $1.0\mu m$. The sheet resistivity of metal is taken to be $0.027\Omega/\square$. Furthermore we assumed the minimum distance between separate power (ground) routes to be $1.2\mu m$ (used to calculate the spacing area A_{sp} in section 5). We supposed the peak (average) current drawn by each module to be $30mA$ ($1mA$). Finally we assumed $|\Pi| = 5$ (see section 2), which means that we selected the best five p/g topologies with respect to our approximation and chose the topology with minimum routing area out of this set.

The table shows that the area requirement for the p/g nets could be reduced for every circuit by application of our method. The improvement was between 5% and 24%. As mentioned above, the widths of all p/g segments contained in the p/g topology, which was generated by the method applying traffic rules, were optimized with regard to p/g routing area. Thus the area reduction is due *only* to the optimization of the *p/g topology*.

7 Conclusions

The minimal routing area of p/g nets depends significantly on the choice of the topology. So it is useful to reduce the total p/g routing area by optimizing the p/g topology. The method presented in this paper tries to generate a p/g topology with minimal routing area. Its application to macrocell circuits yielded p/g routing area improvements of up to 25%.

Acknowledgement

The first author would like to thank R. Dachauer of Siemens AG for his encouragement and valuable discussions.

References

[CB85] S. Chowdhury and M. A. Breuer. The construction of minimal area power and ground nets for VLSI circuits. *ACM/IEEE Proc. 22th DAC*, pages 794–797, June 1985.

[CB86] S. Chowdhury and M. A. Breuer. Minimal area sizing of power and ground nets for VLSI circuits. *Proc. 4th MIT Conf. on Advanced Research in VLSI*, April 1986.

[CB87] S. Chowdhury and M. A. Breuer. Minimal area design of power/ground nets having graph topologies. *IEEE Trans. on Circuits and Systems*, CAS-34(12):1441–1451, December 1987.

[CB88] S. Chowdhury and M. A. Breuer. Optimum design of IC power/ground nets subject to reliability constraints. *IEEE Trans. on CAD*, CAD-7(7):787–796, July 1988.

[Cho87] S. Chowdhury. An automated design of minimum-area IC power/ground nets. *ACM/IEEE Proc. 24th DAC*, pages 223–229, June 1987.

[D'H71] F. M. D'Heurle. Electromigration and failure in electronics: An introduction. *Proc. IEEE*, 59:1409–1418, October 1971.

[DMS89] R. Dutta and M. Marek-Sadowska. Automatic sizing of power/ground (p/g) networks in VLSI. *ACM/IEEE Proc. 26th DAC*, pages 783–786, 1989.

[GN72] R. S. Garfinkel and G. L. Nemhauser. *Integer Programming*, pages 354–366. Wiley & Sons, New York, 1972.

[IWL90] International Workshop on Layout Synthesis. Microelectronics Center of North Carolina, 1990.

[KSJ88] J. M. Kleinhans, G. Sigl, and F. M. Johannes. GORDIAN: A new global optimization / rectangle dissection method for cell placement. *IEEE International Conf. on CAD, ICCAD-88*, pages 506–509, November 1988.

[KSJA91] J. M. Kleinhans, G. Sigl, F. M. Johannes, and K. J. Antreich. GORDIAN: VLSI placement by quadratic programming and slicing optimization. *IEEE Trans. on CAD*, CAD-10:356–365, 1991.

[Mou83] A. S. Moulton. Laying the power and ground wires on a VLSI chip. *ACM/IEEE Proc. 20th DAC*, pages 754–755, July 1983.

[RM81] H.-J. Rothermel and D. A. Mlynski. Computation of power and ground supply nets in VLSI layout. *ACM/IEEE Proc. 18th DAC*, pages 37–47, 1981.

[SG82] Z. A. Syed and A. El Gamal. Single layer routing of power and ground networks in integrated circuits. *Journal of Digital Systems*, 6(1):53–63, 1982.

VLSI 91
A. Halaas and P.B. Denyer (Eds.)
Elsevier Science Publishers B.V. (North-Holland)
1992 IFIP.

On Distributed Logic Simulation Using Time Warp

HERBERT BAUER, CHRISTIAN SPORRER, AND THOMAS H. KRODEL

Institute of Electronic Design Automation
Department of Electrical Engineering
Technical University of Munich
W-8000 Munich 2, Germany

Bell-Northern Research
P.O.Box 3511 Station C
Ottawa, Ontario
Canada K1Y 4H7

In this paper we discuss issues related to distributed logic simulation using the promising Time Warp mechanism based on the Virtual Time paradigm.

To overcome the extensive memory requirements of Time Warp, we employ an incremental state saving technique instead of checkpointing; state saving itself requires only a negligible amount of overhead.

A new asynchronous algorithm for the determination of Global Virtual Time is presented. The *GVT* process runs in parallel to the simulation processes and guarantees efficient *GVT* determination without affecting the ongoing simulation. On the average, less than N messages are needed to calculate a new *GVT* thereby reducing communication costs for *GVT* calculation to a minimum.

Experimental results obtained with a prototype simulator are included in the paper.

1 Introduction

Logic simulation, being the number one verification tool for digital designs, currently encounters serious problems with respect to the ever increasing complexity of integrated circuits. These problems are due to excessive cpu-times as well as tremendous memory requirements for simulation. Current solutions comprise a higher level of abstraction, partitioning of the circuit into smaller simulatable portions, employing a hardware accelerator or using a distributed simulation approach. High-level simulation is fine if only functional verification is needed, however, this approach is unable to verify critical timing conditions. Also, a completely different (high-level) model of the circuit has to be available. The major drawback of separately simulating small portions of the circuit, the most widely used technique, is that the interface of a partition cannot be simulated realistically. The designer is forced to define input vectors on his own that he thinks are typical for the interface. Hardware accelerators, though offering a substantial speed up, are often unflexible with respect to modeling capabilities, have limitations in the amount of memory, are expensive and go out of date soon after purchase.

Although digital circuits are inherently parallel, the event driven simulation algorithm which is very efficient for high precision delay simulation, strangely enough, lacks concurrency. It seems that optimistic distributed simulation could be an answer to these problems, since it changes the event driven algorithm slightly so that it can exploit the circuit's parallelism better. Distributed simulation likewise accelerates simulation by concurrent computations and addresses the memory problem by distributing the whole circuit over many processing nodes (which are assumed to have local memory).

Optimistic distributed simulation (as opposed to conservative approaches) abandons the notions of synchronous, universal simulation time and allows each simulation process to have its own and possibly different local time. Therefore, special synchronization mechanisms are needed to guarantee results that obey the law of causality.

A promising mechanism for synchronization that is based on the Virtual Time paradigm, called Time Warp, was presented by Jefferson in 1985 [Jef85b]. The basic idea is to enable simulation processes to return to a previous time step, called *rollback*, by saving previous simulator states. As simulation time advances, each simulator assumes stable inputs as long as no external events are received. This assumption however can turn out to be wrong quite late, namely when an external event is received with an execution time that lies in the past. Then, a rollback to that previous time step has to be performed.

Besides the conventional event driven algorithm, Time Warp incorporates mechanisms for rollback, state saving and the determination of *global virtual time*, which is indispensable for memory management.

Section 2 of this paper will give a short overview of the Time Warp mechanism. Several adaptations that are needed for logic simulation will be discussed. In section 3 we will report on our experiences of using a memory efficient incremental state saving technique which can be performed with negligible computational effort.

An important issue of Time Warp is the continuous determination of *global virtual time* (GVT), which is the smallest local time of all processes (and the smallest time a rollback may return to). All saved states of time steps prior to GVT can be discarded (freed) and all waveforms up to GVT can be printed out. However, a distributed environment has no notion of simultaneousness and it would not be possible to get a freeze-frame of all processes without stopping them, which would be a major irritation in concurrent computing. On the other hand, consulting the simulators sequentially is not sufficient since simulation is no longer monotone in time—an already consulted simulator might return to an earlier time afterwards, due to a rollback.

In section 4 we will present an asynchronous GVT algorithm with minimal overhead, that does not need to halt any processes. Nevertheless, the estimated GVT is quite close to the real GVT.

Experimental results and a short comparison with existing GVT algorithms are given in section 5 and 6, respectively.

2 Time Warp

One method to speed up simulation consists of the distribution of the simulation task onto several processing nodes simulating in parallel. In Time Warp [Jef85b], a distributed discrete event simulation method, each node progresses as if it was independent from the others, no direct synchronization exists. Therefore, each processing node, or simulator, has its own *local virtual time* (LVT) and time "warps" across the simulators, so to speak. However, the simulators are not independent from each other and may receive external events that affect their past. In this case, the affected simulator has progressed too far in the future, erroneously assuming that receiving no external events characterizes stable conditions.

Because the simulators are "hoping" their assumptions prove to be right, Time Warp is classified as an optimistic distributed approach. However, when wrong, a simulator has to return to an earlier time (*roll back*) to allow correct processing of the external event.

For logic simulation the digital circuit will be partitioned and distributed onto several simulators. We assume structural modeling; the circuit will be represented as a graph with edges corresponding to the signals, and nodes corresponding to the elements. Primary in- and outputs will be modeled as nodes, too.

Events occurring in logic simulation can be described as a quadruple

$$E(s_i, T_{gen}, T_{exe}, a_j),$$

where s_i denotes the state variable (signal) which is going to change, T_{gen} and T_{exe} the generation and the execution time of the event and a_j the action that has to be performed on s_i. In our case, a_j will be the instruction to set the state of a signal (s_i) either to 0 or to 1.

In order to capture the timing behavior of digital circuits, the execution of events is delayed, i.e. $T_{exe} = T_{gen} + T_{delay}$, where T_{delay} denotes the time span between the logic value change at an input signal and the output signal s_i of an element el_n. Since we assume $T_{delay} > 0$ for all elements el_n of a circuit, it follows that $T_{exe} > T_{gen}$.

During simulation, events will be generated and executed either locally or sent to another simulator. Within one simulator, all the actions a_j are performed in the order of their corresponding event execution time T_{exe}. This still leads to correct results in the distributed environment as long as external events are received prior to their execution time (i.e. $T_{exe} \geq LVT$). But receiving an external event with $T_{exe} < LVT$ renders the simulation between T_{exe} and LVT more or less invalid due to the wrong assumption that s_i would not change state.

Simulation must then be brought back to some time prior to T_{exe}. This involves restoring the old states as well as dealing with potentially erroneous events that have already been sent to other simulators.

Therefore, simulation states have to be saved periodically during normal simulation. However, since simulation will never roll back to a time prior to GVT [Jef85a], those states with $LVT < GVT$ can be discarded, thus freeing some of the limited memory resources. Having k simulators, the *global virtual time* can be calculated as

$$GVT = \min_{k} LVT^k \tag{1}$$

assuming that every message is received instantly after sending.

We employ one processor node for the GVT calculation and the collection of simulation output. It also sets up the environment and provides the simulators with the circuit description and the input vectors.

3 Incremental State Saving and Rollback

As mentioned above, a simulator must be provided with the ability to restore an earlier state. This is normally done by *checkpointing* which means periodically dumping the current state on disk. In the case of a rollback to time T_r ($= T_{exe}$ of the external event that caused the rollback), the state with the highest LVT smaller than (or equal to) T_r gets loaded. Let this time be T_s. Then, the external event will be scheduled and simulation restarts at T_s. Redundant resimulation from T_s to T_r can be reduced by checkpointing more often.

Although this procedure will work correctly, it has the major disadvantage of consuming vast amounts of memory by storing mainly redundant information. On one hand, the state of a simulator comprises not only all the signal values but also internal states of memory elements (like flip-flops) as well as the event queue where events are stored for future execution. On the other hand, usually only small portions of the state of a simulator change at a time step. Statistics on simulation runs reveal that simulation activity per time step decreases rapidly with increasing circuit size and timing accuracy. [BS88], for example, show that when simulating a 32-Bit RISC Processor, on average only 0.06 % of the elements are active during one time step, leading to some 6 elements to evaluate per step.

For these reasons we employ an incremental state saving technique, saving only changes. Since changes are caused by events, the old state (os = value of s_i prior to the execution of a_j) will be stored together with the event:

$$E(s_i, T_{gen}, T_{exe}, a_j, os).$$

Upon event execution the event will simply be left in the event queue for state saving purposes.

This is sufficient for combinational circuits but not so for sequential circuits where elements can hold an internal state which has to be saved as well. To have a unified saving mechanism, we introduce *phantom events* for this purpose

$$PE(el_n, T_{gen}, os),$$

where el_n identifies the element, T_{gen} denotes the time of change and os holds the old state.

Every time an element evaluation leads to a change of the element's internal state, a phantom event will be created and stored in the event queue at the current time step (as mentioned earlier, the event queue also serves as a data structure for state saving). In case of a rollback, the phantom event will restore the old internal state and then be deleted as time steps back over T_{gen}. There is no T_{exe} and a_j associated with phantom events since their only purpose is to store old states.

When a rollback becomes necessary, the old state is restored by *reverse simulation* which performs the following three tasks while scanning the event queue (containing old events) in reverse:

- restore the signal states
- restore the internal states of sequential elements
- restore the event queue (and, finally, add the newly arrived external event)

Reverse simulation only serves to restore old states, therefore no element evaluations are necessary.

For a detailed description of the rollback phase we have to distinguish between local events (events that were generated within the same simulator), external events and phantom events (which are local events as well).

Let L_{ev} denote the event queue (event list), L_s the list of signal values and L_{el} the list of element state values. Let T_r be the time to roll back to.

Whenever the rollback algorithm reaches an event that has already been executed, the previous state will be restored (i.e. the old signal value for an event or the old element state for a phantom event). If the event was an external one, it will be kept in L_{ev} since the local rollback does not affect its validity. If the event was a local one and its T_{gen} is greater than T_r, it will be removed from L_{ev} (these events were not part of L_{ev} at time T_r). At this point we have to consider the fact that this local event might also have been sent to another simulator where it should be canceled as well. However, the *antimessage* announcing cancellation will not be sent right now, instead *lazy cancellation* is employed [Jef85a], i.e. the antimessage will only be sent if the following forward simulation does *not* generate the same event any more. This avoids an unnecessary rollback at the other simulator. To keep track of potentially invalid events that have been sent out, the corresponding local events will not be deleted but moved from L_{ev} into a special list L_{sev} awaiting their final approval during the next forward simulation. Finally, phantom events are always deleted. The complete rollback algorithm used with incremental state saving is given in Fig. 1.

With incremental saving we keep memory requirements to a minimum, storing no redundant information and by introducing phantom events we are able to use the regular event queue for state saving, thus reducing the state saving overhead drastically.

4 Calculating the Global Virtual Time

It is a fundamental characteristic of Time Warp that simulation might not proceed monotonously in time as individual simulators step back in time during a rollback. However, despite this fact, simulation will eventually advance into the future and terminate.

```
set T := T_exe of E(s_i, T_gen, T_exe, a_j, os) with E being the last event in L_ev
while T > T_r
    for each E with T_exe = T
        if T < LVT      /* old LVT */
            move os back into L_s
        if T_gen > T_r
            remove E from L_ev
            if E has been sent, move E to L_sev else delete E
    for each PE(el_n, T_gen, os) with T_gen = T
        move os back into L_el
        delete PE
    set T := T_exe of the previous event E
endwhile
set LVT := T_r      /* new LVT */
```

Figure 1: Rollback algorithm for incremental state saving

Jefferson [Jef85b] has shown that at any given real time t during simulation, there exists a *global virtual time* GVT which is the smallest simulation time that a simulator might have to roll back to. Basically, the GVT is equal to the smallest T_{exe} of any not yet executed event in the system.

Knowing the GVT is important because all the simulation up to GVT can be considered valid and the corresponding simulation results can be written to a file. Furthermore, it is not necessary to keep any states older than GVT since no rollback would need them any more. The Time Warp approach requires relatively large amounts of memory so it is important to release memory as soon as possible.

There are several reasons why finding the event with the smallest T_{exe} is not a trivial task in a distributed environment. The first problem arises from the fact that it is not sufficient to look at all the local times alone since the event in question might currently be "hidden" in some communication channel.

Given the following notation

- $k \in S$: simulator k is an element of the set S of simulators
- $m^{b \triangleright a}$: message sent from simulator b to simulator a, containing an event E
- $T_m^{b \triangleright a}$: T_{exe} of the event contained in message $m^{b \triangleright a}$
- $T_{ch}^{b \triangleright a}(t)$: minimum $T_m^{b \triangleright a}$ of messages already sent by b but not yet (at real time t) received by a (hidden in the channel from b to a)

we consider the communication channels by defining the *global virtual time* at any real time t as

$$GVT(t) = \min\left(\min_{k \in S} LVT^k(t), \min_{a,b \in S} T_{ch}^{b \triangleright a}(t)\right). \tag{2}$$

In other words, GVT is the minimum of any LVT and any message that has already been sent but was not received up to now. As can be seen in (2), all values have to be taken at the same real time t. Imposing an additional problem, this is usually impossible to achieve in a distributed environment other than by stopping all simulators and then gathering necessary information about them and the channels. Since it is desirable to have the GVT value up to date, this procedure would have to be repeated periodically, thereby degrading simulation performance significantly.

We will now present an algorithm that asynchronously determines a lower bound of the *global virtual time.* The simulators autonomously send information messages at arbitrary intervals to a dedicated central processor, which continuously determines GVT based on received information and in turn distributes the result periodically back to the simulators. Since the central processor is also employed for collecting simulation output, the information messages are just appended to the regular traffic. The whole process of GVT determination does not interfere with the simulators at all. Of course, it is not imperative that the GVT task be executed on an extra processing node, it could also be put on any simulator node since it doesn't consume much cpu-time.

We assume error-free communication channels where messages are received in the same order as they were sent. All messages through a certain channel will be numbered, so we can use the following notation:

- $n^{b \triangleright a}(t)$: serial number of a message $m^{b \triangleright a}$ sent from b to a at real time t, increasing with each sent message (equals the number of messages sent during the time interval $[0, t]$)
- $r^{a \triangleleft b}(t)$: number of messages received by a from b during the time interval $[0, t]$
- $T_m^{b \triangleright a}(n)$: T_{exe} of the event contained in message $m^{b \triangleright a}$ with serial number n

We develop our approach starting from (2). First, we define an *extended local virtual time ELVT* for a simulator. Considering the channels leading to a simulator as part of that simulator, we define

$$ELVT^a(t) = \min\left(LVT^a(t), \min_{b \in S} T_{ch}^{b \triangleright a}(t)\right). \tag{3}$$

Using $ELVT$, we can now express (2) as

$$GVT(t) = \min_{k \in S} ELVT^k(t). \tag{4}$$

Since $T_{ch}^{b \triangleright a}(t)$ is the minimum $T_m^{b \triangleright a}$ of events already sent to but not yet received by a,

$$T_{ch}^{b \triangleright a}(t) = \min_{r^{a \triangleleft b}(t) < n \leq n^{b \triangleright a}(t)} T_m^{b \triangleright a}(n). \tag{5}$$

Equation (5) covers all messages that are underway in a channel at some real time t. Consider a message that has just been received at some real time t^a by simulator a (i.e. with number $r^{a \triangleleft b}(t^a)$) coming from simulator b. Let us assume this message was sent at the earlier real time t^b, i.e. $n^{b \triangleright a}(t^b) = r^{a \triangleleft b}(t^a)$. This means, we can rewrite (5) for real time t^a, looking at the messages only from the sending side, but during a time interval:

$$T_{ch}^{b \triangleright a}(t^a) = \min_{n^{b \triangleright a}(t^b) < n \leq n^{b \triangleright a}(t^a)} T_m^{b \triangleright a}(n). \tag{6}$$

To explicitly refer to the two times involved in (6), we will use the abbreviation

$$T_{min}^{b \triangleright a}(t_1, t_2) = \min_{n^{b \triangleright a}(t_1) < n \leq n^{b \triangleright a}(t_2)} T_m^{b \triangleright a}(n) \tag{7}$$

and can now rewrite (3) for real time t^a

$$ELVT^a(t^a) = \min\left(LVT^a(t^a), \min_{b\in S} T_{min}^{b\triangleright a}(t^b, t^a)\right). \tag{8}$$

Equation (8) states that we can determine $ELVT^a$ at real time t^a, if we know $LVT^a(t^a)$ and the value of $T_{min}^{b\triangleright a}$ during the interval $[t^b, t^a]$ (with $n^{b\triangleright a}(t^b) = r^{a\triangleleft b}(t^a)$) from each simulator $b \in S$ that sends messages to a. Determining the minimum of all $ELVT$'s at real time t^a yields $GVT(t^a)$ (according to (4)).

The only remaining problem is that we probably won't have all the information messages from exactly the times t^a and t^b. In the following part, we will estimate a lower bound for GVT, using information from arbitrary real times t.

The constraint on t^b (the time when messages were sent) can be relaxed quite easily by covering a wider time interval for $T_{min}^{b\triangleright a}$. Since

$$T_{min}^{b\triangleright a}(t^b, t^a) \geq T_{min}^{b\triangleright a}(t, t^a), \quad t < t^b \tag{9}$$

we just have to extend the interval to the next available information message prior to t^b.

The explanation of how to deal with information messages from different times t^a (the time for which to determine GVT) will be carried out by determining GVT for only one pair of simulators (two channels). This procedure has then to be applied for every possible pair of simulators (which are connected by a channel). Since GVT is a value that can only increase with time and all information messages stem from the past, the minimum of all the partial GVT's is a lower bound of the current GVT.

Let us assume that the latest information messages from simulator a and b stem from times t_2^a and t_2^b, respectively. Simulator a has just received a message that was sent by b at real time t_1^b (b's received message was sent by a at t_1^a). See Fig. 2 for illustration.

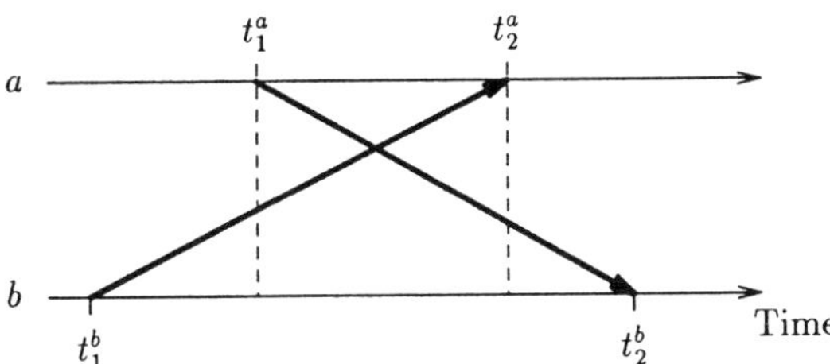

Figure 2: Time axis showing the relation between sent and received messages

We have chosen $t_2^b > t_2^a$, but since the problem is totally symmetric, this doesn't constitute any loss in generality. We will now determine a lower bound for $GVT(t_2^b)$.

According to (8),

$$ELVT^a(t_2^a) = \min\left(LVT^a(t_2^a), T_{min}^{b\triangleright a}(t_1^b, t_2^a)\right). \tag{10}$$

Since

$$T_{min}^{b\triangleright a}(t_1^b, t_2^a) \geq T_{min}^{b\triangleright a}(t_1^b, t), \quad t > t_2^a \qquad \text{(e.g. } t = t_2^b\text{)} \tag{11}$$

(similar to (9)), and LVT^a will either increase during the interval $[t_2^a, t_2^b]$ or might step back to a time not smaller than $T_{min}^{b\triangleright a}(t_1^b, t_2^b)$ (receiving a rollback event that was sent somewhere in the interval $[t_1^b, t_2^b]$), we can predict a lower bound for $ELVT^a$ during the interval $[t_2^a, t_2^b]$:

$$ELVT^a(t) \geq \min\left(LVT^a(t_2^a), T_{min}^{b\triangleright a}(t_1^b, t_2^b)\right), \quad t_2^a \leq t \leq t_2^b. \tag{12}$$

Since a cannot send a message with $T_{exe} < ELVT^a$, (12) also constitutes a lower bound for $T_{min}^{a\triangleright b}(t_2^a, t_2^b)$ (which we will need later on):

$$T_{min}^{a\triangleright b}(t_2^a, t_2^b) \geq \min\left(LVT^a(t_2^a), T_{min}^{b\triangleright a}(t_1^b, t_2^b)\right). \tag{13}$$

Focusing now on b, we get with (8) (analogous to (10))

$$ELVT^b(t_2^b) = \min\left(LVT^b(t_2^b), T_{min}^{a\triangleright b}(t_1^a, t_2^b)\right). \tag{14}$$

If t_2^a would have been smaller than t_1^a (not shown in Fig. 2), we could employ the relation

$$T_{min}^{a\triangleright b}(t_1^a, t_2^b) \geq T_{min}^{a\triangleright b}(t_2^a, t_2^b) \tag{15}$$

(similar to (9)).

On the other hand, if $t_2^a > t_1^a$ (Fig. 2), we split the interval $[t_1^a, t_2^b]$ at t_2^a and get

$$T_{min}^{a\triangleright b}(t_1^a, t_2^b) = \min\left(T_{min}^{a\triangleright b}(t_1^a, t_2^a), T_{min}^{a\triangleright b}(t_2^a, t_2^b)\right). \tag{16}$$

Replacing $T_{min}^{a\triangleright b}(t_1^a, t_2^b)$ in (14) with (15), (16) and (13), we get

$$\begin{aligned} ELVT^b(t_2^b) &\geq \min\left(LVT^b(t_2^b), T_{min}^{a\triangleright b}(t_1^a, t_2^a), T_{min}^{a\triangleright b}(t_2^a, t_2^b)\right) \\ &\geq \min\left(LVT^b(t_2^b), T_{min}^{a\triangleright b}(t_1^a, t_2^a), LVT^a(t_2^a), T_{min}^{b\triangleright a}(t_1^b, t_2^b)\right). \end{aligned} \tag{17}$$

It should be mentioned that in case $t_2^a \leq t_1^a$, we set $T_{min}^{a\triangleright b}(t_1^a, t_2^a) := \infty$ (the neutral element for the *min*-operator) since there are no messages sent by a before t_2^a that b would not already know of at t_2^b.

With (12) and (17) we have all the necessary information to determine a lower bound of GVT at real time t_2^b despite the fact that we have not received any information from simulator a after t_2^a:

$$\begin{aligned} GVT(t_2^b) &= \min_{k\in\{a,b\}}\left(ELVT^k(t_2^b)\right) \\ &\geq \min\left(LVT^a(t_2^a), T_{min}^{b\triangleright a}(t_1^b, t_2^b), LVT^b(t_2^b), T_{min}^{a\triangleright b}(t_1^a, t_2^a)\right). \end{aligned} \tag{18}$$

In practice, the information message from a simulator k consists of its LVT^k, $r^{k\triangleleft a}$ for each incoming channel and $n^{k\triangleright a}$ and $T_{min}^{k\triangleright a}(t_{i-1}^k, t_i^k)$ for each outgoing channel (t_i being the current time and t_{i-1} the time of the previous information message). The GVT processor then applies (18), determining the t_1- ("message sent-") times by comparing $n^{a\triangleright b}$ and $r^{b\triangleleft a}$ for each channel ($n^{a\triangleright b}(t_1^a) \leq r^{b\triangleleft a}(t_2^b)$ will give a valid t_1-time, cf. (9)).

Since an idle simulator (with an empty event queue) cannot *cause* a rollback by itself, it will transmit $LVT = \infty$. Eventually, when every simulator has sent $LVT = \infty$ and all channels are empty, it can be concluded that the end of simulation has been reached.

Whenever the estimate on GVT has increased, it will be distributed to the simulators, which in turn can delete obsolete state-saving information.

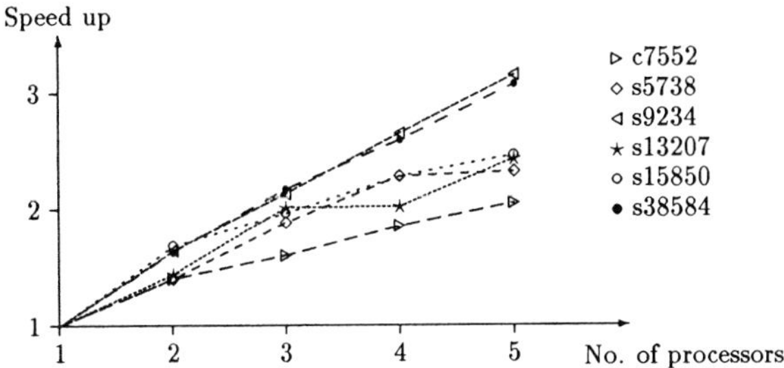

Figure 3: Speed up of parallel simulation

Since each simulator can decide on its own how often it will send *GVT*-finding information, this gives a tuning parameter: the more often it is sent, the closer will the estimated *GVT* be to the real *GVT*. On the other hand, this leads to a higher degree of communication. We obtained good results by sending *GVT*-finding information after each simulation cycle together with simulation output results. It is important to note that the only activity the simulators are involved in for calculating *GVT* is sending information data; this is quite a small amount of data when compared to the simulation output data. Moreover, they are completely autonomous in deciding when to send the information and are thus able to choose a suitable occasion.

5 Experimental Results

We have implemented a distributed simulation system by taking a 6-valued ambiguity delay logic simulator [Kro90] and applying the algorithms described above. Currently the system is running on a Sequent Symmetry. Although this machine provides shared memory, we employ a message based communication protocol since our targeted hardware architecture is a distributed environment without shared memory (like hypercubes or workstations connected via a local area network).

Circuit	No. of gates	No. of flip-flops	No. of processors				
			1	2	3	4	5
c7552	3512	—	42.7	30.7	26.7	23.1	20.8
s5378	2779	179	366.8	260.0	194.6	160.2	158.4
s9234	5597	228	484.0	295.2	227.6	182.3	153.5
s13207	7951	669	296.8	205.7	147.4	146.8	122.3
s15850	9772	597	531.8	316.1	271.9	233.3	215.9
s38584	19253	1452	568.7	349.0	261.6	219.8	184.7

Table 1: Simulation times in seconds on a Sequent Symmetry

The results for some well-known combinatorial and sequential benchmark circuits are given in Table 1 and the respective speed up in Fig. 3. All times are elapsed times, obtained by simulating with random input vectors. Partitioning algorithms are still under investigation; for these examples we employed a cluster algorithm that groups gates around flip-flops.

6 Related Work

The main problem for GVT calculation are messages that are hidden in communication channels (transient messages).

Bellenot [Bel90] proposed a GVT estimating algorithm which tackles this problem by employing acknowledgement messages, but this doubles the network traffic. An additional $4N$ messages (N denoting the number of simulators) are needed to calculate a new GVT.

Lin and Lazowska [LL90] showed that it is possible to solve the transient message problem without an explicit acknowledgement after every sent message. Their approach is based on START-messages, which are broadcasted to all N simulators to initiate the calculation of GVT. In addition, in the worst case another $N^2 + N$ messages are needed to obtain a new GVT.

With our algorithm running asynchronously, we observed that less than N messages were needed for computing an updated GVT, which results in cutting the communication costs for GVT calculation by a factor of 2 when compared to [LL90].

7 Conclusion

A memory and computationally efficient *incremental state saving* mechanism and the associated rollback algorithm has been presented.

We have developed a new GVT calculation algorithm that is running asynchronously and in parallel to the simulators. This process does not interfere with the simulators and on the average only needs N messages to determine an updated GVT.

Obtained results indicate that distributed simulation gives good speed up if the circuit partitions are large enough so that a sufficiently high percentage of the generated events are local events (depending on the communication costs).

Future work will focus on further optimizing the rollback algorithm and especially on new partitioning schemes to minimize communication overhead.

Acknowledgement

The authors would like to thank Professor Dr. K. Antreich for his encouragement and valuable discussions. This work was supported by the German National Science Foundation (Deutsche Forschungsgemeinschaft) under Grant An 125/8.

References

[Bel90] Steven Bellenot. Global virtual time algorithms. *Distributed Simulation 1990*, pages 122–127, 1990.

[BS88] Mary L. Bailey and Lawrence Snyder. An empirical study of on-chip parallelism. *25th ACM/IEEE Design Automation Conference DAC*, pages 160–165, 1988.

[Jef85a] David R. Jefferson. Implementation of time warp on the caltech hypercube. *Proceedings of the SCS Distributed Simulation Conference*, pages 70–74, January 1985.

[Jef85b] David R. Jefferson. Virtual time. *ACM Transactions on Programming Languages and Systems*, 7(3):404–425, July 1985.

[Kro90] Thomas Krodel. An accurate model for ambiguity delay simulation. *Proceedings of the 1990 European Design Automation Conference*, pages 563–567, March 1990.

[LL90] Yi-Bing Lin and Edward D. Lazowska. Determining the global virtual time in a distributed simulation. Technical Report 90-01-02, Department of Computer Science and Engineering, University of Washington, Seattle, WA 98195, January 1990.

VLSI 91
A. Halaas and P.B. Denyer (Eds.)
Elsevier Science Publishers B.V. (North-Holland)

An integrated environment for the design and simulation of self-timed systems

E. L. Brunvand[a] and M. Starkey[a]

[a]Department of Computer Science, University of Utah, Salt Lake City, UT 84112, U.S.A

Abstract

Self-timed systems can offer significant improvements over traditional globally clocked systems in such areas as robustness, ease of design, incremental improvement, and even speed. At present, however, there is a lack of tools available to help the designer explore the promise of such systems. We present an integrated environment for the design, simulation, and testing of self-timed systems based on OCCAM and LISP. Systems described as programs written in a subset of OCCAM are simulated by running the OCCAM code using a LISP-based interpreter. Parts of the system not included as standard OCCAM primitives may be described using LISP. These programs can then be automatically translated into self-timed circuits using syntax-directed techniques. The resulting circuits can be simulated at the gate level with any special features now instantiated as transistor-level models. Finally, the system is sent to an automatic place and route tool to prepare for fabrication.

1. INTRODUCTION

As systems become larger, faster, and more complex, timing problems become increasingly severe and account for more and more of the design and debugging expense. A major culprit is the traditional synchronous design style in which all the system components are synchronized to a global clock signal. One solution is to use asynchronous self-timed techniques. *Self-timed* circuits avoid clock-related timing problems by enforcing a simple communication protocol which is insensitive to delays in circuit components or the wires that are used to connect them.

The design of asynchronous or self-timed circuits and systems has long been considered too difficult by most circuit designers. One reason for this perception is that a long history of designing synchronous circuits has resulted in a great deal of knowledge, and a large variety of tools, to aid in that style of circuit design. The same level of support for asynchronous design does not exist at present. Because there is no great demand for suitable circuits, semiconductor manufacturers do not offer commercially available parts that are suitable for asynchronous design. In addition to being difficult to design, individual asynchronous circuit modules are usually larger than their synchronous counterparts. Historically, because space on an integrated circuit is limited, the overhead involved in implementing asynchronous circuits has been considered too great to consider that form of implementation.

Despite these difficulties however, asynchronous design is currently attracting renewed interest as a method for coping with some of the problems associated with improved VLSI technology. The lack of commercially available parts can be largely overcome simply by designing the needed parts directly. The size penalty for individual circuits is becoming less severe as device size shrinks and die size increases, and while the individual circuit modules may still be more difficult to design than their synchronous counterparts, the more serious problem of system integration can be made much easier.

Faced with a large complex system, and an unfamiliar circuit design style, the complexity of the design process itself becomes a major stumbling block. One method for taming this complexity is to use automatic methods for generating circuits from behavioral descriptions. This allows the designer to abstract away details of the low-level circuits and think of system behavior in terms of high level programs. In addition to making system design easier, formal proof techniques can be used to verify that the program meets the system specification. Because the generated circuits faithfully mimic the behavior of the program, the resulting circuits are correct by construction. At the same time, if a programmer is to design efficient systems in this way, then there must be a way for the programmer to reason about the resulting circuit based on the program text. The translation process must be sufficiently transparent to give the programmer some idea of how different program alternatives will affect the compiled circuit.

In this paper, we present an integrated environment for designing, simulating, and testing self-timed systems. Systems are described as programs written in a subset of OCCAM[8] and developed in a LISP-based environment. The system can be simulated by running the OCCAM code directly [17], or by using the automatic translator to generate a self-timed circuit from the OCCAM description and simulating the circuit at the transistor level [4, 3]. The transistor level netlist is sent to a place and route tool to prepare for fabrication. Test programs written to drive simulations may also be used to drive the tester when the chip is returned from fabrication.

2. SELF-TIMED CIRCUITS AND SYSTEMS

One main reason that self-timed design is attractive is that it separates timing from functionality. Timing has always been a major concern in synchronous circuits. As transistor sizes decrease, and system sizes increase, the problems involved in distributing a global clock signal to the entire circuit become severe. Self-timed circuits localize the timing information to individual circuits rather than synchronizing the entire system to a global clock and thereby avoid the standard problems of clock distribution and clock skew. Also, in a complicated pipelined system such as a processor, the necessity of balancing the worst case path through all the pipe stages requires analyzing literally thousands of paths and iterating a large number of times.

The greatest benefit of designing with self-timed parts may be composability. Systems may be constructed by connecting components and assembling subsystems based only on their functionality and not on their timing characteristics. Timing is now a separate issue related to performance. Increasing the speed of the individual modules can improve performance without affecting functionality. Because of this, large self-timed systems can be built from subsystems that operate at widely different speeds, or that will continue to operate over a wide range of voltage or temperature conditions. Increased composability of subsystems can also greatly decrease the time required to assemble a new system.

A benefit related to composability is incremental improvement. In a properly designed asynchronous system it is possible to improve the performance or functionality of a system by incrementally improving or replacing individual subsystems without changing or retiming the whole system. Incremental improvement of large synchronous systems is usually difficult, if it is possible at all. This notion of incremental improvement also extends to improvements in the process in which the circuit is fabricated. A large synchronous design is not likely to be able to take advantage of incremental improvements in the manufacturing process because these improvements may upset the timing relationships necessary for the synchronous system to operate correctly. A change in the timing of a single component may have system wide effects.

Performance itself is another issue where asynchronous designs offer improvement. Synchronous systems usually exhibit worst-case behavior while asynchronous systems tend to reflect the average-case. Because the results from a computation must be available at the next clock tick, the time between clock ticks in a synchronous system must reflect the worst case delay in any subcircuit. A self-timed system, on the other hand, starts computation when data are available, and signals that the result has been computed however long or short a time this takes. In systems where the latency through parts of the system depends on the data values, this can allow the system to run much faster than the worst case in most instances. The subsystem will slow down to the worst case speed only in the rare case where that delay is actually necessary.

Finally, as it becomes possible to integrate more and more onto a single chip, it becomes attractive to think of the system as a collection of independent, cooperating systems rather than one large system. This allows the designer to deal with large systems hierarchically. Each subsystem can be designed and tested independently with confidence that, because a self-timed communication protocol is used, they will operate correctly when assembled into a larger system.

3. RELATED WORK

Early work in asynchronous circuits was concerned with state machine synthesis where a state machine is implemented as a combinational network with delayed feedback [11, 21]. Some more recent work [6, 12] has continued to focus on implementing asynchronous state machines in this way. Macromodules, developed at Washington University around 1970 [14], demonstrated the usefulness of flexible system building blocks that use distributed asynchronous control. Modules from a set of around 15 different types were used to design and implement parallel computing structures of diverse kinds.

Recently there has been interest in using concurrent programs as specifications for asynchronous circuits. Martin [5, 10] has had good success compiling programs written in a language based on CSP[7] to gate-level circuits, and a group at Philips Labs [13] has also compiled a similar CSP based language into an abstract circuit. Sutherland [18] has proposed a paradigm he calls "micropipelines" which is designed to be used with asynchronous circuits.

With increasing knowledge about self-timed circuit design, it has been possible for researchers to build larger systems. The first design of an asynchronous microprocessor was published by Martin [9]. It was fabricated and tested and has thus provided a solid existence proof of the possibility of using asynchronous circuits for general processor design.

4. SOURCE LANGUAGE

Entia non sunt multiplicanda praeter necessitatem. — William of Occam

Since VLSI circuits are inherently concurrent at a low level and can be used to implement

```
PROC buffer (CHAN A, B) =          (Process buffer ((Chan A B))
  WHILE TRUE                         (While True
    VAR temp:                          (Seq ((Var temp<8>))
    SEQ                                  (? A temp)
      A ? temp                           (! B temp))))
      B ! temp
```

Figure 1: An example of OCCAM written in the traditional style and our LISP-like format

systems that are also concurrent at a higher level, the language used to describe these systems should allow explicit description of concurrent as well as sequential computation. This explicit control of concurrency also allows the system designer to exchange increased circuit speed due to parallel execution for the extra hardware that such parallel execution requires. If circuit area is at a premium, then sequential execution on a single hardware component may be required.

OCCAM[8] is a programming language based on CSP[7] and designed, in accordance with Occam's razor, to be a simple language for expressing concurrent programs. Control over concurrent and sequential aspects of the computation is explicit and processes may not communicate through shared data. In particular, OCCAM is useful for describing small-grain processes communicating over fixed, synchronized links called channels. OCCAM requires that the channels between the processes, as well as the processes themselves, be fixed at compile time.

The syntax for OCCAM code used in this system is very similar to that in [8] except that parentheses are used for block structuring instead of indentation alone. This allows OCCAM programs to be parsed easily by the LISP-based interpreter, and also allows the system description to use LISP functions when appropriate. Figure 1 shows a simple OCCAM program written in standard OCCAM and our LISP-like syntax. The complete syntax can be found in [3, 17].

The variant of OCCAM we use departs from the standard language in several ways. Arrays of variables and channels, and replicated commands are omitted, as are commands that deal with resource allocation and priority such as *pri-alt*, *placed-par*, *pri-par*, and *allocate*. Commands that deal with timing such as *now* and *after* are also omitted. A weaker version of *pri-alt* called *lex-alt* is added to the command set of basic OCCAM.

The only data type available in this dialect of OCCAM is a simple word consisting of a fixed number of bits. Variables and channels may be defined with any number of bits, with a *Subseq* operator used to access contiguous subparts of a variable or channel. The number of bits in a variable is specified by giving the variable width in angle brackets after its definition. If no width is given, the variable is assumed to be a single-bit Boolean value. The width of the data path of a channel is specified in the same way. If a channel is used for synchronization only with no data being passed it must have a width of zero. Note that although there is an explicit *Block* process used to create a new computational block, *Seq* and *Par* processes implicitly create a new block with possible variable and channel declarations.

5. A LISP-BASED OCCAM INTERPRETER

Although there is a large body of theoretical work investigating properties of CSP and OCCAM programs, the system designer's task still involves running the program to test its behavior. An environment that uses an interpreter to execute the program is ideal for the type of iterative refinement and development that characterizes system design. While theoretical reasoning can determine many interesting properties of concurrent programs, running the program on sample input is still a good way to determine if the program exhibits the desired behavior.

A notion of hierarchy is also important to ease system design. OCCAM programs can be developed a piece at a time with small programs of known behavior being combined together to produce larger systems. Since OCCAM uses a purely message passing model of process communication, hierarchical development is greatly simplified. System pieces can easily be modeled as separate processes using the channels as the interface to the rest of the system.

5.1. Additions to OCCAM Syntax

Simulation of a complete system requires that all the various parts of the entire system are present. For our purposes, some of the pieces of the system may be implemented directly as circuit modules, while others will be described using OCCAM to be translated into circuits. Still others may represent the environment in which the system operates. The data path, for example, is often implemented directly as a circuit module rather than using an OCCAM description. The circuit module is designed to obey the same protocol as the synthesized circuits, but can be carefully designed by hand to improve performance.

When simulating the system by running the OCCAM program, it is essential that the parts of the system that are implemented directly as circuits are included. In this environment, any pieces that are not directly described in OCCAM may be described using a LISP function. To integrate these auxiliary functions with the OCCAM program a LISP function must be written to perform the appropriate behavior. This function is then registered with the system and may be used wherever the programmer desires. If the function must maintain some state between invocations, a closure can be used.

5.2. Providing User Interaction

While debugging or running a program, it is sometimes useful to display certain variables or to direct the program with different user inputs. Printing messages or prompting the user for data can be done in three ways. One method is to define a behavioral model of a process in LISP which has the side effect of printing or obtaining data. This method is very flexible, since the data can be formatted by the associated LISP code as desired.

Another method of transferring data to and from the user is to use two special predefined channels. Sending data to the OUTPUT channel causes data to be displayed on the screen. The channel definition is modified slightly by allowing not just bit strings of data but also character strings for debugging purposes. Receiving on the INPUT channel will prompt the user for data.

The third method is to use a pair of special gates. These gates help to monitor and affect expression evaluation. The GET gate gets data from the user and substitutes that data in the expression. The PUT gate displays the value of an expression. These operations are transparent to expression evaluation and do not affect program execution.

```
(Block ((Chan input<8> output<8>))
  (Par ((Chan mid<8>))                  ; Parallel composition
    (buffer input mid)                  ; of two single-place
    (buffer mid output))                ; buffers
```

Figure 2: OCCAM Code for a Two Place FIFO Buffer

```
(Process environment ((Chan in out))    ; Define an environment for the buffer
  (Par ((Var temp<8>))                  ; Parallel composition
    (! out 1)                           ; Send some values to the buffer
    (! out 2)
    (? in temp)                         ; Get some values from the buffer
    (? in temp)))
```

Figure 3: An Environment for a FIFO Buffer

5.3. Interpreting An OCCAM Program

A simple example of an OCCAM program to implement a simple single word buffer was shown in Figure 2 as the macro *buffer*. This macro simply reads words from its input channel into a temporary variable, then sends the value of that variable to its output channel. A parallel composition of two of these buffers, shown in Figure 2, implements a FIFO buffer with a two word capacity. This example will serve not only to show how the interpreter is used, but also to demonstrate the translation method used to synthesize a circuit.

One way to simulate the entire system is to write a separate OCCAM program to implement the environment in which the system will operate. A simple example is shown in Figure 3. This process is combined using a *Par* construct with the FIFO process to allow the buffer to accept input and produce outputs. This environment process can also be translated automatically into code to drive the transistor level simulation once the circuit has been synthesized. Alternatively, the user can write such a program in LISP which will access the channels of the system directly.

6. TRANSLATION INTO A SELF-TIMED CIRCUIT

The source for the translation procedure is the OCCAM program that describes the desired system. The target for the translation is an asynchronous circuit. In particular, the target circuits are delay-insensitive control circuits using two-phase transition signalling [15] combined with bundled data paths. The main step in translating OCCAM programs into circuits is the initial substitution of circuit structures for OCCAM language constructs. The circuit modules used in the initial substitution are like those described in [2, 18, 19] and are implemented as a set of standard cells using the MOSIS scalable CMOS design rules. The modules used include a *Merge* module that implements the "or" function for transitions, a *C-Element* that is the "and" function for transitions, a *Call* module that acts as a hardware subroutine call allowing multiple access to a shared subcircuit, a *Select* module that steers a transition to one of two outputs based on the value of a "select" signal, an *Enable* element that enables bundled data at the output in response to transition control signals, and a *Register* that latches bundled data in response to transition control signals.

Control modules are connected with signals that use transition signalling, in which a transition

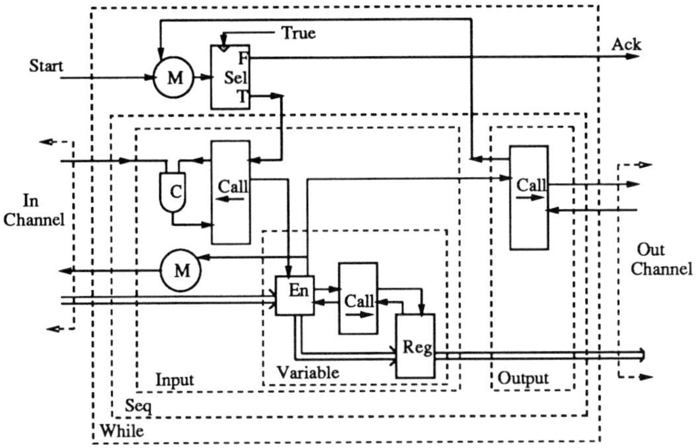

Figure 4: Initial Transformation of Buffer Macro

from low to high or from high to low signals an event. Each module responds to a global *clear* signal by forcing its outputs to a known state and clearing any internal state. The environment issues a transition signal called *start* after a global clear to get things started. This is connected to the request input of the process obtained by translating the single top-level OCCAM process.

The resulting circuit is improved by applying optimizing transformations automatically. These transformations are similar to peephole optimization used in software compilers and are described as an initial circuit and a replacement circuit. When a circuit topology is matched with a template it is replaced with a new structure that retains the behavior of the original but improves its performance or reduces the circuit size.

7. EXAMPLE: A SIMPLE FIFO BUFFER

As a simple example of the entire process, consider the simple FIFO buffer shown in Figure 2. After simulating the FIFO and its environment by running the OCCAM code using the interpreter, the program is automatically translated into a circuit. The translation of the OCCAM program for the single *buffer* macro is shown in Figure 4. First the repetitive construct is expanded. The component process of the repetitive construct is a sequential construct with two component processes: an input and an output statement. Notice that the *buffer* macro begins operation upon receipt of a *start* signal and signals completion with an *ack* signal. There are two channels, one input and one output. The C-Element that performs the synchronization at the channel is included as part of the circuit for the input channel.

Optimizing the circuit removes a great many components. This, along with the parallel composition of two of the simple *buffer* macros, yields the final result shown in Figure 5. Most of the optimizations are readily apparent in Figure 4: a *Call* module with only one client can be removed and replaced by wires; an *Enable* module whose outputs are the only drivers of a data bus can be removed; a *Merge* module with one input and a *Select* module whose select condition is *true* can be removed. These steps, along with a few others, yield the final circuit,

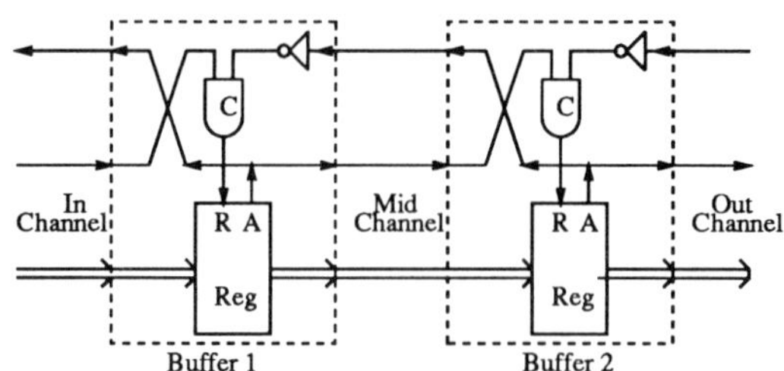

Figure 5: Optimized Two-Place FIFO Buffer

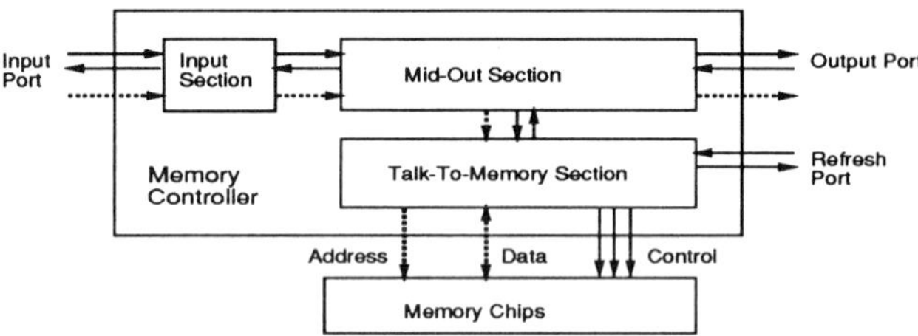

Figure 6: Memory Controller Block Diagram

which is the best circuit known for a FIFO using bundled data paths and transition signalling.

The result of generating the circuit is a netlist of standard modules and their connections. This netlist will contain the names of each of the extra functions or gates described in the code with the assumption that a transistor, and ultimately, layout, description of the module exists. This netlist can be compiled into standard *sim* format and simulated using a variety of transistor level simulators. The environment process used previously can be translated into a program to drive this simulation which now includes transistor models of every part in the system.

After simulation, the netlist may be passed to a place and route tool in preparation for fabrication. The same driver programmer may then be used to test the completed system part.

8. EXAMPLE: A SELF-TIMED MEMORY CONTROLLER

One of the larger example systems that has been generated using this system is an asynchronous memory controller [3]. This circuit is intended to operate with several DRAM parts to make an asynchronous memory system and is based on a memory control circuit designed and built previously at Carnegie Mellon University[16].

The memory controller is a chip that allows asynchronous system parts that use a two-phase bundled data protocol to interface with standard DRAM parts. The block diagram is shown in Figure 6. The bundled data paths at the right and left of the figure are eight bits wide. The memory chips are shown connected to the bottom of the block diagram. Timing for the memory chips is implemented using timing models external to the memory controller chip.

Sequences of bytes are delivered to the input port of the memory controller that contain command and address information. Based on the command and function specified in the input packet, the memory controller performs some function on the memory chips and may deliver data to the output port. In a single command the memory controller can do either a read cycle, a write cycle or both to implement a read-modify-write cycle. Data can be delivered to the output port on any cycle, even a write cycle, and the data written to the chip can be combined with data read from the chip through a full ALU. The data path to the memory is 16-bits wide, and the current controller uses 18-bit addresses (although three bytes of address are consumed). There is also a control channel used to initiate refresh events to the memory chips.

The ALU used by this system is an example of a function implemented directly and described in LISP while running the program. The ALU accepts four inputs: two operands, a function code, and an adder select. The function of the whole ALU is (*If* addsel then (A + *f*(A,B)) else *f*(A,B)). The LISP code used in the simulation simply enumerates the cases for each function of the ALU and returns the result. The actual circuit module is implemented using a carry-completion sensing adder to provide an acknowledge signal.

The OCCAM code that describes the memory controller contains approximately 200 lines of code and has been simulated by running the program, and then translated into a circuit by the prototype compiler. The initial circuit included 236 circuit modules, and optimization reduced this number to 114. The resulting circuit was simulated completely using the the RNL simulator [20], and when simulation was successful, the circuit was automatically assembled using the MOSIS Fusion place and route service [1]. The result from the Fusion router is 5254 by 6056 microns and fits into the MOSIS 64PC69x68 64-pin standard pad frame. The optimized version sent for fabrication includes $\approx$22,000 transistors using the current cell library.

9. CONCLUSIONS

We have found that OCCAM is an excellent way to describe self-timed systems, especially systems organized as pipelines or queues. The OCCAM programs are clear, easy to manipulate, and easy to simulate by executing the programs using our OCCAM interpreter. The OCCAM style of programming with collections of small concurrent processes mirrors exactly the style of asynchronous system design we have been exploring in hardware. Because there is a clear relationship between the OCCAM programs and the resulting circuits, the programmer or system designer has a good idea how modifications will affect the performance or size of the result.

The prototype silicon compiler we have built suggests that a principal weakness of asynchronous systems—the difficulty of their design—can be largely overcome by automatic translation from a clear specification such as an OCCAM program. We have found that an integrated environment that allows the system to be designed as an OCCAM program with LISP additions and then simulated at both the program level and circuit level in the same environment is an extremely effective method for developing self-timed systems.

10. REFERENCES

[1] Ronald F. Ayres. Completely automatic completion of VLSI designs. *IEEE Transactions on Computer-Aided Design*, 9(2):194–202, February 1990.

[2] Erik Brunvand. Parts-R-Us: A chip aparts ... Technical Report CMU-CS-87-119, Carnegie Mellon University, 1987.

[3] Erik Brunvand. *Translating Concurrent Communicating Programs into Asynchronous Circuits*. PhD thesis, Carnegie Mellon University, 1991.

[4] Erik Brunvand and Robert F. Sproull. Translating concurrent programs into delay-insensitive circuits. In *ICCAD-89*, pages 262–265. IEEE, November 1989.

[5] Steven Burns and Alain J Martin. Synthesis of self-timed circuits by program transformation. Technical Report 5253:TR:87, California Institute of Technology, 1987.

[6] Alan B. Hayes. Stored state asynchronous sequential circuits. *IEEE Transactions on Computers*, 1981.

[7] C. A. R. Hoare. *Communicating Sequential Processes*. Prentice-Hall, 1985.

[8] Inmos. *Occam Programming Manual*, 1983.

[9] Alain Martin, Steven Burns, T.K. Lee, Drazen Borkovic, and Pieter Hazewindus. The design of an asynchronous microprocessor. In *Proc. Cal Tech Conference on VLSI*, 1989.

[10] Alain J. Martin. Compiling communicating processes into delay insensitive circuits. *Distributed Computing*, 1(3), 1986.

[11] R.E. Miller. *Switching Theory*, volume 2, chapter 10. Wiley, 1965.

[12] C. E. Molnar, T.P. Fang, and F.U. Rosenberger. Synthesis of delay-insensitive modules. In *Chapel Hill Conference on VLSI*. Computer Science Press, 1985.

[13] Cees Niessen, C.H. (Kees) van Berkel, Martin Rem, and Ronald W.J.J. Saeijs. Vlsi programming and silicon compilation; a novel approach from philips research. In *ICCD*, Rye Brook, NY, Oct 1988.

[14] S. M. Ornstein, M. J. Stucki, and W. A. Clark. A functional description of macromodules. In *Spring Joint Computer Conference*. AFIPS, 1967.

[15] C. L. Seitz. System timing. In *Mead and Conway, Introduction to VLSI Systems*, chapter 7. Addison-Wesley, 1980.

[16] Robert F. Sproull, Ed Frank, and Ivan E. Sutherland. Memory control circuit. Technical Memo 3619, Sutherland, Sproull, and Associates, May 1985.

[17] Mike Starkey. A LISP based OCCAM interpreter. Technical Report UUCS-91-002, Computer Science Department, University of Utah, 1990.

[18] Ivan Sutherland. Micropipelines. *CACM*, 32(6), 1989.

[19] Ivan E. Sutherland, Robert F. Sproull, and Ian Jones. Standard asynchronous modules. Technical Memo 4662, Sutherland, Sproull and Associates, 1986.

[20] Chris Terman. RNL user's guide. In *VLSI Design Tools Reference Manual*. Northwest LIS, University of Washington, 1988.

[21] S. H. Unger. *Asynchronous Sequential Switching Circuits*. Wiley-Interscience, 1969.

VLSI 91
A. Halaas and P.B. Denyer (Eds.)
Elsevier Science Publishers B.V. (North-Holland)

A General Purpose Network Solving System

T.J. Kazmierski, A.D. Brown, K.G. Nichols and M. Zwolinski

Department of Electronics and Computer Science, University of Southampton
Hampshire SO9 5NH, UNITED KINGDOM

Abstract

The complexity of the circuit simulation problem is well known, as are the inadequacies of conventional simulators. To reduce simulation times, systems are developed employing mixed- or multimode simulation techniques, but there are many real situations that still require inordinate simulation times. This paper presents a new and powerful technique for the solution of general purpose networks, of which the electronic circuit is a subset. By supporting an almost arbitrary variety of levels of system abstraction, the user can describe a circuit in any form (circuit and/or functional), and this ability to hide complexity means that solution times for extremely complex systems can be cut to orders of magnitude less than that of conventional simulators. As examples, we analyse the behaviour of four 'difficult' systems, a Schmitt trigger, a bipolar circuit with thermal runaway, a phase-locked loop with a frequency multiplication factor of 100, and the Lorenz Chaos Equations. These systems either require vast CPU times, or are incapable of solution by conventional SPICE-like simulators.

1. Introduction

ALFA (Algorithmic Language of Functional Analysis) is a general purpose object-oriented network description language, designed specifically with the requirements of VLSI analysis in mind. Although capable of the time-domain solution of an arbitrary network, its main use is in the solution of electronic systems, which it allows to be described at almost any level of abstraction. The solution techniques used by ALFA are governed by the description methodology employed by the user. Explicit evaluation of logical signals and behavioural descriptions are combined with a conventional circuit simulator to enable the system to keep simulation times to a minimum.

In addition, we observe that systems that will get normal simulators into difficulties (difficult phase locked loops, etc) are usually those that are more appropriately modelled at some level of abstraction other than the circuit level. For example, in our phase locked loop, the multiplier is described behaviourally very simply. A corresponding four-quadrant discrete subsystem would slow any conventional analysis down markedly.

ALFA is a network description language, primarily designed to make multi-mode circuit simulation powerful, flexible, and cheap.

One of the key ideas in ALFA is the **model**. A model is a user defined entity, similar to, say a **class** in C++ or SIMULA. Models provide data encapsulation and initialisation, topological details and descriptions of the circuit (terminal) equations. Model libraries contain conventional

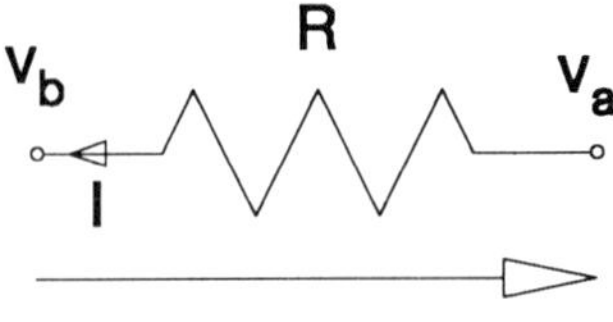


```
model   res(real R, node a,node b)
flowvar I(a,b);
FLOW {
 "flow equation is Ohm's Law"
 I: a-b-I*R = 0
 "partial derivatives of flow eqn"
 " wrt to system varaibles a,b,I "
 {@a := 1,   @b := -1,   @I := -R}
}
```

Figure 1. *ALFA model of resistor.*

SPICE-like devices (resistors, capacitors and so on), but in situations where these are inadequate, it is easy for the user to create his or her own. The basic concept in defining a new model is to separate the internal details of the behaviour from the properties essential to those who use the model - see, for example, the Schmitt trigger of figure 3. A model can be a quite complex abstract entity that comprises both the underlying data and actions to be performed on that data. Instances of models can be declared and interconnected to define a circuit or system.

ALFA descriptions do not lend themselves to one-at-a-time von Neumann style of programming and it is not possible, for example, to derive a control flow of the operations performed by the simulator. Programming models based on parallel structured operations but without a control flow have been proposed elsewhere [1] and have very attractive properties.

An ALFA network is a graph, consisting of **nodes** connected by **flows**. For example, a resistor can be defined as a simple model with one **flow** and two **nodes**, where the nodes represent the potentials at the at the resistor terminals and the flow is a model of Ohms Law relating the current to the potential difference, as in figure 1. User created models may be arbitrarily complex: if two devices in a circuit have a thermal or optical interaction, this can easily be included as an extra flow and set of nodes in the model - see, for example, the thermal runaway simulation example shown in figures 5 and 6.

2. The ALFA Simulator

Before demonstrating the power of ALFA modelling in the next section, we must first emphasise the fundamental differences between SPICE [2] and ALFA.

- SPICE uses a fixed set of built-in models of devices and input waveforms, while ALFA allows new model definitions, which can be exotic semiconductor devices, complex waveforms, logic blocks or even abstract differential equations that have no direct electronic equivalent.
- The underlying simulator in SPICE interprets the data structure resulting from the circuit description, while ALFA compiles its input data to produce both a database and an instruction set to solve the circuit equations. This latter approach is far more efficient and flexible as actions required by the user are not limited by the structure of the simulator itself.

The structure of the ALFA simulator is shown in figure 2. The problem of multi-mode time-domain simulation of complex systems is stated as follows: the simulator must generate time waveforms of the set of circuit variables, which usually comprises the node voltages, some branch currents, and, in ALFA, may also include various other quantities, at the discretion of the user. The system is a set of stiff ordinary differential as well as algebraic and logic equations which may be solved numerically by applying a stiffly-stable multi-step predictor corrector algorithm [3] combined with iterative evaluation of the FLOW sections - see figure 2. We report elsewhere [4] how this can be implemented on a multiprocessor system by exploiting parallelism within both the equation building and the solution phase.

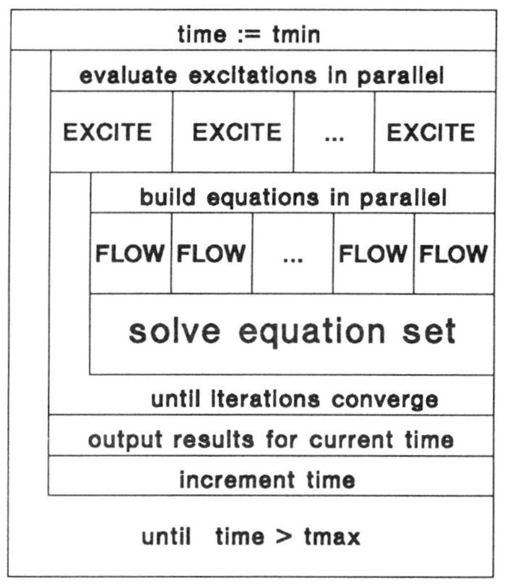

Figure 2. Parallel algorithm of circuit simulation.

It is important to realise that the processes labelled 'EXCITE' and 'FLOW' in figure 2 are not the simple evaluation of primitive components as supplied in SPICE. They can be extremely complex behavioural descriptions, and the 'EXCITE' processes can remove a great proportion of the computational load from the innermost loops of the simulation process.

Rather than formally describe the ALFA language, we will demonstrate its use in the following sections, where fragments of the descriptive language accompany each example.

3. Simulation examples using ALFA models

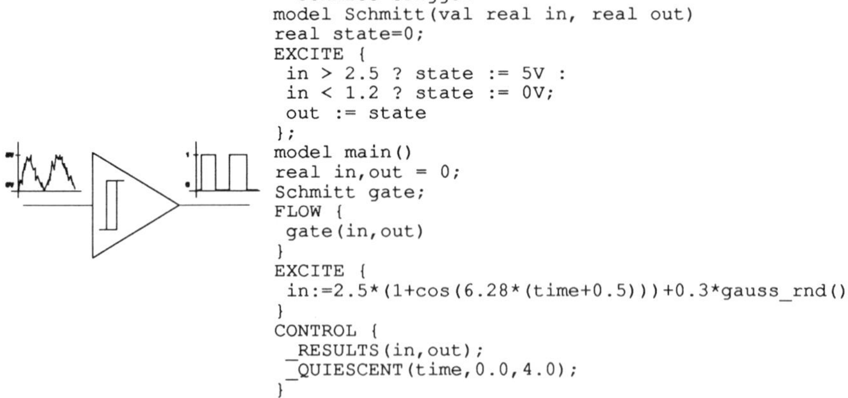

```
 "Schmitt trigger"
model Schmitt(val real in, real out)
real state=0;
EXCITE {
 in > 2.5 ? state := 5V :
 in < 1.2 ? state := 0V;
 out := state
};
model main()
real in,out = 0;
Schmitt gate;
FLOW {
 gate(in,out)
}
EXCITE {
 in:=2.5*(1+cos(6.28*(time+0.5)))+0.3*gauss_rnd()
}
CONTROL {
 _RESULTS(in,out);
 _QUIESCENT(time,0.0,4.0);
}
```

Figure 3. ALFA model of Schmitt trigger.

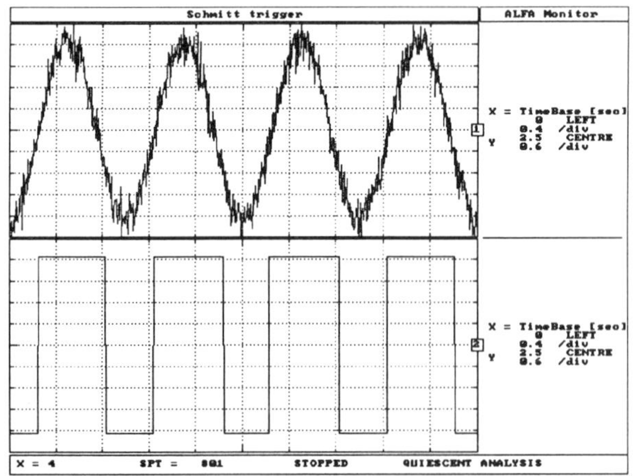

Figure 4. Schmitt trigger simulation results.

The following sections describe the use of ALFA in modelling the behaviour of four systems, which are extremely difficult to model with SPICE-like simulators.

The Schmitt trigger

The structure , response of the system and the entire ALFA text needed to generate this simulation and response of this system are shown in figures 3 and 4.

The model 'Schmitt' is defined first, with a definition of

its own internal behaviour, which is hidden from the outside world by ALFAs object-oriented philosophy. This is followed by the definition of the main module, which contains one instance of the Schmitt, called 'gate', and the excitation required to produce the simulation results.

Thermal runaway

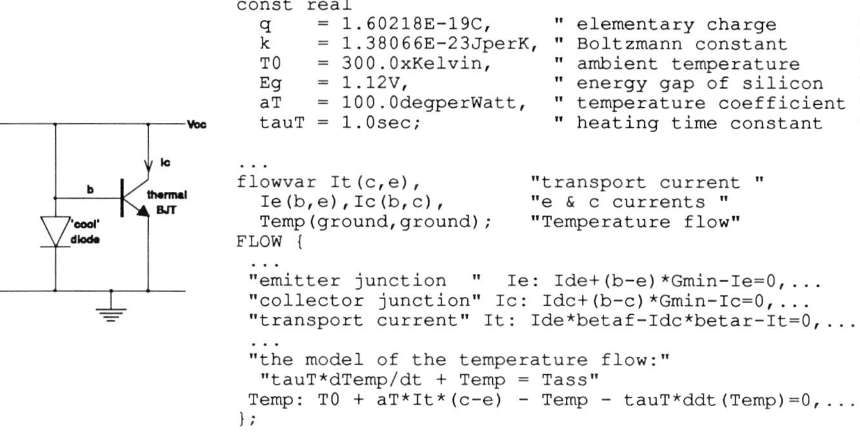


```
model T_NPN(node c,node b,node e)
const real
  q    = 1.60218E-19C,       " elementary charge         "
  k    = 1.38066E-23JperK,   " Boltzmann constant        "
  T0   = 300.0xKelvin,       " ambient temperature       "
  Eg   = 1.12V,              " energy gap of silicon     "
  aT   = 100.0degperWatt,    " temperature coefficient   "
  tauT = 1.0sec;             " heating time constant     "

...
flowvar It(c,e),           "transport current "
  Ie(b,e),Ic(b,c),         "e & c currents "
  Temp(ground,ground);     "Temperature flow"
FLOW {
 ...
 "emitter junction  "  Ie: Ide+(b-e)*Gmin-Ie=0,...
 "collector junction" Ic: Idc+(b-c)*Gmin-Ic=0,...
 "transport current" It: Ide*betaf-Idc*betar-It=0,...
 ...
 "the model of the temperature flow:"
  "tauT*dTemp/dt + Temp = Tass"
 Temp: T0 + aT*It*(c-e) - Temp - tauT*ddt(Temp)=0,...
};
```

Figure 5. ALFA model of thermal runaway.

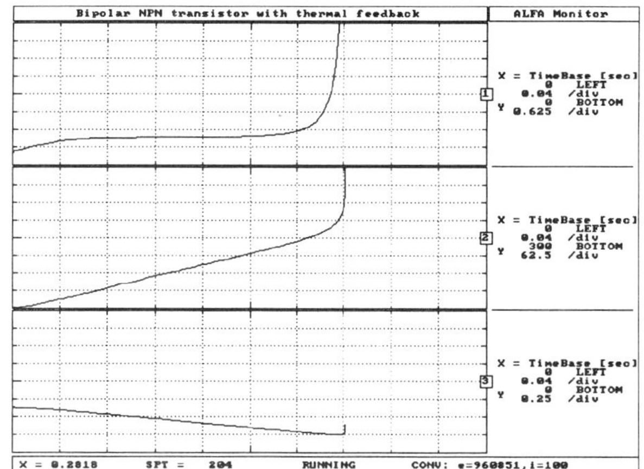

Figure 6. Thermal runaway simulation results: 1. Collector current I_c (t), 2. Temperature T(t), 3. Base-emitter voltage V_{be} (t).

The circuit to be modelled is shown in figure 5, and comprises simply a conventional diode in parallel with the base-emitter junction of a bipolar transistor. The power dissipated by the transistor causes it to heat up and hog the drive current, which in turn causes the device to heat up more. The model assumes a constant, linear thermal impedance between the junction and 'thermal ground', and the transistor eventually destroys itself. In the example, the junction temperature is a primary system variable, and is controlled by the energy dissipated by the device as a whole. The transistor temperature $T(t)$ as a function of time obeys the following differential equation:

$$T(t) = T_0 + a\, I_{ce}\, V_{ce} - \tau \frac{dT(t)}{dt}$$

where $T_0 + a\, I_{ce}\, V_{ce}$ is the asymptotic temperature, T_0 is the nominal temperature, a and τ are coefficients and $I_{ce}\, V_{ce}$ is the power dissipated at the transistor.

As the results in figure 6 show, the analysis effectively halts when the value of the collector current exceeds the computer's floating point range and a loss of convergence occurs.

It is not possible to perform this kind of experiment using SPICE, as one of the limitations imposed by the program is that the temperature, although controllable by the user, must remain constant during the simulation period.

```
"Lorenz Chaos Equations  "
const s = 10.0, b = 2.667, r = 28.0;

model main()
flowvar x(ground,ground)=0,y(ground,ground)=5,z(ground,ground)=25;
FLOW  {
 x: ddt(x)+s*(x-y)    = 0,    {@x := sdt(x) + s,     @y := -s },
 y: ddt(y)+y+x*z-r*x = 0,    {@x := z - r, @y := sdt(y) + 1, @z := x},
 z: ddt(z)+b*z-x*y    = 0,    {@x := -y,     @y := -x,      @z := sdt(z) + b }
}
CONTROL {
  _RESULTS(x,y,z);
  _TRANSIENT(0.0,30.0);
}.
```

Figure 7. ALFA model of Lorenz Chaos Equations.

The Lorenz chaos equations:

The fact that ALFA is capable of analysing arbitrary networks means, of course, that its use is not restricted to electronic circuits. In this example, a coupled set of first order differential equations are solved:

$$\frac{dx}{dt} = s\,(y-x)$$

$$\frac{dy}{dt} = rx - y - xz$$

$$\frac{dz}{dt} = xy - bz$$

These are the famous Lorenz Chaos Equations [5] derived from a fluid convection model, that are known to have no steady-state or periodic solution. The ALFA description and $z(x)$ plane for a given s, r and b are shown in figures 7 and 8.

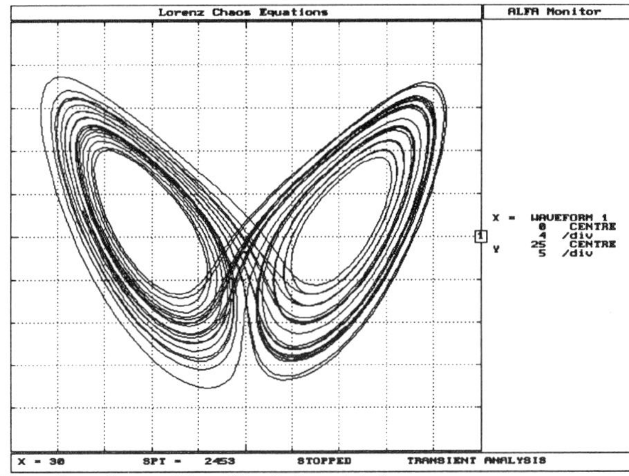

Figure 8. Lorenz Chaos Simulation; z(x) plane.

The phase locked loop

The circuit of figure 9 is an example of a mixed-mode PLL-based frequency multiplier. The input waveform of 10kHz is compared with the output from the counter C by the phase detector (multiplier M and filter F), which produces the error waveform that drives the VCO. The output frequency of the VCO is 1MHz, which is divided by 100 before being fed back to the phase detector.

Circuits of this type are extremely difficult to simulate by SPICE-like simulators for the following reasons:

- The maximum step size must be kept much smaller than the period of the VCO waveform whereas the simulation must run over thousands of VCO periods.
- SPICE models of some circuit blocks, such as the VCO or counter are themselves extremely complex due to the limited macromodelling facilities available.

The (digital) model of the counter keeps track of the input period count and modifies the internal logic state every time the count exceeds some preset value (in this case 100). Models of this kind do not require much processing power to implement.

Figure 10 shows the results of simulating the phase locked loop frequency multiplier circuit. The driving waveform starts with a frequency of 10kHz, and this changes abruptly to 11 kHz at 5ms into the simulation. The bottom trace (channel 5) of figure 10 shows the filtered error voltage (VCO control voltage) displaying the two transients arising from the startup and sudden shift in drive frequency. The channels 1,2 and 3,4 display the waveforms $in1(t)$, $in2(t)$ and $Vx(t)$ during

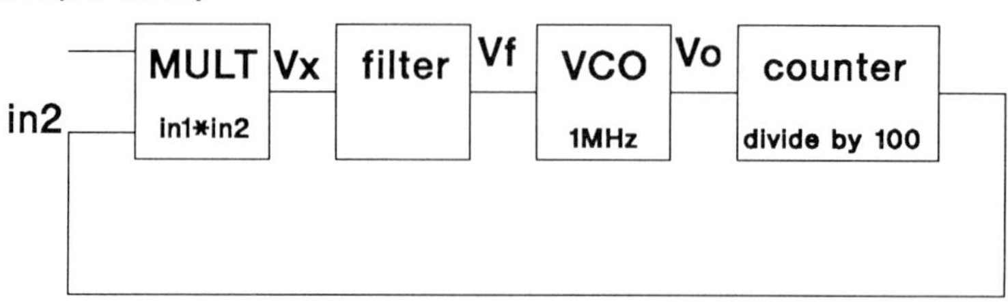


```
" mixed-mode PLL frequency multiplier x100 with a frequency jump "
 const real      phase  = 0.6us,       twopi  = 6.283185307;
model VCO(val real in, real out)
 const real  fc = 1MHz,     "centre frequency"
            df = 1MHzpV, "slope "
            Vc = 0V;       "centre frequency voltage"
 real phase_old = 0,time_old = 0.0,freq=fc,phase=0;
EXCITE {
 " evaluate frequency"  freq := max(0.5MHz,fc+(in-Vc)*df);
 "integrate frequency to obtain phase"
 phase := phase_old + (time-time_old)*freq*twopi; phase := phase %
twopi;
 "update state"  phase_old := phase;  time_old := time;
 "output voltage (sine wave)" out := 2.5V*(1.0 + sin(phase))
};
model counter(val real in, logic out)
 const int MaxCount = 50; "divide-by-100 counter"
 logic state=0,in_old=0,count=0;
EXCITE {
 (in  2.51V) & (in_old=0) ? {in_old := 1, count := count+1} :
  (in  2.49V) & (in_old=1) ?  in_old :=0;
 count = MaxCount  ? {count := 0, state := !state}; out := state
};
model main()
 real period = 100us; "initial input waveform period - 10kHz"
 real  Vx=0,Vf=0,Vo=0,t=0,in1=0,in2=0;
 logic out;
 MULT    X;
 filter  F;
 VCO     V;
 counter C;
FLOW {
 X(Vx,in1,in2),          " multiplier"
 F(Vx,Vf),               " filter    "
 V(Vf,Vo),               " VCO       "
 C(Vo,out)               " counter   "
} "end of FLOW section"
EXCITE {
" reduce period when time>8ms " time>8ms ? period := 91us;
"input wave  in1" t := (time + phase) % period;
 in1 := 2.5V*(1+sin(twopi*t/period));     "0..5V"
 "D-A conversion of counter output" in2 := 5V*out
}
CONTROL {
```

Figure 9. ALFA model of PLL frequency multiplier.

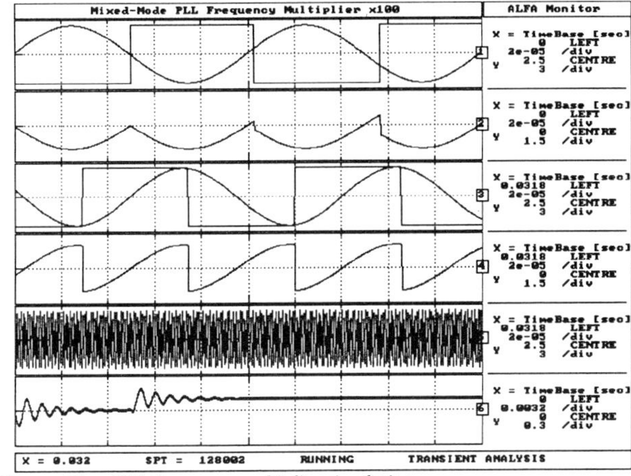

Figure 10. *PLL frequency multiplier simulation.*

the first and last 200μs of the simulation correspondingly, clearly showing the system moving towards steady state.

4. Run times

All the simulations were run on a 25MHz 386/387 IBM PC clone and the run-time requirements of the four simulations discussed here may be summarised below:

	CPU time	Simulated time	Time points	CPU time per time point
Schmitt trigger	4 sec	4 sec	801	5 msec
Thermal runaway	13 sec	3 sec	204	63.7 msec
Lorenz Chaos	34 sec	30 sec	2448	13.9 msec
PLL frequency multiplier	16 min	32 msec	128002	7.5 msec

5. Final remarks

ALFA is a system simulator of exceptional power and flexibility. The flexibility comes about by adopting an object-oriented approach with its descriptive input language, it allows true description hierarchy with complete data encapsulation where necessary. The power of the simulator comes from its unique ability to handle a multitude of abstract levels of modelling, enabling simulations of circuits difficult or impossible to analyse by means of conventional methods. Also, much of the numerically intensive processing can be transferred to the outer loop of the simulation process so that, for some models, CPU requirements are vastly reduced.

6. References

[1] Chandy, K.M. and J. Misra, "Parallel Program Design: a foundation",1988,Addison-Wesley Publ. Co. Inc. Reading, MA ISBN 0-201-05866-9.

[2] Quarles, T., A.R. Newton, D.O. Pederson, A. Sangiovanni-Vincentelli, "SPICE 3A7 user's guide", 1986, Dept. of Electrical Engineering and Comp. Sci., Univ. of California, Berkeley, CA, USA.

[3] Chua, L.O. and P.M. Lin. "Computer-aided analysis of electronic circuits", 1975, Prentice-Hall Inc., Englewood Cliffs, New Jersey, USA.

[4] Kazmierski, T.J. and Y. Bouchlaghem, 'Hierarchical solution of linear algebraic equations on transputer trees', Proc ECCTD, September 1989, Brighton, U.K.

[5] Sparrow, C. "The Lorenz equations: bifurcation, chaos and strange attractors", 1982, Springer-Verlag, ISBN 0-387-90775-0.

VLSI 91
A. Halaas and P.B. Denyer (Eds.)
Elsevier Science Publishers B.V. (North-Holland)

ON-CHIP CMOS SENSORS FOR VLSI IMAGING SYSTEMS

Peter B. Denyer, David Renshaw, Wang Gouyu, Lu Ming Ying, Stuart Anderson

Department of Electrical Engineering, University of Edinburgh, Mayfield Road, Edinburgh, EH9 3JL, Scotland

Abstract

We present techniques for the integration of vision sensors and processors on common CMOS substrates. This establishes the feasibility of single-chip vision systems which offer reductions in size, power and cost over contemporary techniques.

1. INTRODUCTION

Potential electronic vision applications are widespread; examples include:-

- bar-code and text readers
- security cameras
- image capture for DTP
- biometric verification; fingerprints, faces, etc.
- fax
- production line inspection
- video telephones
- vision subsystems in robots and autonomous vehicles
- consumer camcorders and still cameras
- automotive applications

Virtually all of these applications, and a host of others, are sensitive to cost, size and power consumption. This applies not only to the camera function but also to the subsequent image processing functions, which are invariably computationally demanding.

Commonly such systems are constructed from commodity camera modules, frame grabbers and PC's, or dedicated image processing hardware. These systems can hardly be described as miniature, and their cost and power consumption severely limit their application. Only where production volumes are very large, as in consumer camcorders, do all parameters fall to attractive levels.

Fortunately the advancing capability of VLSI, especially CMOS, technology permits the integration of video-rate A/D conversion and the implementation of powerful custom image processing architectures. Challenging imaging applications, such as fingerprint verification, have become possible within a few tens of cubic inches, consuming a few watts of power. This is still far from ideal and the dominant limiting factor is usually the camera module itself.

It is well known that silicon can act as an excellent photoreceptor over the visible spectrum. The majority of solid-state cameras today use CCD technology (a variant of MOS) which, over two decades, has been highly refined to optimise this function. Some sensor manufacturers use a variant of single-channel MOS technology in which only doping levels are altered to optimise optical performance parameters such as anti-blooming.

Other workers [1,2] have recognised the attraction of implementing sensors in an unmodified CMOS process, permitting the inclusion of the sensor with other control and processing functions on the same chip. Despite encouraging results this technique has never been developed to the point at which the sensor performance matches that of CCD cameras. The purpose of our work is to realise, in unmodified CMOS technology, array image sensors which match the performance of CCD cameras. We have succeeded in this aim and report here the circuit techniques and results that we have achieved, including demonstrator single-chip vision systems.

2. TECHNIQUES

In common with others [3] we use a photodiode sensor comprising an array of MOS transistors, one per pixel, Figure 1. The photodiode is implemented by extending the source region of the transistor. This may be reset and then isolated under control of the MOS transistor gate. All of the gates in each row are driven in common.

Once reset, the reverse-biased (photo)diode converts incident light into a small photocurrent which gradually discharges the photodiode capacitance. The pixel is read by opening the gate, connecting the photodiode to the MOS transistor drain. All of the drain in each column are connected in common and only one row is read at any time.

Commonly, the column lines are gated through an analogue multiplexer to a single external charge sense amplifier. The requirements of this amplifier are daunting considering that high-speed and wide dynamic range must be achieved from a charge packet in the pixel which may be of the order of 1fC. Accordingly we have replaced this scheme by providing charge sense amplifiers at the top of each column. These need not work so quickly, since their activation frequency is equal to the line rate rather than the pixel rate, and they are situated as close as possible to the pixel array. Their sole constraints are the need to achieve a good dynamic range and to be realised within the pixel pitch (of the order of 10-20μm). We use a single-ended differential charge integrator which gives a low impedance 1v analogue representation of the pixel charge with a theoretical dynamic range of 70dB. The read time is approximately 500nsec.

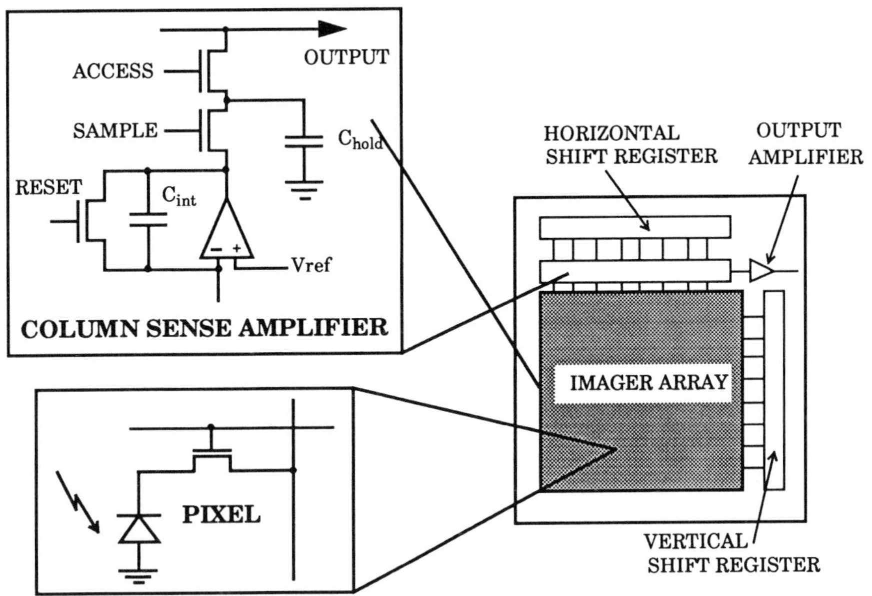

Figure 1. Architecture of a CMOS image sensor with column charge sense amplifiers.

The voltage representation at the output of these sense amplifiers is sampled and stored on a row of capacitors and the information on these is multiplexed out in the conventional manner, with the exception that we implement the output charge integrator on-chip, including a sample-and-hold stage. For applications requiring a composite video waveform it is relatively easy at this stage to include an analogue multiplexer to switch in blanking and sync. levels at appropriate times.

Serially-scanned operation is achieved by adding vertical and horizontal digital shift registers at the periphery of the array and these also must match the pixel pitch. The vertical register successively activates the row lines, whilst the horizontal register controls the sequential pixel read-out within each line. The performance of the array is quite insensitive to these control waveforms and amplitudes, in contrast to CCD, and this is a distinctive advantage of the approach.

Prototype CMOS arrays using this architecture give remarkably good results. They operate over very wide margins of temperature and supply voltage. The single parameter of concern is fixed pattern noise from two sources; threshold variations in the MOS pixel access transistors causing speckles, and mismatches between the column sense amplifiers causing vertical stripes. Without compensation these effects have an rms value around 1% of saturation. In later designs we have eliminated these effects by:-

(i) reducing the applied pixel reset voltage to make the actual reset value independent of the gate potential and gate threshold.

(ii) implementing an offset compensating phase in the common sense amplifiers during idle periods, such as line and frame synchronisation.

These circuit techniques successfully eliminate the fixed pattern noise and overcome a traditional objection to the potential of this approach.

Figure 2. A demonstration CMOS CCTV camera.

Figure 2 shows a completed CCTV camera chip fabricated on a standard 2 micron CMOS ASIC line. The chip contains an array of array of 312 x 287 pixels with timing and automatic exposure control on-chip.

(a) (b)

Figure 3. Comparison of CCD and CMOS camera performance under identical conditions:- (a) CCD (b) CMOS

Figure 3 compares the picture output of this device with an existing CCD camera module. The results are subjectively indistinguishable.

3. AUTOMATIC EXPOSURE CONTROL

By electronically controlling the integration period we can proportionately decrease the sensitivity of the array. We can achieve this through the vertical shift register by controlling the duration, in cycles, of a 'reset' pulse entered at the top of this register. This varies the integration time in steps equal to the line period. We further gate this signal with one of short duration to reduce the exposure time in steps equal to the pixel period, down to a minimum time constrained by the read time of the column sense amplifiers. This is approximately 500nsec, giving a total exposure range of 40,000:1.

If we now alter the exposure time in response to the monitored video output we can implement fully automatic electronic exposure control and avoid the need for mechanical iris control on the lens.

Figure 4 shows a simple scheme for such a control algorithm. The video stream is internally histogrammed in three bins:- very white, average and very black. The exposure is increased or decreased according to whether the image content is judged to be too bright or too dark.

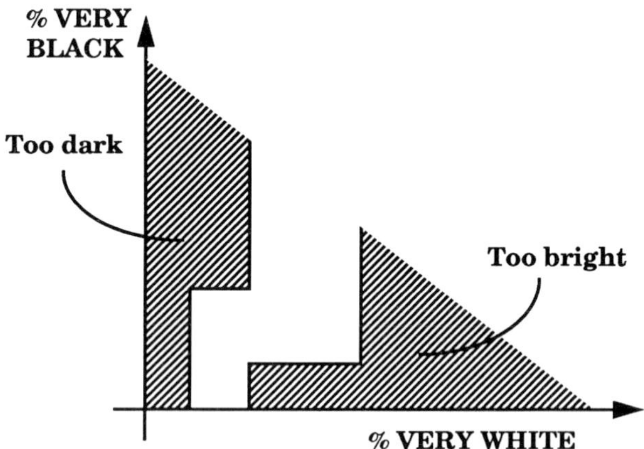

Figure 4. Automatic exposure control decision diagram

We have implemented this scheme (which costs approximately 1000 gates) on several CMOS sensor arrays and obtained satisfactory performance.

4. RESULTS

Over a series of prototypes we have improved the performance of CMOS sensor arrays to match or out-perform typical monochrome CCD performance in most respects. A detailed comparison is given in Table 1. The CMOS data is measured from the CCTV camera demonstrator device shown in Figure 3. The CCD data is compiled from manufacturers data sheets.

	CCD Camera Module (typical)	Integrated Edinburgh CMOS Camera
voltage	12v	5v
power	1W	200mW
minimum pixel size for 1.5μm process	10μm x 10μm	16μm x 16μm
saturation level	20 lux	20 lux
s.n.r.	52dB	51dB
output	composite video 1v p-p	composite video 1v p-p
integration time range	300:1	40,000:1
dark current as fraction of saturation at room temperature, 20msec integration time	0.005	0.0004
antiblooming factor	100x	100x

Table 1. Comparison of CCD and CMOS sensor performance

5. Application Examples

We report here two examples which illustrate the potential of this technology for integrated vision applications.

Our first example is a low-resolution camera for use in intruder-alarm verification. This device is installed with a Passive Infrared detector and, upon detecting an alarm, transmits a short sequence of video to a control unit which compresses and transmits this data to a remote monitoring station. Within a few seconds of an alarm event a remote observer can deduce its cause and take appropriate action. As greater than 95% of such alarms are false, the provision of video verification will eliminate much unnecessary police action.

Passive alarm units are very low cost items and cost is a primary constraint on this device.

Figure 5 shows the complete intruder-alarm camera module. This includes a 1.5μm CMOS camera chip measuring 5.8mm x 4.0mm combining a 156x100 pixel sensor array with all timing and control electronics on-chip. The device is customised for this application, and this includes drivers which automatically trigger lamp or flash devices to assist in dark conditions. A further novel feature of this device is the use of a miniature glass lens which is bonded directly to the sensor chip surface. This enables a 90° field of view and assists in improving the robustness and cost of this small unit.

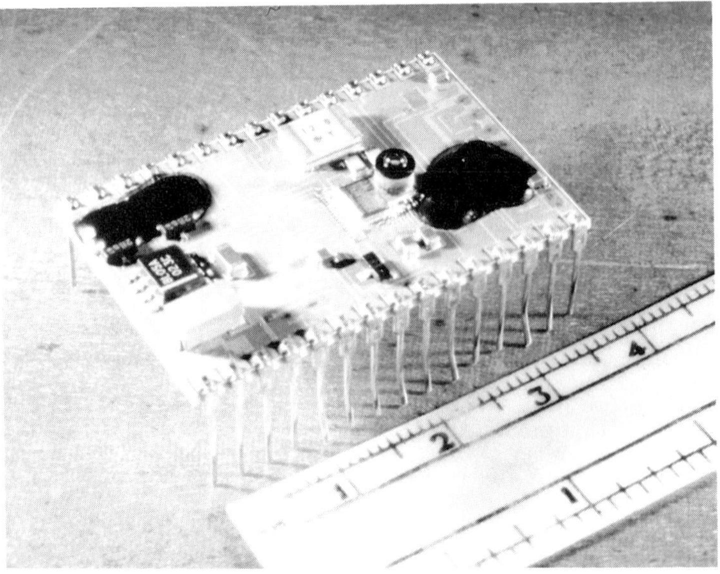

Figure 5. Complete alarm verification camera module using a custom CMOS sensor with a miniature chip-mounted lens

The assembly shown is on a ceramic hybrid substrate and includes a 5v regulator, a clock crystal and approximately one dozen passive components. A simple bipolar stage provides sufficient output impedance to drive 200ft of co-axial cable to the modem unit.

Our second example shown in Figure 6 is a very substantial single-chip vision processing system. This includes:-

. 258 x 258 pixel array.

. Image preprocessing and quantisation to form a normalised binary image.

. 64-cell 2000 Mops/sec correlator array.

. Post-correlation decision hardware.

. 16k bits RAM cache.

. 16k bits ROM look-up table.

With the aid of two external devices (one 64kbit RAM and one 8051 microcontroller) this device performs all of the image sensing and processing functions necessary to capture and verify a fingerprint against a stored reference print within one second.

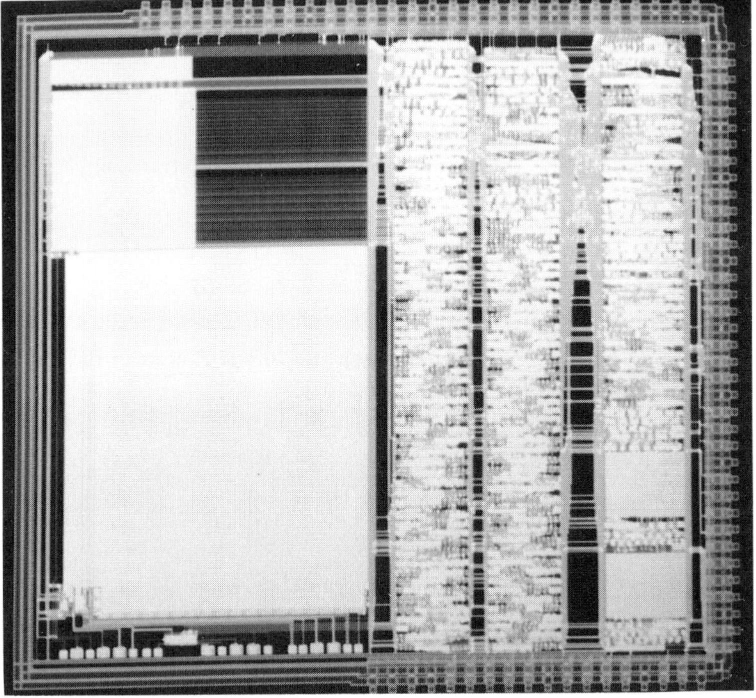

Figure 6. An integrated vision system 258 x 258 sensor array and image processor for fingerprint capture and verification

This is perhaps the best example of our goals; the integration of a sensor and powerful image processor on one substrate.

We have demonstrated several other devices in both 2 micron and 1.5 micron ASIC CMOS technologies and claim that these techniques are portable to any commodity CMOS process.

6. CONCLUSIONS

The aim of 'smart' vision devices, incorporating image sensors and processors in one chip has been substantiated. These integrated devices can be implemented today in unmodified commodity CMOS technology. Vision systems implemented in this way enjoy unprecedented reductions in size, cost and power consumption.

7. ACKNOWLEDGEMENTS

This work was supported by SERC award GR/F 36538, Automated Security Holdings Plc., De La Rue and VLSI Vision Ltd.

8. REFERENCES

1. Lyon, R.F., "The Optical Mouse and an Architectural Methodology for Smart Digital Sensors", VLSI Systems and Computations, 1981, Springer-Verlag, ISBN 3-540-11251-0, pp. 1-19.
2. Tanner, J.E. and Mead, C., "A Correlating Optical Motion Detector", 1984, Conference on Advanced Research in VLSI, MIT, pp. 57-64.
3. Renshaw, D. and Denyer, P.B., "ASIC Vision", CICC 1990, §7.3.
4. Anderson, S. et al., "A Single Chip Sensor and Image Processor for Fingerprint identification", CICC 1991, §12.1.

VLSI 91
A. Halaas and P.B. Denyer (Eds.)
Elsevier Science Publishers B.V. (North-Holland)

A Customizable Neural Processor for Distributed Neural Network

J. Ouali[a], G. Saucier[a], P.Y. Alla[a], J. Trilhe[b], L. Masse-Navette[b],

[a]Institut National Polytechnique de Grenoble / CSI, 46 Avenue Félix Viallet, 38031 Grenoble - France, Tel: (33) 76 57 46 87, Fax: (33)76 50 34 21

[b]SGS THOMSON, avenue des martyrs, 38019 Grenoble France, Tel: (33)76 58 50 00, Fax: (33)76 58 56 47

Abstract :
This paper proposes a distributed, synchronous architecture for artificial neural networks. A basic processor is associated with a neuron and is able to perform autonomously all the steps of the learning and the recognition phases. Data circulation is implemented by shifting techniques. Customization of the network is done by fixing customization data in dedicated memory elements. The neuron has been implemented on silicon. It is shown that in a silicon compiler environment dedicated network can be easily generated by cascading these elementary blocks.

Introduction

Neural computing relying on non algorithmic information processing has led in the recent last years to a tremendous research effort all over the world [1, 2, 3]. It appears clearly that neural computing will be practical and efficient if dedicated, fast architectures exist. Usually, neural computing is simulated on existing Von Neumann or parallel computers (CRAY2, GAPP, Transputer). Attempts have been made to build dedicated architectures with commonly available components and some other studies concentrate on the design of either analog or digital specific components [4..7].

This paper focuses on the design of a distributed and customizable neural architecture, based on a processor which can be easily cascaded. The processor structure is well fitted to an automatic generation in a silicon compiler environment. Modification of parameters such as precision and size of data or neuron output codes must not result in any additional design effort. In the neural network, a basic processor is associated with a neuron and is able to perform autonomously all the steps of the learning and recognition phases. Data circulation is implemented by shifting techniques. The customization of the network is done by setting identification data in dedicated memory elements. According to these parameters it is then possible to perform different calculations with the same data-path structure.

In section 1, the global network architecture will be presented. Subsection 1-1 will describe the connections between neurons. In subsection 1-2, the way calculations are achieved with the proposed architecture will be detailed on an example. Subsection 1-3 will discuss communications in a layered network.

Section 2 presents the neural processor architecture, and subsection 2.2 focuses on the hardware requirements.

Finally, a customizable data-path is described in section 3. A brief overview of future work will conclude this paper.

1. General frame of the architecture.

We consider a neural network which may be made up of one layer (Hopfield network) or several layers (multilayered network) of neurons. As previously mentioned, a physical entity (neural processor) is associated with each neuron. During a recognition phase, the neural processor realizes the classical operations shown in figure 1. The processor number i first compute a ponderated sum V_i called "potential". The weight $C_{i,j}$ associated with the input data are called "synaptic coefficients". A function f, called activation function, is then applied to the potential to calculate the final output O_i (usually called "state" of the neuron). This section presents the global interconnections between the processors and an example illustrates the way data circulate throughout the network.

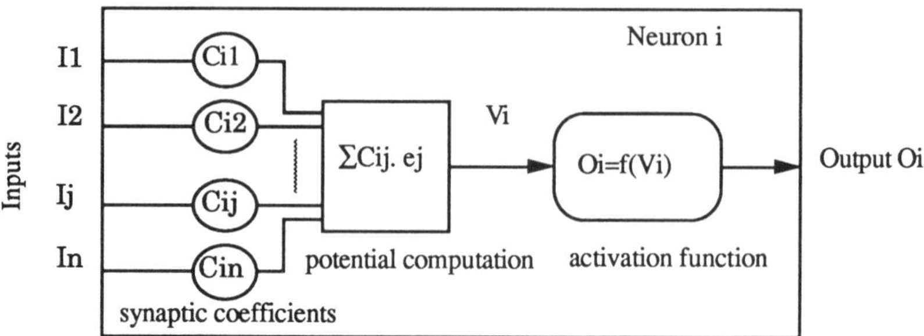

Figure 1 : Function performed by the neural processor during the relaxation phase.

1.1. Network organization

It has been pointed out frequently that a large number of connections between the neurons is a major drawback for their implementation on silicon.Since a processor cannot be connected to a large number of neighbours or external inputs, each neural processor has to seize his input data sequentially and must pick them up when they are at hand. Every neuron of a given layer will process the same input data. It is therefore natural to make both the input and output data circulate through the network via registers connected through buses in a circular stack-like structure (figure 2). Spatio-temporal virtual connections thus replace spatial connections [9].

The proposed architecture is shown figure 3. The neurons are organized in a 2-D array of processors. To ensure communications, they will be placed in rows between two buses. These buses are made up of bus segments which can be connected by software programmable switches when required during the phases of neuro-computing, thus implementing the stack structure.

A chip on which a network is integrated, has this same topology. Most of the time, however, a whole network will not be implemented on a single chip, due to the size of current applications. But it is possible to build large networks by assembling such basic blocks.

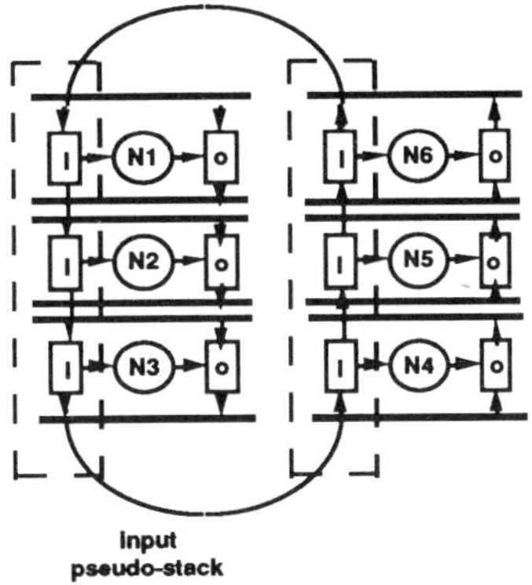

Figure 2: implementation of the shifting technique (Ni = basic neuron).

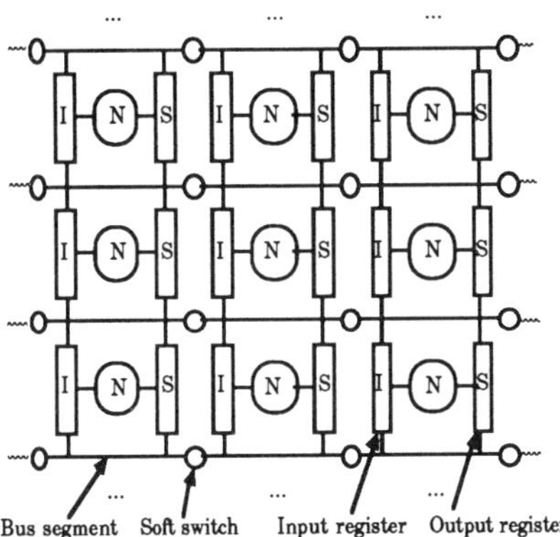

Figure 3: Global architecture (partial view)

As mentionned above, many classical neural architectures divide a network in several layers. In a Hopfield network, there is only one layer, and the neurons process their own outputs. In a multilayered network, the layers are disjoint and process the ouput states of the neuron of the previous layer. It is interesting to implement a given layer or part of layer on a single chip. The following section concentrates on the way classical calculations are performed in a layer with the proposed topology.

1.2. Layer folding

This global architecture is now applied to the example of 8 processors realized on a 4 by 2 array. First, let us show how the computation of potentials is performed ; the equations of potentials Vi are given below.

$$V_1=\mathbf{C_{1,1}.I_1}+C_{1,2}.I_2+C_{1,3}.I_3+C_{1,4}.I_4+C_{1,5}.I_5+C_{1,6}.I_6+C_{1,7}.I_7+C_{1,8}.I_8$$
$$V_2=\mathbf{C_{2,1}.I_1}+C_{2,2}.I_2+C_{2,3}.I_3+C_{2,4}.I_4+C_{2,5}.I_5+C_{2,6}.I_6+C_{2,7}.I_7+C_{2,8}.I_8$$
$$V_3=C_{3,1}.I_1+C_{3,2}.I_2+\mathbf{C_{3,3}.I_3}+C_{3,4}.I_4+C_{3,5}.I_5+C_{3,6}.I_6+C_{3,7}.I_7+C_{3,8}.I_8$$
$$V_4=C_{4,1}.I_1+C_{4,2}.I_2+\mathbf{C_{4,3}.I_3}+C_{4,4}.I_4+C_{4,5}.I_5+C_{4,6}.I_6+C_{4,7}.I_7+C_{4,8}.I_8$$
$$V_5=C_{5,1}.I_1+C_{5,2}.I_2+C_{5,3}.I_3+C_{5,4}.I_4+\mathbf{C_{5,5}.I_5}+C_{5,6}.I_6+C_{5,7}.I_7+C_{5,8}.I_8$$
$$V_6=C_{6,1}.I_1+C_{6,2}.I_2+C_{6,3}.I_3+C_{6,4}.I_4+\mathbf{C_{6,5}.I_5}+C_{6,6}.I_6+C_{6,7}.I_7+C_{6,8}.I_8$$
$$V_7=C_{7,1}.I_1+C_{7,2}.I_2+C_{7,3}.I_3+C_{7,4}.I_4+C_{7,5}.I_5+C_{7,6}.I_6+\mathbf{C_{7,7}.I_7}+C_{7,8}.I_8$$
$$V_8=C_{8,1}.I_1+C_{8,2}.I_2+C_{8,3}.I_3+C_{8,4}.I_4+C_{8,5}.I_5+C_{8,6}.I_6+\mathbf{C_{8,7}.I_7}+C_{8,8}.I_8$$

In a first computation step, bus segments are connected as shown in figure 4. The switches of the second column are open. The input of neuron N1 is available on the bus segment of the second row for both neurons N1 and N2, the input of neuron N3 is available similarly on the next bus segment for neurons N3 and N4 and the same for N5, N7. The computations performed during this step are in bold characters in the above equations.

The input data I_i can be either primary inputs or outputs of another layer, or outputs of the current layer. If the network is of the Hopfield type, for instance, e1 will be the current state of neuron N1, e3 that of N3, and so on.

In the second step of the computation, a circular shifting is performed in the first layer as indicated in figure 5. The previous input of neuron N1 has been shifted on the next bus segment and is now available for both neurons N3 and N4.

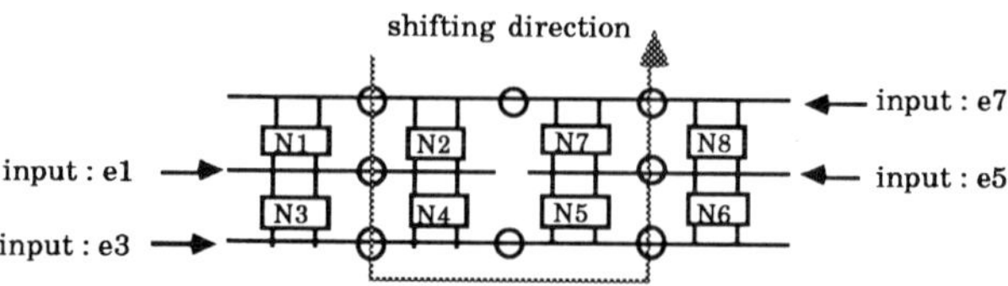

Figure 4 : first computation step.

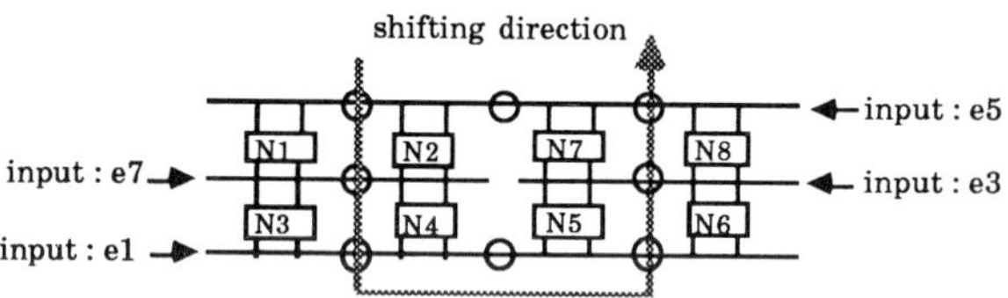

Figure 5 : Second computation step

In this example, it is clear that 4 elementary shiftings - a complete circular shifting - will be necessary to perform half of the complete computation. Two complete circular shiftings (8 elementary shiftings) will be required to end up the computation. Once this stage is over, the controller initializes the calculation of the activation function, which does not result in any communication between the neurons.

Generally speaking, a layer of N processors will be implemented on a (N/k by k) array. Since each input data is transmitted to N/2k neurons (half of a line) when it is on a bus, 2k elementary shiftings are required to communicate the data to every neuron of the block. If each neuron must consider N' inputs, the data have to be sent in N'/2k different waves. That makes a total of N' elementary shiftings to complete a standard potential computation. In the case of a Hopfield network there must be N elementary shiftings.

It is clear in the above example that the localization of the open switch(es) sets the topology of the network. The column where the switches are open corresponds to an inversion of the shifting direction, and can be seen as a folding of the layer or part of layer. A given layer folding implies a fixed path of the data in the network.

If the different synaptic coefficients must be loaded from ouside the chip, there will be an initialization phase when the coefficients will be stored in a local memory. The above remark is also valid for this operation. That means the coefficients will be listed in this memory in the same order as the input data will be sent. Therefore, a sequential access to the local memory will provide the neuron with the right coefficients. That makes this approach especially fitted to silicon compiler requirements, since no significant modification is induced by a change in the network topology.

1.3. Communication between layers.

This section will discuss the way layers can exchange data with the proposed architecture. This will be illustrated on an example with two layers.

When computing the potentials and activation function, the network layers will be insulated from each other by open switches, as shown in figure 6.

If there is no pipe-line between the layers, the external input data of the first layer can be entered as explained in the previous sections, provided that all the switches of the following layers are closed. But for the other layers, the input data will come from only one side. The network will therefore be oriented from inputs to outputs, the inputs being the bus segments of the first layer , and the outputs that of the last layer. If we want to keep for each layer the ring structure mentioned above, each wave of input data must be sent by the previous layer in two clock cycles (one additional clock cycle for each wave). Between each wave, there will be a communication phase when each layer will read the next wave of input data from its predecessor, and pass chosen outputs on to the next layer.

In the case of a pipe-lined processing, the switches between two layers must be set open to provide independant functionning. Therefore, it is not possible to feed the input data of the first layer from both sides of the network anymore. The computation of a wave of data constitutes an elementary task of the pipe-line. In a steady-state functionning, all the layers must communicate with their two neighbours at the end of each wave processing. All the switches must be closed in both sending and receiving layers. Since we do not want any hand-shaking between the neurons, some layers must be left idle during an appropriate number of clock cycles to ensure synchronisation. A communication phase can be achieved in four clock cycles, provided that the rule below is followed: "if a given layer receives its inputs before sending its ouputs, then the following one will first send and then receive, and vice-versa". In a network with more than two layers, odd rank layers can for instance be of the "send-receive" type and even rank layers of the "receive-send" type (figure 6).

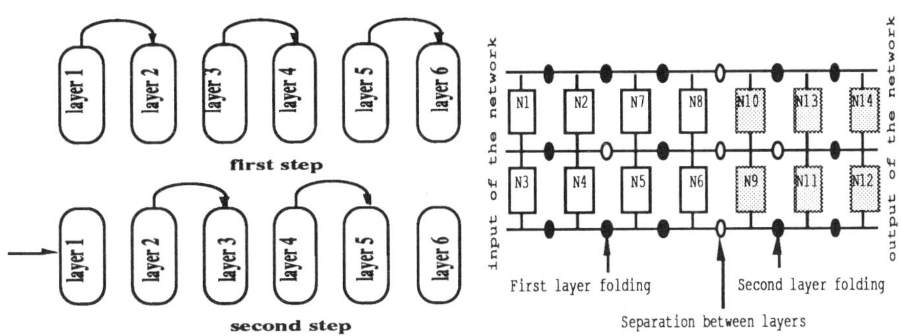

Figure 6: communication phase.

Figure 7: Organization of a two-layer network.

In the example below, the eight neurons of the previous section now constitute the first layer in a multi-layered network, and they have 16 inputs (2 waves). It is supposed that the second layer is made up of six neurons (figure 6), with an arbitrary layer folding.

With this example the calculations performed by the network can be described as follows:

First wave:
Step 1: Processors N1 and N2 read their first input on bus medium. Processors N3 and N4 read their first input on bus low.
Step 2: Processors N7 and N8 read their first input on bus high. Processors N5 and N6 read their first input on bus medium
Step 3: The neurons of the first layer process the first wave. Second layer is inactive.

Second wave :
Step 1: Processors N1 and N2 read their first input on bus medium. Processors N3 and N4 read their first input on bus low .

Step 2 :Processors N5 and N6 read their first input on bus high. Processors N7 and N8 read their first input on bus medium.
Step 3: The neurons of the first layer process the second wave. Second layer is inactive. The inputs of second layer are available at the end of this step in the output buffers of the first layer processors. The first layer can begin another series of computations.

First wave:
Step 1: Processor N1 communicates his state to neuron N10 on bus medium. Processor N3 communicates his state to neuron N9 on bus low.
Step 2: Processor N7 communicates his state to neuron N13 and N14 on bus high. Processor N5 communicates his state to neuron N11 and N12 on bus medium.
Step 3: Processors N1 and N2 read their new input on bus medium. Processors N3 and N4 read their new input on bus low.
Step 4 : Processors N7 and N8 read their new input on bus high Processors N5 and N6 read their new input on bus medium
Step 5: The neurons of the first layer process the first wave of the second series. Second layer processes the first wave of the first series. The switches separating the two layers are open.

Second wave :
Step 1: Processor N2 communicates his state to neuron N10 on bus medium. Processor N4 communicates his state to neuron N9 on bus low.
Step 2: Processor N8 communicates his state to neuron N13 and N14 on bus high.Processor N9 communicates his state to neuron N11 and N12 on bus medium.
Step 3 : Processors N1 and N2 read their new input on bus medium Processors N3 and N4 read their new input on bus low
Step 4 : Processors N5 and N6 read their new input on bus high Processors N7 and N8 read their new input on bus medium
Step 5: The neurons of the first layer process the second wave of the second series. The second layer processes the second wave of the first series.

This process can be repeated indefinitely if there are more waves. In the case of non fully-connected layers, this same algorithm can be brought into play, thanks to masking techniques, without important modification.

2) Architecture of the neuron processor.

This sections outlines the proposed neural processor architecture.

2.1. Characteristics of the architecture.

The architecture is distributed and synchronous, which means that there is no communication between the different processors in the course of any neural processing. Each neural processor has autonomous control and is able to perform all the processes of neuro-computing (potential and activation function computations, learning phase). The neuron processor contains a set of identification registers which store the necessary information about the global architecture of the network. An initial loading of these registers will therefore customize the network. The local memory is a RAM for a chip implementing both learning and recognition phases and an EPROM for an optimized chip dedicated to the recognition phase of chosen applications. For an optimized recognition application, the ROM technology should be used for both the identification registers and memory coefficients. The first implemented prototype uses a 64 by 8 bit RAM to simplify the design problem. External control is just required to feed in the data. It can perform both potential and activation function computation.

Integrated on silicon using a 1.5µ technology, the total core area of the prototype is 6mm^2 with approximately 8000 transistors and a 20Mhz clock. A single layer network with 8 neurons, 16 full-connected inputs and sigmoidal activation functions will calculate its outputs in 300 clock cycles. The activity of each neural processor in such a configuration will be a little over one million connections per second.

This approach is quite different from other ones based on powerful multipliers without integrating any of the other functions [5,9]; the proposed method can sometimes be more time-consuming than the latter, but it is far more flexible and therefore more suitable for silicon compilation and fast generation of ASICs.

2.2. Hardware description.

The neural processor (Figure 8) is made up of :

1) 8 identification registers (ID_R):
- N1 representing the number of bits encoding the state.
- N2 representing the number of inputs sent in a wave.
- N3 representing the number of coefficients stored in the memory.
- N4 representing the shifting direction of the inputs in the layer.
- N5 representing the number of inactive state in the communication phase (for multilayered uses).
- N6 to N8 containing formatting parameters that will be clarified in section III.

2) a local memory in which the coefficients $\{C_{ij}\}$are stored.
3) a data path able to perform mutiplications and additions/subtractions.
4) a controller supervising the computation of the potential, the state of the neuron and the data transfer with the other neurons.
5) an input register to store and shift the input value.
6) an output register to store and shift the state of the neuron.
7) three counters C1, C2, C3.

C1 is initially loaded with the value of the register N1 and is decremented during the multiplication. C2 is used on each shifting state when it is initially loaded with the value of one of the registers N2, N5, N6, N7 and is decremented during the shifting.

C3 is initially loaded with the value of the register N3 and is decremented after each read cycle in the memory.

3. Customizable data-path.

The set of basic operations to be performed in the configurable neuron processor will be now detailed, in order to determine the relevant parameters and conventions for silicon compilation, concerning the data size and precision.

3.1. multiplication and detailed structure of the data-path.

The two main operations involved in the recognition phase can be splitted up in elementary arithmetic calculations.

For the computation of the potential ($V_I=\Sigma_j C_{I,j} e_j$) multiplications, additions and subtractions must be carried out. Multiplications are implemented using a Booth algorithm to avoid a costly combinatorial multiplier. The prototype's architecture presented in [11] has been improved to reduce the hardware. Only one 2's complement adder/subtractor is thus needed in the data-path. The minimum size for the adder/subtractor must be the number of bits encoding the input data of the Booth multiplier, that is to say the external inputs and the RAM coefficients. If their sizes are not identical, the maximum is chosen. The width of this ALU will determine the width of every memory element in the data-path. If the ALU comprises n bits, and if there are i coefficients (hence i inputs), the theoritical upper bound for the number of bits of the potential is roughly equal to n*i. Therefore, the size of the register needed to store the result of the potential is a multiple of the ALU size. It is then possible to describe the detailed architecture of the optimized data-path (figure 10) in a bit-slice format.

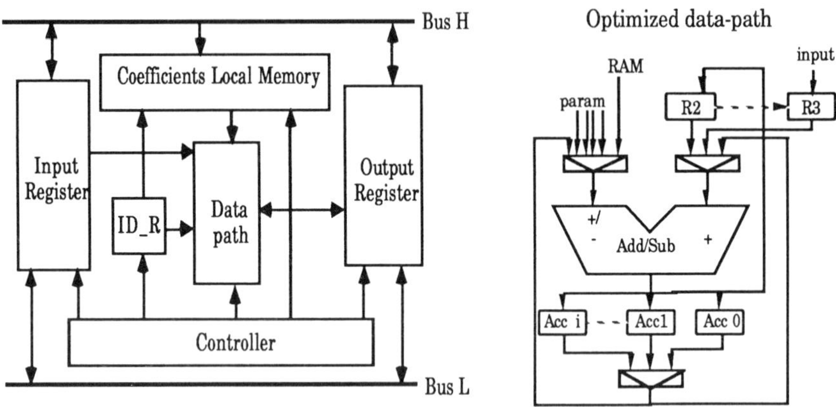

Figure 8 : Neural processor

Figure 10 : customizable data-path features.

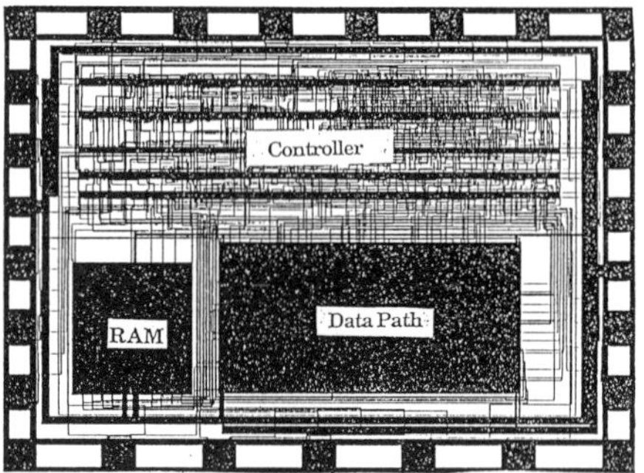

Figure 9 : Layout of the neuron

3.2. Accumulation.

The result of a multiplication will be stored in the two registers R2 and R3, and must be added to the current state of the partial sum ($\Sigma_j^k C_{I,j} e_j$, k<i) contained in the accumulator $acc_0..acc_i$. Since we cannot add two values whose decimal point is not at the same place, it is sometimes necessary to shift the result of the multiplication to superimpose the two decimal parts. If n1 is the desired length for the decimal part of the potential, n2 that of the external input, and n3 that of the coefficients, n2+n3 will be the length of the decimal part in the result of the multiplication. To superimpose the two decimal points, n1-(n2+n3) shifts will be

required ; these are right shifts if n2+n3>n1, left shifts otherwise. The number of necessary shifts and their direction will be stored in identification register N6.

Most of the time the shift will be performed right, because n1 will be equal to n2 and n3. Moreover, the final precision of the potential will be determined by the precision of the accumulated multiplications, hence by the precision of input data and coefficients. It would not be of much use, nor very rigorous (because $\Delta(xy)=x\Delta y+y\Delta x$) to widen the decimal part of the potential beyond that of the multiplication results. In the case of a right shift, the sign bit must be fed right, and left constant.

In any case, notice that the addition cannot be done in a single clock cycle because both R2 and R3 must be accumulated, and the subsequent carry propagated. This operation is described by the following algorithm (figure 11). It has been called accumulation.

```
Acc 0 = Acc 0 + R3;
Acc 1 = Acc 1 + R2 + C0;
for i:=2..b-1 do
    Acc i = Acc i + Ci-1+sgn(R2) *I(n);
    (where I(n) is 11...1 n bits)
```

figure 11 : algorithm used for accumulation.

3.3. Activation function.

Once the potential has been computed, the neuron calculates the associated state given by the activation function. It is clear that any function which can be approximated by a low order polynomial can be calculated by our processor. Only additions and multiplications are involved. The type of the activation function to be used is coded in identification register N7.

However, if a N*J-bits by N*K-bits multiplication has to be made, J*K n-bits multiplications are required, and additional memory elements are necessary. It can then be sometimes worthwile to truncate the calculated potential. The truncation is again performed via shiftings of the accumulator. The number of shifts is deduced from the values of the desired number of significant bits and is stored in identification register N8. Another truncation is sometimes performed when the result of the activation function is available, to take into account the number of bits asked for the next layer inputs. Its size will be stored in identification register N9.

In many cases, however, the activation function is far simpler. In the case of hand-written character recognition, this same architecture has been successfully implemented with a threshold activation function.

3.4. Silicon compiler strategies and future work.

It has been demonstrated for the recognition phase that the same data-path structure enables a large number of distinct uses.

Modifications for the learning phase are under consideration, but it is a bit more complicated, given the large number of learning techniques currently employed. Improvements of the general architecture, proposed by [10], are also being studied.

The important fact is that only the functioning of the controller will be modified by the values of the identification registers, so it is the only element of the customizable processor subject to changes.

We can then sort out two radically opposite strategies concerning a possible network compilation:

-A truly general controller, which can execute as many different neural behaviours as possible, is used. The generality of the controller is limited by silicon integration contingencies, and there is a waste of hardware for simple applications. But a compilation is then simply performed by setting the chip's topology (layer folding) and loading the proper identification registers.

-The compiler includes a library of partial controllers. A compilation is done by picking up parts of control graphs and setting registers. Each compilation is a bit more complex, because a Finite State-Machine compiler must be used.

The best solution probably lies in between. It might be a good compromise to have a standard controller that could be used in the most common cases. The second strategy could then be brought into play only to deal with more marginal architectures.

Conclusion

A fully digital implementation of a neural processor has been presented and the flexibility of this approach has been demonstrated. Via modifications of relevant customization registers a set of neural processors can be tailored to fit multiple data formats and connectivity patterns. It is possible to take advantage of these good properties in a silicon compiler approach, and different strategies have been outlined to reach this goal.

Acknowledgements

The authors are grateful to L. Personnaz, G. Dreyfus from ESPCI (Paris) and M.Weinfeld from Ecole Polytechnique (Paris) for helpful discussions.

References

[1] "Disordered systems and biological organization", Ed. by E. Bienenstock et al., Springer, 1986.

[2] "Neural information processing systems", Ed. by D.Z. Anderson, American Institute of Physics, 1988.

[3] "Neural networks : from models to applications", Ed. by L. Personnaz and G. Dreyfus, IDSET, Paris 1989.

[4] "A programmable binary synaptic matrix chip for electronic neural networks" A. Moopenn, H. Langenbacher et al., Neural information processing systems, Ed. by D.Z. Anderson, American Institute of Physics, 86.

[5] "A digital VLSI module for neural networks", M. Duranton, J. Gobert, N. Mauduit, in "Neural networks : from models to applications", Ed. by L. Personnaz and G. Dreyfus, IDSET, Paris 1989.

[6] "A VLSI asynchronous cellular architecture dedicated to multilayer neural networks". B. Faure, G. Mazaré, Neural networks : from models to applications, Ed. by L. Personnaz and G. Dreyfus, IDSET, Paris 1989.

[7] "A new VLSI architecture for neural associative memories", M. Verleysen, B. Sirletti, P. Jespers, in "Neural networks : from models to applications", Ed. by L. Personnaz and G. Dreyfus, IDSET, Paris 1989.

[8] "Réseau neuromimétique sur silicium : mythe ou réalité ?", J. Ouali, G. Saucier J. Trilhe. Int. Workshop Neuro-nîmes'88, Nimes, France, November, 88.

[9] "A fully digital integrated CMOS Hopfield network including the learning algorithm". M. Weinfeld, VLSI Computer Architecture and Digital Signal processing. Ed. by J. G. Delgado-Frias, 1988.

[10] "A Comparison Between Digital Neural Network Architectures". Yaron Kashai, Yair Be'ery, Department of Electronic Communication, Control and Compute Systems, Tel-Aviv University. Private communication.

[11] "Neural Processor for Neural Networks on Silicon" . J. Ouali, G. Saucier, European Conference Specific Integrated Circuits Design, EURO ASIC 90, Paris, France, May 1990.

VLSI 91
A. Halaas and P.B. Denyer (Eds.)
Elsevier Science Publishers B.V. (North-Holland)
1992 IFIP.

A VLSI Module for Analog Adaptive Neural Architectures

Daniele D. Caviglia, Maurizio Valle and Giacomo M. Bisio

Department of Biophysical and Electronic Engineering
University of Genoa - via Opera Pia 11/a - I-16145 Genoa - ITALY

Abstract

This paper describes a CMOS VLSI module for the implementation of adaptive analog neural architectures with Long Term Memory functionalities. This module is part of a Multi Layer Perceptron architecture and features on-chip learning on the basis of the Back Propagation algorithm. The analog approach followed to implement this module can be also used in the implementation of other types of neural architectures.

1. INTRODUCTION

The research on VLSI Artificial Neural Networks (ANNs) first considered ways to implement efficiently a simple threshold model of neuron and a multiplicative linear synapse. Digital realizations of neural primitives need a considerable amount of silicon area; in particular if an high computational accuracy, which is the main advantage of digital realizations, is required, multiplications and integrations claim complex circuit implementations. In order to save silicon area, simple models of neural computations have ben developed: 1) synapses assuming values in the set {-1, 0, +1}; 2) synapses assuming power of two values; 3) limited set values for weight update; 4) staircase or piecewise linear neuron activation functions.

A different approach has been proposed by C. Mead [1, 2] who recognizes that computational power of ANNs derives not only from massive parallelism but also from analog processing. He emphasizes the role of neural network modeling for learning and self-organization, but does not present schemata for a broad domain of application, considering sensorial systems.

In this paper we present the architectural and circuital features of an analog VLSI module to be used as a building block of adaptive neural systems in wide domains of application. The module is based on analog computational primitives that are complex in their functionality, but that can be implemented very efficiently in silicon respect to both power consumption and area occupation.

This work has been partially funded by "Progetto strategico Reti Neurali" CNR, Roma, Italy.

2. THE ANALOG ADAPTIVE NEURAL MODULE

The definition of a neural computational model involves the specification of: 1) the feed-forward algorithm; 2) the learning algorithm; 3) the types of I/O signals. These computations should be expressed by a network of massively interconnected neurons. This structural schema, that works well for biological networks, when transferred in silicon should be supplemented by various modules taking care of coding and decoding computation and memorization.

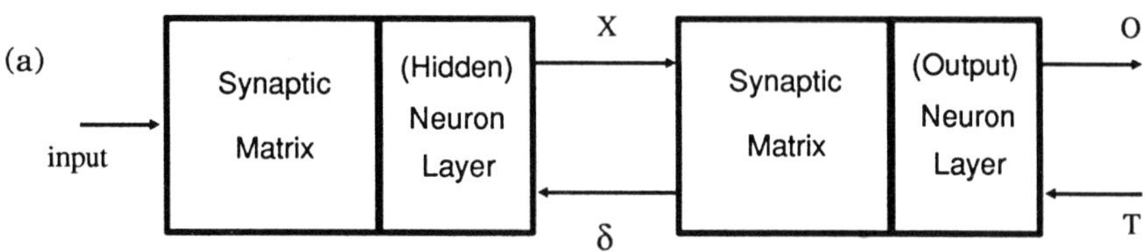

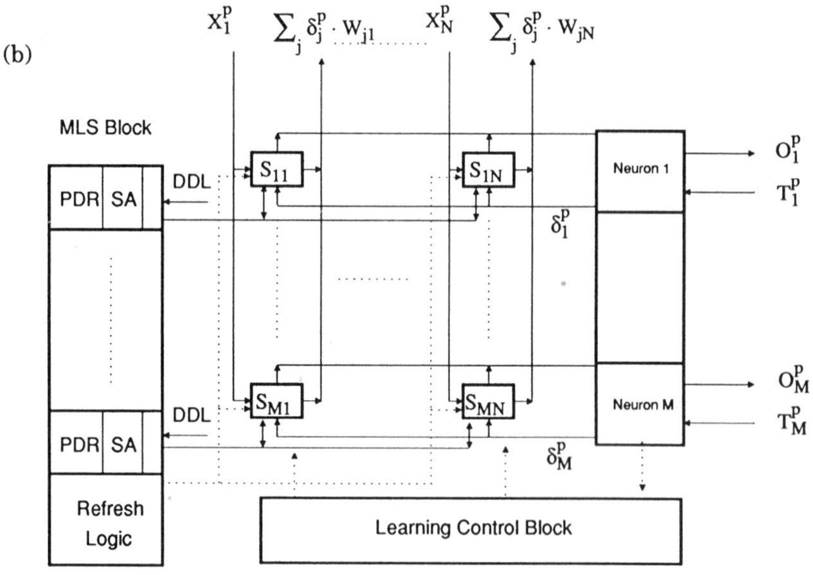

Figure 1. (a) Architecture of a Multi Layer Perceptron (2-Layer) trained with the Back Propagation algorithm.(b) Architecture of the single layer module (the case of the output layer is illustrated).

Basically the feed-forward phase needs multiplications and sums, the learning phase needs integration of weights (summation over time of the weight updates) and multiplications in the synapse module and computations of the error contribution in the neuron module (e.g. computation of the derivative of the neuron transfer function in the Back Propagation learning algorithm).

In a Multi Layer Perceptron (MLP) adaptive architecture each layer consists of an array of M neurons, each of which is connected through N adaptive weights to the input vectorX_i^p . The output of each neuron O_j^p is compared to a signal, which corresponds to some desired response T_j^p for that neuron, and an error signal is generated. The learning algorithm uses these errors to adjust synaptic weights (W_{ji} , $1 \le i \le N$,$1 \le j \le M$) so as to match neuron output and desired response in a (statistically) meaningful way [3]. In Fig. 1 the output layer of a MLP is illustrated. It is represented as a regular matrix of synapse cells $\mathbf{S}^{\mu}$ connected to an array of neuron cells in a crossbar arrangement.

The analog adaptive architecture is based on three main components:

- the synaptic cell S_{ji}^{μ} ;
- the neuron cell;
- the Long Term Memory circuitry distributed at each synaptic site, which stores the weight voltages.

Moreover:

- the *Learning Control Block* controls the learning (backward) phase, performed locally at each synaptic site.
- the Long Term Memory, which make use of a dynamic refresh technique, is controlled by the *Multi Level Storage (MLS) Block* (see Section 4).

A detailed diagram of a synapse is illustrated in Fig. 2a: block F performs the feed-forward computation, block B1 the error back propagation, block W_{ji} the dynamic weight storage, block B2 calculates the weight update following a single pattern presentation, block Σ calculates the weight update following the presentations of all exemplary patterns (*epoch*) and performs the weight update and W_{ji} stores the weight value. The block W_{ji} is accessible to/from the MLS Block.

The basic neuron module sums up the output synaptic currents ($I_j = \sum_i I_{ji} = \sum_i W_{ji} \cdot X_i$) and applies a quasi-linear transfer function with saturation to produce the direct output voltage O_j (normalized to the saturation value) and the inverting output voltage $1 - O_j$.

The error quantity to be propagated is:

$$\delta_j = (T_j - O_j) \cdot \frac{\partial O_j}{\partial I_j}$$

where the derivative $\frac{\partial O_j}{\partial I_j}$ can be approximated by $O_j \cdot (1 - O_j)$ if the transfer function is quasi linear with saturation (e.g. sigmoid).

(a)

(b)

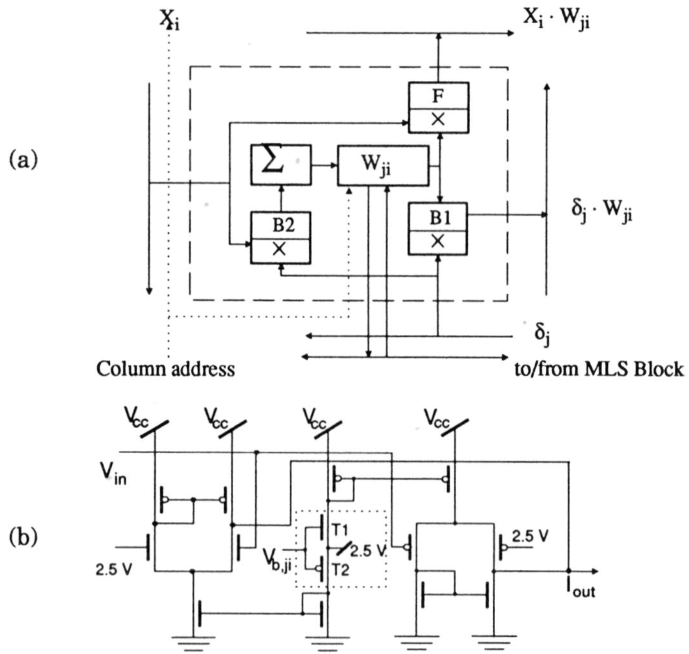

Figure 2. (a) Functional schema of the synapse S_{ji} of the MLP in Fig. 1. (b) Detailed circuit of blocks F and B1.

3. CIRCUIT IMPLEMENTATION OF SYNAPSES AND NEURONS

Synaptic weights are controlled through analog voltages as charge stored locally at each synaptic site on the capacitors constituted by the gates of MOS transistors to maintain the locality of computations.

The circuit for the synaptic multipliers F and B1 (see Fig. 2b) consists of a pair of complementary transconductance amplifiers, and a cell for the weight storage (transistors T1 and T2): V_b is the weight controlling voltage. V_{in} is the input voltage and I_{out} is the output current.

The neuron circuit is depicted in Fig. 3: it is an inverting amplifier in closed loop whose gain is controlled through the variable feedback resistor R. One can distinguish two main blocks: the gain stage and the differential output stage. The gain stage is constituted by two CMOS inverters.

The wide range four quadrant multiplier multiplies the neuron output voltages O_j and $(1 - O_j)$ and generates the output current $\hat{I}_j \approx O_j \cdot (1 - O_j)$. The current buffer uses $\hat{I}_j$ to bias the TA whose input voltages are T_j and O_j and whose output current is $I_j^{out} \approx (\hat{I}_j + I_z) \cdot (T_j - O_j)$. The transistor M_z generates

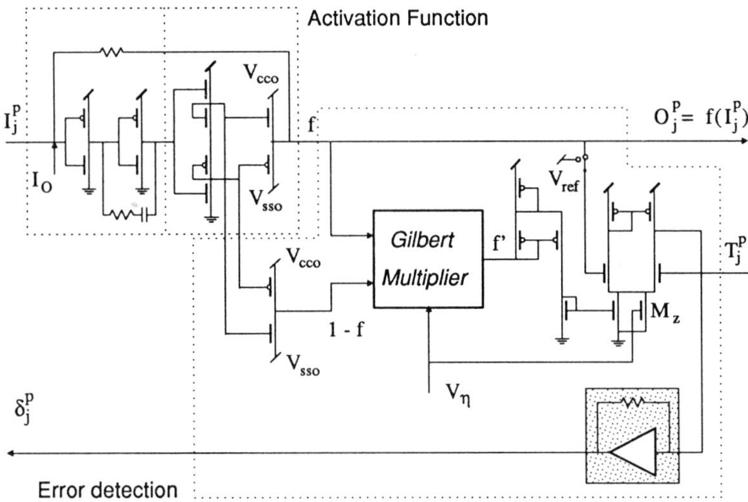

Figure 3. Circuit implementation of the neuron block.

the extra current I_z to set I_j^{out} greater than zero when the output O_j is opposite to the target and the derivative is close to zero. The wide range four quadrant multiplier and the TA can be controlled adapting the learning rate voltage V_η . The current-to-voltage converter generates the error voltage δ_j .

The neuron output stage operates as a class AB amplifier with a quiescent current lower than 700μA. I_O adjusts the neuron output voltage around V_{ref} compensating the offset. Since the ouputs of the neurons should be connected to the inputs of the synapses of the following layer they must fit into the operating range of the synaptic multiplier; consequently a low power supply should be adopted for the output stage of the neuron (e.g. V_{sso} = 2.4 V, V_{cco} = 2.6 V for V_{DD} = 5 V). In such a restricted range the differential signal can be easily obtained with the complementary output stage of Fig. 3.

4. LONG TERM MEMORY (LTM) IN SYNAPTIC CELLS.

One of the major problems of the analog implementation of neural networks, is the storage of synaptic weights. The various solutions that can be envisaged depend on the specific desired functionality and on the technology available [2, 4, 5, 6, 7].

Learning algorithms require the continuous updating of synaptic weights, while long term storage is required in the following feed-forward computation. Various authors have proposed a flexible solution, which uses the gate of a standard MOS transistor as charge storage capacitor, and controls the charge through a properly driven transfer gate [8, 9]. The maximum storage time depends on the required precision, on the affordable capacitance value and on the leakage current. This technique requires some kind of refresh circuitry. To this purpose, we adopt a Multi Level Storage (MLS) technique, which extend the concept of refreshing standard dynamic RAM's [10, 11, 12]. This solution implies that the synaptic weights can vary over a set of discrete values. We alreaday analyzed the effects of weight discretization on the functionality of the Back Propagation (BP) learning algorithm [13], showing that the coarser level of approximation still remains below the minimum requirements of the learning algorithm and the overall behaviour of the network can be maintained. In this Section we present the circuit implementation of the LTM.

The memory array is distribuited through the synaptic matrix (blocks S_{ji}, see Fig. 1). The Multi Level Storage (MLS) block manages (in a column-wise fashion) the refresh of the memory array, while neurons and synapses perform independently neural computations.

The functional schema of MLS Block is illustrated in Fig. 4. The MLS Block is organized in M slices (one for each row of the synaptic array). Each slice is composed by a Sense Amplifier (SA), a Dummy Data Line (DDL) and a Programmable Digital Register (PDR). The Refresh Logic contains a Comparison Block (CB), a Programmable Digital Counter (PDC) and a Digital-to-Analog Converter (DAC). The PDC cyclically addresses the memory cells in column wise fashion. The DAC writes a discrete value into the cell W_{ij} through the Write Data Line (WDL); the SA reads the analog value of W_{ij} from the Read Data Line (RDL).

In the READ phase of the refresh operation, the RDL is charged to the voltage value stored in the addressed memory cell; then the PDC starts counting and a staircase waveform from the DAC is applied to the DDL. When the voltage value of the DDL equals the voltage value of the RDL, the SA enables the storing of the PDC output data into the PDR; this data is the digital code of the analog value stored in the memory cell.

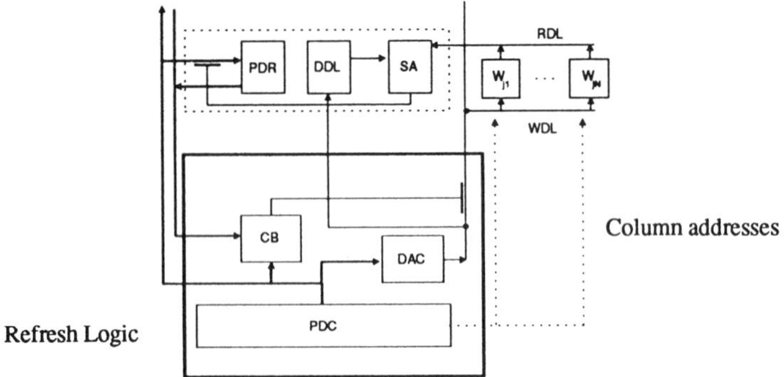

Figure 4. Functional schema of the Multi Level Storage (MLS) Block (see Fig. 1) and of the jth row of the synaptic array.

In the WRITE phase of the refresh operation, the digital data in the register PDR has to be transferred into the memory cell. The PDC starts counting; when the count reaches the value coded in PDR, the CB enables the DAC, that, through WDL, drives the addressed memory cell at the corresponding staircase voltage level.

In this way, each memory cell stores, in a dynamic way on a capacitor, a set of discrete voltage values {Vi}, $2 \le i \le M$. The number M of discrete voltage values is user programmable. For a specified voltage range the maximum value of M depends on the height of the minimum detectable voltage step (ΔV_{min}). The voltage step is determined by the voltage gain and the precision of the SA; using a standard CMOS process, a 15mV voltage step could be the lower limit, corresponding to 256 storable levels. Pratically a 50 mV step (corresponding to about 80 storable levels) derives from our present simulations. Since the adopted synaptic multipliers operate during most of the time in subthreshold to lower power consumption, the two voltage ranges (for negative and positive values) where weights are meaningful, are about 1.0 V wide each. For each range, 20 discrete voltage values (ΔV = 50 mV) are sufficent not to affect the functionality of the Back propagation algorithm as it has been demonstrated in [13]. The circuit solution proposed here meets this limit.

The SA is constituted by two complementary input stages (transconductance amplifiers), and by an output stage. Input stages are active alternatively during RDP and RDN periods. The SA detects voltage values in a lower range [F_n ÷ 2.5 V], $F_n \approx 0.8$ V and in an upper range [2.5 V ÷ (5 V - F_p)], $F_p \approx 0.6$ V.

Adopting the synaptic multipliers of Fig. 2b, positive weights correspond to voltage values in the range [F_n ÷ 2.5 V], while negative values correspond to voltage values in the range[2.5 V ÷ (5 V - F_p)]. The synaptic weight is 0 when the corresponding controlling voltage is 2.5V.

The number N of bits in the PDR satisfies the constraint: $N \ge \ln_2 \frac{M}{2} + 1$.

The basic circuit diagram of a memory cell is illustrated in Fig. 5 and a typical timing diagram is illustrated in Fig. 6: a seven levels (2-bit plus the sign) operation timing is shown for simplicity.

The READ mode operation is performed during two time periods: for detecting voltage values in the range [F_n ÷ 2.5 V] (period RDN) and in the range [2.5 V ÷ (5 V - F_p)] (period RDP), respectively. During RDN, the DDL is discharged to GND and the RDL is precharged at 2.5V (signal P_{nj}). Afterwards the cell output voltage is put on the RDL (signals R_{nj}) and the DAC output is applied to line RM (signal D_{nj}). During the second period (RDP), the DDL is pre-charged at 2.5V and the RDL is pre-charged at 5V (signal P_{pj}). Then the cell voltage is put on the RDL (signal R_{pj}) and the DAC output is applied to signal RM (signal D_{pj}).

Due to data lines (RDL and DDL) pre/dis-charge periods, the DDL voltage value is always increasing. When the SA input voltage values are equal, the stored value is recognized and the SA enables the storing of the output of the counter into the digital register.

In the WRITE mode operation the output of the DAC is put on the WDL and the counter starts the count. When the output of the counter equals the value stored in the digital register, signals T_{ij}^{n} and T_{ij}^{p} are valid and the actual output voltage value of the DAC is stored in the cell.

Each READ operation is always followed by a WRITE operation (REFRESH cycle).

The READ operation is not destructive because the cell storage capacitance does not drive directly the RDL; the discrete voltage value is thus always valid.

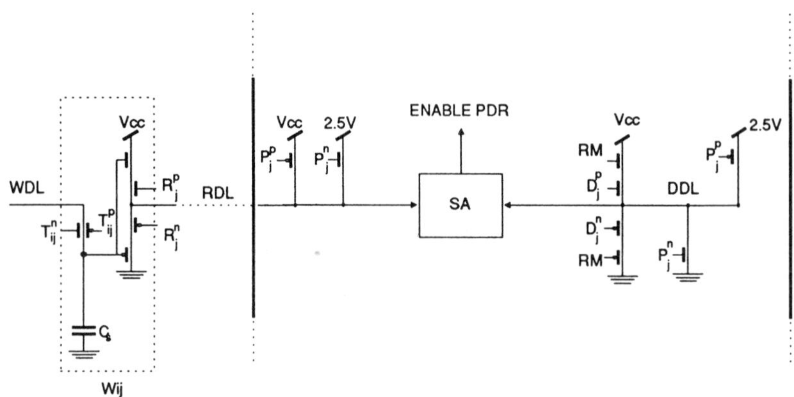

Figure 5. Basic circuit diagram of a memory cell (W_{ij}) and of the connections with the Read Data Line (RDL), the Write Data Line (WDL) and the Dummy Data Line (DDL). The output of the SA enables the writing of the Programmable Digital Register (PDR). C_S represents the gate capacitance of the two weight transistors (T1 and T2, Fig. 2b). The timing of the signals is detailed in Fig. 6.

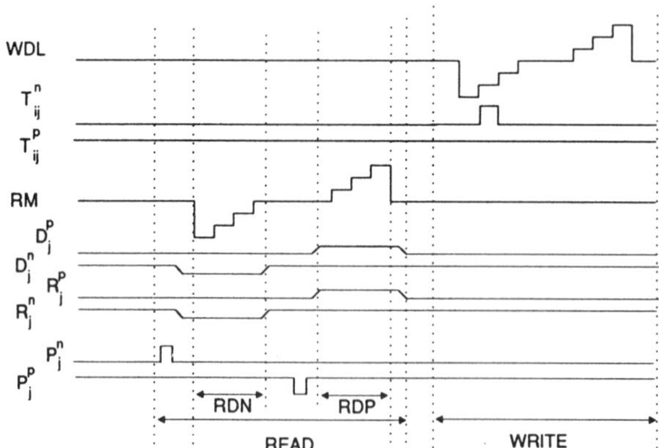

Figure 6. Timing diagram for read and write operations for signals in Fig. 5 in the case of seven levels. During RDN and RDP the voltage values in the lower range [F_n ÷ 2.5 V] and in the range [2.5V ÷ (5 V - F_p)] are red respectively. In the example shown , during writing cycle the second negative level is written into the cell.

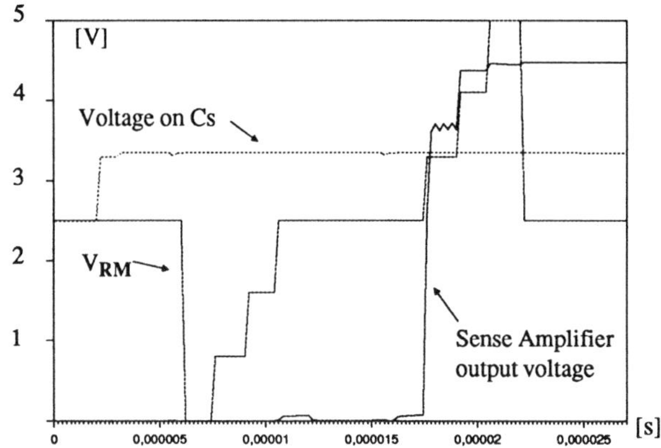

Figure 7. Simulation results for the seven levels case. The sixth voltage value is stored on C_S at time t=2.5 μs. At time t = 5.0 μs the read operation starts. When voltage V_{RM} reaches the sixth level (time t= 17.5 μs), the Sense Amplifier recognizes the stored value.

In Fig. 7 the simulation results in the case of seven levels are illustrated.

During the refresh period the voltage decay should be less than 50% of the minimum voltage step ($\Delta V_{min} = 15\ mV$). From [11], for a standard 3 μm CMOS process, the estimated refresh period, for a storage capacitance of 100 fF is about 0.3 s, at room temperature, and $0.5 \cdot 10^{-3}$ s, at 90 ºC.

The area of the MLS circuitry with respect to the rest of the synapse is less than 10%.

5. CONCLUSIONS

In this paper we presented an analog adaptive VLSI module for neural systems. The module is equipped with Long Term Memory capabilities for weight voltage refreshing through a Multi Level Storage dynamic memory. Though we focussed our attention on the MLP trained with the Back Propagation (BP) learning algorithm, our approach can be easily applied to neural architectures which exhibit regular synaptic arrays like the Kohonen's self-organizing maps [15].

A learning iteration consists in the presentation to the architecture of a set of exemplary patterns and the corresponding targets, and in one update for all the weights; its duration is usually much smaller than the refresh period. Consequently, during learning, MLS refresh can be suppressed.

During forward computation (e.g. classification) MLS refresh is activated and weights assume discrete values: weights precision decreases but it still remains within the constraints of calculations [14]. Furthermore, we showed [13] that, also using a discretization technique, the weight precision still satisfies the requirements of the learning algorithm; if necessary, MLS refresh may be interleaved with learning iterations with marginal effects on the learning performances.

A neural chip implementing the proposed analog architecture is currently being fabricated.

6. REFERENCES

1 Mead C.A., "Analog VLSI and Neural Systems", Addison-Wesley, 1990.
2 Mead C.A., "Neuromorphic Electronic Systems", Proc. IEEE, Vol. 78, No. 10, Oct. 1990, pp. 1629-1636.
3 Ruhmelhart D.E. and McClelland J.L. (Eds.), "Parallel Distribuited Processing", MIT Press, Cambridge, Mass., 1986
4 Vittoz E.A., "Analog VLSI Implementation of Neural Networks", Proc. Int. Symposium on Circuits and Systems, New Orleans, 1990, pp. 2524-2527.
5 Schwartz D.B., et al., "A programmable Analog Neural Network Chip", IEEE Journal of Solid-State Circuits, Vol. 24, No. 2, Apr. 1989, pp. 313-319.
6 Pasero E., "Floating Gates as Adaptive Weights for Artificial Neural Networks", Proc. of IFIP Workshop on Silicon Architectures for Neural Nets, St. Paul de Vance (F), Nov. 1991.
7 Boonyanit K., et. al., "Neural Network Circuit Implementation with Analog Weight Storage via EPROM Technologies", Tech. Rep., EE 478, Stanford University.
8 Tzividis Y, and Satyanarayana, "Analogue Circuits for Variable-Synapse Electronic Neural Networks" Electr. Letts. Nov 1987, Vol. 23, no. 24, pp. 1313-1314.
9 Hochet B., "Multivalued MOS Memory for Variable-Synapse Neural Networks", Electronics Letters, 11th May 1989, Vol. 25, No. 10, pp. 669-670.
10 Rich D.A., "A Survey of Multivalued Memories", IEEE Transactions on Computers, Vol. C-35, No. 2, Feb. 1986, pp.99-106.
11 Masakazu A. et al., "A 16-Level/Cell Dynamic Memory", IEEE Journal of Solid-State Circuits, Vol. SC-22, No. 2, Apr. 1987, pp. 297-299.
12 Masashi H., "An Experimental Large-Capacity Semiconductor File Memory Using 16-Levels/Cell Storage",IEEE Journal of Solid-State Circuits, Vol. SC-23, No. 1, Feb. 1988, pp. 27-32.
13 Caviglia D.D., Valle M., and Bisio G.M., "Effects of Weight Discretization on the Back Propagation Learning Method: Algorithm Design and Hardware Realization", Proc. Int. Conference on Neural Networks, San Diego, 1990, pp. II631-II637.
14 Hollis P.W. et al., "The effects of Precision Constraints in a Back-Propagation Learning Network", Proc. 1989 International Joint Conference on Neural Networks, pp. II-625.
15 Kohonen T., "Self Organization and Associative Memory", Springer-Verlag, Berlin, 1984.

VLSI 91
A. Halaas and P.B. Denyer (Eds.)
Elsevier Science Publishers B.V. (North-Holland)

Has CAD for VLSI Reached a Dead End?

A. Richard Newton

Department of Electrical Engineering and Computer Sciences
University of California at Berkeley
Berkeley, CA. U.S.A.

Abstract

While the microelectronics industry has grown enormously over the past forty years, the design technology industry — representing all aspects of the application of computers to design, including CAD — continues its boom-to-bust cycles, with innovation seemingly driven only by new technologies or by changing end-user market requirements Not only are fundamental algorithms and techniques seemingly "reinvented" by the research and development communities in each cycle, but these ideas are often embodied in EDA products that only differ from the previous generation products in terms the color of the wrapper. The answer to the question I pose is no, CAD for VLSI has not reached a dead end yet. However, we cannot be left behind, concentrating our efforts on refining solutions to traditional CAD problems at the expense of important, new problems in the area of design technology. We must step back and take a broader look at the problems designers face today and reset our research priorities and investments. Perhaps more important, researchers must work together with industry to provide a robust technological and industrial foundation for design technology, one we can count on to be there through the ups and downs of the commercial markets.

1. INTRODUCTION

It's been almost forty years since the first computer aids were used to help design electronic systems. In that time, we've seen the complexity of ICs grow more than six orders of magnitude and we've seen major IC companies grow successfully in conditions of fierce international competition. We've seen considerable innovation in the area of CAD for integrated circuits, reported in part at this meeting in years past. However, it seems that both our research thrusts and the industrial developments that flow from them have a cyclic nature, pushed largely by new technological developments or, more recently, pulled by user and market requirements. What seems to be missing in the design technology area is a lasting infrastructure. This lack of infrastructure cuts across all aspects of our profession, including teaching and research, our professional societies, both captive and commercial EDA developments, and our symbiotic relationships with the semiconductor industry.

I use the term Design Technology here to refer to all software and infrastructural support aspects of the design of electronic systems. This includes tools for verification and synthesis, from behavior to layout, and all aspects of the integration of such tools with other technologies. Software for storing and managing design programs as well as design data, user interface technology, special-purpose hardware accelerators and interfaces, and design decision support systems (e.g., ASIC gate and megacell library support) are all key elements of an overall solution to todays VLSI design problems. All of these comprise the components of the design technology industry.

From an industrial point of view, why is it that no electronic design technology company has been successful for more than a decade? Why is it that many of them seem to grow rapidly for a few years and then just seem to implode? With the industry track record of the past fifteen years, it's no wonder customers don't really trust their design technology vendors today.

I believe that unless some fundamental infrastructural changes are made in both our research and development practices as well as in industry very soon, we will to repeat the boom-to-bust cycle once again, disappointing customers and the financial community, as well as ourselves.

The use of computers in IC design has evolved significantly from the simple drafting tools of the 1970s and the rudimentary CAE systems of the early 1980s. Still, the electronics industry has yet to come to grips with the fact that today's ASIC industry is driven much more by design technology than by IC processes and production. The 1970's saw the emergence of the first major wave of commercial design technology for electronics systems. These were mainly "point tools;" each separate program would read its input from a file or from the user, process its data and produce a well-defined output file for printing or plotting. Companies like Calma, Applicon, NCA and Phoenix Data Systems emerged to lead the industry in custom IC design. Programs such as SPICE and Tegas defined the industry *de facto* standards for netlist information at the circuit and logic levels, as did Calma Stream format for layout.

The second wave was designated Computer-Aided Engineering (CAE) and from a software point of view it was characterized by an emphasis on tool integration. In the early 1980's, companies such as Daisy Systems, Mentor Graphics and Valid Logic Systems delivered multiple, integrated tools, emphasizing schematic capture coupled to simulation and standard component libraries. Dissatisfied with the *ad hoc* and proprietary formats used for moving data between tools and CAE systems, users and CAE vendors organized to develop integration-oriented standards like the Electronic Design Interchange Format (EDIF) and, more recently, the CAD Framework Initiative (CFI).

In each of these waves, companies grew rapidly to become household names — the promise for the future. Even more rapidly, many of them fell from favor, some disappearing without trace but most leaving some unfortunate remnant of a support headache for their customers.

Many people believe we are entering the third wave of design technology today, characterized by design-task-oriented user interfaces, higher-level design entry and automatic synthesis for some components, and a strong coupling to manufacturing by way of libraries and design decision support systems. Integration with mechanical design, improved support for network-based design teams, and practical support for software/firmware co-design may also be prominent features of this next wave.

What changes have taken place in our industry that will allow the present leaders to make the necessary investments and develop the relationships needed to remain strong? Not very much that I can see. Certainly, once a customer chooses a design technology vendor and makes a large investment in design systems, it is difficult to replace those systems quickly. Certainly the investments are larger and the technology is used more broadly in the design process than a decade ago. When Sun Microsystems was selling a billion dollars worth of workstations per year, Digital equipment was still taking in almost as much in annual PDP-11 sales. Does that mean the PDP-11 is here to stay? No design technology company has developed customer relationships so strong that they cannot be displaced over time.

2. INDUSTRIAL OPPORTUNITIES

Like similar software industries, the design technology industry can be viewed as comprising five major segments: Component Suppliers, Solution Suppliers, Sales Channel, Consulting Services, and Research.

The Component Suppliers provide technology that does not stand alone. Examples of components include CAD tools, like simulators, synthesis tools, or a layout-rule checker, as well as simulation libraries, ASIC libraries, or an engineering design data management system.

The Solution Supply business involves selecting and packaging components into a design system that addresses a specific task, as seem from the user's perspective. For example, an ASIC design system today might include schematic capture, a simulator, synthesis, one or more libraries, and the framework technology needed to integrate all the components seamlessly. The ASIC design solution would be packaged with ease of use and design efficiency as high customer priorities.

In our industry, a company cannot become a truly large only selling components. Most component companies are doing well with $20M-$40M annual sales and the most successful ones can grow to perhaps $100M-$200M revenues at most. A company selling solutions, which adds its own value to the components it sells through integration, can grow somewhat larger provided it can identify and follow (even encourage, sometimes!) major trends in technology and in electronic system design methodology.

As the industry grows, design technology has become more complex, the customers' ability to make an informed decision has become both increasingly difficult and expensive, and so another industry segment is just beginning to become significant — Consulting Services. Specialized consulting organizations work with large corporations to help plan corporate-wide design-technology investments, including the choice of appropriate products, standards, and services. This is particularly important when the customer's needs cannot be satisfied by a single design technology vendor, which is usually the case today. These consulting groups may also play a follow-on role to help implement and maintain the design technology, much as the role played by the EDS company in the MIS area, or may even provide consulting in the design of electronic circuits for large system houses as well. While such activities are not broadly successful today, they represent a major opportunity in coming years.

As a design technology company becomes successful, either supplying components or solutions, the value of its installed base — its customers and their allegiance — soon become its major asset. At this point, the company begins to play the role of a Sales Channel and satisfying the many varied requests of its customers; delivering the best technology it can find through the channel; becomes its primary objective. By adding continuously to its technology portfolio and by working closely with customers, such companies have the chance to become significantly larger and more stable than component or solution companies alone. To date, it is this transition — from component or solution supplier to becoming a customer-oriented channel — that has been the main reason for the lack of stability in our industry. I believe there are three major areas where infrastructural changes are needed before companies can make that transition successfully: integration standards, customer relations, and professional organizations and activities. I deal with each of these in more detail below.

Finally, changes are needed in our approach to teaching and research in design technology, including a more rigorous approach to the evaluation and reporting of research, and the development of core teaching curricula that emphasize fundamental techniques, illustrated by their application in specific problem domains.

3. INFRASTRUCTURAL REQUIREMENTS

In the transition from component or solution supplier to sales channel, a design technology company must be able to integrate and support new components or solutions obtained from third parties. Because of the lack of component integration standards, system suppliers find they must often develop and integrate their own components, rather than being able to act as a value-added re-seller for third-party products. Alternately, they feel compelled to buy the component supplier outright so that they can integrate at the program source level — usually a very expensive undertaking.

In the personal computer industry, the suppliers of the principal integration standards (e.g. DOS, Windows, Macintosh toolkits) also produce some components (e.g. Microsoft Word, Excel, MacDraw). But unlike the design technology industry, Microsoft and Apple do not provide special access and hooks for their own developers that are not available to their competitors in the component business. They understand the importance to the health of the entire industry of providing an open, practical integration standard. The standard becomes a base that all players in the industry, including the financial community, can rely on.

The CAD Framework Initiative (CFI) is designing a variety of integration standards for electronic design technology. Over forty design technology vendors and customers are members of the CFI and most of them are represented on one of the eight technical subcommittees. However, without a significantly larger commitment from design technology companies and especially design technology users, I do not believe the CFI will be able to deliver on its promise. History has shown us repeatedly that practical integration standards cannot be developed by a committee, or even by a small team of developers isolated from real users. The committee's job is to ratify the standard, not to develop it. It is up to the major design technology suppliers, working with each other and working closely with their major customers, to develop a set of integration interfaces. Once they are satisfied that the technology is practical, they should present their prototype solution for ratification by a standards body like the CFI.

Not all areas of integration technology are ready for standardization today. In some areas, like design flow management, there has been much University research activity but the industry has not had sufficient practical experience with these ideas to be sure they will meet users' needs. On the other hand, we have a great deal of accumulated experience in areas like storage management and design representation. In these areas, I have no doubt that it is possible to develop practical core standards today. Even simple standards in these areas will permit many third-party developers, including universities, to supply new programs and ideas to design technology users for evaluation. Universities as well as companies will be able to count on such standards and should be willing to develop their own instructional and research programs around them.

Why is it that design technology companies, with a few notable exceptions, have not invested heavily in infrastructural technology to date? A major factor has been the sensitivity of the financial markets to such long-term investments and the impact that they might have on the short-term performance of the company. As Ivan Sutherland stated in a recent presentation, it seems that investors in US public corporations "rent" companies today rather than buy them. For example, every outstanding share of both Cadence and Mentor has changed hands in the last eight months. These fickle investors penalize companies heavily for even the slightest changes in their short-term financial performance. Under such conditions it is no wonder that the management of public design technology companies feels obligated to organize for short-term profits!

It would be unfair to imply that all the infrastructural problems in the design technology industry are caused by the financial community or the design technology companies themselves. A significant portion of the problem lies in the hands of the consumers of design technology; the IC and system designers and manufacturers alike. Since the delivery of the first computer programs for design, the term "CAD" has stuck in the throat of most users. Promises the developers could not keep and user expectations that could never be met have left users rightly cynical about new claims and promises in the design technology area.

But while users cry out for standards and a design technology infrastructure on which they can depend, most large corporations have not been willing to invest the significant resources needed to make it happen. Most customers continue to play one vendor against another and provide only minimal participation in standards activities. Design technology is driven by its users; it requires ongoing and active leadership by major IC and electronic system developers. Corporations that invest tens or even hundreds of millions of dollars annually in design technology must be willing to invest significantly in the development of industrial standards and strong partnerships with their design technology suppliers. They must develop long-term, cooperative relationships. If the leading design technology vendors of today fail, as leaders have in the past, users must shoulder as much of the blame as any other group.

Finally, there are three other forums that could play a more significant role in stabilizing and developing our profession: professional societies, trade organizations, and national and international conferences and trade shows.

It is certainly a responsibility of our professional societies to represent the interests of the design technology profession as a whole. Unfortunately, design technology has evolved as a component of a variety of organizations and activities rather than as a single body that can represent issues common to all design technology professionals. This fact only accentuates the "second cousin" attitude users often take to their design technology colleagues. Design technologists play important roles in the IEEE Computer, Circuits and Systems, Electron Devices and Microwave Theory and Techniques Societies, the IEEE Solid-State Circuits Council, the ACM, and IFIP, just to mention a few.

Unfortunately, in most of these groups the design automation aspect has evolved as a separate activity, often secondary to the principal role of the organization and its management. To be effective in the future, the design technology profession must organize its own professional organization, involving both developers as well as users. This organization must work closely with all related professional organizations, including those listed above. The many conferences, workshops, and trade shows held world wide should also be consolidated under this professional organization.

From an industrial point of view, the recent formation of the EDAC trade organization, whose membership includes the majority of design technology companies, is an important step in providing a common ground for a variety of industry-wide infrastructural issues, including standards. However, EDAC must find a way to involve users more actively in its decision-making process if it is to achieve its goals.

In all of these areas, it is critical that the users, the developers, and the providers of design technology play an integrated role. Without cooperation and balance in all areas, we will continue to promote a divisive relationship and the industry will never be able to achieve its full potential.

4. SUMMARY

Today, design technology represents the critical link between electronic system designers and the manufacturers of electronic components. Both of these groups must recognize the true value of design technology and the importance of a sound infrastructure in stabilizing the industry. The design technology companies themselves must be willing to make the investments in infrastructure needed for the future, rather than concentrating primarily on products and services that deliver short-term revenue advantages. The financial community must learn to reward companies for making these investments, rather than penalizing them in the marketplace. Researchers must take a more critical look at their contributions and should build teaching curricula around a generally-accepted minimal set of fundamental techniques and concepts. Finally, the design technology profession must accelerate the development of its own professional infrastructure, separate from and in cooperation with those professional organizations that represent its primary customers. If we don't achieve at least some of these goals soon, we'll miss the critical turn and run head-on into the dead end that is almost upon us.

VLSI 91
A. Halaas and P.B. Denyer (Eds.)
Elsevier Science Publishers B.V. (North-Holland)

Partitioning-Based Allocation of Dedicated Data-Paths in the Architectural Synthesis for High Throughput Applications*

Werner Geurts[1] Stefaan Note[1]†Francky Catthoor[1,2] Hugo De Man[1,2]

[1]IMEC Laboratory, Kapeldreef 75, B-3001 Leuven, Belgium, Tel:+32/16/281201
[2]ESAT Laboratory, Katholieke Universiteit, K. Mercierlaan 94, B-3001 Leuven, Belgium

Abstract

In this paper we will present a new approach towards the allocation of data-paths in the high-level synthesis for high throughput signal processing applications. This domain contains applications such as image and video processing, front-end audio applications and high sample rate linear filters. The computational bottleneck that is encountered in these applications is tackled by the use of data-paths whose composition is dedicated to the application at hand. We will give a short overview of the synthesis script that is used for this application domain. The main contribution of this paper is a technique which minimises the area of the allocated data-paths for a given throughput requirement. The technique is initially developed for applications that are characterised by flat (non hierarchical) data-flow graphs. In addition an extension is presented which enables the optimal allocation for hierarchical graphs and which performs this optimisation across the loop boundaries.

1 Introduction

High-level synthesis [1] is the computer aided transformation of the behavioural description of an algorithm into an equivalent register transfer description. In this paper we will concentrate on high-level synthesis systems which are particularly suited for some classes of high throughput applications. The methodology that we present will be restricted to the real time signal processing domain. Examples of applications in the target domain are medium level and front-end image and video processing modules, front-end audio and linear filters with a high sample rate. This application domain features two important characteristics. The first characteristic is the presence of multidimensional signals of a substantial size, which will have to be stored in bulk memories or in medium size distributed memories. The optimisation of the memory size and the number of access ports is an important task in the compiler script. The second characteristic is that the number of operations that have to be performed within a time frame (which is typically a few hundred MOPS), is much larger than the achievable clock frequency (e.g. 20 MHz). This will require the allocation of a large number of resources. The cost of the interconnections that are required to let these resources work in parallel will become an important optimisation criterion.

In this paper we will use the term *maximum hardware sharing factor* $HSF = T_{EVAL}/T_{CLOCK}$. T_{EVAL} is the amount of time - specified by the user - that is available for the evaluation of one instance of the algorithm (for DSP applications this is usually the sample frequency). The HSF

*This research was sponsored by the ESPRIT Basic Research Action 3281 (ASCIS) of the E.C.
†Currently with Philips, International Technology Centre, Leuven, Belgium.

	HSF	adders	mults	mux	regs
Park [2,5]	3	5	3	23	42
G.E. [3]	2	8	4	20	18
CATH-III [11]	2	8	4	3	4

Table 1: Comparison between the result of the traditional approach and the CATHEDRAL-III approach.

expresses how many times a resource can be re-used for the evaluation of one instance of the algorithm. It corresponds to the potential degree of multiplexing or sharing of any hardware resource. For high throughput applications this number will be low (typically in the range 1...20).

The traditional approach towards synthesis for high throughput applications is to perform pipelined scheduling - which minimises the number of FUs [2,3,4], followed by a binding step [3,5] - which aims to minimise the register and interconnect cost. The result of this "scheduling first" approach is that a large number of operations have been distributed over a large number of operators and it has become very difficult to route the intermediate variables from their source operator to the destinations. This is reflected in the large number of multiplexers and busses that are required. Table 1 illustrates this at the hand of the results that where published for the 16 tap symmetrical FIR filter benchmark [2] are shown. The irregularity introduced by the scheduling also results in an unacceptable growth in the size of the control unit. In section 2 we will introduce a synthesis methodology that preserves the inherent regularity of the algorithm. The result for the FIR filter that is obtained by this methodology is also shown in table 1.

The synthesis paradigm that will be used throughout this paper is the minimisation of the total area consumed by all resources - i.e. memories, operators and interconnect - under a user specified throughput constraint (e.g. the sample rate). The methods explained in this paper are part of the CATHEDRAL-III synthesis system that is currently under development. The synthesis script of CATHEDRAL-III is presented in section 2. In the remainder of the paper we will present a solution for the data-path allocation task as it is handled in CATHEDRAL-III .

2 The Synthesis Script

The architectural model that is used for CATHEDRAL-III was presented in [6]. In this model all operations are performed on dedicated data-paths whose structure strongly reflects the structure of the data-flow graph. In the remainder of this text these data-paths will be referred to as ASUs or *application specific units* [7]. The fixed grouping of operators strongly reduces the required inter-ASU interconnections. The intra-ASU interconnections are fixed and hence no multiplexers are required. Data routing inside an ASU can in most cases be done by making use of certain properties of the operators (e.g. an adder can be placed in pass mode if one of the inputs is set to zero). The critical delay of an ASU is dominated by the delay of its longest ripple path and is therefore not much larger than the critical delay of a single FU. The validity of this architectural style has been proven in a number of manual designs of real life applications (see e.g. [6]).

The starting point for the synthesis environment is an applicative (parallel) description of the signal flow graph in the SILAGE language [8]. This description can contain a complex combination of nested iterations and conditions. A precise specification of the word lengths of each variable is also included. Furthermore the required throughput is specified in terms of T_{EVAL} and an initial estimate is made for the clock frequency f_{CLOCK_ESTIM}. For some applications the clock frequency can be used as an optimisation parameter, for others it is dictated by the system of which the chip will be a part. The sample frequency is a characteristic of the particular application.

Before entering the synthesis script, the SILAGE description is translated into an internal data model called the DSFG [9] or *decorated signal flow graph*. During the synthesis process this data model is gradually annotated with the synthesis information. At the same time the structural information is stored in the *architecture net-list* or ANL.

The first synthesis task is the high-level memory management (see figure 1). The overall control flow structure is transformed in such a way that the area needed to store multidimensional variables is minimised. A decision is made on the number size and type of background[1] memories and on the number of ports for each memory. Also the assignment of multidimensional signals to specific instances of background memories is determined.

After the memory management there is still some freedom left to modify the DSFG. Typically this can incorporate those control flow transformations on the loop structure which do not violate the constraints imposed by memory management, optimisation of arithmetic expressions and data-flow transformations such as the expansion of expensive operations into a number of less expensive operations (e.g. expansion of constant multiplication into adds and shifts, selection of multiple precision operations for long words, etc...). This is a crucial step which has to be considered in conjunction with the data-path mapping task following it. For each of the operations in the optimised DSFG a list of possible operators or ABBs[2] are selected from the synthesis library. E.g. a compare operation can be performed on a comparator or on a subtractor or even on an adder/subtractor. The postponement of a unique selection provides freedom for subsequent optimisation tasks.

In the data-path mapping part it is the objective to define a set of ASUs with minimal cost (in terms of the total ASU area) dedicated to the application at hand. This is done in two steps. First the DSFG is partitioned into a number of clusters (step 2.2) of tightly interconnected operations, e.g. the taps of an FIR filter or an adapter section in a recursive filter. The complete set of clusters is then partitioned into a number of disjoint subsets, each of which will correspond to an ASU. In the second step (step 2.3) these ASUs are constructed using the constructive matching algorithm of [7]. For this purpose the number of clusters in each subset should not exceed the HSF. At this point each operation is uniquely mapped onto a specific operator or ABB. It is now possible to optimise the hardware in the data-path with respect to timing. At this point an HBB[3] is selected for each ABB and pipeline registers can be inserted in the data-path in order to meet the estimated clock frequency that was put forward.

At this point the clusters can be represented by a *superoperation* for which there exists a binding to one or more ASUs and which is characterised by a latency (in number of clock cycles) due to the internal pipelining of its ASU. The scheduling of the superoperations is done by the ATOMICS tool [10] which constructs an urgency based list schedule and also performs loop folding.

A number of design feedback loops are provided in order to: (i) optimise the ASU-partitioning

[1] A distinction is made between background or bulk memories and foreground memories such as registers. Background memories (e.g. RAMs which can either be on or off chip) are usually used to store large multidimensional signals and scalars which have a long lifetime. Foreground memories are used to store short lived intermediate variables.

[2] ABB = Abstract Building Block. This is a primitive operator in the synthesis library for which only the functionality is specified, e.g. an adder or an incrementer.

[3] HBB = Hardware Building Block. An HBB corresponds to a possible hardware implementation of an ABB, e.g. the adder ABB can be implemented as a ripple adder or as a carry select adder.

in conjunction with the control/data flow transformations, (ii) refine the initial estimate of the HSF based on the clock frequency and (iii) redistribute the complete cycle budget over the different loops in the algorithm.

The low level mapping incorporates detailed memory and data-path organisation. A detailed discussion of this synthesis scipt can be found in [11].

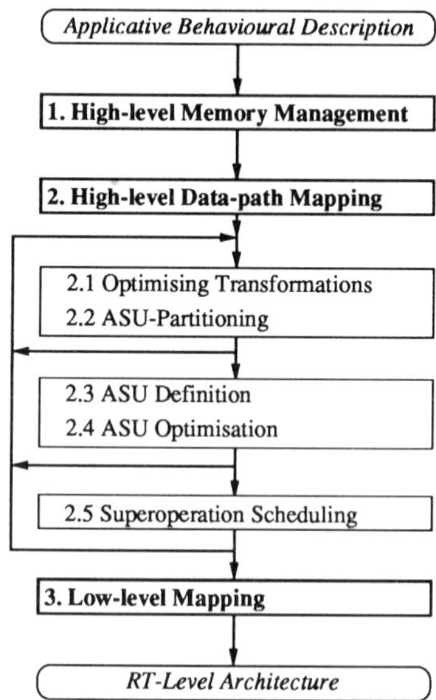

Figure 1: The script of the CATHEDRAL-III synthesis system.

3 The ASU-Partitioning Task

Definition 1 (*The ASU-Partitioning Problem*) *Given a* DSFG, *partition this graph in a number of clusters of highly interconnected operations in such a way that the total cost (area) of resulting ASUs that can execute these clusters within the given time constraint is minimal.*

We will proceed in solving this problem in two steps:

- **Step 1:** [Operation Clustering] In this step the most suitable clusters are identified in the DSFG.
- **Step 2:** [ASU assignment] A number of ASUs are defined in such a way that each cluster can be executed by exactly one ASU.

3.1 Operation Clustering

Definition 2 (*Cluster*) *A cluster C of operations - represented by the graph $G_C(V_C, E_C)$ - is a connected subgraph of the* DSFG *($G_{DSFG}(V, E)$). The vertices $v \in V$ represent arithmetic, rela-*

tional and logical operations. The edges $u \rightarrow v \in E$ represent scalar signals and are characterised by a signal type.

The goal for the clustering task is the identification of those clusters which can be efficiently mapped on a number of ASUs - i.e. those clusters which have a high degree of compatibility (see section 3.2.1). The main constraint is that the critical delay of a cluster may not exceed the proposed clock period (a cluster corresponds to a pipeline section). At this moment we are studying heuristic approaches as well as statistical global optimisation techniques to solve this problem. In this paper we will assume that the clustering step has been performed already.

Example: In the remainder of this paper we will use the example of figure 2. The DSFG and its clusters are shown in figure 3.

```
#define W8  fix<8,0>   /*  8 bit fixed point type */
#define W12 fix<12,0>  /* 12 bit fixed point type */
func main(A, B : W8[16]) X : W12[16], Z : bool[16] =
  begin
    S[0] = W8(0);    D[0] = W12(0);
    (i:0..15):: begin
        S[i+1] = S[i] + (A[i] + B[i])/16;
        D[i+1] = D[i] + (A[i] - B[i]);
      end;
    (j:0..15):: begin
        X[j] = abs(A[j] - B[j]) - D[16];
        Z[j] = (A[j] + B[j]) > S[16];
      end;
  end;
```

Figure 2: SILAGE code for the example.

3.2 ASU Assignment

At this point we have partitioned the DSFG into a number of clusters of tightly connected operations which now have to be assigned to an ASU. The main objective will be to assign

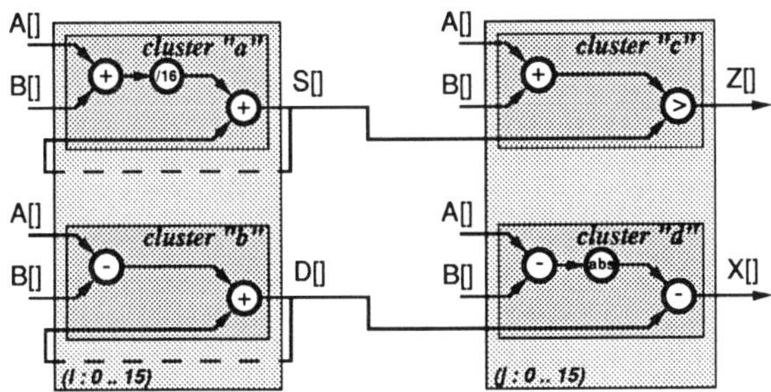

Figure 3: Data flow graph for the example. The four clusters - a and b for the i-loop and c and d for the j-loop are indicated.

clusters which are highly compatible to the same ASU. We will therefore need a quantitative measure which expresses this compatibility.

3.2.1 A Compatibility Measure for Clusters

In view of the computational complexity it is impossible to evaluate all possible assignments of N_{cl} clusters to N_{ASU} ASUs. We will instead perform the assignment on the basis of a pairwise measure for the compatibility of two clusters.

For each cluster c we compute the cluster cost C_c as the sum of the area of the ABBs needed to implement the cluster.

$$C_c = \sum_{\forall ABB_i} Area(ABB_i) \tag{1}$$

Since there may still be a number of possible ABB choices in the list provided for a given operation, the area of the smallest ABB alternative is chosen. If two clusters a and b - with costs C_a and C_b - are matched on one ASU, the cost of this ASU is represented by C_{ab} (see section 3.2.2). For *perfectly compatible* clusters - i.e clusters whose graphs are isomorphic, where corresponding vertices represent identical operations and where corresponding edges have the same signal type - we have: $C_a = C_b = C_{ab}$. If these conditions are not satisfied then $C_{ab} > C_a$ and/or $C_{ab} > C_b$.

Definition 3 (*Matching Cost*) *The matching cost*

$$C_M(a,b) = 2C_{ab} - C_a - C_b \tag{2}$$

represents the overhead cost for the merging of two clusters on the same ASU.

Only if the two clusters are perfectly compatible $C_M = 0$, otherwise $C_M > 0$. Based on the matching cost we can define the compatibility graph

Definition 4 (*Compatibility Graph*) *The compatibility graph $CG(V,E)$ is a complete graph in which a vertex represents a cluster. There is an edge between any two vertices a and b and this edge $a-b$ has a weight equal to $C_M(a,b)$.*

The construction of this graph has a complexity $\mathcal{O}(|V|^2)$. The matching cost gives an exact measure for the closeness of two clusters of operations based on the operations in the clusters. There are no terms which reflect potential sharing of interconnect or potential parallelism. At this point we believe that it is not necessary to incorporate estimates with respect to the interconnect optimisation because this problem is largely solved by the chosen architectural style. Moreover the global interconnect problem is revisited in the low level mapping stage of out high-level synthesis script. Since a cluster is always part of a loop body the desired degree of parallelism can be obtained by folding this loop body.

3.2.2 Computation of the ASU Cost

The computation of the matching cost requires the cost C_{ab} of the resulting ASU of the matching of two clusters a and b. Two ways to compute this cost are proposed:

1. The ASU cost can be determined by calling a constructive matching tool [7] for every pair of clusters. In view of the complexity of the matching this approach is only advocated for the final design iterations.

2. For the first design iterations only a good estimate of the ASU cost C_{ab} is required. The cost of the two cluster ASU is approximated by the cost of the minimum set of required ABBs (this is the component allocation problem presented in [12]).

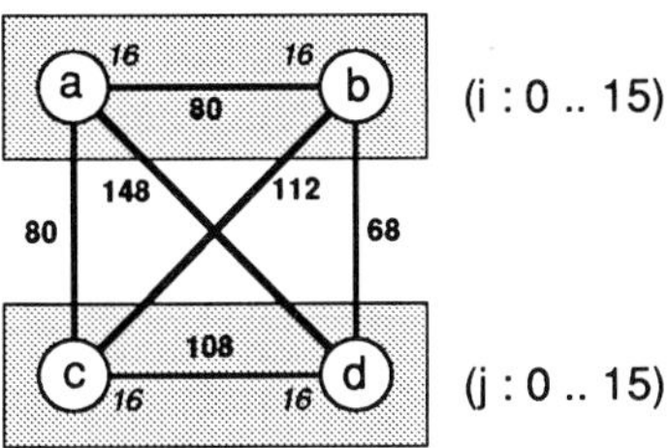

Figure 4: Compatibility Graph for the example. The edge weights are printed bold, the vertex weights (see section 3.2.4) are printed italic.

3.2.3 Partitioning of the Compatibility Graph

First we will consider the case where there are no loops. The minimum number of ASUs that have to be allocated is given by $N_{ASU} = \left\lceil \frac{N_{cl}}{HSF} \right\rceil$ (N_{cl} is a result of the clustering step). The vertex set V of the compatibility graph has to be partitioned into N_{ASU} disjoint subsets in such a way that the most compatible vertices are in the same subset. This can be achieved by minimising the total weight of all edges that do not cut across partitions:

$$\min(\sum_{s=1}^{N_{ASU}} C_s) \tag{3}$$

The accumulated cost of the edges in each subset can be computed by:

$$C_s = \sum_{\forall i-j \,:\, i \in S_s \wedge j \in S_s} C_M(i,j).x_{i,s}.x_{j,s} \tag{4}$$

in which

$$x_{i,s} \;,\; x_{j,s} = \begin{cases} 1 & \text{if } i \in S_s \\ 0 & \text{otherwise} \end{cases}$$

is a binary variable which indicates that vertex i of the compatibility graph is in subset S_s. The *quadratic term* $x_{i,s}.x_{j,s}$ selects all edges between vertices in the same set.
The constraints are:

1. The number of clusters assigned to each subset cannot exceed the available number of cycles (HSF):

$$\sum_{v=1..N_{cl}} x_{v,s} \leq HSF \;\;,\; s = 1..N_{ASU} \tag{5}$$

2. Each cluster has to be assigned to exactly one subset:

$$\sum_{s=1..N_{ASU}} x_{v,s} = 1 \;\;,\; v = 1..N_{cl} \tag{6}$$

This 0/1 programming problem with quadratic cost function is solved exactly by means of an implicit enumeration algorithm [13].

3.2.4 Extension for DSFGs which Contain Loops

A DSFG which contains hierarchically nested loops exhibits the same loop structure that can be found in programming languages. A loop L of the DSFG can contain:

- Operations in the form of clusters belonging to that loop $c \in L$
- Nested loops $l \in L$.

We will use the symbol L to denote a loop at a certain hierarchical level and the symbol l to denote a loop that is nested one level deeper: $l \in L$. A loop is characterised by a *multiplicity* m_l ($m_l > 1$) equal to the number of iterations of that loop. The total cycle budget HSF is specified for the outermost loop of the DSFG. The number of cycles for a nested loop $N_{cyc,l}$ should be determined during synthesis. We will assume that all loops at a certain depth of nesting are executed sequentially[4].
For a hierarchical DSFG the compatibility graph has to be modified:

Definition 5 (*Compatibility Graph for Iterative Algorithms*) *The compatibility graph for Iterative Algorithms $CG(V, E)$ is a complete graph in which a vertex represents a cluster. There is an edge between any two vertices a and b.*

- *Edges have a weight equal to the matching cost between the two vertices they are incident to: $w(e_{a,b}) = C_M(a, b)$*
- *Vertices have a weight equal to the multiplicity of the loop that encloses the corresponding cluster: $w(a) = m_l$.*

The partitioning of this modified compatibility graph can again be formulated as an integer programming problem. The cost function (3) remains unchanged. The constraints have to be modified in the following way:

1. The constraints of type (6) still hold.
2. The constraints of type (5) however have to be extended: for each loop L the sum of the number of cycles consumed by (i) nested loops $l \in L$ and (ii) operations of that loop $c \in L$ (represented by $N_{cyc,cL}$) should not exceed the available number of cycles for L.

$$\sum_{\forall l \in L} m_l \cdot N_{cyc,l} + N_{cyc,cL} \leq N_{cyc,L} \quad , \forall L \in DSFG \tag{7}$$

   The constraints of this type can be found by a recursive traversal of the loop structure of the algorithm.

3. For the operations of loop L the constraint of type (5) can be rewritten as

$$\sum_{\forall c \in L} x_{c,s} \leq N_{cyc,cL} \quad , s = 1..N_{ASU} \tag{8}$$

[4]If there are no data precedences between two loops at a certain nesting depth, they could in principle be executed in parallel and controlled by separate controllers. In this paper we only consider single thread of control machines.

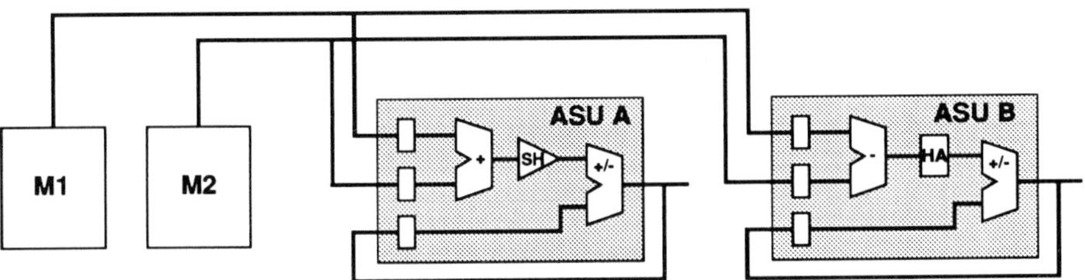

Figure 5: Resulting architecture for the example. The two memories M_1 and M_2 are used to store the vectors $A[]$ and $B[]$. ASU A implements clusters a and c and ASU B implements clusters b and d.

4. If for a certain loop body, represented by loop L there exists a lower bound CB_L for the achievable number of cycles - e.g. due to memory or I/O bandwidth limitations - then this constitutes an additional constraint for $N_{cyc,L}$.

$$N_{cyc,L} \geq CB_L \tag{9}$$

   CB_L can be computed from the number of memory ports that where allocated during high-level memory management and from the number of read/write operations to these ports.

For hierarchical DSFGs it is not possible to compute the minimum required number of ASUs beforehand. The partitioning problem will therefore have to be solved for a number of reasonable values, e.g. $N_{ASU} = 2 \ldots 5$, so that the best solution can be chosen.

Example The compatibility graph for the example is shown in figure 4. We assume that there are 32 cycles available to execute the algorithm and that two values ($A[i]$ and $B[i]$) can be fetched from background memory each cycle. We will allocate two ASUs, A and B. Under the constraint of type (7): $16 \cdot N_{cyc,i} + 16 \cdot N_{cyc,j} \leq 32$ and the constraint of type (6): $x_{a,A} + x_{a,B} = 1$ etc... , the only valid distribution of the cycle budget is: $N_{cyc,i} = N_{cyc,j} = 1$. The constraints of type (8) permit the following partitionings:

$S_1 = \{a, d\}$ and $S_2 = \{b, c\}$ with a cost $C = 148 + 48 = 196$
$S_1 = \{a, c\}$ and $S_2 = \{b, d\}$ with a cost $C = 80 + 68 = 148$

Clearly the latter one will yield the cheapest ASUs. The block diagram of the architecture is shown in figure 5.

4 Conclusions

In this paper a technique has been presented for the allocation of dedicated data-paths in the architectural synthesis for high throughput DSP applications. The technique - based on an integer programming problem - minimises the total area consumed by the data-paths and performs this optimisation across the loop boundaries of a hierarchical data flow graph. The allocation task has been situated in the design script of the CATHEDRAL-III compiler and its relation with the other synthesis tasks has been indicated. In the future more research is needed to automate the clustering step - especially for those cases for which the clustering imposed by the hierarchical structure of the algorithm does not lead to satisfactory results. Furthermore the influence of data flow transformations on the data-path allocation will have to be investigated.

References

[1] M. C. McFarland, A. C. Parker R. Camposano "The High-level Synthesis of Digital Systems" in *Proceedings of the IEEE* vol. 78 no. 2, Feb. 1990, pp. 301-318.

[2] N. Park, A. Parker "Sehwa: A Software System for the Synthesis of Pipelines from Behavioral Specifications" is *IEEE Transactions on the Computer Aided Design of Integrated Circuits and Systems* vol. 7 no.3, March 1988, pp. 356-370.

[3] K. S. Hwang, A. E. Casavant, C. T. Xhang and M. A. d'Abreu "Scheduling and Hardware Sharing in Pipelined Data-paths" in *Proc. IEEE International Conference on Computer Aided Design*, Santa Clara, Calif., Nov. 1989, pp.24-27.

[4] D. J. Mallon and P. B. Denyer "A New Approach to Pipeline Optimisation" in *Proc. 1st ACM/IEEE European Design Automation Conference*, Edinburgh, Schotland, 1990, pp. 83-88.

[5] N. Park, F. J. Kurdahi "Module Assignment and Interconnect Sharing in Register-Transfer Synthesis of Pipelined Data Paths" in *Proc. IEEE International Conference on Computer Aided Design*, Santa Clara, Calif., Nov. 1989, pp. 16-19.

[6] F. Catthoor, H. De Man, "Application-Specific Architectural Methodologies for High-Throughput Digital Signal Processing and Image Processing" in *IEEE Transactions on Acoustics, Speech and Signal Processing*. vol. ASSP-38 no. 2, Feb. 1990, pp. 339-349.

[7] S. Note, F. Catthoor, J. Van Meerbergen, H. De Man, "Definition and Assignment of Complex Data-Paths suited for High Throughput Applications." in *Proc. IEEE International Conference on Computer Aided Design*, Santa Clara, Calif., Nov. 1989, pp.108-111.

[8] P. Hilfinger, J. Rabaey, D. Genin, C. Scheers, H. De Man, "DSP Specification using the SILAGE language", in *IEEE International Conference on Acoustics, Speech and Signal Processing*, April 1990, pp.1057-1060.

[9] D. Lanneer, G. Goossens, F. Catthoor, M. Pauwels, H. De Man "An Object-Oriented Framework supporting the full High-Level Synthesis Trajectory" in *Proc. COMPUTER HARDWARE DESCRIPTION LANGUAGES and their Applications*, Marseille, France, April 1991 pp. 281-300.

[10] G. Goossens, J. Rabaey, J. Vandewalle and H. De Man "An Efficient Microcode Compiler for Application Specific DSP Processors" in *IEEE Transactions on the Computer Aided Design of Integrated Circuits and Systems* vol.9 no.9, Sept. 1990, pp.925-937.

[11] S. Note, W. Geurts, F. Catthoor, H. De Man "CATHEDRAL-III: Architecture-Driven High-Level Synthesis for High Throughput DSP Applications" in *Proc. 1991 Design Automation Conference*, San Francisco, Calif., 1991.

[12] P. Marwedel "Matching System and Component Behavior in MIMOLA Synthesis Tools" in *Proc. European Design Automation Conference*, Edinburgh, Scotland, March 1990, pp. 146-156.

[13] C. McMillan *MATHEMATICAL PROGRAMMING: An Introduction to the Design and Application of Optimal Decision Machines*. J. Wiley, 1970.

VLSI 91
A. Halaas and P.B. Denyer (Eds.)
Elsevier Science Publishers B.V. (North-Holland)

A New Approach to Multiplexor Minimization in the CALLAS Synthesis Environment[1]

N. Wehn[2] J. Biesenack M.Pilsl
Siemens AG Corporate Research and Development

Abstract

In this paper we present a new approach to the multiplexor minimization problem in the CALLAS synthesis environment. Instead of flattening multiplexor cascades into their logic equations we perform the minimization on if-then-else DAGs and show how this structure can be efficiently manipulated to minimize multiplexor cascades in a global view. We discuss the efficiency of the new strategy on different examples.

1 Introduction

The synthesis of digital circuits from an algorithmic specification of the behavior is an important research and development topic [1, 2, 3, 4]. Currently, no commercial high-level synthesis tools are available, and designers do not seem to believe in this approach to design automation. Contacting our circuit designers showed two major problems for high-level synthesis. First, designers think in terms of register transfers rather than in sequential algorithms (most circuit specifications we receive from our customers for test purposes already have a structural nature). Second, often designers are not satisfied with the results of the synthesis process due to restrictions imposed by the particular target architecture. To overcome the first problem there is an ongoing activity in our group to show that the design time can be reduced drastically by the use of the high-level synthesis system CALLAS [5, 6, 7]. For solving the second problem we are working on evolutionary improvements of the approach implemented in CALLAS. We are convinced that, due to our recent work on RT-level optimization and timing specification capabilities [8], CALLAS now provides a powerful tool for designing control dominated ASIC's which are satisfying designer's requirements. In addition we are investigating new ideas for high-level synthesis [9] in cooperation with other universities. The subject of this paper is a new approach to multiplexor minimization that is well suited to our target architecture.

The paper is organized as follows: Section 2 gives a brief overview of CALLAS, Section 3 explains the multiplexor minimization problem arising in CALLAS. In Section 4 the new approach is presented to solve this problem. In Section 5 it is explained how the

[1]This work was partly supported by the BMFT under the JESSI contract NT2869
[2]N.Wehn is also with Darmstadt University of Technology

new approach fits into the CALLAS system, and Section 6 gives some results obtained by this method. Finally, Section 7 presents some concluding remarks.

2 High-level Synthesis with CALLAS

CALLAS is a comprehensive synthesis toolset that supports the design process starting from the behavioral specification of a circuit using a subset of VHDL [10].

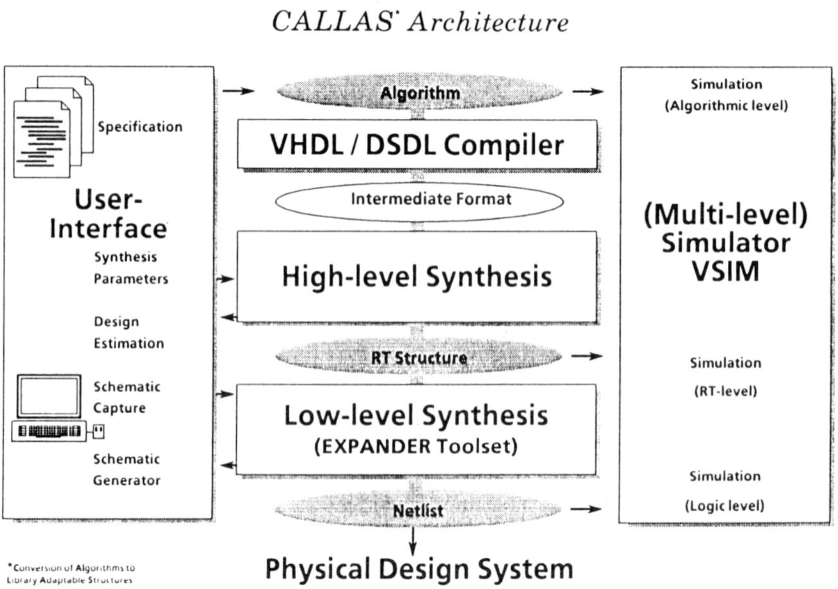

Figure 1: CALLAS system

The synthesized results can be handed over to state-of-the-art CAD systems for physical design (refer to [7] for details). The overall structure of CALLAS is shown in Figure 1. Since VHDL interfaces on all levels of abstraction are provided, any VHDL simulator may be used for validation. The VHDL subset assures that there are no inconsistencies between synthesis and simulation results. The synthesizer in CALLAS maps the algorithm on a multiplexed architecture. A finite state machine with MEALY behavior controls the data path by switching multiplexors and loading registers. Optimization tools such as the method described in this paper, local transformations on Boolean and arithmetic components, lifetime analysis and register sharing, partitioning into blocks and multi level logic optimization are all used to minimize the RT-structure. The control is generated in an AFAP-like manner [8, 11]. The attributed abstract components of the synthesized circuit structure are mapped onto a cell library by a tool called EXPANDER. Schematics can be generated to visualize the results. Besides the algorithmic entry to the system, CALLAS supports also a schematic entry on block- and logic-level as well as state-table and truth-table input.

3 The Problem of Multiplexor Minimization

Conditional statements ,e.g., *if* and *select* are frequently used in behavioral descriptions. A common way to transform these statements onto an appropriate RT-structure is the use of multiplexors to resolve conditional assignments. Further, multiplexors are generated to switch between the initial and the feedback path for variables used in loops. The sharing of functional units and registers with disjunct lifetimes causes the generation of additional multiplexors in the data path. Since CALLAS improves an initial synthesized RT-structure by iterative cycles, complex multiplexor cascades containing redundancies are generated in front of register inputs and functional units. The following piece of code

```
if c1 then
  begin if c2 then c<=a else c<=b end
else
  begin if c2 then c<=b else c<=a end;
```

can directly be mapped into an RT-structure (see Figure 2).

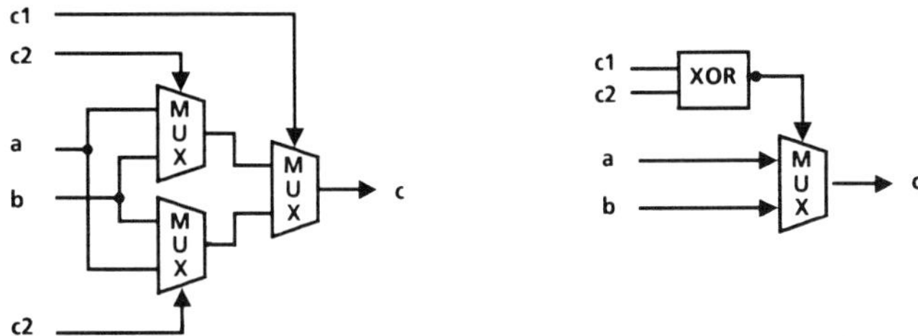

Figure 2: Example of a multiplexor cascade and its optimized structure

However, a careful examination of this example shows that the same behavior can be implemented with a single 2:1 multiplexor which control input consists of a *nexor* gate. Obviously, the second solution is more efficient.

There are two common methods that solve the problem of multiplexor minimization:

- local transformations
- global logic minimization strategies

The local transformations in CALLAS directly work on th RT-level structure. Unfortunately, these transformations are restricted to simple multiplexor structures and produce only local minima, e.g., a typical transformation replaces a 2:1 multiplexor with a constant "1" input by a *or* gate. Only high sophisticated look-ahead techniques as described in [12] could overcome local minima. However, for most of the RT-level local transformations look-ahead techniques do not apply. Thus, a more complicated method is necessary for removing redundancies in multiplexor cascades. First, the multiplexor structures are flattened into their equivalent logic equations. Then, the resulting logic is minimized with well known algorithms [13, 14]. However, this strategy implies several problems: due to the NP-complexity of the global logic minimization problem, these algorithms are based on heuristics and the quality of the solution is a trade-off between computation time and minimization result. Since multiplexor inputs are often busses the equivalent logic may explode and the logic minimization itself becomes inefficient.

In the CALLAS system the logic optimization is done in the EXPANDER, which performs also the technology mapping. Examples have shown that the logic minimization is often not able to preserve multiplexor structures. At the moment we need the flattening process and the logic minimization to remove the redundancies in the multiplexor cascades, but as a result we obtain a minimized glue logic instead of multiplexor structures. This involves some drawbacks: the data path structure becomes difficult to survey which complicates the designer controlled iterative design cycle. In addition multiplexor structures are often more efficient when comparing their layout sizes with glue logic realizations. The EXPANDER is able to map multiplexors, thus able to benefit from this fact.

4 The New Approach

Our goal in the new approach is to minimize multiplexor cascades on the RT-level by removing redundancies while preserving the multiplexor structures. In the next subsections we discuss a natural representation for multiplexors that can be easily manipulated by a set of routines. The routines are able to remove redundancies and to optimize the cascades.

4.1 An Efficient Multiplexor Representation

The idea of an efficient multiplexor representation is motivated by *if-then-else DAGs*. These DAGs are based on the *if-then-else* operator:

Definition: *The* **if-then-else** *operator is a ternary Boolean function, with* (**if** a **then** b **else** c) *defined as* $ab + a'c$ *or, equivalently* $(a + c) * (a' + b)$.

All binary Boolean functions can be defined with the if-then-else operator. Using this definition an *if-then-else DAG* is defined:

Definition: *An* **if-then-else DAG** *is a ternary directed acyclic graph in which each inner node has three successors, an* if-, a then- *and an* else- *subDAG. Each leaf is labeled* *with true or false or a literal. The meaning of a leaf node is the label on the node, and the meaning of an internal node is defined recursively as* (**if** *meaning* (**if**-part) **then** *meaning* (**then**-part) **else** *meaning* (**else**-part)).

If-then-else DAGs have a long history [15, 16]. If the *if*-subDAG is restricted to a leaf, we obtain binary decision diagrams (*BDDs*) that are the most popular representatives and have often been used for logic verification [17, 18]. An efficient implementation of if-then-else DAGs was recently published [19]. For further discussions a depth-first traversal of the DAG is assumed. The outgoing edges of a node are visited in the order if-, then-, else-edge. Thus, we can represent the DAG as a string, i.e., (**if** a **then** b **else** c) is represented by (a, b, c).

A multiplexor is defined by two sets: a set of control-inputs s_v, denoted in the following *select* variables. Consequently, a multiplexor has $2^{|s_v|}$ inputs. The variables in s_v are ordered: the first element is the highest order bit, the last element the lowest order bit. The second set v_v represents the *data* inputs. This set is ordered too: the first element represents the input that is selected if all control inputs are "1", the last element represents the input that is selected if all control inputs are "0". Inversions of select and data variables are considered as part of the multiplexor.

The DAG for a single multiplexor is obtained by applying the Shannon decomposition procedure. The select variables are the decomposition variables corresponding to the if-leaves. E.g., a 4:1 multiplexor with $s_v=(c_1, c_0)$ and $v_v=(i_3, i_2, i_1, i_0)$ is equivalent to the DAG $(c_1, (c_0, i_3, i_2), (c_0, i_1, i_0))$.

The DAG model can be used to represent multiplexor cascades. A multiplexor cascade is a set of multiplexors obtained as follows: starting with a multiplexor which output has a fanout > 1 or which output is connected to an element that is not a multiplexor, a recursive search is started along its inputs for multiplexors which are connected to the actual one. We stop the search if the fanout of a multiplexor input net is > 1 or the net does not belong to a multiplexor. Thus, if a DAG leaf represents a multiplexor output, this leaf is recursively substituted by its corresponding DAG. Obviously, if-then-else DAGs are a natural representation of multiplexor structures.

4.2 DAG Optimization

In this section we discuss some routines to manipulate and optimize the DAG.

The first routine removes all redundancies that are caused by the appearance of the *if* leaf in the *then/else* successor of a node, i.e., $(c, then(c), else(c))$: *known select variables are substituted by its values in the subDAGs.* Only the processing of the *then()* case is discussed (processing of the *else*() subDAG is done in a similar way). There are three possibilities:

- c is a *if* leaf, i.e., there is a subDAG in $then(c)$ with (c, a, b). This subDAG is substituted by the single node a. To visit the *then* subDAG, c must be "1". Thus, we have the situation $(1, a, b)$ that can be simplified to a.

- c is a *then* leaf, i.e., there is a subDAG in $then(c)$ with (a, c, b). In this case c is substituted by "1": $(a, 1, b)$.
- c is a *else* leaf. This is processed as the case before, i.e., c is substituted by "1": $(a, b, 1)$.

The second routine simplifies the DAG in the case of constant *truc* and *falsc* lcaf nodes:

- the subDAG $(dag\_if, 1, 0)$ is substituted by dag_if.
- $(dag\_if, 0, 1)$ is substituted by dag_if'.
- $(1, d\_then, d\_else)$ is substituted by d_then.
- $(0, d\_then, d\_else)$ is substituted by d_else.

The third routine places *inversions in an unique way.* This is done by removing recursively the inversions in if and $then$ subDAGs. Inner nodes of the DAG are allowed to be inverted:

- An inversion of an if subDAG can be deleted by exchanging the *then* and *else* subDAG, i.e., (c', t, e) is substituted by (c, e, t).
- An inversion of a *then* subDAG can be removed by shifting this inversion, i.e., (c, t', e) is substituted by $(c, t, e')'$.

The last routine removes *identical subDAGs*, i.e., there is a subDAG $(c, f(), f())$. This subDAG is substituted by $f()$. It can be shown that two subDAGs represent the same function if they have exactly the same structure. Thus, the test if two subDAGs are identical is quite simple: it must be checked if the two subDAGs are equivalent in their structure and labeling.

An if-then-else DAG that satisfies these conditions can be shown to be canonic under the assumption that there is a unique order in the select variables [16, 20]. A single multiplexor always satisfies the unique order restriction. However, DAGs derived from multiplexor cascades do not always satisfy this requirement. Without restricting our further considerations, a global order is assumed.

The canonic DAG representation has some nice advantages: the multiplexor cascade is simplified since the complexity (in terms of DAG nodes) of the canonic representation is always less or equal the complexity of the original DAG, e.g., given is a multiplexor cascade with three 2:1 multiplexors: $mpx1 : s_v{=}c0,\ v_v{=}(a, b)$; $mpx2 : s_v{=}c0',\ v_v{=}(b, a)$; $mpx3 : s_v{=}c1,\ v_v{=}(mpx1\_out, mpx2\_out)$. The if-then-else DAG of this cascade is $(c1, (c0, a, b), (c0', b, a))$. The canonic representation is $(c1, a, b)$ that corresponds to a single 2:1 multiplexor.

The canonic representation permits a fast check and extraction of Boolean functions, e.g., $(c, d, 0)$ represents a *and* function, $(c, 1, d)$ represents a *or* function, (c, d, d') represents a *nexor* function etc. The following example illustrates this advantage: given is a 4:1 multiplexor which multiplexes two busses and their inverse values. The DAG form of the multiplexor is $(c1, (c0, a, b), (c0, a', b'))$. The canonic representation is $(c1, (c0, a, b), (c0, a, b)')$. Thus, the 4:1 multiplexor can be substituted by a single 2:1 multiplexor which is controlled by $c0$. The output of the multiplexor is input to a *nexor* gate with $c1$ as second input. Obviously, this solution is more efficient than the 4:1 multiplexor structure.

The canonic representation is *prime* and *irredundant* in the context of the DAG, i.e., no subDAG can be replaced by "0" without changing the meaning of the multiplexor cascade (*prime*) and no subDAG can be replaced by a "1" without changing the meaning of the multiplexor cascade (*irredundant*). The canonic representation of the corresponding multiplexor cascade does not contain any redundancies.

However, there is a problem not yet addressed. The complexity of the canonic representation (number of DAG nodes) depends on the select variable ordering [16, 18], e.g., the canonic representation of a 4:1 multiplexor $(c1, (c0, a, b), (c0, a, c)$ differs from the canonic representation if $c1$ and $c0$ are reordered: $(c0, a, (c1, b, c))$, which is a better solution than the first one. To overcome this problem the select variables are globally reordered. Due to the often small numbers of select variables, a permutation procedure is justifiable to find the best solution. If the number is too large (e.g. > 4), a heuristic ordering strategy is applied [21]: we traverse the DAG from the root to the leaves by depth-first order to find the longest path. Select variables visited during this traversal are ordered first. Then, the next longest path is searched. If its select variables are not yet in the order list, they are appended. This procedure is continued until all select variables are processed.

In the considerations discussed so far, the *if* subDAGs were assumed to be leaves, thus preserving the similarity to *BDDs*. From now we ignore this restriction. It can be shown that this extended DAG is also canonic [22]. New transformations are introduced that make benefit of this extended structure. The goal of these transformations is a further reduction of the complexity of the multiplexor cascades, i.e., the number of leaves that represent data inputs is further reduced:

- Local reordering of select variables:
  - $(c_i, (c_j, f_a, f_b), (c_j, f_a, f_c)) \rightarrow (c_j, f_a, (c_i, f_b, f_c))$
  - $(c_i, (c_j, f_a, f_b), (c_j, f_c, f_b)) \rightarrow (c_j, (c_i, f_a, f_c), f_b)$
- Identical *then*, *else* subDAGs:
  - $(c_i, (c_j, f_a, f_b), (c_k, f_a, f_b)) \rightarrow ((c_i, c_j, c_k), f_a, f_b)$
  - $(c_i, (c_j, f_a, f_b), (c_k, f_b, f_a)) \rightarrow ((c_i, c_j, c_k'), f_a, f_b)$
- One common subDAG:
  - $(c_i, (c_j, f_a, f_b), f_b) \rightarrow ((c_i, c_j, 0), f_a, f_b)$

- $(c_i, (c_j, f_a, f_b), f_a) \rightarrow ((c_i, c_j, 1), f_a, f_b)$
- $(c_i, f_a, (c_j, f_b, f_a)) \rightarrow ((c_i, 1, c_j'), f_a, f_b)$
- $(c_i, f_a, (c_j, f_a, f_b)) \rightarrow ((c_i, 1, c_j), f_a, f_b)$
- $(c_i, (c_j, f_a, f_b), (c_k, f_a, f_c)) \rightarrow ((c_i, c_j, c_k), f_a, ((c_i, c_j, 1), f_c, f_b))$
- $(c_i, (c_j, f_b, f_a), (c_k, f_c, f_a)) \rightarrow ((c_i, c_j, c_k), ((c_i, c_j, 0), f_b, f_c), f_a)$
- $(c_i, (c_j, f_a, f_b), (c_k, f_c, f_a)) \rightarrow ((c_i, c_j, c_k'), f_a, ((c_i, c_j, 1), f_c, f_b))$
- $(c_i, (c_j, f_b, f_a), (c_k, f_a, f_c)) \rightarrow ((c_i, c_j, c_k'), ((c_i, c_j, 0), f_b, f_c), f_a)$

Although these transformations increase the number of select variable leaves, the number of leaves representing data variables decreases. Select variables are one bit and data variables are busses, consequently, an improvement results. It can be shown that the DAG after these transformations is still canonic if the DAG before was canonic and the DAG traversal order is fixed. The efficiency of these transformations is illustrated on the example of the introduction. Transformation 4 directly provides the optimum structure.

A further advantage of the DAG representation is the efficient use of *don't care* information. *Don't cares* occur frequently in multiplexor structures (e.g. a case statement in which not all possible cases are specified). A don't care can be recognized on the RT-level by a multiplexor input that is not connected. Thus we introduce a special leaf node that represents a don't care input. Don't cares (x) can be efficiently processed in the DAG reduction process: (i, x, e) is transformed to e, (i, t, x) is transformed to t and (x, t, e) is transformed to t *or* e.

The retransformation of the optimized DAG into structure is very easy and not explained explicitly. Conventional gates as described previously can be extracted and are favored. Due to the permutation procedure for finding the best select variable ordering, a reduced DAG for each permutation is obtained. Thus each DAG is weighted by a cost function which takes into account number of data leaf nodes, number of select leaf nodes, number of multiplexors, and complexity of select logic. The DAG with the cheapest cost is selected as final solution.

5 Integration in the CALLAS Environment

The multiplexor minimization approach is integrated in CALLAS. First the RT-level data structure is scanned for occurrences of multiplexor cascades. After a multiplexor cascade has been detected the new minimization algorithm is applied. The final DAG is retransformed into structure consisting of multiplexors and some basic logic gates, which replaces the old cascade. The above steps are repeated until all multiplexor cascades have been minimized. Obviously, the new multiplexor minimization method behaves like a local transformation rule. In fact we integrated it as a 'super' transformation rule, replacing a set of older local transformations. Thus, the RT-level and logic-level optimizations integrated in the CALLAS system can be classified as follows:

- Local transformations consisting of a set of sequential rules, a set of arithmetic rules, a set of logic rules and one transformation for multiplexor minimization.
- Global logic optimization using MIS: interconnected abstract components of a specific type, e.g. logic gates, small arithmetic components are combined into combinatorial submodules, which can be optimized and mapped by tools like MIS. The decision which types of abstract components should be flattened into submodules is interactively guided by the designer.

Since the complete synthesis cycle is interactively controlled, the best optimization method can be chosen with respect to the design to be synthesized.

6 Results

In this chapter we present some results obtained with the new minimization technique. Multiplexor cascades originating from real examples are used as test vehicles. In a first step, the multiplexor cascades were minimized by the traditional strategy: applying of local transformations, flattening, logic optimization with MIS, and technology mapping. In a second step the cascades were minimized with the new strategy and directly technology mapped without any logic optimization. The following table shows the results. The values are only referenced to the multiplexor cascades. The solutions are compared in terms of area A (without routing overhead), equivalent gates E_G, cells to be placed C_P and number of interconnections I_C obtained by the EXPANDER.

old strategy versus new strategy								
name	A^{old}	E_G^{old}	C_P^{old}	I_C^{old}	A^{new}	E_G^{new}	C_P^{new}	I_C^{new}
c1	76374	50.5	33	115	58116	50.5	17	67
c2	155598	101.5	70	256	109432	88	32	112
c3	74664	47	38	122	66115	55	22	80
c4	76374	51	33	116	58135	51	17	68
muxdc	47306	31.5	21	73	33057	27.5	11	40
c5	62125	41	29	104	61555	53	19	73

Obviously, in all cascades there is a gain in layout size compared to the old strategy. Since routing overhead is not considered a further improvement is expected since the number of nets to be interconnected is drastically reduced.

Nevertheless, there are structures in which the old strategy is superior to the new one. The minimization method as presented in this paper is still a local minimization technique that can not always reduce identical busses fed to different inputs in a cascade. If the inputs of the cascade are constant values, flattening with logic optimization probably produces better results. Our strategy is well suited if we have large data busses at the inputs of multiplexor cascades. At the moment we apply the new strategy first and then dependent on the new cascade structures some multiplexors are still flattened and logic minimized interactively guided by the designer. In larger examples, up to 70% of the initial multiplexors could be removed making the RT-structure more understandable and the logic optimization more efficient.

7 Conclusion

In this paper we presented a new approach to the multiplexor minimization problem in the CALLAS synthesis environment. The new algorithm works on the RT-level and simplifies the logic minimization step. Mainly two aspects make the new approach attractive. First the new approach substitutes a set of non optimal local transformations and thus improves the multiplexor optimization considerably. In fact, the new approach can be extended to a complete logic minimization algorithm based on an if-then-else DAG representation of Boolean functions. The second effect is that the hierarchy of the synthesized RT-structures is retained, since the results of the multiplexor minimization dispends us in many cases from the need of flattening multiplexors into logic. Thus after minimization the data flow remains visible in the generated schematic diagrams. This simplifies understanding and simulation of synthesized circuits for designers and so increases acceptance of high-level synthesis.

References

[1] J. V. Meerbergen and H. D. Man, "A true silicon compiler for the design of complex ics for digital signal processing," *Philips Technical Review*, vol. Vol. 44, No. 7 1989.

[2] G. D. Micheli and D. Ku, "Hercules - a system for high-level synthesis," in *Proc. of 25th Design Automation Conference*, 1988.

[3] D. Thomas, E. Dirkes, R. Walker, J. Rajan, J. Nestor, and R. Blackburn, "The system architects workbench," in *Proc. of 25th Design Automation Conference*, 1988.

[4] H. Krämer and W. Rosenstiel, "System synthesis using behavioural descriptions," in *Proc. of EDAC 1990*, 1990.

[5] M. Koster, M. Geiger, and P. Duzy, "Asic design using the high-level synthesis system callas: A case study," in *Proc. of ICCD Conference 1990*, 1990.

[6] S. März, K. Buchenrieder, P. Duzy, R. Kumar, and T. Wecker, "Callas - a system for automatic synthesis of digital circuits from algorithmic behavioral descriptions," in *Proc. of EURO ASIC Conference 1989*, 1989.

[7] P. Duzy, H. Krämer, M. Neher, P. Pilsl, W. Rosenstiel, and T. Wecker, "Callas - conversion of algorithms to library adaptable structures," in *Proc. of IFIP VLSI Conference 1989*, 1989.

[8] A. Stoll, H. Vollmer, and P. Duzy, "High-level synthesis with timing constraints," in *submitted to ICCD 91*, 1991.

[9] J. Scheichenzuber, W. Grass, U. Lauther, and S. März, "Global hardware synthesis from behavioural dataflow descriptions," in *Proc. of 27th Design Automation Conference*, 1990.

[10] W. Glunz and G. Umbreit, "Vhdl for high-level synthesis of digital systems," in *Proc. of 1st European Conference on VHDL Methods*, 1990.

[11] R. Camposano and R. Bergamaschi, "Synthesis using path-based scheduling: Algorithms and exercises," in *Proc. of 27th Design Automation Conference*, 1990.

[12] D. Gregory, K. Bartlett, A. D. Geus, and G. Hachtel, "Socrates: A system for automatically synthesizing and optimizing combinational logic," in *23rd Design Automation Conference*, 1986.

[13] R. Brayton, G. Hachtel, C. McMullen, and A. Sangiovanni-Vincentelli, *Logic Minimization Algorithms for VLSI Synthesis.* Kluwer Academic Publishers, 1984.

[14] R. Brayton, R. Rudell, A. Sangiovanni-Vincentelli, and A. Wang, "Mis: A multiple-level logic optimization system," *IEEE Transactions on Computer Aided Design*, vol. Vol. 6, Nov 1987.

[15] S. B. Akers, "Binary decision diagrams," *IEEE Trans. on Computers*, Jun 1987.

[16] R. Bryant, "Graph-based algorithms for boolean function manipulation," *IEEE Trans. on Computers*, Aug 1986.

[17] R. Bryant, "Graph based algorithms for boolean function manipulation," in *1985 Chapel Hill Conference on VLSI* (H. Fuchs, ed.), 1985.

[18] S. Malik, A. Wang, R. Brayton, and A. Sangiovanni-Vincentelli, "Logic verification using binary decision diagrams in a logic synthesis environment," in *Proc. of IEEE ICCAD-88*, Nov 1988.

[19] K. Brace, R. Rudell, and R. Bryant, "Efficient implementation of a bdd package," in *Proc. of 27th Design Automation Conference*, 1990.

[20] J. Madre and J. Billon, "Proving circuit correctness using formal comparison between expected and extracted behavior," in *Proc. of 25th Design Automation Conference*, 1988.

[21] N. Fujita, H. Fujisawa, and N. Kawato, "Evaluation and improvements of boolean comparison methods based on bdds," in *Proc. of 25th Design Automation Conference*, 1988.

[22] K. Karplus, "Using if-then-else dag's for multi-level minimization," in *Decennial Caltech Conference on VLSI*, Mar 1989.

VLSI 91
A. Halaas and P.B. Denyer (Eds.)
Elsevier Science Publishers B.V. (North-Holland)

Meta VHDL for Higher Level Controller modeling and synthesis

Ahmed Jerraya[1], Pierre G. Paulin and Simon Curry

BNR, PO. Box 3511, Stn. "C", Ottawa, Canada K1Y 4H7

Abstract

This paper presents Meta VHDL (MV), a hardware description language based on VHDL with the addition of primitives to support an efficient specification of hierarchical and communicating Finite State Machines. MV supplements VHDL with three new paradigms: hierarchical composition of FSMs, exception handlers and global transitions between FSMs. An algorithm that translates MV specifications into VHDL models is also presented. First experiments show that the use of these new paradigms for the specification of high level controllers allows for a fourfold reduction of the description size when compared to the VHDL model.

1. Introduction

While many behavioral synthesis systems [McPC90] support complex high-level data path machines, until very recently [RoGi90] there has been very little work in the area of high level controllers and protocol machines. These are often modeled as a set of hierarchical and communicating state machines.

1.1. Motivations

This paper presents a new modeling language called MV (Meta VHDL). MV is based on VHDL with the addition of new primitives that simplify the specification of hierarchical and communicating FSMs by hiding all the details related to process control and exception handling.

It is often easier to structure a large system as a collection of cooperating sub-systems rather than as a single sequential process. An MV description is organized as a set of hierarchical and cooperating Finite State Machines (FSMs), similar to the process organisation in CSP [Hoar78] but with a simpler communication scheme. The specification of the control of such processes in VHDL leads to a verbose description. In VHDL the user is asked to provide all the activation and de-activation conditions of a process in order to control its progress. MV hides these details by introducing an inter-FSM transition primitive.

In almost all complex controller systems, it is the case that the normal course of actions can be disrupted at any point in time by events signalling exceptions (interrupts, resets, context

---

[1]The present address of the author is: IMAG/TIM3, 46 Avenue Felix Viallet, 38031, Grenoble CEDEX, France

switches, timeout). The specification of such exceptions in VHDL induces a lot of duplication in the code. MV introduces facilities for exception specification.

1.2. Previous work

None of the existing HDLs, in our experience, provides all the required facilities for the specification of such systems. Some existing HDLs tried to provide facilities for complex controllers specification. HSL-FX[HoKN85] and UDL/I [Kara89, Udli90] provide the notion of tasks and control transfer between tasks for FSMs specification. The model of control transfer is based on an extended model of co-routines that restricts the specification of concurrent FSMs. The authors in [BuMa88] present a language for system level specification. This language is based on CSP [Hoar78]. The authors use a communication scheme (handshake) that restricts the modeling power of CSP. None of these languages provides facilities for exceptions specification. Exceptions (reset, initialization, emergency situations, etc.) induce a lot of duplications in the code description.

Most of the paradigms listed above are handled by system level description languages oriented towards protocols, and real time systems. Some of these languages are: ESTELLE[ISO87], LOTOS [Loto89], the HAREL's statecharts[Hare90], SML [Clar89], ESTERELLE[BeCo84].

Several recent papers report on the link between system-level languages and hardware specification in order to allow mixed software/hardware systems design. The automatic generation of hardware descriptions starting from system-level specifications is the backbone of such a link. The translation algorithm reported in [Tier88] produces a VHDL model composed of a unique sequential process even when the controller specification includes parallel FSMs. Another translation algorithm that produces a hardware model from a HAREL's Statechart description is reported in [DrHa89]. This algorithm produces a model composed of a large number of parallel FSMs and a large number of signals. This amount of processes and signals decreases the efficiency of the simulation and synthesis. [NaVa90] uses a similar algorithm for the translation of SpecCharts (SpecCharts is a new system-level description language partly based on HAREL's Statecharts) into VHDL. A more recent algorithm is described in [Dutt91]. It translates a hierarachical state table (with no parallelism) into a set of VHDL guarded blocks. In fact, the guards act like a global state register.

MV is a different method of linking system level and hardware design. It provides a set of primitives that simplifies system modeling and synthesis using VHDL. The algorithm developed for the automatic generation of VHDL starting from MV, preserves the organization of the initial description and produces an efficient and accurate VHDL model. The VHDL description is verbose when compared to the MV specification — our experiments indicate that MV allows for more than a fourfold reduction in the size of the description when compared to VHDL. This is caused mainly by the lack of facilities for the specification of process control within hierarchical FSMs (e.g., to stop or restart a VHDL process that contains wait statements).

1.3. MV: A Meta VHDL for High level controllers specification

Despite VHDL's growing acceptance and use in academia and industry, facilities around VHDL for specification and synthesis are desirable for usability — much like the addition of X-windows and motif to UNIX or the addition of C++ to C.

Meta VHDL (MV) is a new modeling language based on VHDL with the addition of new primitives that simplify the specification of high level controllers. These new facilities allow to reduce the size of hierarchical and communicating FSMs specification by hiding all the details related to process control and communication.

MV is intended to be integrated into a system-level synthesis environment. Fig. 1 shows a global view of MV. The hierarchical controller modeling and synthesis process starts with an MV description. This MV description can be compiled in one of two methods. The first produces a VHDL model which is used for the validation (by simulation) of the initial specification. The compilation algorithms are described in greater detail in section 3. The second method produces a SIF model. SIF is a synthesis intermediate form developed at BNR, and is described in [JePa91]. SIF is the entry point of a larger system aimed at system-level synthesis. The target of this synthesis process is a VHDL model accepted by the existing synthesis tools such as those provided by SYNOPSYS[Carl90]. The MV primitives have been defined with a synthesis scheme in mind.

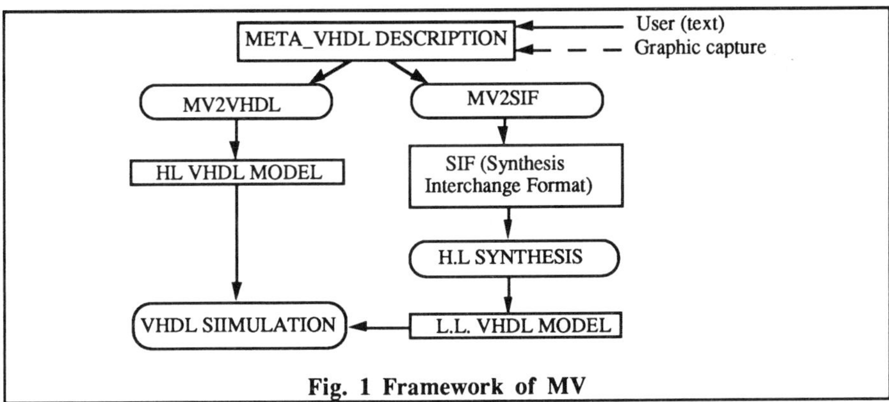

Fig. 1 Framework of MV

2. THE DESCRIPTION LANGUAGE: MV

This section provides an overview of MV, a more detailed presentation of which can be found in [JePa91].

2.1. The syntax

Fig. 2 gives the syntax of the MV primitives. There are four main constructs: HFSM, FSM, EXCEPTION and TRANSITION. A controller is described as an HFSM. The use of the HFSM primitive follows the same rules that apply to block statements in VHDL. There are two kinds of HFSMs. They are defined by the composition mode of the components. An HFSM which is composed of a set of FSMs running in parallel is called a PFSM. An HFSM composed of a set of sequenced FSMs (only one of the components can be active at the same time) is called an SFSM. In the last case an initial state must be defined.

The leaf FSMs are described by a primitive called FSM. Its use follows the same rules that apply to the VHDL process statement. No assumptions are made about the complexity of the states. A state may hide a complex controller which is described by a sequential program with implicit states.

The exception primitive has two parameters:

1 An event: which can be any signal visible in the HFSM that contains the exception.

2 An action that will be executed when the event is true (high value).

The global transition primitive "next_state" specifies an exit from the current FSM state (or HFSM) and an entry to another specific FSM (or HFSM) state. It has two parameters:

1 An FSM name: it specifies the FSM to enter. If omitted, the default value is the current FSM or HFSM.

2 An entry point: in the case of an SFSM the entry point specifies which component to start. If omitted, the initial state of the specified (H)FSM will be started.

The semantics of the next_state primitive assumes that an FSM has two predefined states called *idle* and *active*. An exit from an FSM forces this FSM to the idle state. Entering an FSM makes this FSM active. The VHDL modeling of these states will be explained below (see section 3).

HFSM and FSM primitives can be combined within a VHDL description as parallel statements. A VHDL architecture statement[2] can then include more than one controller description. The other primitives can only be used in accordance with the syntax specified in Fig. 2.

```
HFSM          ::= label: HFSM_TYPE [VHDL_BLOCK_DECLARATION]
                      begin  {EXCEPTION} {COMPONENT} end ;
HFSM_TYPE     ::= PFSM | SFSM (initial_state => FSM_NAME)
COMPONENT     ::= FSM | HFSM
FSM           ::= label: FSM [VHDL_PROCESS_DECLARATION]
                  begin {VHDL_PROCESS_STATEMENT | TRANSITION} end;
EXCEPTION     ::= exception (Event =>SIGNAL_NAME; Action => FUNCTION)
FUNCTION      ::= TRANSITION | VHDL_FUNCTION_CALL
TRANSITION    ::= next_state [([fsm=> FSM_NAME;] entry => FSM_NAME);]
```

Fig. 2. MV Syntax

2.2. A description example

The MV primitives will be explained using a simple description representing a request manager [JePA91]. This request manager acts between a host and a server (Fig 3). It accepts requests from the host and transmits them to the server. The host and the server work at different speeds. The three components are synchronized through control signals. The host can reset the buffer through the signal RESTART. REQIN is high when the host is transmitting DATAIN. INQUIRE is high when the request manager is transmitting DATAOUT. RDY is

---

[2] In VHDL a design unit is described by an entity statement that specifies the interface and an architecture statement that specifies the function of the design as a set of parallel statements.

high when the server is ready to receive a request. The request manager uses a circular buffer to store requests. This buffer is addressed through two indexes: BUFFIN provides the address where the next request from the host will be stored and BUFFOUT provides the address of the next request to be sent. When the buffer is full, the host is alerted through the signal B_FULL. The buffer is full if after a receive action BUFFIN becomes equal to BUFFOUT. The buffer is empty if after a send action BUFFOUT becomes equal to BUFFIN. In order to simplify the example, we suppose that each request is composed of one data bit.

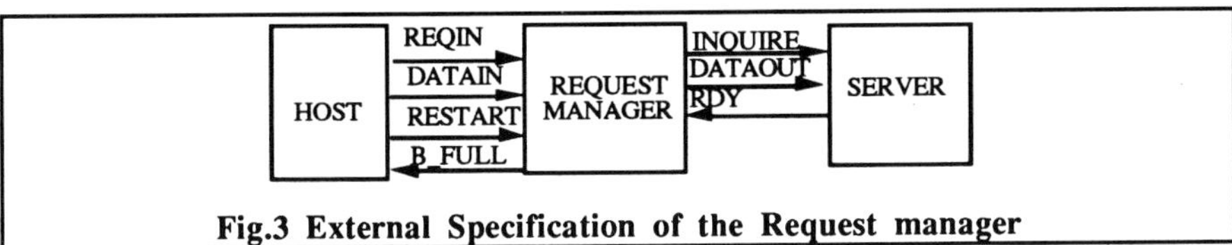

Fig.3 External Specification of the Request manager

An MV description of the controller is given in Fig. 4. The MV primitives (in bold) are mixed with VHDL statements. The description has the following organization. The controller is an HFSM composed of two sequenced components: the FSM INIT and the PFSM SEND_RECV. INIT resets the buffer and transfers the control to SEND_RECV. The latter is composed of two parallel components SEND and RECV. RECV receives requests from the host and queues them in the buffer. It is composed of two sequenced FSMs WAIT_REQ and BUF_FULL. The first accepts the requests. BUF_FULL is activated when the buffer is full. The FSM SEND is in charge of sending the requests to the server. The transition between FSMs is controlled by the next_state primitives. The restart action is specified as an exception in the SEND_RECV HFSM.

The exception primitive is a powerful construct which minimizes redundant transition declarations in the specification. In this case, it expresses *"Regardless of the state of SEND_RECV, If restart = '1', then exit SEND_RECV and enter INIT."*

Each FSM (or HFSM) description uses the signals provided by the environment: ports of the entity that contains the FSM, and all the signals visible from the FSM as well as the internal signals and variables declared inside the FSM.

3. MV to VHDL

During the compilation process, the MV primitives are translated into standard VHDL statements with no effect on the rest of the description. Some of the VHDL statements contained in the FSM bodies are also transformed during the translation process. The VHDL *wait* statements are transformed in order to allow real-time handling of exceptions. The statements that perform assignment of shared variables are transformed in order to handle the conflict access. All these transformations are based on an abstract FSM model .

```
ENTITY REQUEST_MANAGER is port
      ( signal RESTART, REQIN, SERVER_RDY : in Bit ;
      signal DATAIN : in bit ; signal DATAOUT : out bit ; ) ;
end REQUEST_MANAGER ;
use synt.extension.all;
ARCHITECTURE SYSTEM_LEVEL_FSM of REQUEST_MANAGER is
      constant N : integer := 10 ;
      type BUF_array is array ( 1 to N ) of bit ;
      signal BUF: BUF_array ;
      signal BUFFIN: X_W_Share integer ; EXCLUSIVE WRITE SHARED
      signal BUFFOUT: integer ;
begin
   MANAGER : SFSM (initial_state => "INIT")
   begin
   INIT: FSM
    begin
      BUFFIN <= 1 ; BUFFOUT <= 1 ;
      NEXT_STATE ( "SEND_RECV") ;
    end FSM INIT;
   SEND_RECV : PFSM
   begin
     Exception (RESTART, NEXT_STATE ( "INIT" ));
     RECV : SFSM ("WAIT_REQ") ;
     begin
         WAIT_REQ: FSM
         begin
              WAIT UNTIL REQIN = '1' ;
              BUF[ BUFFIN ] <= DATAIN ;
              If BUFFIN + 1= BUFOUT
                   then B_FULL <= 1; NEXT_STATE ("BUF_FULL");
                   else BUFFIN := BUFfIN + 1;
              end if;
          end FSM WAIT_REQ;
          BUF_FULL: FSM
          begin
                wait until BUFFIN /= BUFFOUT ;
                BUFfIN := BUFFIN + 1
                NEXT_STATE ("WAIT_REQ");
          end FSM BUF_FULL;
     end EFSM RECV;
     SEND : FSM
     begin
       INQUIRE <= '0';
       if BUFFIN = BUFFOUT then wait until BUFFIN /= BUFFOUT ; end if;
       send_loop: while BUFFIN /= BUFOUT
                INQUIRE <= '1'
                wait Until SERVER_RDY = '1' ;
                DATAOUT <= BUF[ BUFOUT ] ; BUFFOUT <= BUFFOUT + 1;
       end loop send_loop
    end FSM SEND;
  end PFSM SEND_RECV;
end SFSM MANAGER ;
```

Fig. 4 MV description of the request manager

3.1. The abstract FSM model

The abstract FSM model used here provides for hierarchical manipulation of an FSM, indepedently of its internal state. It introduces two predefined states called idle and active for each FSM. An FSM is idle if none of its internal states are active. In the case of an HFSM, the idle state implies that all the component FSMs are idle. A state is active when it is executed. An FSM is active when one of its states is active. An HFSM is active when one of its components is active.

This abstract FSM model corresponds to a VHDL process with the addition of two control signals (enter_FSM, exit_FSM). The process is composed of two loops (idle loop and active loop) that models the two states of the abstract FSM. The "active" loop contains the body of the FSM. The "idle" loop emulates the idle state. The transition between the two loops is controlled through the two control signals (see Fig 5). When the exit signal is high, the process exits the active loop and enters the idle loop. The process enters the active loop when the exit signal is low and the enter signal is high.

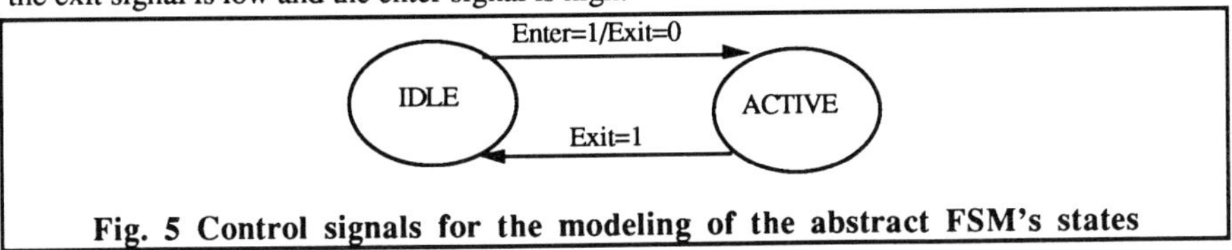

Fig. 5 Control signals for the modeling of the abstract FSM's states

An exit from an FSM forces this FSM to the idle state. If the FSM is hierarchical, then all of its components are also forced to the idle state. Entering an FSM makes this FSM active. In the case of an SFSM, either the default entry state is entered, or an explicit sate can be specified. In the case of a PFSM, all the components are activated.

The MV model is delay independent. The FSMs are synchronized by the wait statements and the transitions between FSMs. The VHDL model produced is also delay independent. The only delays are introduced by the sequential execution of the waits and the transitions. In fact, the execution of each next_state primitive induces a delay of one basic time unit (1 femtosecond in VHDL). This delay is introduced in order to avoid races.

3.2. Algorithms

The translation algorithm starts from an MV description and produces a VHDL model where all the MV primitives are expanded. The file also includes the declaration of the control signals. The translation of (H)FSMs and wait statements are syntax directed. Examples are given below to illustrate the transformations. More details are given for the next_state primitive translation.

Fig. 6 shows the translation of an HFSM. The HFSM request_manager is translated into a VHDL block statement. The exception is translated into a process that executes the action (specified by the exception) when the event is true.

A leaf FSM causes the declaration of two new control signals (enter_fsm, exit_fsm). It is translated into a VHDL process composed of an Idle loop and an Active loop. The two loops corresponding to the FSM "WAIT_REQ" are shown in Fig. 7. The wait statements are transformed in order to allow real time handling of exceptions. The translation process combines the wait with a loop where the exception signals and the control signals are checked. The different forms of the VHDL wait statement need different translation schemes. Figure 8 shows the the translation result of a "wait until" statement.

The translation of the next_state primitive needs an extra computation in order to determine which FSM to exit from and which FSM to enter. The translation algorithm is given in the rest of this section.

```
REQUEST_MANAGER: block -- REQUEST_MANAGER: SFSM
        signal enter_INIT : wor bit register := '1' ;
        signal exit_INIT  : wor bit register := '0' ;
        ...
        SEND_RECV : block -- SEND_RECV : PFSM
                --b1: Exception (RESTART, NEXT_STATE ("INIT"));
                SEND_RECV_B1_EXCEPTION : process
                begin
                        if now = 0fs then BUFFIN <= null ; ....
                        if RESTART = '1' then  [EXPANSION OF THE ACTION]
                        wait on restart;
                end process SEND_RECV_B1_EXCEPTION ; ...
                RECV : block -- RECV : SFSM
                ...
                end block RECV; -- end SFSM RECV;
                ...
        end block SEND_RECV; -- end PFSM SEND_RECV;
end block REQUEST_MANAGER ; -- end SFSM REQUEST_MANAGER ;
```

Figure 6. Example of a HFSM Expansion

```
WAIT_REQ: process -- WAIT_REQ: process
begin
        if now = 0fs then
                BUFFIN <= null ;
                enter_WAIT_REQ <= null; ...
                if exit_WAIT_REQ = '1' then
                        IDLE_LOOP: while true loop
                                wait on enter_WAIT_REQ, exit_WAIT_REQ ;
                                if exit_WAIT_REQ = '0' and enter_WAIT_REQ ='1';
                                        then exit IDLE_LOOP;
                                end if;
                        end loop IDLE_LOOP;
                end if;
                ACTIVE_LOOP: while true loop
                        wait for 1fs;
                        B_FULL <= '0' ;
                        -- WAIT UNTIL REQIN = '1' ;
                        WAIT_2_LOOP: while true loop
                                wait until REQIN = '1' or restart'event or
                                                        exit_WAIT_REQ'event ;
                                if restart = '1' then wait for 1fs; exit ACTIVE_LOOP;
                                elsif exit_wait_req = '1' then exit ACTIVE_LOOP;
                                elsif REQIN = '1'        then exit WAIT_2_LOOP ;
                                end if;
                        end loop WAIT_2_LOOP;
                end loop ACTIVE_LOOP;
end process WAIT_REQ;
-- end FSM WAIT_REQ;
```

Figure 7 example Of an FSM and a wait Expansion

The main data structure used by the translation algorithm is a tree representation of the hierarchy contained in the MV description. The root of the tree is the controller. The leaves are FSMs. Each HFSM is a vertex in the tree. Each vertex contains a record of the exceptions specified at the corresponding level. Figure 8 shows an algorithm that uses this tree (called CHT) in order to compute a transition between two FSMs. Figure 9 depicts an application example of this procedure. The next state primitive is translated to a set control signal assignment statements and an exit from the active loop. In order to avoid conflicts, the control signal assignments are guarded by the higher priority transitions.

```
procedure compute_transition (CURRENT, NEXT : FSM_name) ;
1.  Make use of the CHT in order to find out the shortest path between the Two FSMs [CURRENT, NEXT],
this path contains a single common parent (X).
2.  split the set of FSMs that belong this path into three disjoint subsets
2.1.    the FSMs that belong to the path [CURRENT, X[ (excluding X) are the set of FSM to exit from.
2.2.    the FSM X
2.3.    the FSMs that belong to the path ]X , NEXT] (excluding X) are the set of FSM to enter to.
3.  generate corresponding assignment statement.
```

Fig. 8. Next_state Primitive Computation Algorithm.

```
-- expansion of NEXT_STATE ("BUF_FULL");
-- Exit WAIT_REQ, Enter BUF_FULL
if NOT (restart = '1' ) then -- NO HIGHER PRIORITY TRANSITION
                -- EXIT WAIT_REQ
        enter_WAIT_REQ  <= '0', null after 1fs;
        exit_WAIT_REQ  <= '1' , null after 1fs;
                -- ENTER BUF_FULL
        enter_BUF_FULL <= '1' , null after 1fs;
        exit_BUF_FULL   <= '0', null after 1fs;
end if;
wait for 1fs;
exit ACTIVE_LOOP;
```

Figure 9 example Of next_state Expansion

4.Conclusion

We have described MV, a new specification language for high level controllers. MV is based on VHDL with the addition of a few selected primitives to simplify the specification of hierarchical and communicating FSMs. This is achieved by hiding all the details related to process control and communication. We also presented a new algorithm that translates MV into efficient and accurate VHDL models. This algorithm is based on a powerful abstract FSM model that allows hierarchical manipulation of FSMs independently of their internal state. Early experiments showed that the use of MV allows for more than a fourfold reduction in the size of the description when compared to VHDL. The VHDL verbosity is mainly caused by the lack of facilities for the specification of process control within hierarchical FSMs (e.g. to stop or restart a VHDL process that contains wait statements) and the lack of exception handlers.

In addition to the automatic generation of VHDL and SIF models, future work includes the extension of MV with new facilities such as time-out for the specification of real time controllers; and channels, for the specification of higher level communication between FSMs.

References

[Agne91] Dave Agnew, "VHDL extensions needed for synthesis and design", CHDL'91, Marseille France, 1991.

[BeCo84] G. Berry & L. Cosserat, "The Esterel Synchronous Programming Language and its Mathematical Semantics", Technical report, Ecole Nat. Superieure de mines de Paris, 1984.

[BuMa88] S. Burns & A.J MArtin, "syntax Directed translation of concurrent programs into selftimed circuits", Proc. of the 5th MIT conference on advanced research in VLSI, 1988.

[Carl90] S. Carlson, "Introduction to HDL-Based Design Using VHDL", Synopsys Inc., 1990.

[Clar89] E.M. Clarke al.,"A language for Compositional Specification and Verification of Finite State Hardware Controllers", CMU-CS-89-110, Jan. 1989.

[DrHa89] D. Drusinsky and D. Harel, "Using Statecharts for Hardware Description and Synthesis", IEEE Transactions on CAD, Vol.8, No. 7, July 1989, pp. 798-807.

[Dutt91] N.D.Dutt al. , "A user interface for VHDL behavioural modeling", CHDL'91, Marseille, France, April 1991 .

[Hare90] D. Harel et al, "Statecharts: A Working Environment for the Development of Complex Reactive Systems", IEEE Trans. on Software Engineering, Vol. 16-4, Apr. 1990, pp.403-413.

[Hoar78] C.A.R. Hoare, "Communicating Sequential processes", CACM, Vol 21, No 8, 78

[HoKN85] T. Hoshino, O. Karatsu, T. Nakashima, "HSL-FX" A Unified language for VLSI design" CHDL85, Aug 1985.

[ISO87] International Standard, ESTELLE (Formal description technique based on an extended state transition model), ISO/DIS 9074, 1987.

[JePa91] A.A. Jerraya & P.G. Paulin, "SIF an interchange Format for the design and synthesis of high-level Controllers", High Level Synthesis Workshop 91.

[Kara89] O. Karatsu, "VLSI design language Standardization in Japan", 26th DAC, 1989.

[Loto89] "LOTOS a formal description technique based on the temporal ordering of observational behavior", ISO, IS 8807, Feb. 89.

[McPC90] M.C. McFarland, A.C. Parker, R. Camposano, "The High -Level Synthesis of Digital Systems", Proc. of IEEE, Vol. 78, No. 2, Feb. 1990, pp.301-318.

[NaVa90] T.S. Narayan & F. Vahid and DD. Gajski, " Translating System Specification to VHDL", submitted to the DAC'91, Nov 90.

[PeLi90] M.A. Perkowski, J. Liu, "Generation of Finite State Machines from Parallel Program Graphs in Diades", Proc. of ISCAS, 1990, pp. 1139-1142.

[RoGi90] A. Rotman, R. Ginosar, "Control Unit Synthesis from a High-Level Language", Submitted to the IEEE Transactions on CAD, 1990.

[Tier88] R.A. Tierney, Modeling Complex Systems, VLSI system design,May 1988.

[Udli90] UDL/I Language Reference, Draft Version 1.0b4, Oct 04 1990.

[Vhdl87] "IEEE Standard VHDL LRM", IEEE Std 1076-1987.

VLSI 91
A. Halaas and P.B. Denyer (Eds.)
Elsevier Science Publishers B.V. (North-Holland)

Towards A Formal Model of VLSI Systems Compatible with VHDL*

Philip A. Wilsey, Timothy J. McBrayer, and David Sims

Computer Architecture Design Laboratory
Dept. of Electrical and Computer Engineering
University of Cincinnati, Cincinnati, OH 45221-0030

Abstract

The work described in this paper is concerned with developing a formal semantic model of VLSI systems compatable with the hardware description language VHDL. Currently, we have developed a formal model sufficiently robust to characterize many key components of VHDL. In particular, we have defined the propagation of signal values up and down the component hierarchy (complete with type conversion functions), bus resolution of signal values, process statements including the wait statement, and transaction list definition. The one main concept of VHDL yet to be introduced into this model is the delta delay.

1 Introduction

Formal models of computing systems are important because they allow the construction of general purpose design automation systems. Unfortunately, such models are often difficult for the designer to manipulate. Therefore, a class of computer languages, called Hardware Description Languages (HDLs) have emerged for machine description (*e.g.*, AADL [7], CONLAN [19], ELLA [6, 18], SBL [11], ACE [22], Verilog [21], and VHDL [16, 17]). These languages can then be used by the architect for formally describing the design and then translated into the internal models needed by the automated design tools. Although these languages are quite powerful for describing architectures across a range of abstraction levels, they usually lack a formal semantic definition and thus, even though an architecture is formally described using an HDL, the resulting description may still be subject to different, possibly conflicting, interpretations.

Although some attempts have been made toward formal definitions (*e.g.*, [4, 5, 9, 10, 15]), the lack of such definitions is not an oversight on the designer of the HDL. The real problem lies with the fact that no readily acceptable mechanism for specifying the formal semantics of temporal based languages really exists. Unfortunately, the traditional

*This work was partially supported by the Defense Advanced Research Projects Agency under order no. 7056, monitored by the Federal Bureau of Investigation under contract number J-FBI-89-094.

```
Interval Relation        Pictoral Example
-------------------------------------------
t before s               ttt sss

t equals s                ttt
                          sss

t meets s                tttsss

t overlaps s              tttttt
                             ssssss

t during s                  ttt
                          ssssss

t starts s                ttt
                          ssssss

t finishes s                 ttt
                          ssssss
```

Figure 1: Relationships of Time Intervals

techniques for semantic definition of programming languages (e.g., denotational semantics [12, 20], axiomatic semantics [14, 13], operational semantics, and so on) are, for one reason or another, inadequate for use in formally defining these languages.

The research described in this paper describes an approach for defining the semantics of VHDL as a formal machine model using an interval temporal logic. The specific temporal logic to be used was originally developed by J. Allen [1, 2, 3] for reasoning about action and time. Thus far, we have successfully defined the semantics of a subset of the VHDL hardware description language. Although we do not consider some significant concepts like delta delays, work is continuing to complete an entire VHDL semantic model.

The remainder of this paper is organized as follows. Section 2 introduces the interval temporal logic that we use for the formal definition. Section 3 presents the temporal elements, object declarations, component instantiations, and process statements. Section 4 provides the formal axioms specifying the process of updating the machine state (variables and signals) from the state transformations (signal and variable assignment statements). Finally, Section 5 contains some concluding remarks.

2 An Interval Temporal Logic

In this work, we will use an interval temporal logic that was originally developed by Allen for reasoning about action and time [1, 2, 3]. Our main reason for selecting Allen's formalism is that it allows us to reason about the relationships of time intervals to one another as well as the relationships of actions to time intervals.

This logic is based upon time intervals rather than time points. This approach is taken so that one can easily reason about the occurrence of an action without concern as to the precise point in time that the action actually occurs. With respect to time intervals, Allen has identified seven relationships, with inverses, that can hold between two time intervals. These relationships are summarized in Figure 1.

Each of the interval relations of Figure 1 has a corresponding predicate in the logic. Let t and s each represent some time interval; then these predicates are:

EQUALS(t,s): t and s represent the same time interval,

MEETS(t,s): t is before s and there is not an interval r that is between t and s,

DURING(t,s): t is fully contained within s,

OVERLAPS(t,s): t begins before s begins, s begins before t finishes, and t finishes before s finishes,

STARTS(t,s): t and s begin together, but t finishes before s finishes,

FINISHES(t,s): t starts after s starts, but they finish together, and

BEFORE(t,s): t finishes before s begins and there is a time interval r that is between t and s.

Allen has defined a set of axioms for the behavior of these predicates. Among these are axioms asserting that each relationship is mutually exclusive of the others; others define the transitivity of the predicates. A detailed discussion of these can be found in [2]. For our purposes, we are concerned only with the fact that for any given interval, t, there exists another interval for each predicate of the logic such that the predicate holds. That is, given an interval t, there is an interval that meets t, that equals t, that starts with t, and so on.

Three additional predicates that will be useful in this work are:

IN(t,s) $\Leftrightarrow$ DURING(t,s) $\vee$ EQUALS(t,s),
BEGINS(t,s) $\Leftrightarrow$ STARTS(t,s) $\vee$ STARTS(s,t) $\vee$ EQUALS(s,t), and
ENDS(t,s) $\Leftrightarrow$ FINISHES(t,s) $\vee$ FINISHES(s,t) $\vee$ EQUALS(s,t).

Informally, the predicate IN states that either t and s represent the same interval or t is fully contained within s. The predicate BEGINS states that t and s begin together. The predicate ENDS states that t and s finish together.

Two other components of Allen's logic relate to *properties* and *events*. A *property* is a time dependent predicate defined on some aspect of the universe under consideration, or a particular state of that universe (*e.g.*, a static value of a program variable). An *event* is something that may cause a state change or the value of a particular predicate to change (*e.g.*, incrementing the value of a program variable). Corresponding to these two notions, Allen defines two predicates: HOLDS and OCCUR.

The predicate HOLDS takes a property, p, and a time interval, t, and is true if the predicate p is true throughout t. One of the most important properties of the HOLDS predicate is that if it is true for some predicate p and time interval t then it is also true for every time interval s contained in t. Formally, we state:

HOLDS(p,t) $\Leftrightarrow \forall$ s : (DURING(s,t) $\vee$ EQUALS(s,t)) $\Rightarrow$ HOLDS(p,s).

The predicate OCCUR takes an event, e, and a time interval, t, and is true if the event e occurs in t and there is no subinterval s of t in which e occurs. Thus, the predicate OCCUR has the property that if it is true for an event e in time t, then it is not true for any subinterval of t. Formally, this property of the OCCUR predicate is defined by the axiom:

OCCUR(e,t) $\wedge$ DURING(s,t) $\Rightarrow \neg$ OCCUR(e, s).

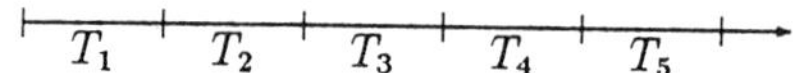

Figure 2: Pictorial View of the Timekeeper

3 The Temporal and Data Spaces of the Model

In Sections 3 and 4, we briefly demonstrate how Allen's logic can be used in the construction of a formal semantic model of VHDL. Throughout this remainder of this paper, a number of simplifying assumptions regarding the semantics of VHDL are made. In most cases, these simplifications occur only because of space constraints. For example, we will ignore concerns for transport or inertial delays and instead assume that if a signal assignment statement schedules a transaction defining a new value of the destination, the new value *will* be assigned to the destination.

The remainder of this section proceeds as follows. First a formal model of time is developed that serves as the basic unit of time of VHDL (femtoseconds). Second, signals are formally characterized. Finally, the propagation of values up and down the component hierarchy is defined.

Before presenting the formal definition, however, we must first introduce some notation and introduce a function. In this model, each element of VHDL is defined as a tuple. Due to the complexity and nesting of references through the tuples, we will always fully qualify references to the elements of tuples. For example, assume that we have the two definitions:

$$P = \{ < M,\ E,\ D,\ fn,\ T > \}$$
$$pa = < F,\ fn_1,\ A,\ fn_2 >$$

where the element F of pa is a member of P. The notation pa.F.M will then be used to reference the element M of F of pa. In addition, we will also use the function VAL(E,t) to indicate the value of the expression E in the time interval t.

3.1 Time

In order to develop a technique for defining the semantic behavior of VHDL, we must first develop a unit time scale. This unit time scale will then denote time in the described machine. In this definition, the unit time scale is defined by an entity called the *timekeeper*. The timekeeper, *TK*, provides a single *universal temporal reference frame,* and is defined as:

$$\text{TK} = \{\text{tk(i)} : \text{i} \geq 0 \wedge \text{MEETS(tk(i),tk(i+1))}\}.$$

The timekeeper defines a set of unit time intervals that are adjacent to one another (Figure 2). Throughout the remainder of this paper, we will use, without qualification, the notation tk(i) to denote the ith element of the timekeeper.

VHDL has a predefined language environment that contains, among other things, the function NOW which returns the current simulation time. Formally, the function NOW is defined as:

$$\forall\, \text{i} : \text{i} \geq 0,\ \text{VAL(NOW,tk(i))} = \text{i}$$

which simply defines the value of the function NOW at every timekeeper unit interval.

3.2 Signals

The set of *signals* (S) in VHDL denote entities that hold values of the machine state. In this definition of VHDL, a distinction between signals declared with a signal declaration and port declarations is made. Port declarations will be discussed in the next section and they are not considered to be in the set S. Formally, a signal, s $\in$ S, can be represented as a 7-tuple:

$s = < E, D, fn, T, TY, K, I >$

where E is the *effective value*, D is the *driving value*, fn:$\{v\} \rightarrow v$ is the *resolution function* mapping a set of values into a single value, T is the *transaction list*, TY is the *data type* of the signal, K is the *signal kind* (`register` or `bus`), and I is the the *initial value*. In the case where no bus resolution function exists in the VHDL description of the signal, the function fn will be a function that simply returns the input value. In order to simplify exposition, the data type of a signal will not be further used or discussed in this paper and we will assume that all signals (ports and variables) have the same scalar type.

The initial value must be placed on the transaction list to update the signal in the initial timekeeper unit interval. This is defined by the axiom:

$\forall$ s :s $\in$ S, VAL(s.T,tk(0)) = s.I

which initialized the machine to a known state.

The resolution function fn is used by the simulation cycle to update the driving value of the signal. The distinction between driving and effective values of a signal is not really necessary, but, it does simplify the definition of the simulation cycle.[1] Because we have included a distinction between driving and effective values, it is necessary to assert that the effective value always reflects the driving value. Formally, we state:

$\forall$ s,t :s $\in$ S, VAL(s.E,t) = VAL(s.D,t)

Distinctions between driving and effective values occur in the formals and actuals of the ports of an instantiated component. This is described in the next section.

3.3 Component Instantiations and Port Connections

The set of *components* (C) in a VHDL description are entities that can be instantiated one or more times for the purposes of describing devices that process and transfer information of the machine state. Components are instantiated with the elements of its formal port clause associated with actual signals (or ports) of the instantiating environment. In the following two subsections, we provide a definition of a formal port clause and then we characterize the propagation of data up and down through the links of the component hierarchy.

---

[1]The VHDL language reference manual commonly refers to ports as signals. In our definition, we completely separate the set of signals from the set of port declarations. Thus, the statement that driving and effective values are the same *is* correct.

3.3.1 Ports

The elements of the set of *formal ports* (P) represent all elements of the formal port clauses found in a VHDL description. In this discussion, we consider only ports of mode IN, OUT, or INOUT.

Formally, a port declaration, p $\in$ P, is represented as a 6-tuple:

$$p = < M,\ E,\ D,\ fn,\ T,\ TY >$$

where M is the *mode* specifying direction for the flow of data through the port, E is the *effective value*, D is the *driving value*, $fn{:}\{v\}\rightarrow v$ is the *resolution function* mapping a set of values into a single value, T is the *transaction list*, and TY is the *data type*. In the case where no bus resolution function exists in the VHDL description of the port declaration, the function fn will be a function that simply returns the input value.

The transaction list and driving values of ports of mode OUT or INOUT are updated by signal assignment statements as defined later in this paper. The effective values of ports of mode IN or INOUT are updated by the propagation of the effective values from the actual designator given in the component instantiation.

3.3.2 Port Associations

In conjunction with component instantiations there are a set of *port associations* (PA) that links the formal port names given in the component definition with the actual signal (or port) names of the instantiating environment. The set PA contains all of the port associations of a complete VHDL description. Formally, a port association, pa $\in$ PA, can be represented as a 4-tuple:

$$pa = < F,\ fn_1,\ A,\ fn_2 >$$

where $F \in P$ is the *formal* port designator, fn_1 is a type conversion function mapping values of the formal's type to values of the actual's type, $A \in S \cup P$ is the *actual* port designator, fn_2 is a type conversion function mapping values of the actual's type to values of the formal's type. In the case where no type conversion function(s) is specified in the VHDL description, fn_1 and/or fn_2 will be a function that simply returns the input value.

In the process of linking the names of the formal and actual parts, instantiating a component initiates the transfer of data up and down the component hierarchy. The data flows up the hierarchy by updating the driving values of the actuals from the driving values of the formals (for ports of mode OUT or INOUT only). Conversely, data flowing down the hierarchy causes the effective values of the formals to be updated with the effective values of the actuals. In either case, data flowing up or down, there may exist type conversion functions between the transmitted formal and actual values.

Formally, the transmission of effective values down the hierarchy (complete with type conversion) is expressed by the axiom:

$$\forall \text{ pa,t :(pa.F.M = IN)} \vee \text{(pa.F.M = INOUT), VAL(pa.F.E,t)} = fn_2\text{(VAL(pa.A.E,t))}$$

Informally this axioms state that for all time intervals, the effective value of the formal will always be equal to the type converted value of the actual.

```
        P1:    process begin
S1                    a <= a1;
S2                    wait on a until c1 for t1;
S3                    if c2 then
S4                           b <= b1;
S5                           wait for t2;
                      else
S6                           b <= b2;
                      end if;
S7                    d <= d1;
               end process;
```

Figure 3: A VHDL Process Statement

$$
\begin{aligned}
&\text{P1} = \langle \text{D, SS, HEAD, SUCC} \rangle \\
&\quad \text{D} = \emptyset \\
&\quad \text{SS} = \{\text{S1, S2, S3, S4, S5, S6, S7}\} \\
&\quad \text{HEAD} = \text{S1} \\
&\quad \text{SUCC(S1,t)} = \text{S2} \qquad \text{SUCC(S4,t)} = \text{S5} \\
&\quad \text{SUCC(S2,t)} = \text{S3} \qquad \text{SUCC(S5,t)} = \text{S7} \\
&\quad \text{SUCC(S3,t)} = \begin{cases} \text{S4,} & \text{if VAL(c2,t) = true} \\ \text{S6,} & \text{otherwise} \end{cases} \qquad \begin{array}{l} \text{SUCC(S6,t)} = \text{S7} \\ \text{SUCC(S7,t)} = \text{S1} \end{array}
\end{aligned}
$$

Figure 4: Formal Representation of a VHDL Process Statement

Similarly, data transmission up the hierarchy is expressed formally as:

$$\forall \text{ pa,t} :(\text{pa.F.M} = \text{OUT}) \vee (\text{pa.F.M} = \text{INOUT}), \text{VAL(pa.A.D,t)} = fn_1(\text{VAL(pa.F.D,t)}).$$

3.4 The Process Statement

Formally a process statement, PR, is defined as a 4-tuple:

$$PR = < D,\ SS,\ HEAD,\ SUCC >$$

where D is a set of *declarations*, SS is the set of *statements* contained in the process statement, *HEAD* is a function to return the statement from SS that is initially evaluated at tk(0), and *SUCC* is a function of two arguments (a statement and a time interval) that determines which statement will be evaluated next.

An example of a process statement and its representation in the model is shown in Figures 3 and 4. For the sake of simplicity, the statements of the process statement in Figure 3 are assigned labels (S1 through S7) which are used in the formal model representation of Figure 4.

In the following three subsections, the wait, signal assignment, and variable assignment will be examined in greater detail. Because of space considerations, the if statement will not be further discussed. Each subsection will derive a logic formula establishing the constraints of the time interval within which the corresponding statement requires for evaluation. The effects of the evaluation of these statements on the machine state will be defined in Section 4.

3.5 The WAIT Statement

Formally, a wait statement, wa, is represented as a 3-tuple:

$wa = < SL, C, TO >$

where SL is a set containing the *sensitivity list* of signals, C is a *boolean expression*, and TO is the *timeout clause* of the wait statement.

We now turn to the problem of establishing the logical relationships on the time interval where a wait statement occurs. For this purpose, we use the following definition:

$$\begin{aligned}\text{WAIT(wa,t)} \equiv\ & \exists\, \text{i,j} : \text{BEGINS(tk(i),t)} \wedge \text{ENDS(t,tk(j))} \wedge \text{j} > \text{i} \wedge \\ & ((\exists\, \text{s} \in \text{wa.SL} : \text{EVENT(s,tk(j))} \wedge \text{VAL(wa.C,tk(j))} \vee \\ & (\text{wa.TO} = (\text{j-i}))) \wedge (\forall\, \text{k,s} : \text{i} < \text{k} < \text{j} \wedge \text{s} \in \text{wa.SL}, \\ & \qquad \neg(\text{EVENT(s,tk(k))} \wedge \text{VAL(wa.C,tk(k))})))\end{aligned}$$

where EVENT is a function (formally defined in Section 4.3) which defines whether a VHDL event occurred. Informally, this definition defines the endpoints of an interval in which a wait statement is evaluated. The first three lines of the formula define the final endpoint of the wait statement. The last two lines indicate that no intermediate time interval exists to satisfy the termination of the wait statement.

3.6 The Signal Assignment Statement

Formally, a signal assignment statement, sa, is represented as a 4-tuple:

$sa = < D, TI, E, d >$

where D is the *destination* (signal name) of the statement, TI is the transport/inertial delay specifier,[2] E is the *expression* defining the new value for the signal, and d is the *delay* time value specified in the after clause. Since delta delays are being ignored, the delay, d, will always be nonzero.

A signal assignment statement is evaluated in a subinterval of a timekeeper unit interval following the proper order of the statements in a process statement. Formally then, we need only define:

$\text{SGNL(sa,t)} \equiv \exists\, \text{i} : \text{i} \geq 0 \wedge \text{IN(t,tk(i))}$.

Informally, this formula merely indicated that evaluation of any signal assignment statement is wholly contained within a timekeeper unit interval.[3]

---

[2]Although we have included this element of the tuple definition, we will hereafter ignore concerns for transport or inertial delays and instead assume that if a signal assignment statement schedules a transaction defining a new value of the destination, the new value *will* be assigned to the destination. Including it in the formal model is relatively straight forward, but it complicates the axioms defining the transaction list (Section 4.2) unnecessarily.

[3]On exception to this is at the very beginning of time where the predicate BEGINS(t,tk(0)) may hold. Hence we use the predicate IN rather than the more restrictive predicate DURING.

3.7 The Variable Assignment Statement

The variable assignment statement will follow from the signal assignment statement. That is, evaluation of a variable assignment statement occurs in a time interval that has the exact same constraints as those of the signal assignment statement. Formally a variable assignment statement, va, is represented as a 2-tuple:

$va = < D,\ E >$

where D is the *destination* (variable name) of the statement, and E is the *expression* defining the new value for the variable.

The boundaries of the time interval within which a variable assignment statement executes is defined by the formula:

$\text{VAR(va,t)} \equiv \exists\ \text{i} : \text{i} \geq 0 \wedge \text{IN(t,tk(i))}.$

Which merely indicates that the evaluation of any variable assignment statement is wholly contained within a timekeeper unit interval.

4 State Transformations

We will now explore the problem of ordering the statements within a process statement and defining the time intervals within which each successive statement is evaluated. Because variable and signal values may be used to to order the statements and because satisfaction of the conditions ending a wait statement depend upon activity on signals, it is necessary to consider the problem of statement order in conjunction with transaction list and signal and variable updates. However, from a presentation standpoint, we can examine only one of these parts at a time. Thus, although they are presented sequentially, the reader must recognize that the set and variable definitions in this section are highly codependent. The remainder of this sections presents each of (i) statement ordering, (ii) transaction list updating, (iii) signal value updating, and (iv) variable updating.

4.1 Ordering the Statements of a Process Statement

Let PR be a set denoting all of the process statements, WA be the set of all wait statements, SA be the set of all signal assignment statements, and VA be the set of all variable assignment statements. In this model each process statement, pr, has an associated set, EV_{pr}. EV_{pr} is a set of 2-tuples denoting a statement and a time interval in which the statement is evaluated. Formally, the sets EV_{pr} are defined as:

$$\begin{aligned}\forall\ \text{pr} \in \text{PR},\ \exists\ EV_{pr} : EV_{pr} = \{ < s_i, t_i > :\ & \text{i} \geq 0 \wedge \\ & \text{BEGINS}(t_0,\text{tk}(0)) \wedge \text{MEETS}(t_i, t_{i+1}) \wedge \\ & s_0 = \text{pr.HEAD} \wedge s_{i+1} = \text{pr.SUCC}(s_i, t_i) \wedge \\ & ((s_i \in \text{WA} \wedge \text{WAIT}(s_i, t_i)) \vee \\ & \ (s_i \in \text{SA} \wedge \text{SGNL}(s_i, t_i)) \vee \\ & \ (s_i \in \text{VA} \wedge \text{VAR}(s_i, t_i)) \vee \\ & \ (s_i \in \text{IF} \wedge \text{IF\_S}(s_i, t_i)))\}.\end{aligned}$$

Thus, the statements, s_i, in a process statement are ordered in successive time intervals according to the successor function SUCC. The initial statement of the process statement begins at tk(0) and the constraints on the boundaries of each statement are defined by the logic formula from Section 3.

Finally, we will call each EV_{pr} an evaluation set and let EV be a set containing the elements from all of the evaluation sets. That is:

$$\mathrm{EV} = \bigcup_{pr \in PR} EV_{pr}.$$

4.2 Transaction List Updates

We must now select elements from the evaluation sets to determine the values of the transaction list for each signal and port. Formally we write:

$$\begin{aligned}&\forall \text{ s,i} : \text{s} \in \text{S} \cup \text{P} \wedge \text{i} > 0,\\ &\qquad \text{VAL(s.T,tk(i))} = \{\text{VAL(sa.E,tk(j)} : \exists \text{ <sa,t>,j} :\\ &\qquad\qquad \text{<sa,t>} \in \text{EV} \wedge \text{sa.D} = \text{s} \wedge \text{j} \geq 0 \wedge \text{IN(t,tk(j))} \wedge \text{i-j} = \text{sa.d}\}.\end{aligned}$$

Informally this axiom states that for each signal or port, s, and every timekeeper unit interval, a set of values is defined. These values are defined by the evaluation of signal assignment statements. The important signal assignment statement evaluations are those that were evaluated in an earlier timekeeper unit interval, tk(j), whose distance (in timekeeper unit intervals) from the current time, tk(i), equals the delay specified on the signal assignment statement (sa.d).

4.3 Signal Updates

Next, the process of updating the signal driving values from its transaction list must be defined. This is accomplished with the axiom:

$$\begin{aligned}&\forall \text{ s,i} : \text{s} \in \text{S} \cup \text{P} \wedge \text{i} > 0,\\ &\qquad (\text{VAL(s.T,tk(i))} \neq \emptyset \wedge \text{VAL(s.D,tk(i))} = \text{s.fn(VAL(s.T,tk(i)))} \wedge\\ &\qquad\qquad \text{EVENT(s,tk(i))} = \text{true}) \vee\\ &\qquad (\text{VAL(s.T,tk(i))} = \emptyset \wedge \text{VAL(s.D,tk(i))} = \text{(VAL(s.D,tk(i-1)))} \wedge\\ &\qquad\qquad \text{EVENT(s,tk(i))} = \text{false})\end{aligned}$$

Which indicates that either the transaction list is non-empty and a new value is defined for the signal's driving value or the transaction list is empty and the signal's driving value is retained.[4] Notice that the function EVENT is defined indicating the occurrence of a signal value change.

---

[4]Which is true only for signals whose signal kind is `register`. For signals of kind `bus`, this axiom would be extended slightly to select a unspecified value from the set of values that the signal can hold. This is a trivial extension to this axiom and a related example can be found in [23].

4.4 Variable Updates

To compute new values of variables, we must inspect the evaluation set that was derived in Section 4.1. Let V denote the set of all variables in the design. Variable assignment statements are then defined by:

$$\forall\ v,<s,t> : v \in V \wedge <s,t> \in EV \wedge s \in VA \wedge s.D = v,$$
$$\exists\ \hat{t} : MEETS(t,\hat{t}\,) \wedge VAL(v,\hat{t}\,) = VAL(s.E,t)$$

which simply updates each variable (v) with new values in a time interval ($\hat{t}$) that is immediately adjacent to the time interval (t) in which a variable assignment statement (va) updating that variable is evaluated.

5 Conclusions

In this paper, we have presented an approach for developing a formal semantic model of VHDL. Probably the most difficult aspect of developing this formal definition has been in understanding the semantics of VHDL. Clearly the formal semantic definition presented in this paper is unambiguous and easy to follow. In addition to providing a clean, unambiguous semantic definition for users and tool builders alike, this definition can facilitate machine verification using techniques such as those developed by [8].

Acknowledgments: We would like to thank Hal Carter for reviewing an earlier draft of this paper. In addition, we would like to thank Dave Barton, Hal Carter, Ron Waxman, Steve Levitan, and Praveen Chawla for their help in clarifying certain points regarding the semantics of VHDL.

References

[1] Allen, J. F. An interval-based representation of temporal knowledge. *Proc. 7th Int. Joint Conf. on Artificial Intelligence* (Aug. 1981), 221–226.

[2] Allen, J. F. Maintaining knowledge about temporal intervals. *Communications of the ACM 26* (Nov. 1983), 832–843.

[3] Allen, J. F. Towards a general theory of action and time. *Artificial Intelligence 23* (1984), 123–154.

[4] Borrione, D. The worker model of evaluation for compter hardware description languages. *Proc. 5th Int. Conf. on Computer Hardware Description Languages* (1981), 3–21.

[5] Brock, B. C., and Hunt, Jr., W. A. The formalization of a simple hardware description language. *Proc. Int. Workshop on Applied Formal Methods for Correct VLSI Design* (Nov. 1989), 83–98.

[6] Computer General Electronic Design Ltd. *The ELLA Language Reference Manual*, 1990. Issue 4.0.

[7] DAMM, W., DOEHMEN, G., MERKEL, K., AND SICHELSCHMIDT, M. The AADL/S* approach to firmware design verification. *IEEE Software 3*, 4 (July 1986), 27–37.

[8] DECAMP, E. Methodology for the specification of temporal behaviours, applied to the formal verification of circuits. Tech. Rep. 580, Laboratoire ARTEMIS IMAG, France, 1986.

[9] GOOSSENS, K. G. W. An operational semantics for a subset of the HDDL ELLA. Version 3.0 Manuscript, April 1990.

[10] GOOSSENS, K. G. W. Semantics for picoELLA. Manuscript, June 1990.

[11] GOPALAKRISHNAN, G. C., SMITH, D. R., AND SRIVAS, M. K. An algebraic approach to the specification and realization of VLSI designs. *Proc. 7th Int. Symp. on Computer Hardware Description Languages* (1985), 131–137.

[12] GORDON, M. J. C. *The Denotational Description of Programming Languages: An Introduction.* Springer-Verlag, New York, NY, 1979.

[13] GRIES, D. *The Science of Programming.* Springer-Verlag, New York, NY, 1981.

[14] HOARE, C. A. R. An axiomatic approach to computer programming. *Communications of the ACM 12*, 10 (Oct. 1969), 576–580, 583.

[15] ISHIURA, N., YASUURA, H., AND YAJIMA, S. NES: The behavioral model for the formal semantics of a hardware design language UDL/I. *Proc. 27th Design Automation Conference* (1990), 8–13.

[16] LIPSETT, R., MARSCHNER, E., AND SHAHDAD, M. VHDL — the language. *IEEE Design & Test 3*, 2 (April 1986), 28–41.

[17] LIPSETT, R., SCHAEFER, C., AND USSERY, C. *VHDL : Hardware Description and Design.* Kluwer Academic Publishers, Boston, MA, 1989.

[18] MORISON, J. K., PEELING, N. E., AND THORP, T. L. The design rational of ELLA, a hardware design and description language. *Proc. 7th Int. Conf. on Computer Hardware Description Languages* (Aug. 1985), 303–320.

[19] PILOTY, R., BARBACCI, M., BORRIONE, D., DIETMEYER, D., HILL, F., AND SKELLY, P. CONLAN report. In *Lecture Notes in Computer Science*, vol. 151. Springer-Verlag, 1983.

[20] STOY, J. E. *Denotational Semantics: The Schott-Strachey Approach to Programming Language Theory.* The MIT Press, Cambridge, MA, 1977.

[21] THOMAS, D. E., AND MRROBY, P. R. *The Verilog Hardware Description Language.* Kluwer Academic Publishers, Boston, MA, 1991.

[22] WILSEY, P. A., AND DASGUPTA, S. Functional and operational specifications of computer architectures. *Proc. 9th International Symp. on Computer Hardware Description Languages* (June 1989), 209–223.

[23] WILSEY, P. A., AND DASGUPTA, S. A formal model of computer architectures for digital system design environments. *IEEE Trans. on Computer–Aided Design of Integrated Circuits and Systems 9*, 5 (May 1990), 473–486.

VLSI 91
A. Halaas and P.B. Denyer (Eds.)
Elsevier Science Publishers B.V. (North-Holland)

Hardware Design using CASE Tools

Wolfgang Glunz and Gerd Venzl

Siemens AG, Corporate Research and Development, Otto-Hahn-Ring 6, D-8000 München 83, Germany, email: glunz@ztivax.siemens.com

Abstract

This paper presents the application of methods and tools known from CASE (Computer Aided Software Engineering) to hardware design. As a particular method and language that is widely used in software design and well supported by tools we use SDL (Specification and Description Language). The problems and benefits of using SDL for hardware design are discussed. The translation of SDL descriptions into the standard hardware description language VHDL is presented. This translation allows for the simulation and synthesis of SDL specifications.

1 Introduction

Through the availability of synthesis methods, hardware can be designed at high levels of abstraction. Synthesis from register transfer level descriptions to the logic level, including the mapping onto a specific cell library, is already widely used. The next generation of synthesis tools will allow one to describe hardware at the behavioral level and translate these descriptions to the register transfer level. The behavior of the hardware is described using a hardware description language (HDL). The most important among HDLs is VHDL. On the one hand, these languages have a lot of similarities to programming languages like C, PASCAL, or ADA, which raises the question of whether tools and methods known from Computer Aided Software Engineering (CASE) can be used for hardware engineering. On the other hand, electronic design automation (EDA) techniques are also well advanced. In particular, simulation methods are well accepted and routinely used by hardware designers which suggests that simulation tools from EDA might be used for CASE applications.

The general objective of our work is to bring together the CASE and EDA worlds. Specifically this means:

- Provide (graphical) specification and design methods known from CASE to hardware designers.
- Build a bridge from abstract system specification to hardware synthesis.
- Provide simulation facilities for EDA *and* CASE users.
- Support system development independent of implementation technology, i.e., support the design of systems in the phases where the partitioning into software, hardware, and possibly firmware is not yet defined.

Interfacing graphical CASE tools to VHDL was already reported in [1] for a CASE tool based on statecharts, in [2, 3] for structured analysis (SA) and recently in [4, 5] for SpecCharts. The primary goal of this work was to provide simulation facilities for CASE applications through VHDL code generation. In this paper we additionally address hardware implementation through synthesis from VHDL descriptions. Hereby it has to be taken into account that all known VHDL based synthesis systems support a subset of VHDL only.

We first give a short overview of SDL which is a widely used and well supported method and language in the Computer Aided Software Engineering (CASE) domain. Then we discuss the usability of SDL in a hardware design process. The next section describes a first version of an SDL to VHDL code generator. Finally we present some ideas for future extensions of this work. As an example we demonstrate the specification of a simple finite state machine and show the automatically generated VHDL code that can be synthesized by a commercially available synthesis system.

2 What is SDL ?

SDL is a Specification and Description Language standardized by the CCITT [6]. The standard defines two different representations of the language, SDL-PR and SDL-GR. SDL-PR is a purely textual format whereas SDL-GR is a corresponding graphical representation. The graphical language is standardized by means of

- a defined set of usable symbols
- the graphical syntax ("which symbols may be combined in what manner?"
- a relation to the SDL-PR representation. This relation implicitly defines the semantics of the graphical language.

A system described in SDL is regarded as a set of concurrently active *processes*. These processes communicate with each other using *messages*. SDL defines three different views of a system.

- *Block diagrams* describe the static structural decomposition of a system into processes.
- *Process diagrams* are used to describe the behavior of the processes and
- *Sequence charts* can be used to clarify typical traces in a system and showing the causal relationship between actions in the different processes.

The behavior of the processes is described in the form of finite state machines (see Figure 1). Extensions to the classical FSMs are:

- Complex *Tasks* may be performed during a state transition.
- The next state does not only depend on the current state and external inputs but also on data local to a process.
- A queue in which arriving signals are stored is assigned to each process implicitly. The default mode of the queue is first-in first-out but the receiving process can influence the order in which signals are consumed from the queue.
- States can be grouped into state groups.

SDL is well supported through tool sets containing editors, analyzers, code generators, transformators (e.g. sequence charts into process diagrams), and test environments. An object oriented extension to SDL is currently proposed standardization. This extensions will support reusability of SDL specifications. More details on SDL can be found e.g in [6, 7, 8]. An overview on available SDL tools is given in [7].

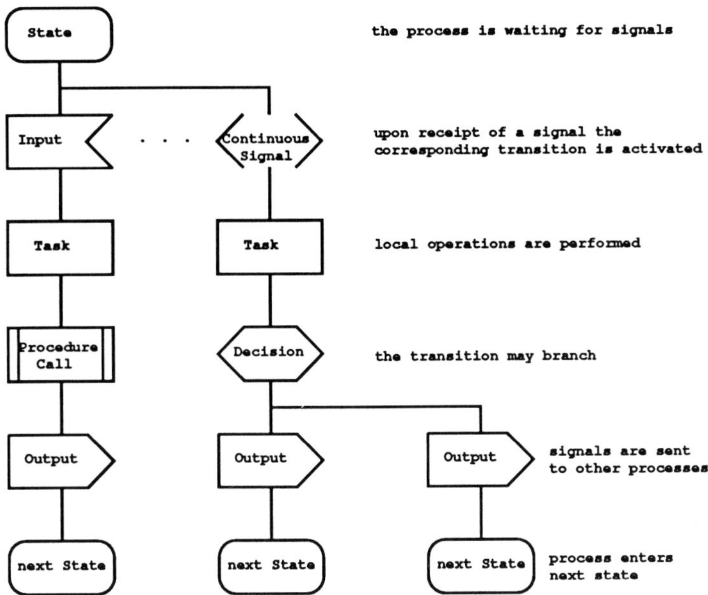

Figure 1: Basic principle of SDL process diagrams

3 Using SDL for Hardware Design

Using SDL in the hardware design process is of interest for the following reasons:

- Hardware is similar to software. Due to the availability of tools synthesizing hardware from behavioral descriptions, hardware can be designed at a level comparable with standard programming languages.
- SDL provides a user friendly graphical representation of behavior, whereas common hardware description languages are just textual representations.
- In the early phases of the system design process it is often open to downstream design decisions which parts and functions of the system will be implemented in hardware or software. SDL has the potential for a method supporting hardware/software codesign.
- The level of abstraction provided by SDL is higher than that of VHDL. As an example communication between processes can be described without considering the communication network and the protocol of the implementation.

A problem in using SDL for hardware design arises from the fact that the language contains features that do not make sense in a hardware context. Thus a "policy of use" has to be worked out in order to insure that SDL specifications will be implementable in hardware without changing the semantics of SDL. Some of the most important questions and proposals for a policy of using SDL for hardware design are discussed in the following.

3.1 What do SDL signals mean in hardware ?

In SDL processes communicate with each other using *signals*. Signals in SDL are abstract events which initiate state transitions. In addition, they may carry data from the sending process to the receiver. In a hardware context there are several possible interpretations of SDL signals. When implementing signals in hardware different signals may be realized differently.

- Signals as static values. Sending a signal like *Reset* to another process may be realized by assigning a certain value, e.g. logical '1', to a wire connecting the two processes. The time the specific value must be kept depends on the receiving process. It must be long enough to assure that the receiver has received the signal. After this time the sender may change the value on the wire to an inactive value, e.g. logical '0'.
- Signals as transition between static values (edges). In this interpretation each change of the value on a wire is regarded as a message. The detection of the changes should be done modulo a clock signal since otherwise the processes would be sensitive to spikes.
- Signals as functions. In this interpretation, signals activate specific functions in the receiving process e.g. `add, sub, ...` in a CPU or `push, pop` in a stack. This view is similar to that known from object oriented languages. Processes here correspond to objects and signals to messages. An interesting point in this context is that SDL allows one to describe the acceptance of functions activations depending on the state of the object (process).

  This interpretation is very attractive for hardware design, in particular at abstract levels. It offers a *function oriented* view to hardware components as opposed to a *pin/port oriented* view that is necessary when using VHDL for example. Due to the lack of this abstract level, several proposals have been made for extensions to VHDL [9, 10].

Another naive interpretation of SDL signals would be to regard them as *clocks* which initiate state transitions in synchronous finite state machine implementations. But this interpretation is of limited use since clocks are usually not sent between processes and addresses implementation issues which are too low level.

3.2 Is the implicit queuing adequate ?

In SDL all processes run concurrently and send signals independently of the state of the receiving process. To deal with this situation SDL defines an implicit input queue for each process. In this queue all arriving signals are stored in a FIFO order by default. The order in which the messages are consumed can be controlled by the receiving process, e.g. by using the SAVE construct of the language.

For hardware design this implicit queuing is of limited interest. Message queues are rarely used in hardware designs, independently of the possible interpretations of signals mentioned above. Implementing a queue for each process would be too much overhead in terms of area and performance. On the other hand, if such a queue were necessary, the designer would probably describe the queue explicitly, e.g. as a separate process, taking care of locking mechanisms and a finite queue length.

Thus when using SDL for hardware design, the designer should not rely on the implicit queuing mechanism but specify the hardware in such a way that implicit queues are not necessary. This policy of use explicitly prohibits the SAVE construct of the SDL language.

3.3 Dynamic creation of processes ?

SDL allows the dynamic (at runtime) creation of process instances. This obviously cannot be supported in hardware and must not be used.

3.4 What do procedures represent in hardware ?

Procedures can be used in SDL to structure process descriptions. For a hardware implementation of procedures and the corresponding procedure calls, the following possibilities exist:

- Each procedure can be mapped onto a single hardware unit. The *CALL* in this case corresponds to transfer of control from the calling unit to the called unit using for example a simple handshake protocol.
- The body of a procedure can be expanded at compile time each time a procedure is called, just like macro expansion. Expanding procedures is mandatory if the procedures contain transitions to other SDL states in which case there is no *return* from the procedure.

3.5 What are continuous signals good for in hardware design ?

Quoted from [6]

> *In describing systems with SDL, the situation may arise where a user would like to show that a transition is caused directly by a true value of a boolean expression. The model of achieving this is to evaluate the expression while in the state, and initiate the transition if the expression evaluates to true. A shorthand for this is called* Continuous signal, *which allows a transition to be initiated directly when a certain condition is fulfilled.*

The semantics of this language construct is quite typical in hardware designs. In finite state machines of mealy type, the output depends both on the current state and the current inputs. Thus any change in the inputs results in an evaluation of the output function. Furthermore an asynchronous reset (boolean expression : reset = '1') causes a state transition to the reset state.

3.6 States in SDL vs. states in hardware

The basic model of SDL processes is a finite state model. This model is also common in hardware designs. The states in a SDL description refer primarily to the *macro states* of the hardware. By macro states we mean the global states of hardware design e.g. `load`, `execute`, `store` of a processor. The tasks that are performed during a state transition between these macro states may result in additional states. We call these *micro states.* For example a series of sequential statements in a task may need a controller with its own states for their execution.

3.7 How can tasks be realized in hardware ?

Tasks in SDL are lists of sequentially executed statements like assignments or operations. The implementation of such statements in hardware is the main goal of hardware synthesis systems like, SYNOPSYS [11] or CALLAS [12, 13]. The implementation may result in a data path controlled by a finite state machine. The states of these finite state machines are the micro states as defined above.

4 Translating SDL to VHDL

SDL can be used as a method and a language to specify and design systems at abstract levels. In order to establish a coherent design environment links have to be provided to the implementation of systems. For the software part various code generators already exist. For the hardware part a corresponding link has been missing up to now and is provided by the translation of SDL to VHDL. The translation of SDL to VHDL is attractive for two reasons.

- Since VHDL may be viewed a special "programming language for an event driven simulator" this allows the use of standard VHDL simulators for simulating systems (hardware and/or software) described with SDL.
- Some available hardware synthesis systems also support VHDL as an input language, e.g. [11, 12]. Therefore the translation of SDL to VHDL makes automatic mapping of SDL descriptions into hardware possible.

We have implemented a first version of an SDL to VHDL translator. In this version only a subset of SDL is supported. Extensions to include more SDL constructs are currently under development. The environment of the code generator is shown in Figure 2. The major goal of the first version was to use SDL as a graphical front-end for the description of hardware units that can be synthesized using a VHDL-based synthesis system. The most important features are summarized below.

4.1 SDL Subset

The SDL-GR symbols of the subset are shown in Figure 2[1]. Based on this subset we developed a methodology for describing hardware modules. Figure 4 shows a specification of a Mealy finite state machine using the subset.

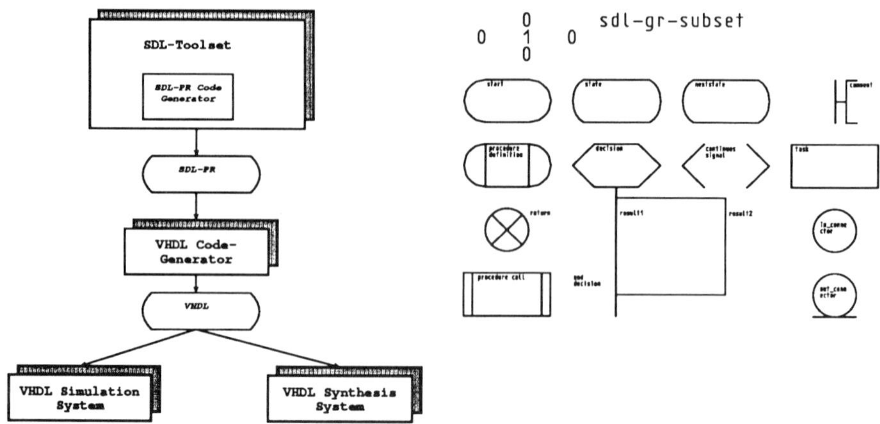

Figure 2: The Tool Environment and the SDL-GR subset

4.2 Processes

Each SDL process is translated to a VHDL entity and an associated architecture. The architecture is built of one or two VHDL processes.

4.3 States

The state symbol may contain single state names, a list (group) of state names, or the "*" state representing all states. The next state name "-" may be used to represent the same state as the current state. The code generator selects all state names from the state symbols and generates a

[1]We concentrate on SDL process diagrams since the mapping of SDL structures to VHDL structures is evident.

corresponding enumeration type plus the declaration of a VHDL-signal of this type as part of the VHDL code. It also automatically expands state lists and collects the transitions corresponding to each state. The code generator generates a state machine description as shown in 5. In this case the basic structure is signal oriented (see Figure 3b). Other options are to generate a state

```
CASE state IS
WHEN ...
  IF    <continous signal 1> THEN
          :
  ELSIF <continous signal n> THEN

  END IF;
:
END CASE;
```

(a) state oriented

```
IF    <continuous signal 1> THEN
  CASE state is
  WHEN ...
  :
  END CASE;
:
ELSIF <continuous signal n> THEN

END IF;
```

(b) signal oriented

Figure 3: State oriented vs. signal oriented code

oriented description (see Figure 3a). This is more closer to the basic model of SDL but is not compatible with the VHDL subset currently supported by VHDL based synthesis tools. A third possiblity would be to create a processes for each state as it was proposed in [5], but this also makes synthesis impossible with currently available synthesis tools.

4.4 Tasks

Tasks are assumed to be written as informal text. This means that sequential VHDL statements have to be used inside the task symbols. Structural or data flow VHDL statements are not allowed. These statements are copied one to one to the target VHDL code.

4.5 Procedure definition

The user may define procedures using the SDL procedure definition symbol. In the body of the procedure the following SDL-GR symbols are allowed: tasks, comments, decisions, decision results, end decisions, return, connectors, and function or procedure calls.

4.6 Continuous signals

As mentioned above, the SDL semantics of continuous signals is typical for hardware. In order to describe FSMs using SDL we use the following methodology:

- The asynchronous part is described using a continuous signal containing the boolean expression "true". The priority of this special continuous signal must be the lowest of all. The next state of the associated state transition must be "-". The tasks in these transitions describe the outputs as functions of the current input values.
- The actual transition to a next state is triggered by a clock edge. This is expressed using a VHDL *clock edge expression*, e.g., "clock'event and clock = '1' ". During the state transition tasks may be performed, e.g., setting some output value or performing operations on local data.
- any number of asynchronously acting reset-like signals may be specified using continuous signals with boolean expressions like, e.g., "reset = '1' ". These signals may be given for the "*" state or for a specific list of states.

4.7 Decisions

Decisions may be used in a state transition and in procedure diagrams. They are translated almost one to one into VHDL decisions.

4.8 Synthesis

To allow the generation of VHDL code synthesizable by a commercially available synthesis tool [11], the clock edge expressions must be the same in all states. Without this restriction imposed by the specific synthesis tool, any mixture of clock edge expressions would be allowed, and might be useful if the VHDL code is used just for simulation. The code generator has an option for checking the compliance with this policy. Other restrictions by specific synthesis systems may impose additional constraints on the statements that may be used in tasks.

4.9 Example

This example describes a Mealy type finite state machine with two states `state0` and `state1`. The SDL specification is shown in Figure 4. The definition of the procedure `shift` that is called during the state transition is not shown here. Figure 4 shows the usage of state lists, the "*" and the "-" state, continuous signals with a clock edge expression, and continuous signals with the "true" expression. The corresponding generated VHDL source code is shown in Figure 5.

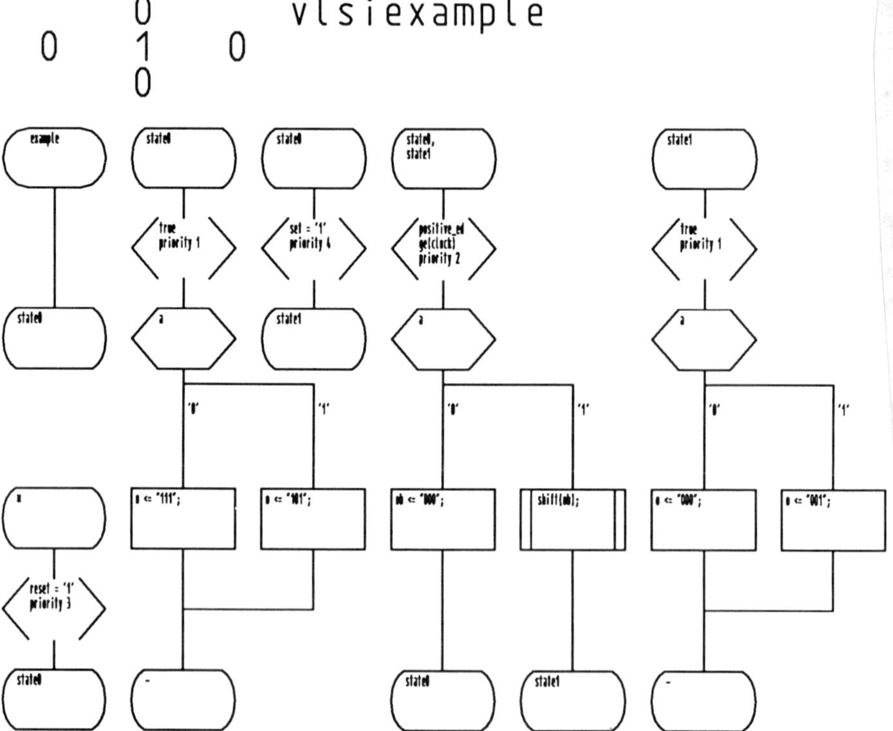

Figure 4: Example

Notice that the transitions corresponding to the continuous signals with the "true" expression have to be put into a separate VHDL process by the code generator in order to achieve the desired semantics and to be synthesizable. This description has successfully been synthesized using the SYNOPSYS VHDL compiler [11].

```
ENTITY example IS
 PORT ( a        : IN     bit;
        reset    : IN     bit;
        set      : IN     bit;
        clock    : IN     bit;
        o        : OUT    bit_vector(1 TO 3);
        ob       : BUFFER bit_vector(1 TO 3));
END example;
ARCHITECTURE synthesis OF example IS
 TYPE      states IS (state0,state1);
 SIGNAL    present_state : states;
 PROCEDURE shift (SIGNAL s : INOUT bit_vector(1 TO 3)) IS
 BEGIN
   FOR i IN 1 TO 2 LOOP
     s(i) <= s(i+1);
   END LOOP;
 END;
BEGIN
  output_function : PROCESS (a,set,reset,clock,present_state)
  BEGIN
    CASE present_state IS
    WHEN state0 =>
      CASE a IS
      WHEN '0' =>
        o <= "111";
      WHEN '1' =>
        o <= "101";
      END CASE;
    WHEN state1 =>
      CASE a IS
      WHEN '0' =>
        o <= "000";
      WHEN '1' =>
        o <= "001";
      END CASE;
    END CASE;
  END PROCESS;

  state_transition : PROCESS (a,set,reset,clock)
  BEGIN
    IF (set = '1') THEN
      CASE present_state IS
      WHEN state0 =>
        present_state <= state1;
      WHEN others =>
          null;
      END CASE;
    ELSIF (reset = '1') THEN
        present_state <= state0;
    ELSIF (clock'event AND clock = '1') THEN
      CASE present_state IS
      WHEN state0 =>
        CASE a IS
        WHEN '0' =>
          ob <= "000";
          present_state <= state0;
        WHEN '1' =>
          shift(ob);
          present_state <= state1;
        END CASE;
      WHEN state1 =>
        CASE a IS
        WHEN '0' =>
          ob <= "000";
          present_state <= state0;
        WHEN '1' =>
          shift(ob);
          present_state <= state1;
        END CASE;
      END CASE;
    END IF;
  END PROCESS;
END synthesis;
```

Figure 5: VHDL code for example in Figure 4

5 Future Extensions

Future extensions of the code generator will be explored in order to support the simulation of systems specified with SDL using a VHDL simulator and to use more powerful synthesis tools for hardware as soon as they are available. Specific issues to be considered are:

- A VHDL process implementing the implicit input queue of each SDL process has to be generated for SDL processes (possibly controlled by the designer). Hereby it has to be considered that SDL signals might be sent from different processes simultaneously. VHDL resolution functions will be useful for dealing with this situation.

- A mapping of SDL signals onto VHDL has to be defined (see section 3.1). The obvious mapping onto VHDL *events* is possible if the system assures that signals are sent only when the receiving process is waiting for the signal. In the general case this assumption does not hold and so another solution has to be found. Also, the selective waiting on signals needs to be considered. However this might not be too difficult for pure simulation purposes.

  In order to allow high level synthesis tools to be used for hardware implementation we must perform a presynthesis step which translates the object oriented interpretation of SDL signals into a description at the algorithmic level. This will allow hardware design to be done at a very high level of abstraction (see above).

- SDL sequence charts may be constructed from results of a VHDL simulation of the SDL process diagrams.

- Support of object oriented SDL (OSDL).

6 Conclusions

We have shown the usability of SDL in the hardware design process. By translating SDL specifications into the standard hardware description language VHDL we have provided a path to hardware synthesis from SDL specifications. Additionally, this translation allows one to use any VHDL simulator for the simulation of SDL specifications, which is interesting both for hardware and software designers. The basic concepts of this translation and a prototype tool have been described. Future work will concentrate on enlarging the current SDL subset supported by the translator both for simulation and synthesis purposes.

Acknowledgements

This work is partially supported by the ESPRIT project ATMOSPHERE, Ref #2565.

The authors would like to thank their colleagues Michael J. Kaelbling and Gerhard Gries of Siemens for their valuable suggestions and constructive criticism during the preparation of this paper.

References

[1] I-Logix Inc, Burlington, MA 01803. The STATEMATE Approach to Complex Systems.

[2] T. Tikkanen, T. Leppänen, and J. Kivelä. Structured Analysis and VHDL in Embedded ASIC Design and Verification. In *Proc. of the European Design Automation Conference*, pages 107–111, 1990.

[3] J. E. Kivelä, T. Leppänen, and M. Sipola. VHDL supported by graphical CASE-tools for high-level system design and implementation. In *Proc. of the 1st European Conference on VHDL*, 1990.

[4] Frank Vahid, Sanjiv Narayan, and Daniel D. Gajski. SpecCharts : A Language for System Level Synthesis. In *Proc. of the IFIP Tenth International Symposium on Computer Hardware Description Languages and their Applications*, pages 145–154, April 1991.

[5] Sanjiv Narayan, Frank Vahid, and Daniel D. Gajski. Translating System Specifications to VHDL. In *Proc. of the European Design Automation Conference*, pages 390–394, February 1991.

[6] CCITT. Functional Specification and Description Language (SDL), Recommendations Z.100-Z.104, Blue Book, October 1989.

[7] Ove Færgemand and Maria Manuela Marques, editors. *SDL '89 The Language at Work, Proceedings of the Fourth SDL Forum.* North Holland, 1989

[8] Birger Møller-Pedersen and Dag Belsnes. Rational and tutorial on OSDL. In *Computer Networks and ISDN Systems*, pages 97–117. North Holland, 1987.

[9] W. Glunz and G. Umbreit. VHDL for High-Level Synthesis of Digital Systems. In *Proc. of the 1st European Conference on VHDL*, 1990.

[10] A. Oczko. Hardware Design with VHDL at a very high level of abstration. In *Proc. of the 1st European Conference on VHDL*, 1990.

[11] SYNOPSYS, Inc. *VHDL/Design Compiler Reference Manual*, 1990.

[12] M. Koster, M. Geiger, and P. Duzy. ASIC Design Using the High-Level Synthesis System CALLAS: A Case Study. In *Proc. of the ICCD Cambridge MA*, 1990.

[13] P. Duzy, H. Krämer, M. Neher, P. Pilsl, W. Rosenstiel and T. Wecker. CALLAS - Conversion of Algorithms to Library Adaptable Structures. In *Proc. of the IFIP VLSI Conference Munich*, pages 197–208, 1989.

VLSI 91
A. Halaas and P.B. Denyer (Eds.)
Elsevier Science Publishers B.V. (North-Holland)

VLSI System Design for the Control of high Performance Combustion Engines

A. Laudenbach, M.Glesner, N. Wehn[1]
Darmstadt University of Technology
Institute of Microelectronic Systems
Karlstr. 15, D-6100 Darmstadt

Abstract

This paper presents a novel VLSI approach for combustion engine control. The approach is based on a real time solution of thermodynamical differential equations. The control system calculates an optimum ignition point by fast measurement and real time processing of signals as temperature, pressure and volume of the combustion chamber. The required computational power cannot be met with standard signal processors. We describe a methodology for the functional mapping of complex algorithms on a parallel architecture. Each design step from the analysis of the heat release algorithm, the optimization of the algorithm and the dataform, the mapping on an architecture, the physical design of the chip set and the chip test is presented. Finally the system integration, the flexibility and the software environment are shown.

1 Introduction

In a long term research project [2] activities are focused on a novel approach for combustion engine control, which is based on the measurement of the pressure in the combustion chamber and on the real time calculation of the heat release.

The first realization of the mechatronic system was implemented with a piezo pressure sensor and a commercial digital signal processor. This design has shown that a commercial signal processor is able to calculate the heat release only with a very simple algorithm in real time. The physical behavior of the combustion is modelled with more accuracy if energy losses over the chamber wall are incorporated and if the combustion dependant gas constant is described more exactly. In this case the heat release is defined implicitely and the calculation is based on the solution of a nonlinear differential equation. The real time solution is not possible with a commercial processor.

[1]also employed by SIEMENS AG Corporate Research and Development

[2]This project is sponsored by the Deutsche Forschungsgemeinschaft under the Sonderforschungsbereich 241 "Integrated mechanical- electronical systems for mechanical engineering"

As a consequence a parallel VLSI architecture tuned to this specific application was developped.

For working with the computer system as an evaluation tool for mechanical engineering some strong requirements had to be met:

- *high computational power* is provided by parallel processing arithmetic units
- *high data throughput* is ensured with a multiplexed and pipelined datapath with several data memory modules on the chip
- *full flexibility* is provided with off-chip program memory. The actual program code is stored in fast static RAM.
- *optimized program code* is automatically generated with a macroassembler that supports arithmetic scalar and vector operations

2 Mathematical model of a combustion engine

State of the art combustion engine control systems [1] determine the ignition point depending on the load of the engine and the revolutions per minute. Thermodynamical parameters of the combustion are not taken into account. In a novel approach for engine control each combustion is analyzed thermodynamically. The fundamental parameter of the control algorithm is the heat release $dQ_B/d\varphi$ as a function of the crank shaft angle φ. The heat release has a great influence on the engine's economy and on the exhaust. The heat release can be calculated with the pressure $p(\varphi)$ in the combustion chamber and the volume $V(\varphi)$ and the differentials of these functions $dp(\varphi)$ and $dV(\varphi)$.

With the law of conservation of energy the heat release becomes:

$$dQ_B = dU + p \cdot dV + dQ_{Wall} \tag{1}$$

with: U : internal energy of the air/fuel mixture
Q_{Wall} : heat due to energy losses over the cylinder walls
$p \cdot dV$: mechanical work on the piston

From the ideal gas equation follows:

$$T = \frac{p \cdot V}{R_g \cdot m} \tag{2}$$

with: T : gas temperature
R_g : gas constant
m : mass of the air/fuel mixture

With an empirical power series expansion [2] the internal energy becomes:

$$U = k_0 + k_1 \cdot T + k_2 \cdot T^2 + k_3 \cdot T^3 \tag{3}$$

The energy loss over the chamber walls which is induced by convection is calculated:

$$\frac{dQ_{Wall}}{d\varphi} = \alpha \cdot A_{wall} \frac{T - T_{wall}}{\omega} \tag{4}$$

with: A_{wall} : combustion chamber surface
T_{wall} : cylinder wall temperature, which is set constant
ω : angular speed

and the average piston speed c_m [3]:

$$\alpha = 130 \cdot V^{-0.06} \cdot p^{0.8} \cdot T^{-0.4} \cdot (c_m + 1.4)^{0.8} \tag{5}$$

At this point the heating value could directly be calculated. However in the burning fuel mixture the ideal gas constant R_g is not constant, but is a function of the combustion rate. The resulting differential equation is of first order and nonlinear. The differential equation is solved with an iterative method which can be processed in parallel. The iterative algorithm calculates a start solution $dQ_B^{(0)}$ with constant R_g and the heat due to energy losses over the cylinder walls set to zero ($dQ_{Wall} = 0$). In the ith iteration we use the old integral heat release

$$Q_B^{(i-1)}(\varphi) = \int_0^{\varphi} dQ_B^{(i-1)} d\varphi' \tag{6}$$

and the old maximum $Q_{B,max}^{(i-1)}$ to calculate the new air/exhaust ratio $r(\varphi)^{(i)}$

$$r(\varphi)^{(i)} = \frac{Q_B^{(i-1)}(\varphi)}{Q_{B,max}^{(i-1)}}(r_{end} - r_{start}) + r_{start} \tag{7}$$

The parameters r_{end} and r_{start} are calculated in the initial phase before the combustion starts. Now the gas constant $R_g = R_g(p(\varphi), r(\varphi), T(\varphi))$ can be calculated

$$R_g(\varphi) = 9.81 \cdot (29.0 + A(\varphi) + p(\varphi) \cdot (\frac{B(\varphi)}{T(\varphi)^{C(\varphi)}} - A(\varphi))) \tag{8}$$

A, B and C are powers of r with a fractional exponent between zero and one. With $R_g(\varphi)$ and (1) - (5) a new heat release $dQ_B(\varphi)^{(i)}$ can be calculated. The analysis is done over a range of 128° crank shaft angle with a stepsize of 1°. The iteration is repeated until the variation of the integral heat release is less than 0.1%:

$$\left| \frac{Q_{B,max}^{(i-1)} - Q_{B,max}^{(i)}}{Q_{B,max}^{(i)}} \right| < 10^{-3} \tag{9}$$

3 Development of algorithm and architecture

Computation time requirements Heat release calculations with sampled input data have demonstrated that with two runs through the iterative loop the variation of the integral heat release is always less than 0.1%, which is the iteration stopping criterion. For generating the start solution and two times running through the iteration loop 218 additions, subtractions and multiplications and 14 divisions are necessary. Since the analysis is done over a range of 128° crank shaft with a resolution of 1°, each operation has to be performed 128 times. A sequential calculation of the heat release must not be longer than the time for sampling one pressure signal. The signals are sampled with a resolution of 1° crank shaft. If the engine is running with the maximum revolution speed of 6000 rpm, the resulting computation time is about 20 μs, which is not sufficient for the sequential calculation of 232 mathematical operations including 14 divisions.

If the algorithm is processed after the sampling period, real time capability is reached, if the computation has finished after 50° crank shaft, corresponding to 1.35 ms at a maximum revolution speed of 6000 rpm. The required computational power can only be reached with parallel processing. Either with a multiprocessor board or with a single chip with internal parallel processing units. Since the goal of the project is an integrated solution the only acceptable way is to develop a single chip parallel processor.

Data format Simulations on behavioural level with different sampled input signals demonstrate that the intermediate results can vary over many orders of magnitude. This result was expected because the input signals can vary in a wide range. The internal data format must be able to represent each value with sufficient accuracy. The simulations have shown that a fixed point representation would require 52 bits, whereas a floating point format with 20 bits provides sufficient accuracy and dynamic for the heat release algorithm. The floating point format was preferred because a 52 bit bus with integer format would result in a big area and power consumption. The optimized floating point data format has 12 mantissa bits, a sign bit and 7 exponent bits.The dynamic of the number representation is 38 orders of magnitude.

Architecture Due to the use of floating point data format the arithmetic units get quite complex, because mantissa and exponent must be processed in a different way and costly error handling becomes necessary. This influences the architecture because the integration of many arithmetic units on a single chip is not possible. The heat release algorithm could easily be mapped on a parallel SIMD (= Single Instruction Multiple Data) architecture with next neighbour connections. But since the SIMD architecture is working efficiently only if a great number of arithmetic units is realized, the integration on a single chip is not possible.

Another parallel processing concept on which the algorithm could easily be mapped

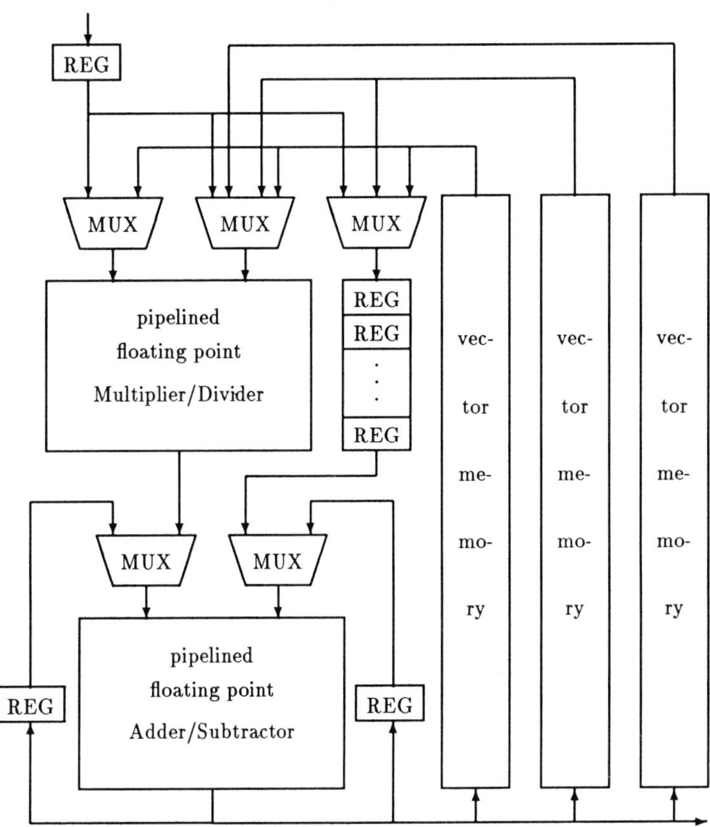

Figure 1: Architecture of the vector processor

is vector processing. Here the data is processed very efficiently with the principle of an assembly line [4], [5]. The implementation of the heat release algorithm on a vector architecture as shown in figure 1, implies that the input signals and intermediate results are processed as vectors. Each degree of crank shaft angle represents one vector component. The vector size is 128. All algorithmic operations except the integration are vector operations. The long vectors make the design of arithmetic units (AUs) with deep pipelines of 4 to 10 stages very efficient. The most frequent operations are of the type multiply-add, multiply-subtract, divide-add or divide-subtract. This is the reason, why the output of the multiplier/divider unit is connected to the input of the adder/subtractor unit. This connection has the advantage of only three instead of four data memory accesses per clock cycle. If the pipeline is filled, two floating point operations per clock period can be processed.

The memory is divided in on-chip memory and off-chip RAM. The need to feed three data into the AUs requires either one high speed on-chip RAM with a high speed bus or several vector memories with a cross bar network.

The cross bar switch was realized with bus multiplexers, the vector memories were realized with shift registers of 128 words times 20 bit size. The shift registers can store three signal vectors or result vectors. The advantage of this solution is a relative moderate clock, while in each clock period two floating-point operations can be processed.

Algorithm The differentiation $df(\varphi)$ is approximated by the secant through the points $f(\varphi+1)$ and $f(\varphi-1)$. The integration is substituted by a summation. The approximation of power and exponential functions is performed with least square optimization techniques. A special tool was developed to automate this task. The user has to specify the functions to be approximated, the range of possible arguments and the allowed error tolerance. The approximation generated by the program is based on orthogonal polynomials. The approximations can easily be calculated in parallel because for each argument the approximation constants are identical. Compared to Spline interpolation which also would be possible, the processor must calculate polynomials of higher order. But this is not a drawback because time consuming memory accesses for fetching Spline coefficients which are different for each vector component, are avoided.

4 Physical design and test

The physical design of the vector processor is based on a macrocell design style. No commercial design tools were used. Two problems originated when designing the macrocell blocks:

- No library was available which contains floating-point arithmetic cells
- All blocks must be parametrizable dependent on the floating point format in order to make possible the optimization of computation accuracy, floorplanning and data formats

To overcome these problems all macrocells were designed in a symbolic design environment [6]. The different blocks were described by a symbolic layout language on a virtual grid. The language offers procedural statements, thus permitting an efficient parametrization. The final layout is obtained by two one-dimensional compaction steps followed by a local two dimensional compaction. After the macrocell generation, the different blocks are placed and routed by a simulated annealing based floorplanning tool. This tool uses a slicing structure as target layout model. The floorplanner and macrocell generator are strongly linked to permit a fast design cycle.

The layout of the vector processor as a single chip in a $1.2\mu m$ CMOS technology has a size of 8.1 mm $\cdot$ 8.6 mm. The whole architecture was split into three different

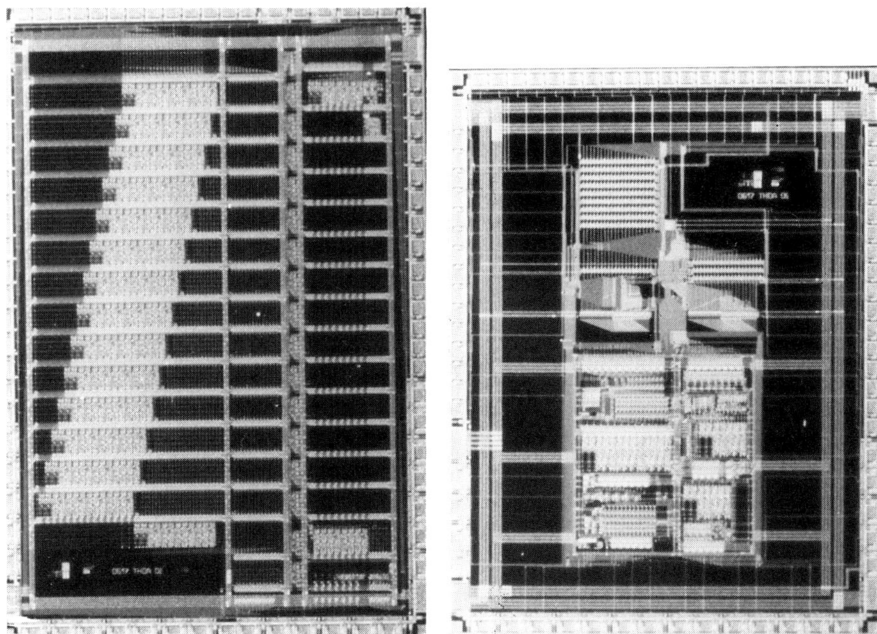

Figure 2: Micro photographs: multiplier/divider and adder/subtractor chip

modules: the multiplier/divider, the adder/subtractor with some glue logic for fast intgeration and maximum search and the vector memory [7]. These modules have been fabricated as test chips. The micro photographs of the multiplier/divider and the adder/subtractor testchips are shown in figure 2.

In the test phase the correct function of each testchip was verified successfully. The sizes of the testchips and the maximum clock frequencies are shown in the following table. The maximum clock frequency of the adder/subtractor is low compared to the other modules. This is caused by power-ground bouncing induced by the buffers in the output pads.

chiptype	macrocell size	chip size	transistors	clock
vector memory	$4098\mu m * 874\mu$m	$5040\mu m * 1584\mu$m	26 800	36 MHz
adder/subtractor	$1703\mu m * 1846\mu$m	$5264\mu m * 4168\mu$m	8 300	15 MHz
multiplier/divider	$5212\mu m * 3462\mu$m	$5928\mu m * 4160\mu$m	31 000	36 MHz

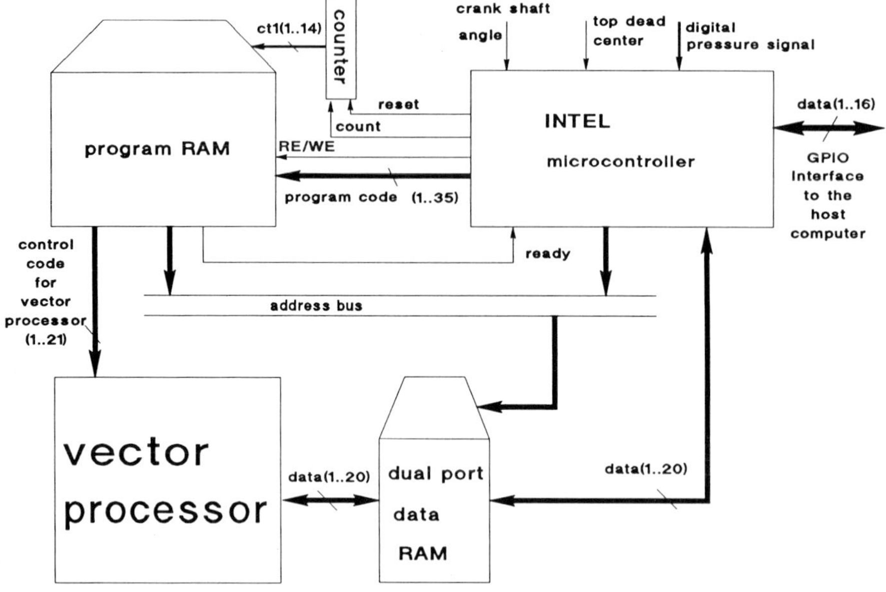

Figure 3: Microelectronic system

5 System aspects

Microelectronic system The microelectronic system for combustion engine control, shown in figure 3, contains a commercial microcontroller and the vector processor board with program and data memory. The microcontroller's tasks are:

- controlling the analog to digital converter board and the vector processor board
- synchronization with the combustion engine
- preprocessing of the input signals, conversion into the 20 bit floating point format and storing in the data memory
- download of the program code for the vector processor into the program RAM
- interfacing to a HP9000/series 300 host computer

At the moment the microelectronic system is built with the testchips. The single chip vector processor is substituted by a board with five testchips, two programmable logic devices (PLD) and several registers. The PLDs substitute the multiplexer crossbar network. With this configuration the system cannot run with full speed of 20 MHz, because the PLDs limit the maximum clock frequency to 5 MHz. Since the slowest testchip, the adder/subtractor can run with 15 MHz, a redesign of this testchip makes no sense. In the next technology run the single chip vector processor will be fabricated. In the design measures will be taken against power ground bouncing, so that a system clock of 20 MHz will be reached.

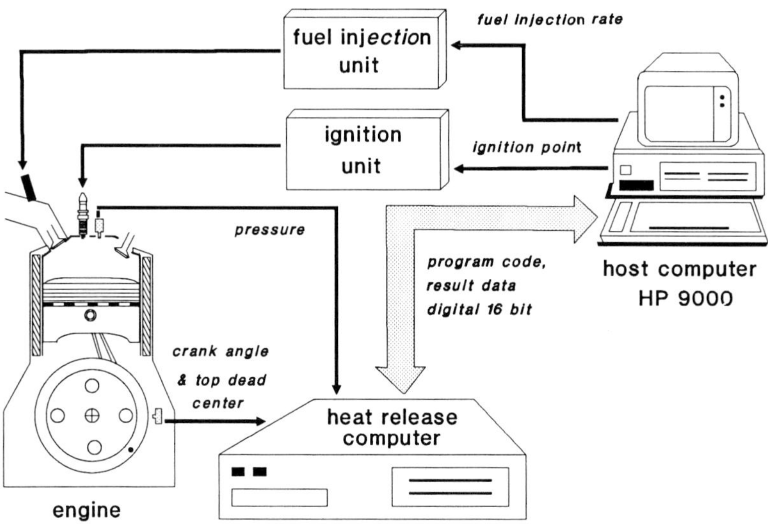

Figure 4: Mechatronic system

Mechatronic system The complete mechatronic system is shown in figure 4. The mechanical part contains the combustion engine, the electronical part cosists of the heat release computer and a HP9000/series 300 computer. Between this two parts the system contains sensors, the ignition unit and the fuel injection unit.

The mechatronic system can be used as an evaluation system for engine control algorithms. The algorithms can be processed very fast in the application specific vector processor. Optimized program code is generated with a macroassembler, which is installed on the HP9000 computer. The macroassembler supports complex arithmetic instructions for scalar and vector operations. With this high level language the heat release algorithm can be described very efficiently. The macroassembler guarantees full flexibility, because variations in the algorithm or in engine constants can be managed by changing the source text.

The HP9000 is used for visual display, operator's console, code generation for the vector processor and interface to the ignition system. On the HP9000 computer the engineer can develop new algorithms which are compiled and loaded down into the program memory of the vector processor. The program memory has 32k lines. The memory can be expanded, if it is necessary. After passing the analog/digital converter board the signals arriving from the engine are converted into the optimized 20 bit floating point format and then they are directly written into the data memory of the vector processor. The data memory can store 4k words of 20 bits. If necessary the data memory can also be expanded. After data processing the vector processor writes the results back into the data memory. Now the microcontroller transfers the result data via GPIO interface to the HP9000 host computer. The HP9000 computer can perform further computations and has control over the ignition unit and the fuel injection unit.

6 Results

A set of testchips has succesfully been fabricated. A board with the testchips for the emulation of the single chip vector processor has been built. The system clock of the emulation board is limited to 5 MHz because of the use of PLDs. The heat release algorithm is calculated with the emulation board in 4.6 ms with a performance of real 6.5 $MFLOPs$. This is nearly the same as the implementation on a DSP32C signal processor from AT&T [8] which needs 5.08 ms. If the algorithm will be processed on the single chip vector processor the computation time will be 1.15 ms, so that the real time requirement of 1.35 ms will be met. The implemented heat release algorithm will run with real 25.8 $MFLOPs$, so that the maximum performance of 40 $MFLOPs$ of the vector processor is occupied to 64%. The vector processor will contain about 130 000 transistors, while the DSP32C has 405 000 transistors. The application specific architecture brings a significant reduction in transistor count and computation time.

The authors would like to express their thanks to INTERMETALL Deutsche ITT Industries company in Freiburg, which has offered a 1.2μm CMOS-technology to fabricate the VLSI chip set.

References

[1] Bosch, *Kombiniertes Zuend- und Benzineinspritzsystem mit Lambda-Regelung: Motronic*, Reihe Technische Untersuchung, 1985

[2] A. Pischinger, *Thermodynamik der Verbrennungskraftmaschine*, Springer Verlag, 1989

[3] G. Hohenberg, *Experimentelle Erfassung der Wandwaerme von Kolbenmotoren*, habilitation thesis, TU Graz, 1980

[4] K. Hwang, *Computer Architecture and parallel Processing*, McGraw Hill, 1988

[5] H. S. Stone, *High Performance Computer Architecture*, Addison-Wesley, 1990

[6] N.Wehn, *Efficient Methodologies for the Physical Design of MOS-VLSI Circuits*, Ph.D.thesis, Darmstadt University, July 1989

[7] A.Laudenbach, M. Glesner, P. Windirsch, J. Plahl, W. Clemens *VLSI Chip Set for Floating Point Vector Processing*, proceedings of the Euro ASIC 91, Paris, May 1991

[8] K. D. Ulery, *The DSP32C: AT&T's Second-Generation Floating-Point Digital Signal Processor*, IEEE Micro, Dec. 1988, pp.30-48

VLSI 91
A. Halaas and P.B. Denyer (Eds.)
Elsevier Science Publishers B.V. (North-Holland)

A Fully Integrated Systolic Spelling Co-processor

P. Frison and D. Lavenier

IRISA - Campus de Beaulieu - 35042 Rennes cedex - France

Abstract

This paper presents a fully integrated spelling co-processor for speeding up the character string comparison process. The algorithm involved gives very good correction rate but has the major drawback of being very time consuming on conventional machines. The chip we propose is architectured around a truncated 2-D systolic array of 69 processors and is able to perform up to 1.3 Gops. Real time spelling correction is then possible on very large vocabularies since dictionaries of 200,000 words can be processed in only 0.1 second.

1 Introduction

The increasing use of computer systems for the handling of textual data has led to a great deal of interest in software for the detection and correction of spelling errors [6]. As a matter of fact, texts are commonly riddled with literals which represent up to 80 percent of the following errors :

- one letter is wrong,
- one letter is missing,
- one extra letter is inserted,
- two adjacent characters are transposed.

These errors are respectively referred as substitution, deletion, insertion and transposition errors. They are mainly due to keyboard mistyping during text edition. If detecting an erroneous word is quite an easy task, proposing suitable corrections requires generally complex processing.

The correction method we use is based on dynamic programming techniques and provides extremely good results compared with other techniques, especially when erroneous words including more than one spelling error are proposed for correction [7]. Key location is one of the most important parameter involved in the correction process. For example, the erroneous word *progessor* will be corrected by *professor* instead of *processor* because the **f** key is nearer to the **g** key than the **c** key on a keyboard.

The major drawback of the algorithm comes from the amount of calculation required. With dictionaries of few hundred of thousand words the computation time on conventional personal computers, and even on recent workstations, can be very long and inadequate for real or acceptable time spelling correction.

Parallel implementations based on dynamic programming algorithms on systolic arrays are very efficient. Many systolic architectures have regularly been proposed to solve similar string correction problems such as speech recognition [1] [2] [3] or nucleic acid sequence comparison [8] [5]. But integrating on a single VLSI chip some tens of quite complex processors was not feasible until these last recent years. Today, the progress of technology makes it possible.

The chip we propose includes a 69 processor array dedicated to spelling correction applications. It can be seen as an hardware accelerator (co-processor) for comparing one string (erroneous word) with a large set of references belonging to a dictionary. The result of the comparison process is a sequence of words which seem adequate for correction.

The goal of the paper is not to present in detail the full chip architecture, but to focus on the original part, namely the processing part which includes the systolic array and the associated data management. This last part is referred to as the *core* chip.

The paper is organized as follows : section 2 and 3 present respectively the method used for spelling correction (string comparison) and the implementation on a 2-D systolic array. The next section is dedicated to the core chip architecture. It is followed by the layout organization. We end by indicating performances and current work.

2 String comparison algorithm

The main idea for comparing two strings of characters is to compute a distance between these two strings. This distance, called the *edit distance*, represents the minimal cost to transform one string into another one by using elementary operations such as substitution, insertion, omission or transposition. They are called *edit operations.*

By using sequences of such edit operations, any string may be transformed into any other strings. It is then possible to take the smallest number of elementary edit operations required to change one string into another one as the measure of the difference between them.

More formally, let $X = (x_1, x_2, ..., x_i, ..., x_n)$ and $Y = (y_1, y_2, ..., y_j, ..., y_m)$ be two strings to compare ; let $d(x,y)$ be the cost of an edit operation to change x into y ; let $\wedge$ represent the null character. It has been shown [9] [7] that the edit distance $D(n,m)$ is obtained from the following recurrence relation :

$$D(i,j) = Min \begin{cases} D(i-1,j-1) + d(x_i, y_j) \\ D(i-1,j) + d(\wedge, y_j) \\ D(i,j-1) + d(x_i, \wedge) \\ D(i-2,j-2) + \gamma_t(x_{i-1}x_i, y_{j-1}y_j) \end{cases} \tag{1}$$

where γ_t is a function such as :

$$\gamma_t(x_{i-1}x_i, y_{j-1}y_j) = \begin{cases} K_t & if\ x_{i-1} = y_j\ and\ x_i = y_{j-1} \\ \infty & if\ x_{i-1} \neq y_j\ or\ x_i \neq y_{j-1} \end{cases} \tag{2}$$

K_t represents the transposition cost. Initial conditions are :

$$D(0,0) = 0, \quad D(i,0) = D(i-1,0) + d(x_i, \wedge), \quad D(0,j) = D(0,j-1) + d(\wedge, y_j)$$

	S	Y	T	O	L	L	I	C
S	0	1	2	3	4	5	6	7
Y	1	0	1	2	3	4	5	6
S	2	1	1	2	3	4	5	6
T	3	2	1	2	3	4	5	6
O	4	3	2	1	2	3	4	5
L	5	4	3	2	1	2	3	4
I	6	5	4	3	2	2	2	3
C	7	6	5	4	3	3	3	2

Figure 1: edit distance computation example

The edit operation $d(x_i, \wedge)$ represents the deletion case while $d(\wedge, y_j)$ represents the insertion case. The elementary edit operation $d(x, y)$ may have different values depending on the characters considered.

The figure 1 is a graphical representation of the edit distance computation between the test string *sytollic* and the reference string *systolic*. For simplicity, insertion cost and deletion cost are constant and equal to one. The edit distance is then equal to 2 since there are one deletion and one insertion. The dash line represents the optimal way to reach the solution.

3 Parallel implementation on systolic array

This section indicates briefly how the algorithm we described in the previous section can be implemented on a systolic array. For sake of clarity, we will only consider model without permutation errors. The general architecture is not affected by this reduction and explanations are greatly simplified.

Systolic implementation is based on associating one processing element to perform the calculation of each value $D(i,j)$. Consider an array of n^2 processors connected as indicated by figure 2. A processor $P(i,j)$ can read data produced by its neighbor processors $P(i-1,j-1)$, $P(i-1,j)$ and $P(i,j-1)$, and propagate its results to processors $P(i+1,j+1)$, $P(i,j+1)$ and $P(i+1,j)$. A systolic cycle represents the elementary distance calculation, that is to say, the data acquisition, the minimization operation described by equation (1) - without transposition errors - and the data propagation.

On such a network, parallel execution of a two string comparison requires $2n-1$ systolic cycles since an elementary distance computation $D(i,j)$ is associated to only one processor $P(i,j)$ and since every processors can run concurrently.

As a complete comparison lasts $2n-1$ systolic cycles and as a processor is used during only one cycle, this last one is then available during the other cycles for other tasks. Furthermore, the edit distance calculation at t_k is computed on a diagonal of processors made by processors $P(i,j)$ such as $k = i+j-1$. On each systolic cycle it is then possible to start a new comparison.

Pipelining the string comparison on this way allows to increase considerably the efficiency of the network since once the array has been started, one comparison result is

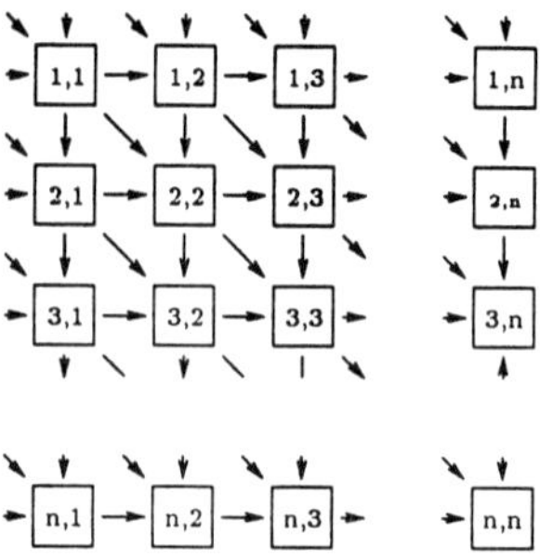

Figure 2: systolic network

produced every systolic cycle.In that way, the speed up in comparison with a sequential machine is given by :

$$\Gamma = \frac{pn^2}{p + 2n - 1} \approx n^2 \qquad with \;\; p \gg n$$

This last assumption is true in the applications envisioned since the number of references can reach hundred of thousands items compared to the low number of characters of the strings (up to 20 !).

4 VLSI implementation

4.1 Systolic array dimensions

Spelling correction implies processing character strings whose length is not very large. The average length of words belonging to french or english dictionaries is about 8 characters. This means that the number of words longer than, saying, 15 characters is quite low and does not require important computation resources. For these words, correction process can be efficiently done on conventional machines.

A 2-D systolic network of 15x15 processors seems a good compromise for this specific application. To be able to process strings of different lengths on a fixed 15x15 processor array, strings shorter than 15 characters are completed with special characters. In that way, results are always picked up on the same processor (see figure 3). Unfortunately, the resulting 225 processor array remains too large to be integrated on a single chip. But, as explained below, the number of processors can be drastically reduced.

As a matter of fact, erroneous words are always corrected by words whose length are similar within a range of ± 2 characters. It means that effective computation is made on a diagonal of the array and that processors which are located on the lower left and upper right corners do not participate actively to the final result. Tests have been made to compute the edit distance without these intermediate distances. They do not have shown appreciable changes and, in most cases, when differences were detected, the proposed corrections were better !

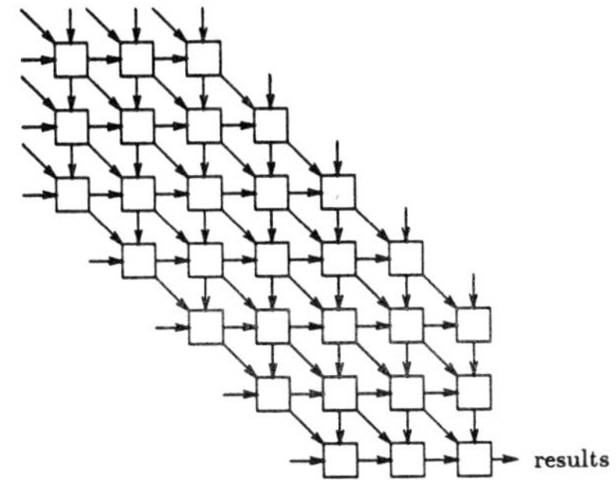

Figure 3: 2-D truncated systolic array

The systolic array we finally implement is shown on figure 3. It's a 2-D truncated array with only five diagonals of processors. The median diagonal includes 15 processors ; this gives a total number of processors equal to 69 ($15 + 2 \times 14 + 2 \times 13$).

4.2 Dataflow and memory management

Remember that one erroneous string must be compared with a large set of references. This means that only data associated to the references are flowing horizontally through the array while data attached to the erroneous string must be considered as constant value during the whole correction process. Consequently, to perform its local computation, a processor $P(i,j)$ needs to receive, on each systolic cycle, three intermediate distances from its neighbors and a character x_{ki} corresponding to the ith character of the kth reference.

Theoretically, costs associated to insertion and deletion depend of the nature of the characters. Practically, they are quite identical and can advantageously be replaced by constant values. So, the insertion and deletion costs $d(\wedge, y_j)$ and $d(x_i, \wedge)$ are respectively changed into constants Ki and Ko. The figure 4 indicates boundary initialization and data propagation through the systolic array.

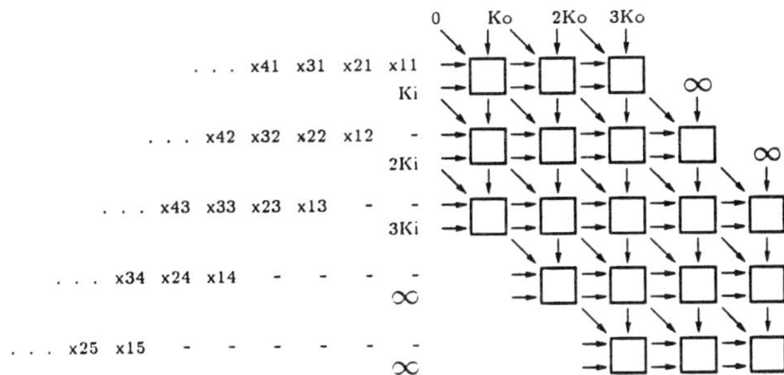

Figure 4: external data management of the network

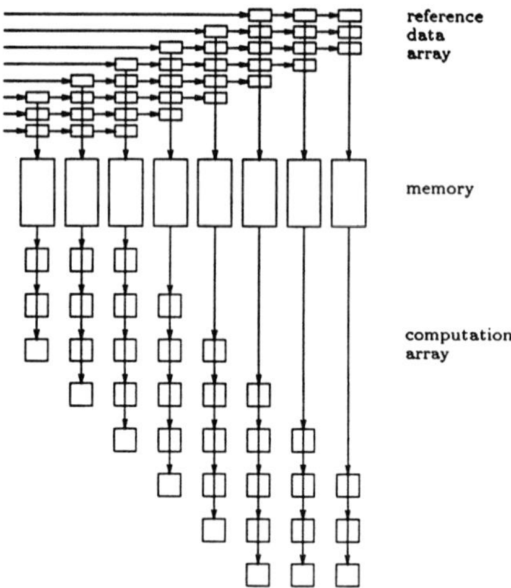

Figure 5: splitting the systolic network into 2 subarrays

This dataflow implies for each processors to have one local table for storing substitution cost in order to be able to compute the first term of equation (1). As the characters of the erroneous string do not move during a complete dictionary comparison, the table stored into each processor can be reduced to only cost associated to a single character. The memory size is then equal to the number of characters of the alphabet. On the other hand, the content of the table inside the processors must be changed each time a new erroneous string must be corrected ; but downloading tables with new values takes a very short time compared to the whole correcting process.

Nevertheless, this solution remains space consuming due to the fact that each processor contains its own table. A less expensive approach relies on two observations. The first one is that the tables of the processors of one column are identical. The second is that, assuming that a systolic cycle needs several clock cycles (at least 5), a single memory element can be shared among them since only one access is needed per processor, per systolic cycle. Therefore, the systolic array can be divided into 2 subarrays operating in parallel as shown on figure 5. The reference data array is a set of registers containing the reference characters and emulating the reference flow through the systolic network. The registers of one column are used sequentially to address the corresponding memory block and obtain the appropriate substitution cost. The computation array is used to compute the equation (1) on each cell, assuming that the distance cost is available in a register updated by the reference data array. This works correctly if the reference data array operates in advance (one systolic cycle ahead).

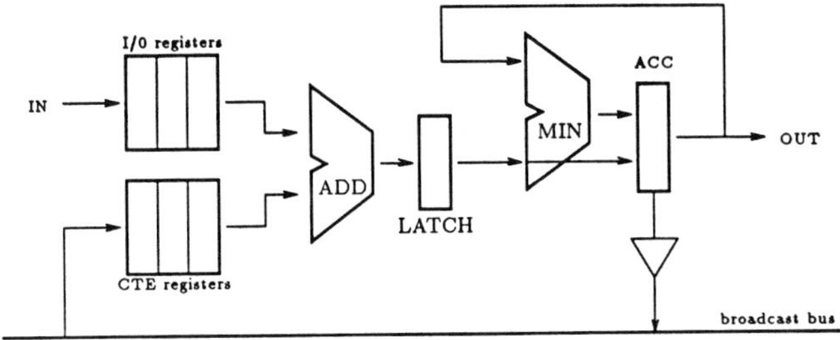

Figure 6: internal processing element architecture

4.3 Memory implementation

As said before, the memory size is equal to the number of characters of the alphabet and its content represents the substitution cost of two characters. This cost is determined by the proximity of the keyboard keys : closer the keys are, smaller the substitution cost is. By examining the cost table associated with a particular key, it can be seen that only a few values are significant (low cost) and that they are associated with neighbor keys. Most of the other have a high value and are quite identical. So, instead of memorizing systematically all values, only interested ones are stored in an associative memory. For the other keys, a default value is proposed. In that way, the memory size is considerably reduced and is independent of the alphabet size.

4.4 Processing element

The basic calculation made by the processors of the computation array is the minimization of terms. Equations to be computed look like :

$$D(i,j) = Min \begin{cases} D(i-1,j-1) + Ks \\ D(i-1,j) + Ko \\ D(i,j-1) + Ki \end{cases} \tag{3}$$

where $D(i-1,j-1)$, $D(i-1,j)$ and $D(i,j-1)$ are results produced by the neighbor processors and Ki, Ko and Ks are respectively insertion cost, omission cost and substitution cost. Ki and Ko are constant value while Ks depends on the reference string character.

The figure 6 gives an idea of the processing element architecture. The **IN** input allows to store, into a dedicated I/O register file data, coming from the other processors (intermediate distances $D(i-1,j-1)$, $D(i-1,j)$ and $D(i,j-1)$). The **CTE** input is connected to a broadcast bus for receiving constant data from the outside world and from the memory. These data are stored on a specific constant register file before to be used.

As the arithmetic operations involved are very dedicated ones (addition and minimization), two specific units are implemented, namely an adder and a minimizer. These two units are pipelined due to the very regular and repetitive structure of the computation.

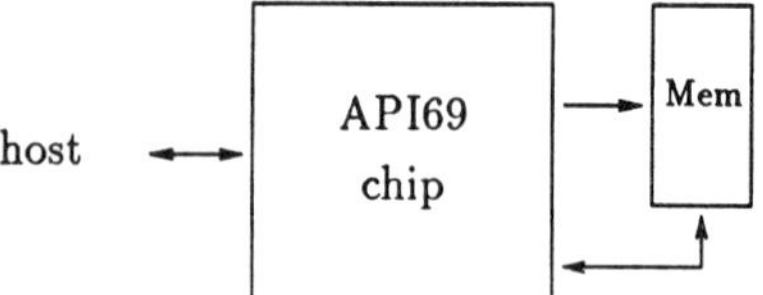

Figure 7: host interface

The accumulator can be loaded either from the adder or from the minimizer. Note that its content can be output on the broadcast bus.

4.5 Chip control

The core chip is programmable in the way that different applications based on the equation (1) can be performed. As a matter of fact, the processors of the array are activated by micro-commands controlling actions such as accumulator loading, I/O and CTE register selection, data acquisition, etc. These actions are specified by an instruction received from the outside and decoded to generate micro-commands. In the same way, the memory actions (as well as data reference actions) can be programmed depending on the calculation being processed.

Processor array, reference memory and data reference array are then operating synchronously since they are executing one instruction every machine cycle and because one instruction specifies actions to realize concurrently on these different units. In that sense, the processor array can be considered to have an SIMD execution mode : all the cells are executing the same instruction at the same time.

4.6 Host Interface

From an external point of view, the chip must be fed with data issued from a dictionary. These data are flowing through the data reference array in such a way that each systolic cycle requires 15 new data. Knowing that a systolic cycle is composed of a few machine cycles, external memory access is not a bottleneck.

Figure 7 shows how the chip is interfaced with the host computer. The dictionary memory is driven by a built-in control mechanism in order to optimize read access during a whole comparison process. This memory can be wrote via the API69 chip for loading, for example, small personal dictionaries.

As the chip can be seen as a peripheral device, a 8-bit data bus with associated I/O commands are provided. From a programming point of view, it means that a data port, a control port and a status port are available.

5 Chip organization

Figure 8 gives the floorplan of the chip. For compactness and suitable VLSI implementation, the systolic array is moved into an orthogonal matrix such as diagonals are laying

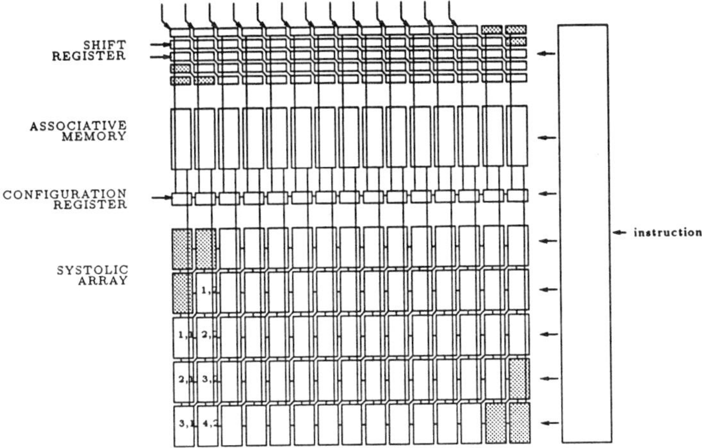

Figure 8: chip organization

down horizontally. Furthermore, to get a very regular structure, extra processors are added at the upper left and lower right corners. They are only used during initialization steps and do not participate to the edit distance computation. For the same reasons, the feeding memory shift register is also transformed into a regular rectangular matrix.

Each column of the processor array is connected to an independant associative memory by a bus, called the *broadcast* bus (Bbus). The processors can input data from this bus and store them in their constant registers. Patterns are provided to the memory by the shift register via another bus called the Xbus.

Before starting a whole comparison process, some constant registers of the processors must be initialized, as well as the memory. A configuration shift register has been added to perform these different tasks. It is connected both to the Bbus and the Xbus and is able to read or write data from/to these two buses. Data are fetched from the outside world on the left side register and can be read on the right side. As an example, loading the memory is realized first by reading 15 different data on the input configuration shift register port and then, by pushing these data to the memory input port. This operation is repeated while the memory is not full.

The configuration register is also a nice way for testing the chip functionality : as the different units are connected to this mechanism, data can be sent and received to/from a specific part of the chip core. To be able to pick up data from every processor, and then to test locally each processor, the micro-command allowing to write its content on the broadcast bus is splitted into the number of processor row. Consequently, SIMD execution mode of the processor array is not respected when performing this action.

6 Performances and conclusion

Due to the highly regular chip structure, full custom cells are designed to get a very compact circuit without any routing connections. The chip core can be seen as a large elementary cell array where cells are placed side by side. This simplifies greatly the design since only a reduced version can be generated for simulation, layout verification or layout/schematic comparison.

The chip core is presently under realization in a 1.5 μm CMOS technology. Operation of the circuit is entirely synchronous and based on a two phase non overlapping clock scheme. The clock frequency is expected to be around 25 Mhz and the number of transistors ranging from 300,000 to 350,000.

Correction process including transposition errors involves a 12 machine cycle systolic step (480ns). Since an edit distance result is available every systolic cycle, more than two millions strings can be processed per second. When all processors are operating, a 1.3 Gops (equivalent to 8 bit addition) peak performance is achieved.

References

[1] J.P. Banatre, P. Frison, P. Quinton, "A network for the detection of words in continous speech," *VLSI81 int. conf.*, J.P. Gray Ed., Academic Press, 1981.

[2] D. J. Burr, B. D. Ackland, N. Weste, "Array Configurations for Dynamic Time Warping," *IEEE Trans. on acoustic, speech and signal processing,* vol. ASSP-32 n° 1, pp. 119-127, feb. 1984.

[3] F. Charot, P. Frison, P. Quinton, "Systolic Architectures for Connected Speech Recognition," *IEEE Trans on ASSP,* vol. 34, n° 4, pp. 765-779, 1986.

[4] P. Frison, P. Quinton, "An Integrated Systolic Machine for Speech Recognition", in *VLSI : Algorithms and Architectures,* Edited by P. Bertolazzi and F. Luccio, North Holland, pp. 175-186, 1985.

[5] M. Gokhale & al., "SPLASH: A Reconfigurable Linear Logic Array", *Technical Report SRC-TR-90-012*, Supercomputing Research Center, Maryland, April 1990.

[6] P. A. V. Hall, G. R. Dowling, "Approximate string matching," *Comput. Surv.*, vol. 12, pp. 381-402, 1980.

[7] D. Lavenier, " MicMacs : un réseau systolique linéaire programmable pour le traitement de chaînes de caractères," *Thèse de l'université de Rennes 1,* Juin 1989.

[8] D.P. Lopresti, "P-NAC : a systolic array for comparing nucleic acid sequences," *IEE Computer,* vol 20, pp. 98-99, July 87.

[9] R. Lowrance, R. A. Wagner, "An extension of the string to string correction problem," *J. Assoc. Comput. Mach.*, vol. 22, n° 2, pp. 177-183, 1975.

VLSI 91
A. Halaas and P.B. Denyer (Eds.)
Elsevier Science Publishers B.V. (North-Holland)

Parallel Architecture and VLSI Implementation of a 80 MHz 2D-DCT/IDCT Processor

W. Liebsch , K. Boettcher

Heinrich-Hertz-Institut für Nachrichtentechnik Berlin GmbH, D-1000 Berlin 10, Federal Republic of Germany

Abstract

A parallel architecture and the implementation in CMOS technology is presented for high-speed forward and inverse two-dimensional discrete cosine transform. These circuits are applicable in advanced television and HTDV data reduction systems working at video sampling rates up to 80 MHz. The architecture utilizes the advantage of parallel and distributed arithmetic to achieve high-speed performance. An IDCT circuit is realized in 1.5 μm CMOS technology as a full-custom VLSI component. It performs 720 million multiplications and 1280 million additions per second at nearly floating point precision for the 8 bit output values. The device demonstrates the transfer of a parallel and regular architecture for high-speed operation into a VLSI full-custom chip layout.

1. Introduction

Transmitting a television video signal through a low bit-rate channel has led to the development of various data compression techniques. These include predictive coding and transform coding. For good quality pictures, predictive coding yields rather low compression rate and is very sensitive to transmission errors. Transform coding based on the discrete cosine transform (DCT) algorithm is one of the most widely used techniques in image data compression and yields better performance, however the computational complexity is much higher. It has been proven to be an almost optimum method, due to the fact that conventional image data has a reasonably high inter-element correlation and image information is mainly concentrated in a few transformed values.

Most published VLSI architectures [1,2,3,4] for two-dimensional transform coding are based on sequential calculation of input sub-blocks, rows and columns to obtain the transform values and require a separate coding block. The procedure to achieve a two-dimensional N-point DCT/IDCT requires 2*N passes through the signal flow graph, additional memory for matrix transposition and comprehensive control logic. These architectures would necessitate a costly and tricky design. For high-speed VLSI implementation the regularity of the data flow and an advantageous circuit structure of discrete cosine transform algorithm has to be taken into account, in order to ease the control, to minimize the number of different stages of the pipeline structure, and to lower storage requirements.

In this paper, the design and implementation of a parallel circuit architecture with direct calculation of two-dimensional DCT/IDCT values will be presented. This realisation does not require any scratch-pad memory for matrix transposition. The circuit design is characterized by a regular structure and utilizes the advantages of parallel and distributed arithmetic to achieve high-speed performance. A one-chip VLSI implementation of the IDCT circuit has been realized in CMOS technology and performs 80 MHz data processing.

2. Algorithm

For an 8x8 input block a direct calculation of two-dimensional inverse discrete cosine transform (2D-IDCT) values is represented by equation (1). The calculation of IDCT values (or DCT with the transposed matrix C) is based on the vector matrix product algorithm and can be implemented using a bank of 64 basic elements. Each basic element calculates one transformed value which is a weighted sum of all the block input data. With expression (1) a distributed arithmetic structure can be derived for a separate calculation for each transformed value.

$$\mathbf{x}(\mathbf{u}) = \frac{1}{8}\sum_{\mathbf{v}=0}^{63} \mathbf{d}(\mathbf{v})\,\mathbf{C}(\mathbf{v},\mathbf{u}) \qquad \text{with } \mathbf{u} = 0 \cdots 63 \tag{1}$$

The input vector d is obtained from the input DCT matrix D by the following rule

$$\mathbf{d}(\mathbf{k}+8\mathbf{j}) = \mathbf{D}(\mathbf{j},\mathbf{k}) \qquad \text{with } \mathbf{k} = 0 \cdots 7,\ \mathbf{j} = 0 \cdots 7 \tag{2}$$

and the output pixel matrix X by

$$\mathbf{X}(\mathbf{j},\mathbf{k}) = \mathbf{x}(\mathbf{k}+8\mathbf{j}) \qquad \text{with } \mathbf{k} = 0 \cdots 7,\ \mathbf{j} = 0 \cdots 7$$

The discrete cosine transform coefficient matrix C has a dimension of 64*64 and its elements are calculated according to

$$\mathbf{C}(\underbrace{\mathbf{l}+8\mathbf{k}}_{\mathbf{u}},\underbrace{\mathbf{s}+8\mathbf{r}}_{\mathbf{v}}) = \mathbf{A}(\mathbf{r},\mathbf{k})\mathbf{A}(\mathbf{s},\mathbf{l}) \qquad \text{with } \mathbf{k}=0\cdots 7;\ \mathbf{l}=0\cdots 7 \qquad (3)$$

$$\mathbf{r}=0\cdots 7;\ \mathbf{s}=0\cdots 7$$

$$\mathbf{A}(\mathbf{n},\mathbf{m}) = \cos\frac{(2\mathbf{n}+1)\pi\mathbf{m}}{16}\mathbf{a}(\mathbf{m}) \qquad \text{with } \mathbf{n}=0\cdots 7;\ \mathbf{m}=0\cdots 7 \qquad (4)$$

$$\mathbf{a}(\mathbf{m}) = \begin{cases} 1/\sqrt{2} & \text{for } \mathbf{m}=0 \\ 1 & \text{for } \mathbf{m}\neq 0 \end{cases}$$

Taking advantage of the special pattern in the discrete cosine coefficient matrix, the number of multiplications and additions can be reduced, according to the degree of decimation. Using the symmetry of columns in matrix C decimation is made by a factor of four. A further reduction of the coefficient matrix is possible, but is not used in order to maintain regularity of the circuit structure. Therefore the number of arithmetic calculations is reduced by a factor of four as well. This is utilized fully to lower the internal clock rate to only one quarter of the pixel sampling rate for chip realization in CMOS technology and for applications of such components in high-speed coder systems. Figure 1 illustrates a block structure of the DCT and IDCT processor. With the reduced coefficient matrix 16 cycles are needed for the calculation of each of the 64 transformed values. The derived architecture enables a simple arrangement of interconnections of the basic elements and a development of an identical basic element circuit structure. This is important for a full custom VLSI implementation. A further symmetry in the rows of the IDCT coefficient matrix can be used for additional reduction of multiplications in the circuit path A,B,C. However, this reduces the regularity of the circuit structure (Figure 2) and the uniformity of basic elements. In this circuit calculation of an 8*8 IDCT block requires 576 multiplications and 1024 additions.

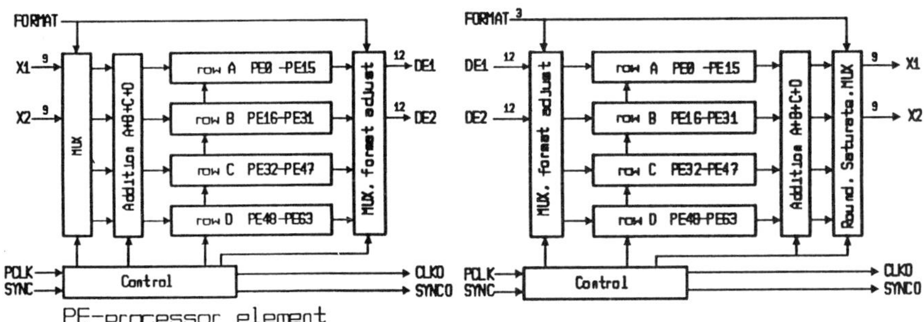

Figure 1. Block diagram of DCT(left) and IDCT(right) processor

3. Architecture and functional description

DCT and IDCT processor

The implementation of a high-speed DCT and IDCT function requires a special-purpose architecture. A block diagram of the two-dimensional DCT and IDCT device based on this approach is shown in Figure 1.

The DCT circuit contains four adder functions and is divided into four data paths. The pixels of an input block are separated into four groups. Four input pixels are processed simultaneously in adders for reducing the internal processing speed. As a compromise between chip pincount and input/output data speed two input values (X1,X2) and two output values (DE1,DE2) are transferred in parallel. In path A,B,C,D the DCT values are calculated concurrently. Each path consists of 16 pipelined basic elements. Each basic element calculates one DCT value. The basic elements are linked to a common output bus system. Four DCT values are obtained for every four input data entered into the processor. A sequential output of the calculated values is produced by controlling the output buffer in the basic elements. 16 cycles are needed to complete a discrete cosine transform of one block. All basic elements can work on a single output bus, but it is convenient to share the bus in order to reduce the load for the bus drivers in the basic elements.

The IDCT circuit has a similar structure with four adders at the output of the circuit. In Figure 2, where the hardware structure for the IDCT is shown, the output bus is divided into four busses and output adders with four inputs are used. The addition A+B+C+D results in the IDCT values (output pixel X1,X2).

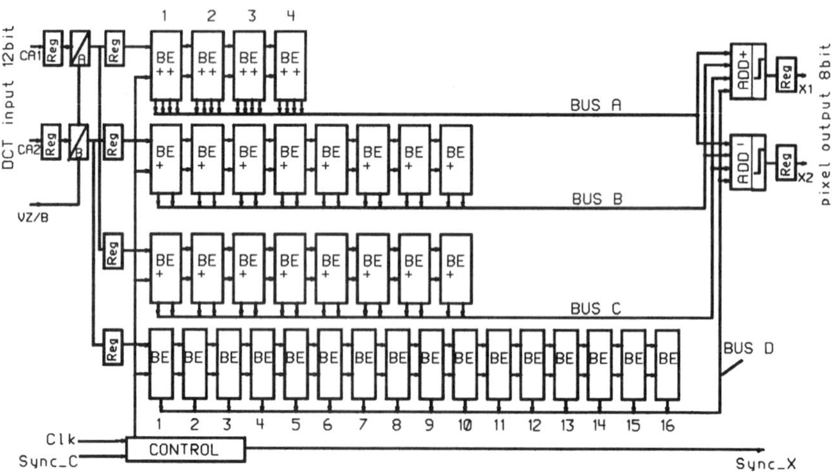

Figure 2. Functional IDCT block diagram (reduced matrix)

In principle it is possible to combine DCT and IDCT functions for a one-chip VLSI implementation. However, as DCT and IDCT require different sets of coefficients, the basic elements must contain both coefficient sets and additional multiplexers and the required chip area would exceed economical magnitudes at these high sampling frequencies (80 MHz). Furthermore, if DCT or IDCT are combined with additional coding or decoding procedures respectively, it is preferable to design separate chips for the two applications. Moreover, for practical use in a data reduction system two components are mostly needed.

Basic elements for DCT and IDCT

A block diagram of the basic element (BE) is shown in Figure 3. The structure of such an element is similar for DCT and IDCT. Each BE calculates one transform value and consists of a multiplier, an accumulator, an output latch with bus driver and a small control unit. The control unit consists of an associated ROM containing the DCT or IDCT coefficients and some control signals to determine the sequence for outputting the DCT/IDCT value via the common bus. This sequence can be made adjustable by selecting one of several control signals contained in the ROM. This is useful for an adaptive coding procedure. All basic elements are synchronized by the main control unit. After 16 cycles, the computation of the transform value is completed and is then transferred to the output latch. An output value will appear at the basic element every 16 cycles. During 15 cycles, the output latch of the basic elements is disabled and the bus is free for the output data of the next basic element values. All arithmetic functions use a sign-magnitude numbering system. Partial sum overflows during accumulation can be tolerated if the final result is in the dynamic range of a output value.

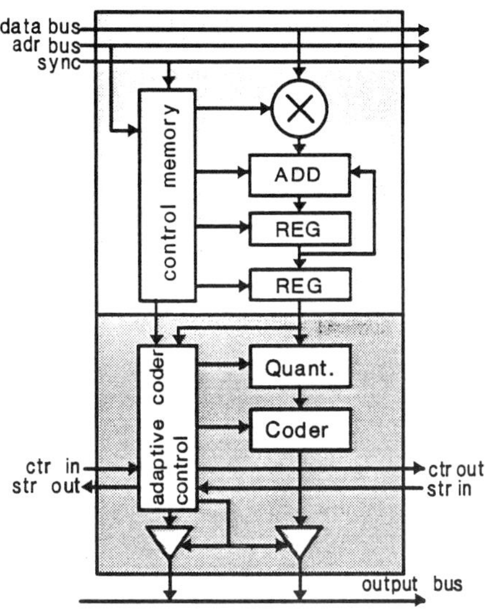

Figure 3. Basic element with coder function

Additional coding functions

The calculation of DCT values in separate basic elements makes it possible to integrate an adaptive coder procedure for bit rate reduction into the processor architecture. In order to adapt the different statistical properties of images different scan paths can easily be fixed by programming the control memory (ROM) in the basic elements without additional hardware. The threshold processing, quantization and further coding procedures of the current DCT values is performed before outputting it of a basic element. The value of the threshold can be adjusted on a block-to-block basis and depends on the fill-status of the output buffer. The DCT values that are above this threshold are quantized, coded and signed for transmission [5,6]. Figure 3 shows a block diagram of additional coding functions in the basic element.

For generation of an output string a new control signal is generated in the active basic element and is fed via a pipeline to the coding module of the next basic element. According to the statistical properties, this procedure begins with the highest order DCT value and the transmission of the block values starts with the first DCT value unequal to zero. Detailed results of the coder part will be published in a further paper.

4. VLSI implementation

In order to investigate the problems of an implementation of such high-speed devices at first we realized the IDCT circuit. The IDCT device, containing 180.000 transistors was designed in a 10*10 mm^2 area using 1.5μm CMOS technology. This layout was shrinked to 8*8 mm^2 and fabricated. Figure 5 shows a photomicrograph of the die. The device consists of input stages with registers for the DCT values, format converter, pipeline registers, clock and control circuits (left), 40 basic elements (center) and output stages with adders, pipeline registers, format converter, limiter and output registers (right).

These 40 basic elements cover 60% of the area and therefore the design of this elements was made very carefully. Sixteen basic elements consist of a multiplier with an output register, a dynamic ROM for coefficients and control bits, one accumulator with pipeline register and one threestate busdriver. All others contain an additional accumulator and busdriver. The algorithm selected for the 11*10 multiplier decodes two neighbouring bits of the 10 bit operand to determine a partial product which is the times zero, one, two or three of the 11 bit operand. Only the times-three operand needs an additional 11 bit adder. The five partial products are selected in parallel by five multiplexers each controlled by two bits. Four carry ripple adders with alternating carry technique perform a 14 bit product in less then 50 ns. Carry ripple technique is also used in the accumulators, format converters and output adders and therefore a five-stage pipeline structure was required.

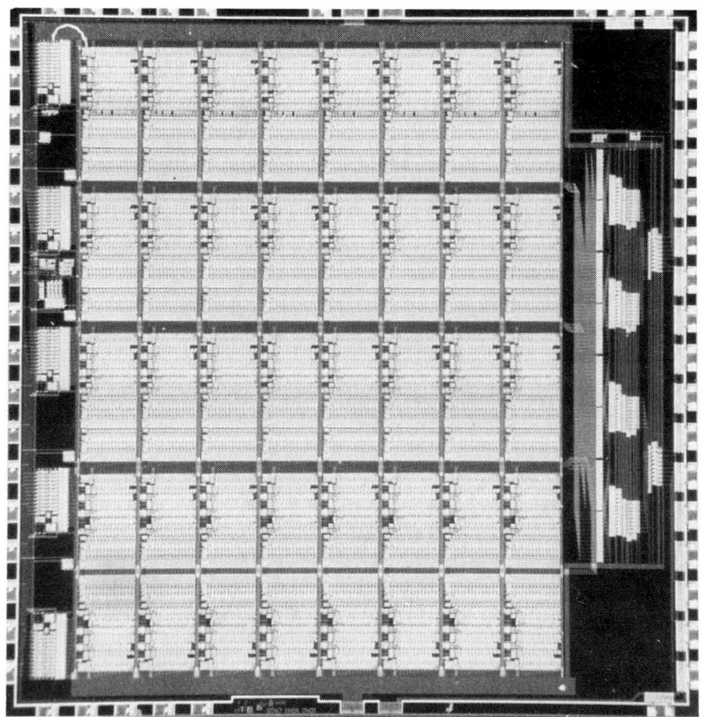

Figure 4. Photomicrograph of IDCT chip

5. Results

Accuracy of the circuit design

For circuit realization, the accuracy of DCT values calculated with this parallel structure has been evaluated for a given 8x8 input sub-block in order to determine the minimum number of bits which does not produce any visible degradation after forward and inverse transform [7]. The overall computational error of this approach results from the rounding operation of the internal values and depends on the selected operand word length. This word length is represented by the word length after precalculation, the word length of coefficients, multiplier output, accumulator output and the final

DCT value at the output of the basic element. The number of bits at the input of a two-dimensional DCT is normally eight. The optimization procedure is carried out by simulation. In order to attain a satisfactory level of accuracy the required number of bits varies from 8 to 12 bits for DCT and 8 to 14 bit for IDCT for internal operations. This is sufficient for nearly floating point precision and for coding and transmission [6]. The IDCT accuracy is compatible with the proposed CCITT standard. With the limited input DCT value word length of 12 bit the mean square error (MSE) is better than 60 dB related to the maximum pixel value of 255.

Due to the statistical nature of the input pixels, many of the ac values may be small compared to the dc value and thus would allow for further reduced word length.

Features of IDCT component

- 8*8 transform block size
- processing of DCT value up to 80 MS/s
- input word length 12 bit (DCT values)
- output word length 8 bit or 9 bit (adjustable)
- block input and output format is adjustable
- fully synchronous interface
- CMOS implementation
- die size 64 mm^2
- single +5V ±10% power supply
- power dissipation $<$ 2.5 W at 50 MHz
- 68-pin PLCC package

6. Conclusions

The architecture proposed here is applicable to high-speed two-dimensional DCT and IDCT working at video sampling rates up to 80 MHz. This architecture utilizes the advantages of parallel and distributed arithmetic to achieve high-speed performance and it can be favourably combined with additional distributed functions for data reduction of video signals. The resulting circuit structure is highly regular with a minimum of control overhead and, because of the widely identical basic elements, is relatively easy to implement. Using this approach, a two-dimensional inverse discrete cosine transform with a block size of 8x8 values has been realized in 1.5μm CMOS technology as a single VLSI component.

Acknowledgements

This project was supported by the German Federal Ministry of Research & Technology

References

[1] Arnould, E., Dugre, J.P., Real Time Discrete Cosine Transform an Original Architecture, Proc. ICASSP, 48.6.1 San Diego, 1984.

[2] Jutland, F.,Demassieux,N.,Concordel, G., A Single Chip Video Rate 16x16 Discrete Cosine Transform, ICASSP 86, 15.8.1 Tokyo, 1986.

[3] A. Artieri, et.al., A one chip VLSI for real-time two-dimensional discret cosine transform, Proc. of ISCAS 1988, pp.701-704

[4] Ligtenberg, A., Wright R.H., O'Neil J.H., VLSI orthogonal transform chip for real-time image compression, AT&T Bell Labs, Visual Communication and Image Processing II, SPIE 1987 Vol. 845-02

[5] Lohscheller, H., and Franke, U., Colour Picture Coding-Algorithm Optimization and Technical Realisation, Frequenz 41 Nr.11/12, 1987.

[6] Chen, W.H., Pratt, W.K.,Scene adaptive coder, IEEE Trans., Com-32, No 3, pp. 225-232, 1984.

[7] Madec, G., A comparison between several fast dct algorithms for hardware implementation, PCS 1987, p.177

VLSI 91
A. Halaas and P.B. Denyer (Eds.)
Elsevier Science Publishers B.V. (North-Holland)

Exact Redundant State Registers Removal Based on Binary Decision Diagrams

Bill Lin and A. Richard Newton

Department of Electrical Engineering and Computer Sciences
University of California, Berkeley, CA 94720[1]

Abstract

We address the problem of removing redundant state registers that are not needed to differentiate the state codes. This can be viewed as a restricted form of re-encoding that can significantly reduce the number of registers (code length) and the complexity of the logic implementation. We show that a greedy approach to this problem is suboptimal and present a new BDD-based algorithm for this problem that is guaranteed to return the maximum set of removable state variables. The size of reachable states that can be handled using our technique is limited by the size of the BDD representation of the reachable set, which enables our technique to be applicable to very large problem instances. We describe effective branching heuristics and a new bounding technique based on the concept of maximal removable set. For the examples tested, we show that the exact solutions can be found in all cases with modest CPU times. We also show how the technique of redundant variable removal can be used to simplify both the BDD representation of the transition relation for verification as well as the size of the gate-level implementation for synthesis. For the examples tested, we have been able to achieve significant reduction in both the BDD size of the transition relation as well as the final implementation cost in terms of area.

1 Introduction

We consider the problem of optimizing multi-level sequential circuits by removing redundant state registers that are not necessary to distinguish the state codes of the reachable states. We assume a sequential circuit description given as an interconnection of combinational logic gates and synchronous registers. A reset register pattern is also assumed to be given.

Given a sequential circuit with N registers, there are 2^N possible state patterns (or simply states). However, only a subset of these states are reachable from the reset state. Using recently developed concepts from sequential verification [7, 11, 5] that are based on the binary decision diagrams and symbolic execution techniques, the set of reachable sets can be computed for sequential circuits with almost arbitrarily large state spaces. The main limitation is the ability of BDD's to implicitly represent the set of reachable states as a characteristic function. We note that the real bottleneck is the regularity of the state space rather than its size.

In many sequential circuits, especially those with a large number of registers, the subset of states that are reachable is often significantly smaller then 2^N (where N is the number of variables). This is the case with many of the circuit examples from the ISCAS-89 sequential test benchmark set. Therefore, it is possible to re-encode the set of reachable states with fewer number of registers. Surprisingly, this property is not uncommon in realistic designs. Also, sequential logic optimization algorithms for area and performance optimization, such as

[1]This project was supported in part by a joint contract between NSF and DARPA under contract number MIP-8719546.

retiming [8] and retiming and resynthesis [10], are known to increase the number of state registers dramatically. However, since the sequential behavior of the circuit remains unchanged, the number of reachable states in the modified circuit tends to remain about the same even with a significant increase in the number of state registers.

One powerful strategy for reducing the number of registers (code length) is to remove state variables that are not necessary to distinguish the different state codes. These state variables are called *redundant* state variables. Consider the following trivial example. Suppose $Q = \{001, 100, 111\}$ is the set of reachable states. The first state variable is redundant since the remaining state variables can uniquely determine the state codes: $\hat{Q} = \{01, 00, 11\}$. In effect, we have re-encoded the states using a subset of their bit patterns. This kind of re-encoding will almost always lead to a more efficient representation or implementation of the sequential circuit. The problem is to find a maximum (or maximal) subset of these variables to eliminate such that the state codes induced by the remaining state variables are unique. Berthet et al. [1] gave conditions for detecting and removing these redundant state variables. However, they did not give any algorithm or results.

A straightforward, but naive, approach is to test and remove one state variable at a time. Unfortunately, we prove this simplistic greedy algorithm is suboptimal. The problem is that the removal of one redundant state variable may preclude the removal of other state variables. We present in this paper a new BDD-based branch and bound algorithm for finding the maximum set of redundant state registers. The algorithm is guaranteed to produce an *exact* solution upon completion. As with most branch and bound algorithms, effective heuristics for branching and bounding are crucial. In our branching heuristic, we try to choose the most *unate* redundant state variable for removal since it can distinguish the least number of state pairs. We have developed an effective bounding technique based on the concept of *maximal removable set*. Empirically, we have found this bounding technique to be extremely effective in pruning the search space. Because our algorithm is BDD-based, and it implicitly examines the states rather than explicitly, it is applicable to any set of reachable states that can be computed and represented using BDD's. Hence, it is applicable to a very broad class of sequential designs.

We have applied our algorithm on a variety of sequential benchmark circuits. After we compute the maximum set of redundant state registers for removal, we can make use of this information to significantly reduce both the BDD representation of the transition relation for verification and the gate-level representation of the circuit for implementation. For design verification, a reduction in the BDD representation of the transition relation can significantly reduced the verification time when checking for multiple design properties. This is important since many design properties may be checked during verification session. Removal of redundant state registers can also lead to much more efficient implementations. This is partly due to the savings in register cost. However, since a subset of state variables are no longer needed, the hardware of the corresponding next state functions can be also discarded. This represents a more significant savings in area. Although some additional logic may be incurred for adapting to the new state codes, the overall area is almost always reduced. Also, with a significant reduction in register count, the resulting sequential designs tend to be more easily testable.

The remainder of the paper is organized as follows. Section 2 presents some background on binary decision diagrams and defines the state machine model that we assume. Section 3 analyzes the redundant state variable removal problem. In section 4, we present our exact algorithm for finding the maximum removable set of state variables. In section 5, we present extensive experimental results on a variety of sequential benchmark circuits using our techniques. We show, in a surprising number of cases, the number of state registers is significantly reduced. We also give results on the effects of register removal on reducing the size of transition relations in BDD form and the area of the circuit level implementation, again with surprising results.

Finally, in section 6, we give some concluding remarks.

2 Preliminaries

2.1 Binary Decision Diagrams

We give in this section a brief description of Bryant's binary decision diagrams (BDD's) [4]. A BDD is a directed acyclic graph (DAG) representation of logic. It is a binary decision graph where each node is associated with a variable and two fanouts. One fanout corresponds to when the variable is set to 0, and the other corresponds to when the variable is set to 1. At the leaves of the graph are two constant nodes representing the constants 0 and 1. A variable ordering is imposed such that all transitive fanouts of node must have a higher ordering index, except for the constant nodes. Also, a variable may not be repeated in a single directed path. The BDD is said to be reduced if there are no isomorphic subgraphs. This representation is unique for a given variable ordering. Standard boolean operators like intersection, union, and negation can be implemented very efficiently with BDD's. Boolean quantifiers such as the existential operator and the universal operator can also be implemented with BDD's as follows. The existential quantification (also called *smoothing*) of a boolean variable v with respect to the boolean formula f can be computed as

$$(\exists v)f \stackrel{\text{def}}{=} f_v + f_{\overline{v}},$$

where f_v is the usual cofactor operation as defined in [4]. Similarly, the universal operator (also called the *consensus* operator) can be computed as

$$(\forall v)f \stackrel{\text{def}}{=} f_v \cdot f_{\overline{v}}.$$

Another commonly used operator is the substitution operator. The substitution of a variable v by a variable w in a formula f, denoted as $f\langle v \rightarrow w\rangle$ can be computed as

$$f\langle v \rightarrow w\rangle \stackrel{\text{def}}{=} (\exists v)[(v \Leftrightarrow w) \cdot f].$$

There are known tricks to make this operation very efficient when renaming over multiple variables.

The advantage of BDD's over other canonical representations (such as a truth table) is that they are usually much more compact and efficient polynomial time algorithms exist to manipulate them. A particularly well suited application of BDD's is the representation of sets as *characteristic functions* defined as follows.

Definition 2.1 *Let Q be a set of n-tuples over $\{0,1\}^n$. Let $A \subseteq Q$. The* characteristic function *of A is the boolean function $\chi_A : Q \rightarrow \{0,1\}$ defined by $\chi_A(\mathbf{x}) = 1$ if $\mathbf{x} \in A$ and $\chi_A(\mathbf{x}) = 0$ otherwise.*

We denote the cardinality of a finite set Q by $\#Q$ or $|Q|$.

2.2 State Machine Models

A *finite state machine* M is a 6-tuple $\langle \Sigma, Q, \mathcal{O}, \delta, \lambda, q_0 \rangle$, where Σ is the input alphabet, Q is a finite set of states, $\mathcal{O}$ is the output alphabet, δ is the state transition function from $Q \times \Sigma$ to Q, λ is the output function from $Q \times \Sigma$ to $\mathcal{O}$, and $q_0 \in Q$ is the initial state of the machine (in general, the initial condition may be a set). An alphabet is a set of symbols. In this paper, we will assume an alphabet is defined over a finite n-dimensional Boolean space $\{0,1\}^n$ where each vertex (minterm) uniquely represents a symbol in the alphabet: e.g. $Q = \{0,1\}^r$, $\Sigma = \{0,1\}^m$, and $\mathcal{O} = \{0,1\}^n$.

If the primary output functions are only dependent on the present state value, then the machine is called a *Moore* machine; otherwise, it is a *Mealy* machine. A machine is said to be *fully specified* if for every combination of primary input and present state, the next state and all primary outputs are defined. Otherwise, the machine is said to be *incompletely specified.*

A state q is said to be *reachable* if there exists an input sequence that will transform the machine from the reset state q_0 to q. The input sequence is called the *justification sequence.* A reachable state is also called a *valid state.* If a state is unreachable, then it is also referred to as an *invalid state.*

A state machine can be implemented at the gate level as an interconnection of combinational logic gates and synchronous registers. Numerous clocking schemes can be devised to synchronize the registers. For simplicity, we assume in this paper a single global edge-triggered scheme. It can be abstractly modeled by a directed graph $G = (V, E)$ where each vertex $v \in V$ represents either a primary input $x_j \in \{x_1, \ldots, x_m\}$, a primary output $z_j \in \{z_1, \ldots, z_n\}$, a logic function f_k, or a synchronous element $l_j \in \{l_1, \ldots, l_r\}$. Associated with each internal logic function f_k is a variable y_k. For a logic function node, its immediate fanins are called its *support.* An arc $e_{u \to v}$ from u to v represents a direct connection between two nodes. For every cycle in the graph, we required that at least one node along the path is a synchronous latch. A combinational network is a special case of a synchronous network where the graph is acyclic and all elements are either a primary input, a primary output, or a logic function.

3 Redundant Encoding Variable Removal

In this section, we consider the problem of removing redundant state variables. For greater generality, we will not assume the set of elements under consideration are states. Instead, we will assume a generic set of already encoded elements A over some finite boolean domain B^n. The lower bound on the number of variables required to re-encode A such that each element is given a unique code is $k = \lceil \log_2 \#A \rceil$. Very often, k is less than n. If this is the case, then it is possible to re-encode A with fewer number of variables. One technique for reducing the number of encoding variables is to eliminate variables that are not necessary to distinguish the codes given to the elements. These variables are called redundant variables.

Definition 3.1 *Let $A \subseteq B^n$ be a set of binary patterns over the boolean space B^n and $a_1 \ldots a_n$ be the corresponding encoding variables. A variable a_i is said to be* **redundant** *if each element after its removal is still unique. It is also called a* **free** *variable.*

Definition 3.2 *Let $A \subseteq B^n$ be a set of binary patterns over the boolean space B^n and $a_1 \ldots a_n$ be the corresponding variables. A* **permissible removal set** *is a set of redundant variables $\{a_i, a_j, \ldots, a_k\}$ such that their removal does not affect the uniqueness of each code.*

Definition 3.3 *A permissible removal set $\{a_i, a_j, \ldots, a_k\}$ is said to be* **maximal** *if it is not fully contained in any other permissible removal set. A permissible removal set is said to be the* **maximum** *if it is the largest permissible removal set.*

Example 3.1 : Let A be $\{0011, 1100, 0000, 1001\}$ and let a_1, a_2, a_3 and a_4 be the encoding variables. Then a_1 can be eliminated since the resulting set of codes are still unique: $\hat{A} = \{011, 100, 000, 001\}$. Since no remaining variables can be eliminated, $\{a_1\}$ is also a maximal removal set. However, $\{a_1\}$ is not a maximum set since $\{a_2, a_3\}$ is also a permissible, but larger, removal set.

The goal is to remove the largest set of variables. The maximum variable elimination problem is stated as follows:

Problem 3.1 : (Maximum Variable Removal Problem) Given a set of binary patterns $A \subseteq B^n$, find the maximum removal set such that the resulting set of codes induced by the remaining encoding variables remain unique.

To handle redundant variable removal for very large sets of elements (e.g. 10^{50} elements and beyond), we must represent the set as a characteristic function in BDD form. Depending on the regularity of the elements, BDD's can represent very large sets. Once captured in BDD form, we can detect redundant variables very easily as follows:

Lemma 3.1 *A variable a_i is redundant if and only if*

$$(\forall a_i)A = 0. \tag{1}$$

The new set of codes after removing the redundant variable a_i can be derived as follows:

$$\hat{A} = (\exists a_i)A. \tag{2}$$

These operations can be efficiently implemented with BDD's.

The use of BDD's to detect and remove redundant variables was proposed by Berthet et al. [1]. However, they did not give any algorithm for finding a removal set. Also, no results were reported. A straightforward approach is to test and remove one variable at a time using the above operations. However, as will be shown, this greedy procedure is suboptimal. Finding the maximum set of variables to remove is a difficult problem in general since the removal of one variable may preclude the removal of other variables. In fact, the problem of finding the maximum removal set is NP-complete so that any algorithm which solves this problem exactly can be expected to have exponential worst-case complexity in the number of variables, even if BDD operations are used. In practice, significantly larger removal sets than a greedy solution can often be found if more intelligent algorithms are devised.

Example 3.2 : Consider the following set of elements:

	a_1	a_2	a_3	a_4	a_5
q_1	0	0	0	1	1
q_2	0	1	0	0	1
q_3	1	1	1	0	1
q_4	1	0	0	0	0

The characteristic function for the above set can be expressed as

$$\bar{a}_1\bar{a}_2\bar{a}_3a_4a_5 + \bar{a}_1a_2\bar{a}_3\bar{a}_4a_5 + a_1a_2a_3\bar{a}_4a_5 + a_1\bar{a}_2\bar{a}_3\bar{a}_4\bar{a}_5$$

In this example, the greedy solution is $\{a_1, a_2\}$. However, $\{a_3, a_4, a_5\}$ is also a permissible removable set. The latter (non-greedy) solution is superior since it removes one additional variable.

Lemma 3.2 *A greedy removal of redundant variables is suboptimal.*

Proof: By examples 3.1 and 3.2. □

4 Exact BDD-based Branch-and-Bound Algorithm

We now illustrate a branch-and-bound algorithm for solving the maximum variable elimination problem. The basic branch-and-bound algorithm for finding the maximum removal set involves the following steps:

1. Determine an upper bound on the number of variables that can still be eliminated. If the size of the current selected set plus the upper bound is less than or equal to a bound (e.g. the size of the best solution seen so far), return from this level of recursion. If there are no more free variables, declare the current solution as the best solution recorded so far.

2. Select a free variable to eliminate. The test for a free variable is given by Equation 1.

3. Add the variable to the selected set and solve the subproblem by obtaining the new set of patterns using Equation 2. Then, solve the subproblem by not selecting this variable.

Choice of Redundant Variable: At each level of recursion, there may be many redundant variables. Good heuristics for choosing the branching redundant variable are crucial in speeding up the branch and bound algorithm. One effective heuristic is to choose the most *unate* variable. We can easily determine which redundant variable is most unate as follows.

$$\text{weight}(a_i) = \text{absolute}(|A_{a_i}| - |A_{\overline{a_i}}|) \quad (3)$$

$$\text{select}(a_i) = \arg\max_{a_i} (\text{weight}(a_i)) \quad (4)$$

The motivation for removing the most unate variables first is because they are the least effective in distinguishing pairs of codes.

Maximal Removable Set: An important feature of the proposed branch and bound algorithm is the use of maximal removable set. This routine is used to compute a simple upper bound that corresponds to the maximum number of removable variables. Here, we use the number of remaining free variables as a heuristic upper bound. A simple bounded look-ahead scheme can be applied for greater accuracy. The purpose of this upper bound is to bound the recursion early so that inferior parts of the search space may be discarded as early as possible.

Constant Variables: If the value of the variable a_i is always 0 or always 1 in every element of A, then we say a_i is a *constant* variable. A constant variable can be removed unconditionally since it cannot distinguish any pairs of codes in A.

Theorem 4.1 *When the above branch and bound procedure terminates, it is guaranteed to return the* **maximum** *removal set of variables that can be eliminated.*

The search space is pruned significantly using the described bounding technique. All operations are performed using BDD's so that the set of elements is never explicitly enumerated. A fast heuristic algorithm can be derived from the above procedure by terminating the recursion early. Experiments show that the first leaf solution is usually quite good. However, we have found a number of test cases where the greedy method is considerably less effective than our exact and heuristic procedures. The effectiveness of the proposed BDD-based algorithms are dependent on the regularity of the set considered. However, using the heuristics described here, we have found substantial reduction in the number of encoding variables for many cases.

circuit	states	registers	exact	reduction	minimum	CPU
s208	17	8	5	37.5	5	0.01
s298	218	14	12	14.3	8	3.40
s344	2625	15	15	0.0	12	0.85
s349	2625	15	15	0.0	12	0.84
s382	8865	21	18	14.3	14	29.44
s386	13	6	6	0.0	4	0.01
s400	8865	21	18	14.3	14	29.45
s444	8865	21	17	19.0	14	34.38
s510	47	6	6	0.0	6	0.01
s526	8868	21	19	9.5	14	30.31
s641	1544	19	14	26.3	11	0.83
s713	1544	19	14	26.3	11	0.84
s820	25	5	5	0.0	5	0.01
s832	25	5	5	0.0	5	0.01

Table 1: Exact Redundant State Register Removal Results.

states	number of reachable states
registers	number of state variables
exact	number of state variables after exact removal
reduction	percentage reduction in the number of state variables
minimum	lower bound on the number of state variables
CPU	times for exact redundant state register removal in seconds on a DEC 5000 Workstation

5 Experimental Results

The techniques described in the previous sections have been implemented using Berkeley's implementation of a BDD package originally described in [2]. In this section we present some preliminary results on experiments based on the exact redundant state registers removal techniques described in this paper. All experimental results presented were measured on a DEC 5000 workstation and the CPU times presented are quoted in seconds. The primary set of test cases used were obtained form the ISCAS-89 sequential test benchmark set.

Table 1 presents the results of exact redundant register removal. The column labeled **states** gives the number of reachable states for each sequential circuit. The set of reachable states is computed using BDD-based implicit enumeration techniques [6]. Particularly, we used Berkeley's implicit enumeration routines [11] and some auxiliary routines from [9]. In most cases, the number of reachable states is much less than 2^N, where N is the number of state registers. In performing exact redundant register removal, we found our maximal removal set bounding heuristic to be very powerful in reducing the depth of recursion. The heuristics for choosing the branching variables were also quite important. The number of registers in the original circuit are shown under the column labeled **registers**. The column labeled **exact** gives the number of registers after exact redundant register removal. Surprisingly, the circuits from the ISCAS-89 benchmark set contain a very large number of redundant registers. The percentage reduction is shown in column **reduction**. Using our exact BDD-based algorithms, we were able to obtain exact results in all cases with modest CPU times.

circuit	original relation	reduced relation	reduction
s208	51	35	31.4
s298	461	413	10.4
s344	377	377	0.0
s349	377	377	0.0
s382	4129	3890	5.8
s386	140	140	0.0
s400	4129	3890	5.8
s444	654	557	14.8
s510	287	287	0.0
s526	462	414	10.4
s641	3280	2917	11.1
s713	3280	2917	11.1
s820	217	217	0.0
s832	217	217	0.0

Table 2: Comparisons of Transition Relation Sizes

original relation BDD size of original transition relation
reduced relation BDD size of transition relation after register removal
reduction percentage reduction in the BDD size of the transition relations

In our second experiment, we examine the effects of redundant register removal on the size of the BDD representation of the transition relation. Reducing the size of the transition relation can have a dramatic impact on the verification, especially design verification, since properties may be checked on the transition relation repeatedly. Table 2 shows the BDD size of the transition relations before and after redundant register removal. For cases where no redundant registers were identified, the size of the transition relation remained the same. However, in a significant number of examples tested, the size of the transition relation reduced by a significant amount. For example, the transition relation of s208 was reduced by over 31%.

In our third experiment, we examine the effects of redundant register removal on the size of the gate level implementation. We used the multi-level logic optimization system misII version 2.2 [3] with the standard script to obtained an optimized result. Under the section labeled **literals only**, we report area measured in terms of literal counts in factored form. **area1** is the area of the optimized original circuit, and **area2** is the area of the optimized modified circuit. The area results reported are in terms of the **total area**, including both the next state and output functions. The area of the original circuit was obtained by running the standard script. The area of the modified circuit was obtained in three steps: first attach necessary re-encoding circuitry to the original circuit; then perform some partial collapsing; then optimize using the standard script.

Depending on the example, significant reduction in area can be achieved (between $20-38\%$ on several examples). Note that much of the reduction comes about because we no longer need to implement certain next state functions if the corresponding register has been removed. However, the procedure cannot guarantee that the area will be reduced since the circuit structure itself

circuit	literals only			with registers		
	area1	area2	ratio	area1	area2	ratio
s208	76	47	0.62	132	82	0.62
s298	112	85	0.76	210	169	0.80
s344	141	141	1.00	246	246	1.00
s349	146	146	1.00	251	251	1.00
s382	152	135	0.89	299	261	0.87
s386	132	132	1.00	174	174	1.00
s400	156	135	0.87	303	261	0.86
s444	150	136	0.91	297	255	0.86
s510	247	247	1.00	289	289	1.00
s526	190	146	0.77	337	279	0.83
s641	189	219	1.16	322	317	0.98
s713	198	219	1.11	331	317	0.96
s820	252	252	1.00	287	287	1.00
s832	258	258	1.00	293	293	1.00

Table 3: Comparisons of Gate Level Implementations

literals only	literals in factored form using misII and standard script
with registers	literals plus register costs (7 literals per register)
area1	optimized area cost for original sequential circuit
area2	optimized area cost of sequential circuit after register removal and simplification
ratio	area cost ratio between modified and original circuit

must be modified. In examples s641 and s713, the area actually increased if the cost of registers is not considered. Since some registers have been removed, it would be more accurate to compare area with the register costs accounted. Under the section labeled **with registers**, we report area results in literal count by giving each register the cost of 7 literals. This corresponds to the relative size of a standard cell implementation versus a similar 7 literal logic gate. With the register cost considered, the total area actually decreased for the examples s641 and s713.

6 Concluding Remarks

In this paper, we addressed the problem of removing redundant state registers that are not needed to differentiate the state codes. We have shown that a greedy approach to this problem is suboptimal. We presented a new BDD-based algorithm for this problem that is guaranteed to return the maximum set of removable state variables. It makes use of new branching heuristics and a new bounding technique called maximal removable set. The size of sequential circuits that can be handled using our technique is limited by the size of the BDD representation of the reachable states, which can be very large dependingly on the regularity of the state space. For the examples tested, the exact solutions were found in all cases with modest CPU times. We have shown how the technique of redundant variable removal can be used to simplify both the BDD representation of the transition relation for verification as well as the size of the gate-level implementation for synthesis. For the examples tested, significant reduction in both the BDD size as well as implementation area were achieved. Currently, we are investigating in BDD-based re-encoding techniques that can guarantee minimum code length without incurring significant area overhead for re-encoding circuitry.

References

[1] C. Berthet, O. Coudert, and J.C. Madre. New ideas on symbolic manipulations of finite state machines. In *International Conference on Computer Design*, October 1990.

[2] K.L. Brace, R.L. Rudell, and R.E. Bryant. Efficient implementation of a BDD package. In *Design Automation Conference*, June 1990.

[3] R. K. Brayton, R. Rudell, A. Sangiovanni-Vincentelli, and A. R. Wang. MIS: A multiple-level logic optimization system. *IEEE Transactions on Computer-Aided Design*, CAD-6(6):1062–1081, November 1987.

[4] R. E. Bryant. Graph-based algorithms for Boolean function manipulation. *IEEE Transactions on Computers*, C-35(8):677–691, August 1986.

[5] H. Cho, G.D. Hachtel, S.W. Jeong, B. Plessier, E. Schwarz, and F. Somenzi. ATPG aspects of FSM verification. In *International Conference on Computer-Aided Design*, November 1990.

[6] O. Coudert, C. Berthet, and J.C. Madre. Verification of synchronous sequential machines based on symbolic execution. In J. Sifakis, editor, *Automatic Verification Methods for Finite State Systems*. Springer-Verlag, June 1990.

[7] O. Coudert and J.C. Madre. A unified framework for the formal verification of sequential circuits. In *International Conference on Computer-Aided Design*, November 1990.

[8] C. E. Leiserson, F. M. Rose, and J. B. Saxe. Optimizing synchronous circuitry by retiming. In R. E. Bryant, editor, *Advanced Research in VLSI: Proceedings of the Third Caltech Conference*, pages 86–116. Computer Science Press, 1983.

[9] B. Lin, H. J. Touati, and A. R. Newton. Don't care minimization of multi-level sequential logic networks. In *International Conference on Computer-Aided Design*, November 1990.

[10] S. Malik, E.M Sentovich, R.K. Brayton, and A. Sangiovanni-Vincentelli. Retiming and resynthesis: Optimizing sequential circuits using combinational techniques. *IEEE Transactions on Computer-Aided Design*, 10(1):74–84, January 1991.

[11] H. J. Touati, H. Savoj, B. Lin, R. K. Brayton, and A. Sangiovanni-Vincentelli. Implicit state enumeration of finite state machines using BDD's. In *International Conference on Computer-Aided Design*, November 1990.

VLSI 91
A. Halaas and P.B. Denyer (Eds.)
Elsevier Science Publishers B.V. (North-Holland)

Resources Restricted Global Scheduling

P.F. Yeung * and D.J. Rees
Department of Computer Science,
University of Edinburgh

Abstract

We describe a scheduling methodology for high-level synthesis and the internal representation which supports it. In this representation, each node in the control-data flow graph is associated with a *condition vector* and a *resource vector*. The interrelation of these two vectors provides a profile describing the resource requirements under different conditions. With this information, we formulate a resources restricted scheduling algorithm which explores the fine-grain parallelism in an aggressive and global sense.

1 Introduction

We now have the capability to put multiple functional units such as ALUs, multipliers, floating point units, memory management, caches etc. on a single chip. In order to maximize the efficiency of a design, resources must be utilized as fully as possible. In the context of high-level synthesis, in order to fit a design within an allowable silicon area, a significant degree of resource re-usage is required. Within the search for an optimal area-time trade-off, area must be used as efficiently as possible. These challenges call for a scheduling methodology which is intelligent enough to extract the grestest timing advantage out of the available resources.

2 Previous Work

Older scheduling algorithms which allocate operations into clock cycles were often applied only to one basic block of the behavioural description at a time. The extraction of potential parallelism from the description as a whole was thereby restricted. The fine-grain parallelism obtained was often not sufficient to keep all the functional units busy. An alternative approach has been adopted in MAHA [PARK86] in which the conditional branching structures are translated into fork-join pairs. The graph is then scheduled as a whole. Unfortunately, MAHA appears not to be able to take advantage of mutually exclusive conditions for resource sharing. Force Directed Scheduling [PAUL87] is, however, able to explore this potential. In both cases, when an operation is placed inside

*financially supported by AEA Technology Ltd., U.K.

a fork-join pair, the scheduler cannot move it out even though there may be abundant resources elsewhere.

The percolation scheduling algorithm described in [POTA90] performs parallelisation across basic block boundaries. Starting with an optimal schedule, semantics-preserving transformations are applied repeatedly to convert the *program flow graph* into a more parallel one. The name 'percolation scheduling' reflects the style of the transformation to move data-independent operations upwards across basic block boundaries towards the top of the program graph. This technique is closely related to our work. However, it does not take into account the effect of resource restrictions.

Simulated annealing [DEVA89] and self-organising [HEMA90] based scheduling algorithms are very good for achieving a generally good solution before final structural and layout synthesis but they are not attractive to the designer who would like to explore the design space quickly. Furthermore, the parameters for controlling the process do not bear any direct relationship with the scheduling problem. This is very undesirable for the designer who wants to have some control over the design strategy.

3 Resource Restricted Scheduling

To overcome these problems, we have developed a *R*esource *R*estricted *Sch*eduling scheme, R^2Sch . Given the input description and the resource constraints, the operations are scheduled into a minimum number of control steps. In order to achieve this, we need to extract as much fine-grain parallelism from the description as possible. This is done by means of an aggressive migration of operations across basic block boundaries. At present, our algorithm can handle:

- Resource restriction, including restriction on abstract buses and memory I/O
- Scheduling with respect to resource restriction
- Scheduling with multicycling and chaining
- Optimisation beyond the limit of basic blocks
- Functional units with different delays and stages of pipelining
- Resource sharing with mutually exclusive conditions
- State overhead (control and steering logic delay)
- Different controller styles
- Simple timing constraints

The main criteria of the scheduling algorithm are:

- Operations should not be restricted by basic block structures or fork-join pairs. They should be allowed to migrate anywhere within the schedule space
- Specification of resource restriction should be allowed for fast area-time trade-off investigation and can be specified either by human designer or by expert system.

In the following sections, the example, *soΠoη* , in figure 1a will be used to explain the various concepts. The internal representation of *soΠoη* takes the form of the **control-data flow graph** (*CDFG*) shown in figure 1c representing the detailed control and data dependencies of each node. Figure 1b depicts its **program flow graph** (PFG) in which local parallelisation is extracted in each basic block.

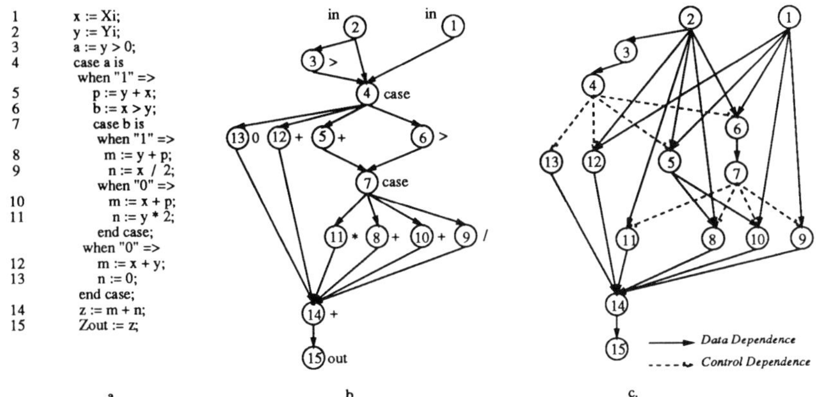

Figure 1: The PFG and $CDFG$ of $so\Pi o\eta$.

In this paper, we will concentrate on the **resource vectors**, the **condition vectors**, the **resource condition matrix** and the formulation of the priority function in sections 4, 5, 6, and 7 respectively. Other aspects of R^2Sch are described in details in [YEUN91].

4 The Resource Vector

When several kinds of resource are used to implement the behaviour, it is necessary to keep track of the usage of each type of resource. To do so, we introduce the **resource vector**(RV) each element of which represents one kind of functional unit.

RV (FU, FU, FU ..., iobus, iomem)

FU	—	functional units which are needed to implement the design
iobus	—	input-output buses
iomem	—	input-output memory ports

For instance, if adder/subtracter and multiply/divide units are required for implementation of the design, then an RV such as: RV ([+,−], [$*$,/], iobus, iomem) can be formulated. Each node in the program flow graph will be assigned an RV representing the quantities of each resource that node requires. For the example $so\Pi o\eta$, in figure 3a, it is assumed that ALUs are allocated for (+, >)s and multiply/divide units are allocated for ($*$, /)s, and concern is on functional units and I/O buses only. As a result, we have the RV: RV([+, >], [$*$, /], *iobus*).

5 The Condition Vector

We consider the conditional branching structure of an acyclic program segment as a tree, **Condition-branch Tree** (*CT*). The condition vector contains a single entry for each branch in the *CT* and the value of this vector at each branch reflects the position of the branch in the *CT*. Within the vector there is a *sub-vector* for each conditional statement with an entry representing each of the branches. In figure 2, the conditional statement with condition c_1 has a sub-vector consisting of the first two entries. Branches reached by following the first branch will have the first entry 1 and the second 0. Unrelated sub-vectors are left blank. The complete condition vector is a

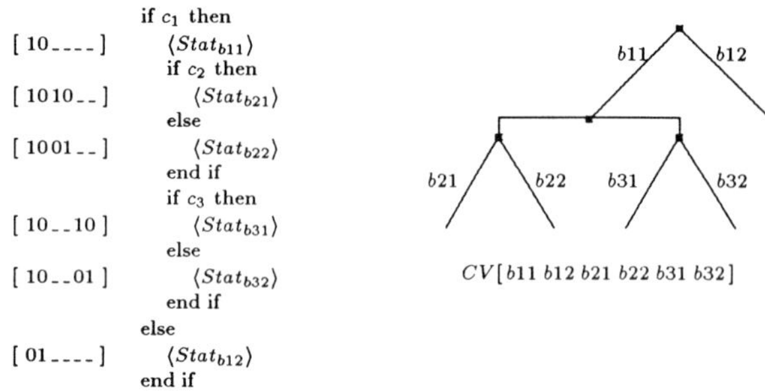

Figure 2: Branch-Based Condition Vector

concatenation of the sub-vectors from all the conditional statements. This *branch-based CV* is different in concept from a path-based *CV*. A path-based *CV* would represent the path in which the node belongs while branch-based *CV* records the conditions under which the node is executed. A branch-based *CV* is better in handling conditional structures. Irrespective of whether they are nested or disjoint, the number of entries will only increase linearly with the number of branches. For exclusivity detection, two statements are **mutually exclusive** if they belong to different branches of a fork. This fact is reflected in the *CV*s. Two vectors are mutually exclusive if they have entries that are different within their common sub-vector i.e. their common ancestry. Referring to figure 2, $\langle Stat_{b21} \rangle$ with *CV*, [*10* 10 _ _], and $\langle Stat_{b12} \rangle$, [*01* _ _ _ _], are mutually exclusive. The same also applies to $\{\langle Stat_{b21} \rangle, \langle Stat_{b22} \rangle\}$ but not, for instance, to $\{\langle Stat_{b21} \rangle, \langle Stat_{b31} \rangle\}$.

In order to formulate a global scheduling algorithm, there is a need to extend the the branch-based *CV* to reflect the global data dependences across basic blocks, i.e. the conditions under which data are defined and used. We call the extended condition vector the **propagated condition vector** (*PCV*). In this case, the *CV* of each data define/use node is propagated forward and backward through the control and data dependence links. The propagated condition vectors are then *bit-wise OR*-ed with the original *CV* to form the *PCV*. This property can also help to enhance mutually exclusive resource sharing. Figure 3b depicts the *CV*s(top) and the *PCV*s(bottom) of

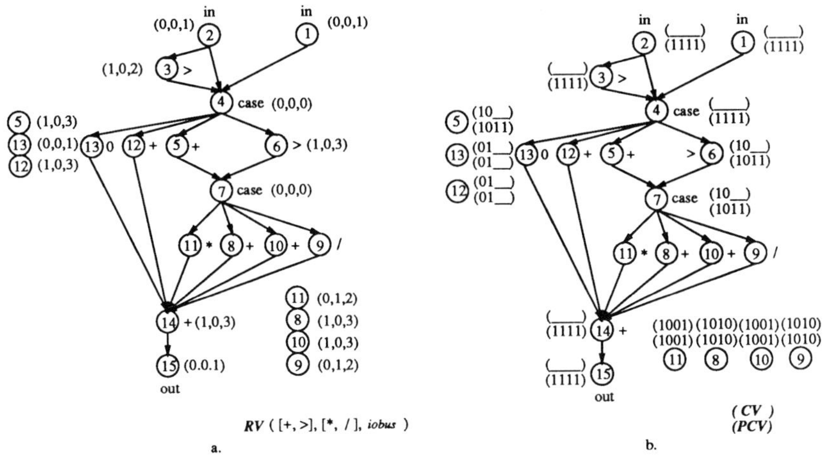

Figure 3: Resource Vectors and CVs, PCVs of $so\Pi o\eta$.

$so\Pi o\eta$ before and after the propagation. The PCV of $node_1$ shows that the definition of x is used in all branches while the PCV of $node_5$ indicates that p is not used in $branch_2$(the branch represented by the second PCV entry).

6 Resource Condition Matrix

The propagated condition vector(PCV) and the resource vector(RV) provide a powerful way to look at the utilisation of resources. By *crossing* the two vectors, we get a 2-dimensional profile, **resource condition matrix**(RCM) which shows the usage of each resource under every condition. Using these matrices, extensive global resource sharing can be carried out. For instance, with the PCV and RV for $so\Pi o\eta$, the 2-dimensional RCM view of it is constructed in figure 4.

RCM	1011	RCM	0000	RCM	0100	RCM	0100
$node_5$	0000	a.	1111	b.	0000	c.	1111
	3033		0000		0300		0300

The RCM of $node_5$ shows that $branch_2$ is uninhabited and no multiplier is required at all(empty second row). Therefore, the execution of $node_5$ can overlap with any node which requires only a multiplier(RCM a) or any node which uses only an ALU or buses in $branch_2$ (RCM b) or both(RCM c).

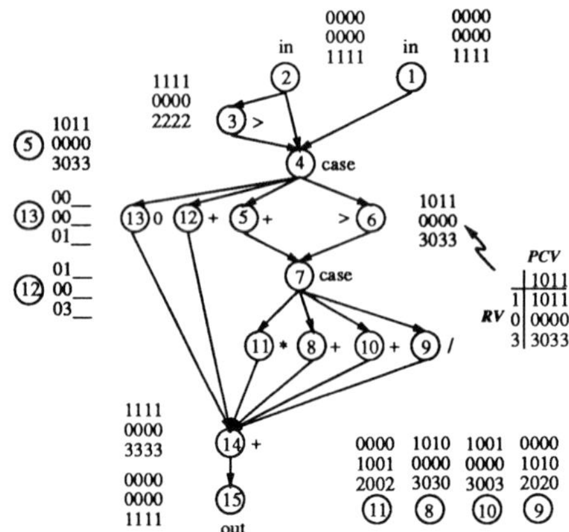

Figure 4: Resource Condition Matrix(RCM) of $so\Pi o\eta$.

7 List Scheduling

The scheduling algorithm, R^2Sch , can be divided into two phases.

1. The first phase scans the flow graph to gather information. Various static parameters like depths, path delays, CVs, PCVs and RCMs of the priority function are computed.

2. In the second phase, a list of candidates ready for scheduling is established. The most urgent one is chosen according to the priority functions(to be described). After a node is assigned to cycles, the candidate list and the resources-used vector will be updated.

During the scheduling process, operations subject to data dependence constraints can migrate to wherever resources are available and with an update of its RV and PCV. Due to lack of space, we have to skip the detail here. Interested readers can refer to [YEUN91].

8 Priority Functions

Three functions are used: *resource priority*, *path delay* and *depth* The resource priority functions of the scheduling candidates are considered first. If they are equal, then the path delay functions are applied. The depth functions will only be used when both the above functions turn out to be insufficient. The user-defined execution probability of the condition branches will also be considered when there are several computations competing for limited resources. The formulation of the priority function can be summarised in figure 5. The axes represent the *conditions*, the *resources*, and the *nodes*.

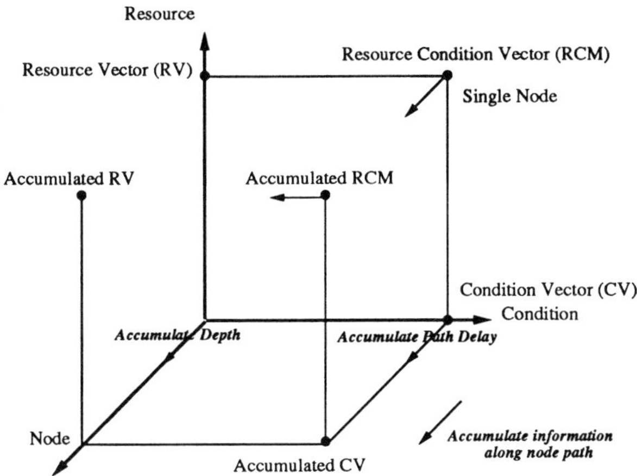

Figure 5: The Formulation of the Priority Functions.

The resource-condition plane represents the information associated with each node. Sliding the plane along the *node* axis, we descend the graph and accumulate information on each node. These accumulated vectors indicate the magnitude of the dependent sub-graph of each node.

The **Depth Function** is defined as the distance (node levels) away from the output. Since this function does not take into account the delay of each operation, it is only used when the other two priority functions prove to be insufficient.

The **Path Delay Function** has been used in Sehwa [PARK88] for pipeline scheduling. The path delay of a node is defined as the longest path delay from the current node to the output. It shows how urgent each node is with respect to the timing constraints. It does not consider the effect of conditional branches, however. We therefore multiply it with the condition vector to take into account the delay of different branches. In figure 3, if the delay of + is 40ns, then the $path\_delay(node_8)$ is [80,40,80,40]. That is the sum of $node_8$:[40,0,40,0], $node_{14}$:[40,40,40,40] and $node_{15}$:[0,0,0,0].

The **Resource Priority Function**. Here we need to define the **accumulated resource condition matrix** (*ARCM*) first. The *ARCM* of a node is the accumulation of its own *RCM* and the *RCM* s from all the nodes in its descendant sub-graph. This sub-graph contains the nodes which form the paths from the current node to the output. The *ARCM* of $so\Pi o\eta$ is shown in figure 6. The *ARCM* provides a view of the

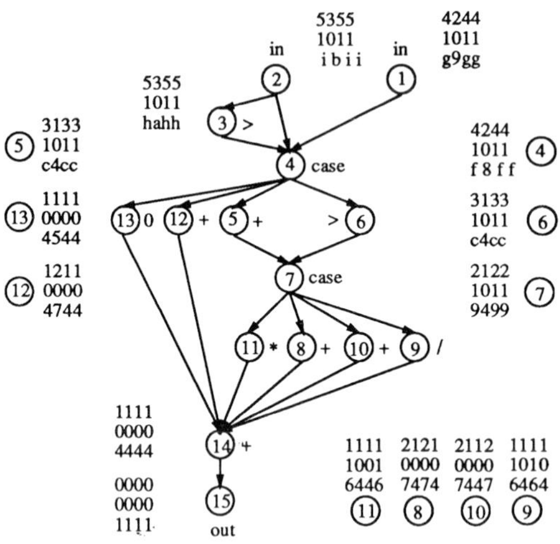

key: a-10 b-11 c-12 d-13 e-14 f-15 g-16 h-17 i-18

Figure 6: The Accumulated Resource Condition Matrix of $so\Pi o\eta$.

foreseeable resource requirement at each node with the requirements on different kinds of functional unit and conditions. By projecting the *ARCM* onto the resource-node plane, figure 5, the **accumulated resource vector** (*ARV*) is formed. For instance, the *ARV* of $node_4$ is [4,1,15]. *ARV* shows the resource usage in the sub-graph of current node. When under resource constraint, it is the very information we needed to determine the priority of a node.

$$\begin{aligned} Available\,Vector &= Constraint\,Vector - Used\,Vector(cycle_c) \\ Differential\,Vector &= ARV(node_k) - Available\,Vector \end{aligned}$$

The **constraint vector** is the initial resource constraints input into the scheduling algorithm. The **used vector**($cycle_c$) represents the amount of resource used so far in the current scheduling cycle, $cycle_c$. The difference, the **available vector**, gives the amount of resource still available in the current cycle. The **differential vector**, represents the *resource shortage* situation with respect to each type of functional unit. This difference from the differential vector can be weighted with the resource costs to determine the priority of a node.

This function gives a global view of the flow graph. The distribution of different kinds of operations and hence the requirement on different kinds of functional units. In addition, as the used vector will be updated constantly as scheduling proceeds, the resource priority function will reflect the priority dynamically.

9 Results

To study the efficiency of the algorithm, several widely used examples have been tested.

a. The example from [PARK86] is chosen first because it contains a substantial number of nested and disjoint condition branches. This helps to establish the effectiveness of R^2Sch in sharing resources under mutually exclusive conditions. As in other published schemes, the cycle length is assumed to be long enough to accommodate two operations. R^2Sch is able to achieve the fastest schedule of 4 cycles with { 2 Add, 1 Sub } which is known to be the optimum resource-delay combination i.e. the resources cannot be reduced further without an increase in number of cycles, and vice versa.

b. Fifth-Order Elliptic Filter — 1988 High-Level Synthesis Workshop Benchmarks
Scheduling with various resource restrictions was performed. The results are summarized in table 1. The results of Force Directed Scheduling(FDS) [PAUL87], Force Directed List Scheduling(FDLS) [PAUL89] and Percolation Based Scheduling(PBS) [POTA90] are summarized for comparison. As in other papers, the functional unit types used are adders with 1 cycle delay, and multipliers with 2 cycles delay. Each schedule takes less than 1 second to compute on a SparcStation 1. The second part of the result makes use of a two stage pipelined multiplier. Since data can be input every cycle, better results are obtained.

	Delay(cycles)					
Resource Requirement	Forward	Backward	Minimum	FDS	FDLS	PBS
3 Add, 3 Mult	17	18	17	17	17	17
3 Add, 2 Mult	18	18	18	18	–	18
2 Add, 2 Mult	19	18	18	19	18	18
2 Add, 1 Mult	21	22	21	21	21	21
1 Add, 1 Mult	28	29	28	–	–	–
3 Add, 2 Pipe	17	18	17	17	17	17
3 Add, 1 Pipe	18	19	18	18	18	18
2 Add, 1 Pipe	19	19	19	19	19	19
1 Add, 1 Pipe	28	28	28	–	–	–

Table 1: Scheduling Result of the Fifth-Order Elliptic Filter.

As a whole, R^2Sch is able to equal the best results published hitherto. These are believed to be the best possible for this heavily studied example. Although R^2Sch performs very well on this example, it is not a good yardstick. The description is a straight line segment of code consisting solely of arithmetic computations with no conditional statements. Nevertheless, it confirms that the underlying algorithm of R^2Sch is sound enough for basic block scheduling.

c. MC6502 — 1988 High-Level Synthesis Workshop Benchmarks
From the original ISP description, we have taken the group 1 instruction decode, address generation, and instruction execution as an example. Most subroutines in the group are expanded. The input description consists of two parts: address generation and instruction execution in two select statements. Each of them consists of 8

branches. Both conventional style and R^2Sch scheduling were carried out. Results show that schedules produced by R^2Sch are 16% to 18% faster. This is described in more detail in [YEUN91]. After studying the schedule graph, it is clear that the performance gain comes from the migration of computations .

10 Conclusions

In this paper, we have presented a new priority function and scheduling methodology to handle dynamic code motion with respect to resource constraints. From the investigations conducted, it is observed that when operations are allowed to migrate to wherever resources are available, better scheduling results can be achieved.

R^2Sch is nevertheless still far from meeting the ideals of high-level synthesis. The main reason is that it considers scheduling as an individual task, while tasks in high-level synthesis are highly interdependent. As a result of its good performance, it is being used for front-end coarse scheduling and back-end cycle length tuning and verification. At the front-end, it helps to evaluate the effect of allocation with different combination of functional units. The flexible allocations and the scheduling results are passed onto the next stage where cycle and module binding are performed. At the back-end, it takes a finished design and refines the cycle length with detail control and interconnection delays.

11 References

[DEVA89] S. Devadas, A.R. Newton "Algorithms for Hardware Allocation in Data Path Synthesis", IEEE Transactions on Computer-Aided Design, pp.768-781, July, 1989.

[HEMA90] A. Hemani, A. Postula "A Neural Net Based Self Organising Scheduling Algorithm", Proceedings of the European Design Automation Conference, pp.136-140, 1990.

[PARK86] A.C. Parker, J.T. Pizarro and M. Mlinar "MAHA: A Program for Datapath Synthesis", Proceedings of the 23rd Design Automation Conference, pp.461-466, 1986.

[PARK88] N. Park and A.C. Parker "Sehwa: A Software Package for Synthesis of Pipelines from Behavioral Specifications", IEEE Trans. on Computer-Aided Design, pp. 356-370, 1988.

[PAUL87] P.G. Paulin and J.P. Knight "Force-Directed Scheduling in Automatic Data Path Synthesis", Proc. 24th Design Automation Conference, pp.195-202, 1987.

[PAUL89] P.G. Paulin and J.P. Knight "Scheduling and Binding Algorithm for High-Level Synthesis", Proceedings of the 26th Design Automation Conference, pp.1-6, 1989.

[POTA90] R. Potasman, J. Lis, A. Nicolau, D. Gajski "Percolation Based Synthesis", Proc. 27th Design Automation Conference, pp.444-449, 1990.

[YEUN91] P.F. Yeung, D.J. Rees "Resources Restricted Aggressive Scheduling", Computer Science Departmental Report, University of Edinburgh, 1991.

VLSI 91
A. Halaas and P.B. Denyer (Eds.)
Elsevier Science Publishers B.V. (North-Holland)

Synthesis of intermediate memories needed for the data supply to processor arrays

M. Schönfeld, M. Schwiegershausen, P. Pirsch

Laboratorium für Informationstechnologie, Schneiderberg 32, 3000 Hannover 1, FRG

Abstract

A systematic synthesis of intermediate memories needed for the data supply to dedicated processor arrays is investigated. It bases on the description of the data streams at the interfaces of the intermediate memory to the data sources and to the processor array. This data description is mapped onto a graph with minimal number of nodes which characterizes the intermediate memory. By graph modifications the total number of arcs is reduced. Then the graph is transformed into a corresponding circuit and the control signals are generated. Thus, the presented strategy is an important step towards the automatic design of intermediate memories. Furthermore, the design of processor arrays is extended by the design of circuitries for the data supply.

1 INTRODUCTION

Dedicated processor arrays provide high computational and throughput rates required for real time realizations of digital signal processing algorithms. Mapping algorithms onto hardware can yield various processor arrays with different structure of the processing elements, interconnections and data requirements [1]–[7]. Nevertheless, the full computational power offered by the processor arrays can only be achieved if the data are supplied due to the requirements of the chosen mapping. The mapping specifies in detail the number of parallel data streams and the data sequences. Often the required data streams are periodical and regular. Repetitions of data, time slots without valid data, and skewed data streams caused by application of pipelining are typically necessary.

Usually, the data sources cannot provide the data as required by the processor array. The demanded adaptation between the data streams of the sources and those needed by the processor array can be realized by an intermediate memory (Fig. 1).

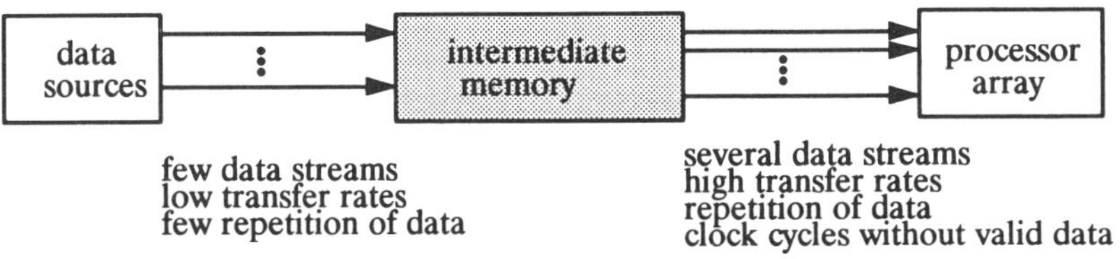

Fig. 1: Data supply to processor arrays

Implementing the processor array as VLSI circuit, the I/O bandwidth is restricted by the number of pins and the maximum transfer rate of external circuits. This often causes a bottleneck of data supply which can be removed by an intermediate memory implemented on the same VLSI circuit as the processor array. This intermediate memory can make use of the recurrence of data and time slots without valid data and decrease the data rates to external circuits if it generates the data repetitions and adapts the data streams to the requirements of the processor array. Thus each single data word has to be provided at minimum only once by the sources.

Until now, there are only a few investigations on data supply to processor arrays. Most of them are limited to special realizations of processor arrays. For the case of communication between exactly two processors a data path synthesis has been presented in [8]. There, only one provided and one required data stream have been considered, whereas the skew of the required data stream can be modified. But in case of data supply to processor arrays it is often not possible to modify the skew of individual data streams without changing the skew of all streams. Therefore in this paper a strategy for the synthesis of dedicated intermediate memories is investigated taking into account all the data streams required by the processor array. It implies that all data streams are synchronous with fixed temporal sequences of the data. Based on the data descriptions at the interfaces to the external data sources and to the processor array appropriate intermediate memories are systematically generated. Two target realizations, register as well as RAM circuits, are supported. The global aim of the synthesis is to minimize the total number of storage cells as well as the number of interconnections between them. This leads to a non–linear assignment problem being NP–complete. Thus the objective outlined in this paper is a suboptimal solution. In the first step the data are assigned to the minimal number of storage cells. In the next step the number of interconnections is reduced by changing the temporal assignment of the data to the storage cells. In order to achieve a higher abstraction level the intermediate memory is described by a directed graph consisting of a minimal number of nodes. The requirements of the data adaptation are mapped onto this graph. Then the number of arcs is reduced by graph modifications in order to minimize the expense of the intermediate memory. The final circuit and the necessary control signals are extracted from the graph representation by transforming nodes and arcs into associated circuit elements. This synthesis strategy offers the opportunity to do major parts of the design automatically. Its implementation as CAD–program is currently under work.

In the following section two fundamental alternatives for the realization of intermediate memories – register or RAM circuits – are discussed. The general synthesis strategy is outlined in section 3 . After explaining the description of the data streams at the interfaces in section 4 a graph representation is introduced in section 5. In section 6 it is shown how the data requirements are mapped onto a given graph. In section 7 some notes about dividing the data adaptation into partial tasks are given. Finally in section 8 the generation of the circuitry and the control signals is mentioned. A conclusion is given in section 9 .

2 REALIZATION OF INTERMEDIATE MEMORIES

In comparison to RAM circuits register circuits need only simple control and no address calculation at all. But there is no random access to data. Furthermore, all registers are simultaneously active controlled by a global clocking which results in a high power consumption. For realizing intermediate memories as RAM circuit two requirements of the data supply are decisive. Firstly, independent read and write access to the RAMs is demanded. Furthermore, several parallel data streams must be provided. Therefore multiport RAMs are necessary. Often several RAMs have to be arranged in parallel because the number of ports per RAM giving the number of simultaneous data accesses is limited by the expense for implementation. A disadvantage of a multiport RAM realization is the high expense for address generation. But the random access to data provides high flexibility by changing the address signals. Moreover, a combination between both realizations is feasible.

3 SYNTHESIS PROCEDURE

The synthesis strategy presented in this paper aims to support both kinds of realization, i.e. circuits with multiport RAMs as well as registers. As shown in Fig. 2 it starts from a detailed description of the data streams at the interfaces of the intermediate memory.

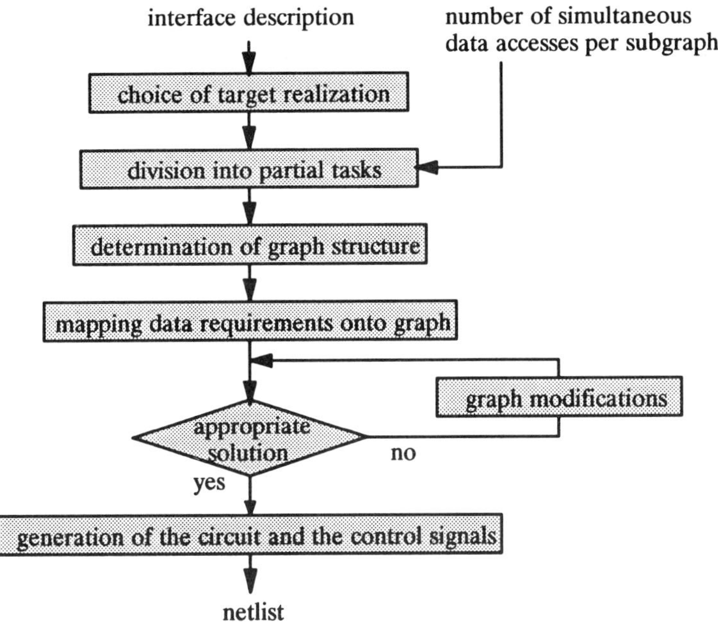

Fig. 2: Synthesis procedure

In the first step the target realization of the intermediate memory has to be chosen. Then the problem of the data stream adaptation is divided into partial tasks, whereas the data are assigned to groups in order to solve conflicts of simultaneous data accesses. In the next step the structure of the basic graph is determined. The number of nodes and arcs depends on the partitioning into tasks. These parameters are extracted from the interface description. Then the requirements of the data stream adaptation are mapped onto the graph structure. By this nodes and arcs are characterized by times at which they are needed for data storage or transfer. Different mapping algorithms are possible resulting in different sets of active times for nodes and arcs. Since nodes and arcs without active times can be removed, the basic graph is transformed into a reduced graph. In order to judge the quality of the solution a cost function is applied to the reduced graph taking into account the number of nodes and arcs, the number of different data stream groups and the number of arcs per node. If no appropriate solution has been found, further graph modifications are applied in order to minimize the number of arcs. Otherwise the graph is transformed into a circuitry of the chosen realization and the netlist of the intermediate memory is extracted. Furthermore, the control signals can be calculated from the active times of nodes and arcs.

4 DESCRIPTION OF THE INTERFACES

Since the intermediate memory shall offer an adaptation between the data streams coming from the data sources and those needed by the processor array, it is necessary to describe the two interfaces at the input and output of the intermediate memory. The dependencies between the two interfaces describe the intermediate memory itself, too. Two aspects are important: the interconnection between the data streams and the amount of delaying for each data word.

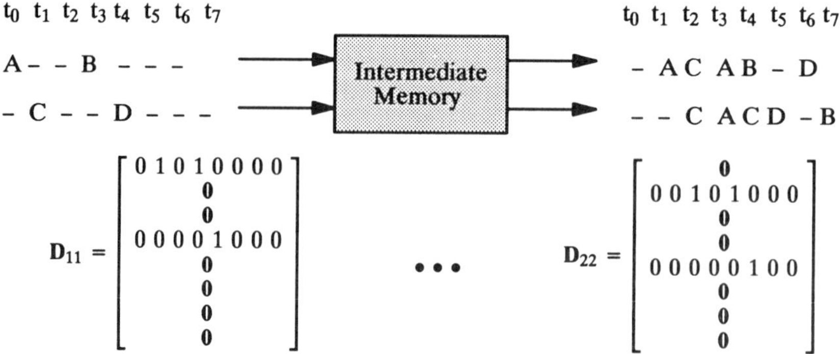

Fig. 3: Example for the interface description

Let $I_1, \ldots, I_M$ be input streams of the intermediate memory coming from data sources, and $O_1, \ldots, O_N$ be output streams providing data for the processor array. Then for each combination of an input stream I_i and an output stream O_j the demanded delays of the data are given by a binary matrix $D_{i,j}$. Each row of $D_{i,j}$ describes the accesses to one data word of the input stream, whereas each column corresponds to a time slot at the output stream. An entry $d_{i,j}(k,l)$ of $D_{i,j}$ is 1, if and only if a data word provided by input stream I_i at time step k is needed on output stream O_j at time step l. Repetitions of data on the same input stream cause identical rows in the matrix.

Let $T_{process}$ be the time period given by the processor array to calculate a whole data set. Furthermore, let the offset time t_{off} describe the delay of each data stream by giving the time period between t_0, the common point of reference for the time axis of all data streams, and the first time slot with valid data. Then the process of the data supply lasts from $t_{\min} = t_0$ until $t_{\max} = t_0 + T_{process} + \max_{i,j} t_{off}$.

Since the matrices have to describe the whole process of data supply, the number of rows of $\boldsymbol{D}_{i,j}$ corresponds to the product of processing time plus the maximum offset time and the clock rate $f(i)$ of the input stream i. The number of columns is given by the product of processing time and clock rate $f(j)$ of the output stream j.

$$n_{row} = f(i) * (T_{process} + \max_{i,j} t_{off}) \qquad n_{colomn} = f(j) * (T_{process} + \max_{i,j} t_{off}) \tag{1}$$

For simplicity in the following it is assumed that input and output streams have the same frequencies and that each data word occurs only once on an input stream. Therefore a data word can be identified by (i,k) giving the input stream I_i and the time k at which it is provided. Fig. 3 gives an example for the interface description with two input and output streams of the intermediate memory assuming $T_{process} * f = 6$.

5 GRAPH REPRESENTATION OF THE INTERMEDIATE MEMORY

An intermediate memory can be described by a graph consisting of a set of nodes connected by directed arcs. The nodes are subdivided into input, output, transmission and storage nodes due to their different functions. Input nodes represent input streams, output nodes output streams. Transmission nodes correspond to multiplexers or demultiplexer according to several arcs pointing towards them or away from them. The function of storage nodes consist in storing data like RAM or register cells. Storage nodes with several arcs pointing towards them combine a transmission and a storage node.

The representation of the intermediate memory starts with a basic type of graph which consists of the minimal number of nodes. The basic graph has a fixed structure which is adapted to the special application by parameters determined from the interface description. The number of input nodes is given by the number of input streams M, the output nodes by the number of output streams N, the input transmission nodes by the maximum number of simultaneous write accesses N_{WR} and the output transmission nodes by the maximum number of simultaneous read accesses N_{RD} of data. The number of storage nodes S_{max} is given by the maximum number of data which have to be stored simultaneously and therefore cannot be further reduced.

$$N_{WR} = \max_k \left| \left\{ (i,k) \mid \exists\, d_{i,j}(k,l) = 1 \; ; \; l > k \right\} \right| \tag{2}$$

$$N_{RD} = \max_l \left| \left\{ d_i(k,l) \mid d_i(k,l) = \bigvee_j d_{i,j}(k,l) \wedge d_i(k,l) = 1 \right\} \right| \tag{3}$$

$$S_{\max} = \max_l \left| \left\{ (i,k) \mid \exists\, d_{i,j}(k,l_1) = 1 \; ; \; l_1 > l \right\} \right| \tag{4}$$

Each input node is connected to each input transmission node (TI) by arcs, each transmission node to all storage nodes and each output transmission node (TO) to all output nodes. In case of input transmission nodes the arcs are directed towards the storage nodes, in case of output transmission nodes towards the transmission nodes. Furthermore, the storage nodes are fully connected to each other (Fig. 4). If a data word is needed on the output stream without delay, the graph has to be expanded by additional arcs pointing from the input to the output nodes. But in the following it is assumed that each data has to be delayed by at least one time step.

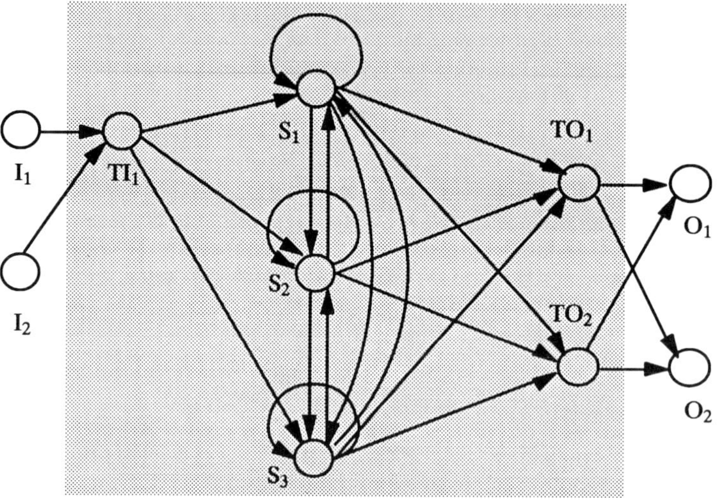

Fig. 4: Basic graph structure for intermediate memories

Arcs, input, output and transmission nodes cause no delay of data. Storage nodes cause a delay of one time step. In Tab. 1 the different node types, their functions, the assigned delay and their labeling in Fig. 4 are listed.

Tab. 1: Set of different nodes

node label	node type	node function	assigned delay
I_1, I_2	input	duplication	0
TI_1, TO_1, TO_2,	transmission	multiplexer, duplication	0
S_1, S_2, S_3	storage	RAM or register cell	1
O_1, O_2	output	multiplexer	0

6 MAPPING THE DATA REQUIREMENTS ONTO A GRAPH

The requirements of the data stream adaptation are mapped onto the basic graph structure whereas nodes and arcs are characterized by times at which they are needed for data storage or transfer. For each time step only one of several arcs pointing towards the same node can be active and transfer data. The active times of nodes and arcs are extracted from the information given by the data description of the interfaces. Different mapping algorithms are possible resulting in different sets of active times.

```
for i = 1 to M , i+1
 for k = 1 to k_max , k+1
   if ∨_{j,l} d_{i,j} ( k,l ) = 1 then
     l_min = k ; l = l_min
     l_max = max_{j,l} { l |d_{i,j} ( k,l ) = 1}
     while l_min ≤ l ≤ l_max
       if l = l_min then
         s = 1, m = 1
         while l ∈ { t ( TI_m ) }
           m = m +1
         end
         while l + 1 ∈ { t ( S_s ) }
           s = s +1
         end
         add l to { t ( I_i )}, { t ( I_i -TI_m), { t ( TI_m) } , { t ( TI_m-S_s)}, l+1 to {t (S_s)}
       else
         for j=1 to N , j+1
           if d_{i,j} (k,l)=1 then
             n = 1
             while l ∈ { t ( TO_n ) }
               n = n +1
             end
             add l to { t(S_s - TO_n )}, { t(TO_n)}, { t(TO_n - O_j)}, {t(O_j)}
           endif
         end j
         if l < l_max then
           if ( l + 1 ∈ { t ( S_{s+1}) } ) ∨ (s = s_max ) ∨ ( realization = RAM ) then
             add l to { t ( S_s - S_s )}, l+1 to { t ( S_s)}
           else
             add l to { t ( S_s - S_{s+1} )} , l+1 to { t ( S_{s+1} )}
             s = s+1
           endif
         endif
       endif
       l=l+1
     end
   endif
 end k
end i
```

Fig. 5: Example for a mapping algorithm

In the following the mapping is explained for an algorithm (Fig. 5) suitable for RAM and register circuits. In case of latter realization the algorithm leads to a unidirectional register chain. A data can allocate different storage nodes in successive time steps. In case of RAM circuits it is assumed that there are no connections between storage nodes because data cannot be transferred between different storage cells within one RAM. Therefore each data allocates exactly one storage node during the whole duration it has to be stored.

The mapping is done for the data of the same input stream in their temporal sequence. Then the next input stream is selected. Therefore the matrix elements of $\mathbf{D}_{ij}$ are processed row by row. In case of a RAM realization the mapping strategy outlined here is to transmit the data which have to be stored to a free storage node and to hold them there as long as required. In case of a register circuit the strategy is to transmit the data to a storage node and then to push them from one storage node to the next one as soon as possible. A node or arc is free for a new data if there is no entry of the current time step in the list of its active times $\{t\}$.

If one of the matrix elements meets the condition $d_{i,j}(k,l)=1; l > k$, the data (i,k) has to be stored and the duration of the storing given by the first time step l_{min} and the last time step l_{max} is determined. In the first time step $l=l_{min}$ the data is transferred from the input node I_i via a free transmission node TI_i to the first free storage node S_s. To find a node which is free to receive a new data the nodes are examined one after the other. Then the lists of the active times of the determined free nodes, except the storage node, and of the arcs between them are expanded by the current time. The active times of the storage node are extended by an entry for the next time step because storage nodes cause a data delay of one time step. For all the next time steps $\{l \mid l_{min} < l < l_{max}+1\}$ the following actions take place. If the matrix element $d_{i,j}(k,l)=1$, the data has to occur at the output j. Therefore the current time is added to the active times of the arcs between the storage and a free transmission node TO_j and between the transmission and the corresponding output node O_j.

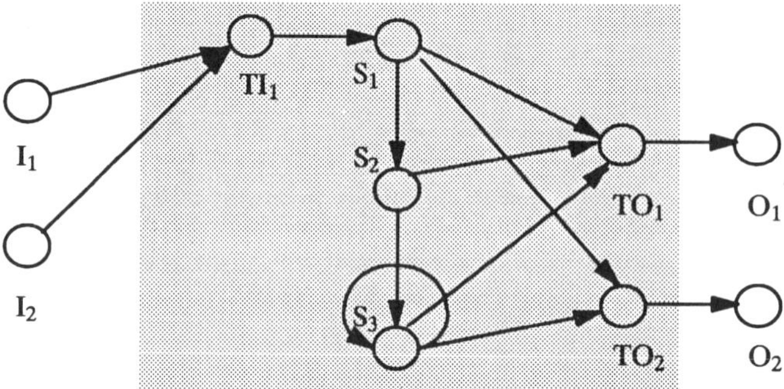

Fig. 6: Example for a graph structure of a register circuitry

If $l = l_{max}$ the data needs not to be stored further on and the next data $k+1$ can be considered. Otherwise if the current data has not yet reached the last storage node s_{max} and if the target realization is a register circuit it is examined if the next storage node $s+1$ in the chain has already an entry for the following time step. If there is no entry, the data is transmitted to the next storage node. Therefore the active times of the corresponding arc are extended by the current time, the active times of the free node by an entry for the next time step. If the next storage node is occupied by another data, if the last storage node s_{max} has been reached or if the target realization is a RAM circuit, the active times of the arc looping back to the same storage node are expanded by the current time and the active times of the storage node are extended by an entry of the next time step. Then the next time step is considered.

Having applied the mapping nodes and arcs without active times can be removed because they have no function. By removing unnecessary arcs the basic graph is transformed into a reduced graph. Therefore different mapping algorithms lead to different structures of the reduced graph. The number of arcs are further reduced by graph modifications. The final reduced graph corresponds to the structure of the later circuit. Because of the correspondence between graph and circuit elements a minimization of the circuit expense can be reached by the choice of the mapping algorithm as well as by additional graph modifications.

7 DIVISION OF DATA ADAPTATION INTO PARTIAL TASKS

In case of a realization with multiport RAMs the number of ports per RAM is limited by the expense for implementation. If the maximum number of ports per RAM is smaller than the number of data accesses which are simultaneously demanded, several RAMs are needed in parallel. Therefore the problem of the data stream adaptation has to be divided onto several RAMs. This results in a graph representation with several subgraphs which consist of separated sets of transmission and storage nodes but common input and output nodes. Since the aim is to obtain the graph representation with a minimum total number of nodes and arcs, the first step is to determine the minimum number of subgraphs which are necessary, if the maximum number of simultaneous data accesses is given. In the next step the solution with the minimum number of nodes is selected among the set of solutions with the minimum number of subgraphs.

8 GENERATION OF THE CIRCUITRY AND CONTROL SIGNALS

The final reduced graph is transformed into the corresponding circuitry. In case of a register circuit transmission nodes with several arcs pointing towards them correspond to multiplexers. Transmission nodes with several arcs pointing away from them correspond to distribution points duplicating the data streams. Storage nodes are transformed into registers and arcs into interconnections between the circuit parts.

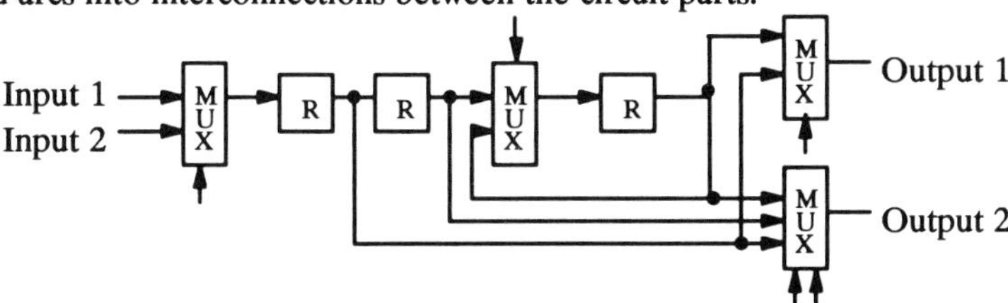

Fig. 7: Resulting register circuitry

If a RAM circuitry is demanded, those subgraphs which consists of a set of transmission and storage nodes where each of the transmission nodes is connected to each of the storage nodes are transformed to multiport RAMs. The transmission nodes correspond to the ports of the RAMs, the storage nodes to the RAM cells. Further transmission and storage nodes which do not belong to this special kind of subgraphs are treated as in case of register circuits. In Fig. 7 the resulting register circuitry for the example of Fig. 6 is shown. The control signals of the circuits can be extracted from the reduced graph, too. The select signals of the multiplexers can be built from the active times of the arcs and the address signals of RAM circuits from the label number of the storage nodes.

9 CONCLUSION

A systematic synthesis of intermediate memories needed to handle the data supply to processor arrays was investigated in case of synchronous and deterministic data streams. Based on the data descriptions at the interfaces the intermediate memory is described by a graph representation with minimal number of nodes. After mapping the requirements of the data adaptation onto the graph the number of arcs is minimized by further graph modifications. Then a transformation into a corresponding circuit is performed. An intermediate memory with minimum number of storage cells is obtained. The netlist and the control signals can be extracted. Thus, the design of processor arrays is extended by the design of circuitries for the data supply. Therefore the expense for the realization of the data supply yields a valuable criterion between alternative architectures of processor arrays.

10 ACKNOWLEDGEMENT

This work is supported by the Deutsche Forschungsgesellschaft (Pi 169/3–1).

11 REFERENCES

[1] H. T. Kung, "Why Systolic Architectures", *Computer Magazine*, Vol. 15, No. 1, pp. 37 – 46, Jan. 1982.

[2] S. Y. Kung, "On Supercomputing with Systolic/Wavefrontarray Processors", *Proc. IEEE*, Vol. 72, No. 7, pp. 867 – 884, July 1984.

[3] B. Wah and J. A. B. Fortes, "Systolic Arrays – From Concept to Implementation", *IEEE Computer*, Vol. 20, No. 7, pp. 12 – 17, July 1987.

[4] S. Y. Kung and S. N. Jean, "A VLSI Array Compiler System (VACS) for Array Design", R. W. Brodersen and H. S. Moscovitz, editors, *VLSI Signal Processing III*, chapter 45, pp. 495 – 508, IEEE Press, New York, 1988.

[5] P. Frison, P. Gachet and P. Quinton, "Designing Systolic Arrays with DIASTOL", S. Y, Kung, R. E. Owen, and J. G. Nash, editors, *VLSI Signal Processing II*, Chapter 9, pp. 93 – 105, IEEE Press, New York, 1986.

[6] D. I. Moldovan, "ADVIS: A Software Package for the Design of Systolic Arrays", *IEEE Trans. on Computer–Aided Design*, Vol. CAD–6, No. 1, pp. 33 – 40, January 1987.

[7] U. Vehlies, "DECOMP – A Program for Mapping DSP Algorithms onto Systolic Arrays", in "Transformational Approaches to Systolic Design", Ed. G. M. Megson, London, Springer, 1991.

[8] J. Decaluwe, J. M. Rabaey, J. L. Van Meerbergen, H. J. De Man, "Interprocessor Communication in Synchronous Multiprocessor Digital Signal Processing Chips", *IEEE Trans. on Acoustic, Speech, and Signal Processing*, Vol. 37, No. 12., Dec. 1989.

VLSI 91
A. Halaas and P.B. Denyer (Eds.)
Elsevier Science Publishers B.V. (North-Holland)

WORKSPACE AND METHODOLOGY MANAGEMENT IN THE OCTTOOLS ENVIRONMENT

Marina Zanella and Paolo Gubian

Dipartimento di Automazione Industriale, Università degli Studi di Brescia, Via Valotti 9, 25060 Brescia, ITALY

Abstract.
This paper describes a research under development whose aim is to endow the Octtools environment with a set of Workspace Management (WM) and Design Methodology Management (DMM) functionalities.

1. INTRODUCTION.

The Octtools system [HMS86] is an advanced VLSI CAD environment developed at the University of California at Berkeley. It consists of a collection of tools and libraries, most of which are fully integrated with the OCT data manager and the VEM graphics editor. OCT performs a centralized data management, i.e. it filters all accesses to data. Applications are insulated from data, they just make calls to OCT procedures. OCT private data structure, the so-called OCT data base, basically consists of sets of records. These records describe a small number of elementary objects, which may be logically interconnected among each other by means of an "attachment" relationship. The same objects and the same attachment relationship are used to create different views of the same circuit, belonging to different abstraction levels. The meaning of each view is established by the policy associated to its "viewtype". A "policy" is a set of rules which interpret the internal structure of a design file. Hence only one data format is used for every circuit representation, be it physical, symbolic, schematic, etc. The correct interpretation of design files is accomplished by policies. In the OCT terminology, the elementary editing unit, i.e. a design file, is called "facet. VEM supports interactive editing and allows the launch of CAD tools by means of a Remote Procedure Call package.
The Octtools system has been chosen as the object of our research since it represents a last generation CAD environment and it is the prototype of electronic CAD frameworks [HNS90].

2. IMPLEMENTATION CHOICES.

The intervention on the Octtools environment has begun by providing a support for work distribution, data exchange and design methodologies.
So far we have skipped the problem of creating and enforcing a taxonomy of object types (versions, alternatives, configurations, ...) and of relationship types (versioning, descending, equivalence, composition). This decision is due to several considerations. First, there are many papers [KAC86][KCB86][KaC87] coming from Berkeley which propose modelling concepts for CAD data. These concepts should eventually bring to the development of a Version Server which should be coupled with the Octtools environment.

Besides, in the VLSI design world there is a stressing need for some support of design methodologies as well as for instruments which are able to archive each design object history. In fact, designers often think in terms of data transformations and identify a design object from its position in the design flow.
In addition, the increasingly larger dimensions of present-day projects are urging to a division of the design space into several workspaces. The goal is to limit the scope of each designer and, at the same time, to control and trace data exchange among designers.
Thus, the first implementation steps have been devoted to the fulfillment of the above outlined requirements.
In our intentions, the WM and DMM services have to alter neither the main characteristics of the Octtools environment nor its interface. However, we are not building an external software architecture that presides over the running of the whole system, instead we are implementing a set of WM and DMM functionalities which co-operate with VEM, OCT and CAD tools. Moreover CAD environment features have been exploited whenever possible. In particular, VEM editing capabilities and the OCT data format have been used so as to describe the special objects which are handled by WM and DMM functionalities. Then the OCT data base has become the repository of all kinds of objects, be they circuits, at various abstraction levels, or WM objects or DMM objects. WM and DMM objects are created by using VEM editing functions and WM and DMM functionalities use OCT procedure calls in order to access data. This choice will give integration and robustness to the final system, since editing commands and OCT calls are quite independent of physical allocation details. This fact has two pleasant consequences. First, it will guarantee that WM and DMM functionalities will continue to work correctly should the file data format be modified in a new Octtools version. Second, WM and DMM functions programmers can ignore details about the file format.
We are working on the Octtools distribution 3.5 [OCT], installed in an ULTRIX environment.

3. WORKSPACE MANAGEMENT.

A "workspace" is a limited logical area where some parts of a project may be created and updated by a designer or a design team.
The implementation of WM functionalities has already been completed and a mechanism is provided which enables an administrator to create a new workspace every time it is necessary, connecting it with existing workspaces within a hierarchical frame. The superuser is the system administrator. There are many authorized OCT designers (individual designers and/or groups). Every user has a default workspace which is popped up every time s/he connects with the system. The operating system is not visible to designers; every interaction between designers and the operating system is filtered by WM functionalities. Designers are asked for username and password at login time. Every designer and/or group has a name as OCT user. There is a workspace for each username, i.e. there is a sub-directory, named from the designer's username, for each OCT user. This sub-directory contains all the design files produced by the concerned user. In addition, every workspace is represented by an OCT cell which is graphically displayed as a box. Many properties of interest, such as the list of valid methodologies, can be attached to the box. All the user sub-directories are placed at the same level in the File System hierarchy.
The superuser's workspace represents the place where all the official data concerning the project are stored.
In case of need, it is possible to create a temporary workspace to which designers working on a particular project variant can turn. Once the variant has been

produced and properly inserted in the main archive, the temporary workspace may be removed or converted to a new use.
The superuser creates an OCT cell comprehending the instances of all the workspaces, which are linked among them in a rooted acyclic graph (Fig. 1). Reciprocal positions and links in the graph establish the scope and reaching rules of each workspace. Every read/write request concerning a design object is disciplined by the scope rules which are implicit in the graph. Every movement or data exchange in the workspace hierarchy is governed by the reaching rules which are implicit in the graph.
By definition, a target workspace is reachable from a source workspace if there is a directed path which connects the source with the target.

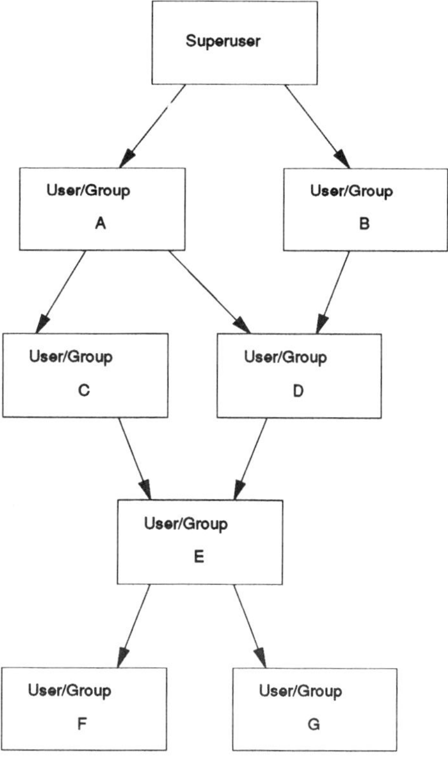

Fig. 1

The following rules express the semantics of the workspace graph.
- A workspace owner has read/write access rights over design objects belonging to his workspace.
- A workspace owner has read access to the design objects which belong to directly connected lower level workspaces.
- A workspace owner has the right to browse the list of all design objects belonging to lower reachable workspaces.
- A user can freely switch to a lower level reachable workspace.
According to the last rule, a designer can open some OCT facets in his/her own workspace and then open some other facets in a lower level reachable workspace, provided he has previously switched to it, and so on. In this way, a designer may act on different files belonging to several workspaces at the same time.
The graph is automatically interpreted by an *ad hoc* function at login time. Note that, since user directories are organized on the same level in the File System, independently of user workspaces' reciprocal positions in the workspace graph, the introduction of a temporary workspace and/or the modification of the graph organization has no effect on the existing File System structure. Any update becomes executive once the new graph has been interpreted.
The user can explore design files and move along the graph arrows according to the above rules. Explorations and movements are driven by browsing facilities and dialog boxes.

The workspace graph pictorial representation is very useful. The superuser can create, move or delete a workspace using the graphics editor, catching the situation at a glance.
Figure 1 shows a workspace graph, as it is displayed on the screen by VEM. The workspace hierarchy has been built by the superuser using the VEM editor itself. According to the above described semantics, every workspace is reachable starting from the superuser workspace. Workspaces A and C are unreachable from workspace B. The superuser can read his own design objects and the design objects belonging to workspaces A and B. At the same time he can switch to every workspace. User B, instead, can read his own design objects and the design objects belonging to D. At the same time user B can switch to workspaces D,E,F and G.

4. METHODOLOGY MANAGEMENT.

A design "methodology" is the specification of an execution path, involving one or more CAD tools.
It is possible either to specify multiple paths, that is different strategies that can be followed in the execution of an action (e.g. the simulation of a complete project), or to force the user to follow a single predefined path (as it is often the case with semicustom designers at customer sites).
DMM services have two major objectives: to discipline designers' behavior and to automate the run of tool sequences.
Proper DMM functionalities act as run-time syntactic controllers of the design paths which are followed by designers. Methodologies are stored in the same way as design data, i.e. they are special OCT cells which are created by (special) users and are handled by DMM functionalities. Hence methodologies can be edited and modified by using VEM editing capabilities. So a centralized and uniform data management is performed for both design data and WM/DMM data.
Since methodologies are represented as OCT cells, then also tools have to be represented as cells so that they can be instantiated within methodologies. The DMM system can control and trace the transformations performed by a tool as well as automatically execute the tool itself only if the OCT description of that tool is available.

4.1 Tool characterization.

From the DMM system's point of view a tool is just a "black box" which is endowed with I/O terminals, representing I/O data. The DMM system is a design flow syntactic controller, therefore it is not interested in the tool's semantic behavior (i.e. in the specific CAD function the tool performs).
There are batch and interactive tools. A batch tool cell consists of several views (Fig. 2), namely one "default view" and a number of "option views". There is an option set associated to each one of the views. The default view describes the tool behavior when specified options (if any) do not influence the number and/or types and/or names of input/output files. Each option view describes the tool behavior when current options change the number and/or types and/or names of I/O files. Every view's contents facet consists of a box. A label containing the tool name and several properties containing active options are attached to the box. Each view's terminal represents a data file (or an homogeneous set of data files). Each terminal has three properties: the first one specifies the data type, the second one specifies the number of associated data files and the last one specifies a parametric name for the file/s.
Interactive tools differ from batch tools in several respects. First, an interactive tool can be run without any input file at all. Second, possible input files may be opened run-time. Third, an interactive execution can be stopped to be restarted

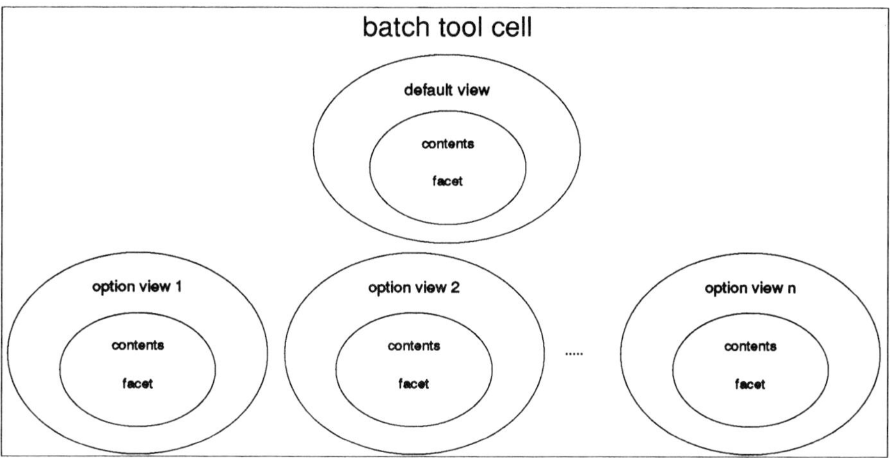

Fig. 2

later. Finally, an interactive execution can be turned into a batch session if a script input file is provided. All these remarks have to be reflected on interactive tool descriptions.

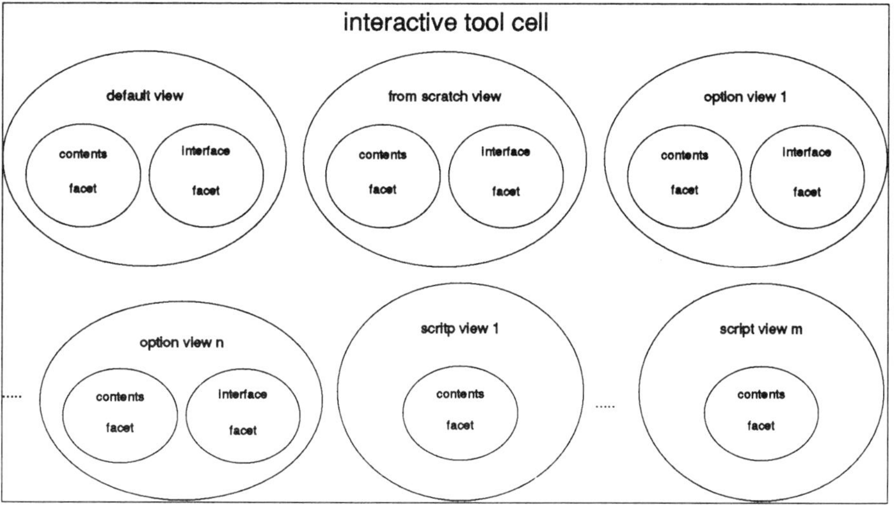

Fig. 3

An interactive tool cell consists of several OCT views (Fig 3). There is the so-called "from scratch view" which has no input terminal. Then there are one "default view" and several "option views", in the same way as for batch tools. Each view has to be put into a cyclic flow (Fig. 4), in order to support the interactive tool cyclic execution, thus creating a (simple) methodology. In the following it is assumed that every interactive tool is always encapsulated into a

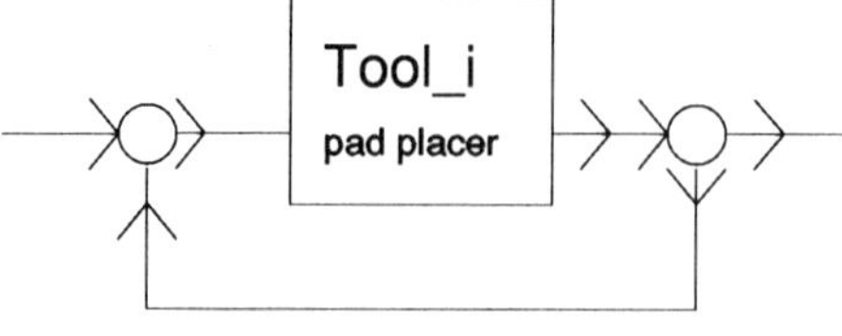

Fig. 4

cycle. Finally, a "script view" has to be introduced in order to describe the interactive tool batch behavior due to a specific script input file. Then there is an interactive tool script view for each script file.
The usage of commands during an interactive session may cause some design files to be opened, updated and saved, independently of the options on the tool invocation line. The commands which have such an effect need to be described singularly, as if they were tools, by means of a "default view".
The characterization of all Octtools RPC applications has been performed following the above directions.

4.2 Methodology description.

A methodology is described by an OCT cell, consisting of one or more views. Each view contains the instances of one or more tool views which are interconnected among each other by means of nets, so as to build a flow (Fig. 5). The DMM system allows the coupling of two terminals only if they are compatible, according to a compatibility rule set. In addition, an execution flow may contain a few characteristic features, such as alternative points, forking points, cycles and feedbacks. An "alternative point" is a point in the flow where the designer can choose one way out of several different ways. A "forking point" is a place in the flow where a file is simultaneously the input of several tools. A "cycle" (Fig. 4) and a "feedback" (Fig. 5) are repetitive executions of one tool and of one sub-methodology respectively.

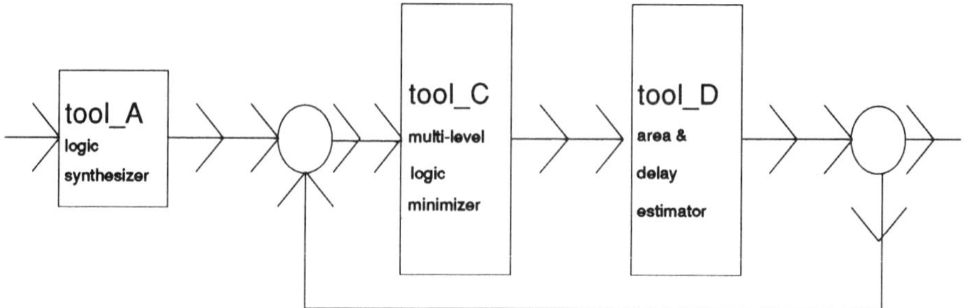

Fig. 5

There exist two kinds of methodology views. By definition, a "white box" view, if it is flattened, contains one tool instance at least. White box views can contain instances of tool views, be they default, from scratch, option or script views, and/or instances of other methodologies' views, which in turn may contain instances of other methodologies' views, etc.

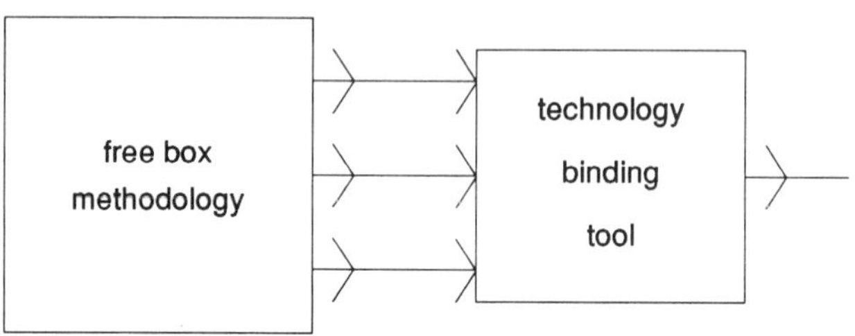

Fig. 6

A "free box" methodology view does not establish any flow to be followed, it just specifies I/O data characteristics and a list of tools which can be used by designers freely. Figure 6 displays a white box methodology view which contains the instance of a free box view, aimed at combinational or sequential network synthesis, and the instance of a view of a tool that accomplishes the technology binding. The final results of the overall methodology will comply with the selected technology.

4.3 Methodology adherence control.

DMM functions have to interpret methodology descriptions in order to monitor designers' actions. For each workspace, there is a set of methodologies which are valid inside that particular workspace. A designer's action is legal if it complies with one valid methodology at least. The same action may comply with different methodologies at the same time.
The methodology adherence control is supported by the concept of file history that will be introduced in the next paragraph. This concept requires that in any methodology view a node is associated to each net coupling two terminals and to each uncoupled I/O terminal. Nodes are numbered univocally. The same node may have several distinct files associated to it.
The methodology uncoupled input terminals are called "primary input" nodes.

4.4 The concept of file history.

The "history" of a file is the sequence of progressive transformations which have led to the creation of the file itself. A file history may be displayed graphically as in figure 7. The same file history may comply with several methodologies which are valid in the workspace the design file belongs to. Figure 8 shows that *file_e*, whose history has been displayed in figure 7, complies with two different methodologies. The DMM system has to track of every design file history. This requirement is due to two main reasons. First, the file history is useful for the designer who wants to analyse his/her past work, who wants to recover a project after a certain amount of time, who imports into his/her own workspace a file belonging to another designer or to a library, who takes over a project from another designer or joins a group when a lot of files have already been produced.

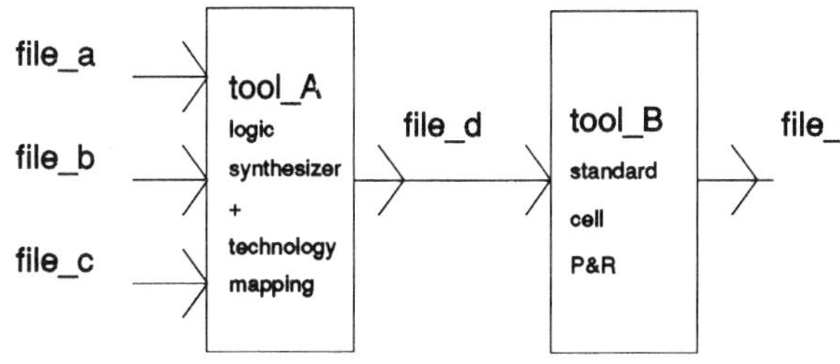

Fig. 7

Second, the knowledge of a file's history enables the DMM system to effectively monitor every operation which will successively be run both on the file itself and on its descendants. This control aims at proving that the actions comply with one methodology at least among those belonging to the current workspace.
Given a methodology and a tool which belongs to it and has only an input node number, a file is a valid input for that tool if the file history certifies that it is placed in the tool's input node. If the tool has several input node numbers, a set of files is a valid input set if each file is placed in the right input node and if all the files have the same ancestors. In fact, not all the files which are placed in the input nodes of a tool can be used together as simultaneous inputs of the concerned tool. In figure 9, the pair (*file_e*, *file_l*) is not a valid input set for

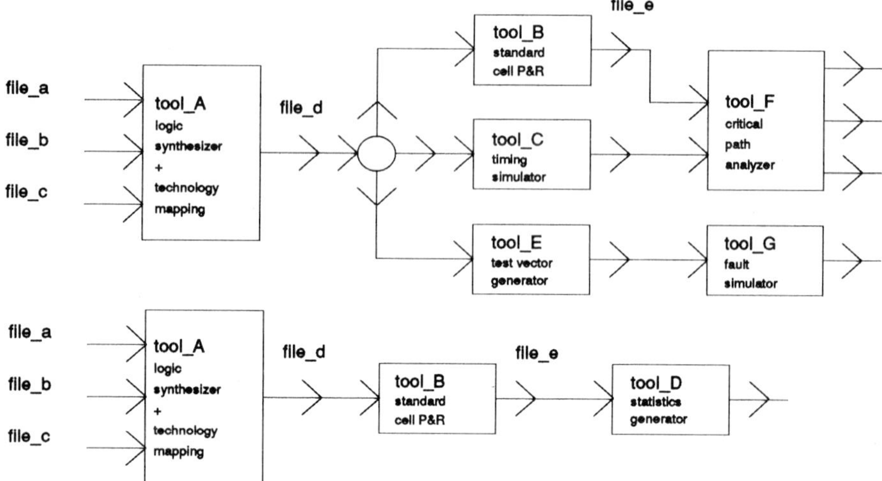

Fig. 8

tool_F, even if each one of the two files is placed in an input node of *tool_F*'s, since the two files have different ancestors.

The past transformation steps of a given file which complies with a methodology may precisely be rebuilt only if there is a distinct file associated to each node number. This is true just in case no file has been overwritten by a tool run. Since

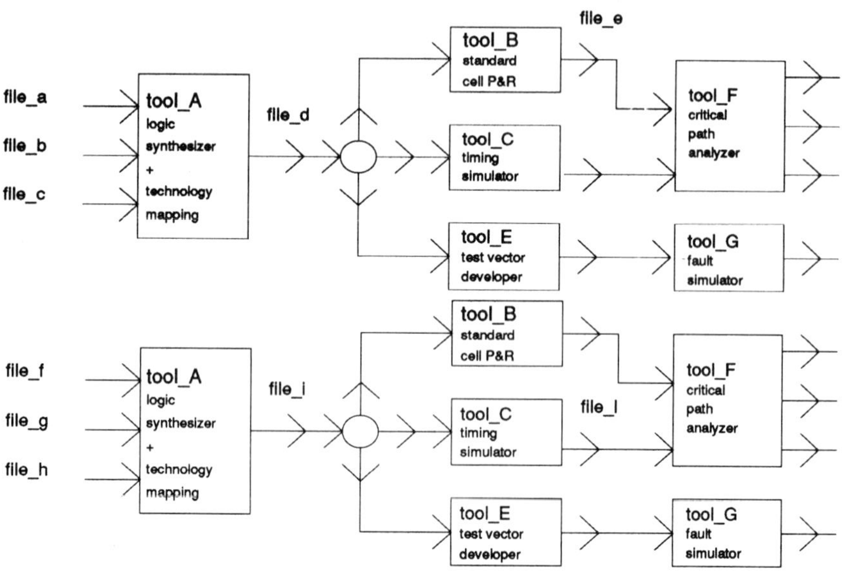

Fig. 9

there are tools (both batch and interactive) which update an input file without modifying its name, DMM functions detect any overwrite action and create a copy of each file which has to be overwritten, asking the designer for a new (unambiguous) name and updating both the name of the file and of its history consequently.
The least information set, necessary and sufficient in order to build the whole past history of a given file and to draw all the methodologies the file complies with, consists of three entities. They are the indication of one (and only one) methodology the file complies with plus the file position (i.e. node number) within that methodology and the names of the input files of the tool which produced the given file as output. All the remaining pieces of information of interest (names of the other possible methodologies the file complies with, names of all the tools which have been progressively run to produce the file, names of all its ancestor files) can be deduced from these few pieces of information, by means of either a small or a great computation effort.
The choice of the information set which has to be stored in order to track every file (i.e. the file history) is the consequence of a compromise between computation time and memory occupation.
Owing to the following remarks, we have chosen to store an information set greater than the least set.
The knowledge of all the methodologies a file conforms to within a workspace is repeatedly used.
Example 1 - Suppose that only the minimum set has been stored, i.e. just the indication of one methodology has been stored for each file. If a designer tries to run a tool assuming a file as input and this action does not comply with the indicated methodology, the DMM system has to find out another methodology the file conforms to and check whether the action is valid according to this methodology, and so on, until an acceptable methodology is found or all the current workspace methodologies have been looked through unsuccessfully.
Example 2 - If a file is exported from a source workspace to be imported into a target workspace, the import can be accomplished only if the file conforms to one methodology at least out of those which are valid in the target workspace. The DMM system has to look over the exchange and check whether this condition is fulfilled. The knowledge of all the methodologies the file complies with in the source workspace may help this test in case the source workspace and the target workspace have some valid methodologies in common.
Since finding out the methodologies a file complies with needs a heavy computation, it is convenient not to repeat it too many times. The computations of the first example can be avoided by storing for each file all the current workspace methodologies the file complies with. This piece of information gives some help even in the second example situation. However, all the computations of the second example are avoided too in case the indications of all methodologies the file conforms to should be stored, be they valid in the current workspace or not (taken it for granted that a file cannot be produced if it does not comply with one methodology at least which is valid in the current workspace).
As it has already been stated, the knowledge of the ancestor files of a given file is quite important in order to check whether declared input files can actually altogether represent an input set for a tool within a specified methodology. Therefore, it is convenient that a file's history contains also the univocal identification of the file's ancestors.

4.5 Methodology automation.

The DMM system provides the user with a set of services which are able to automatically run flows that belong to white box methodologies. A "flow" is a sequence of tool invocations which is included in a methodology, called "mother

methodology". Methodology automation services involve two operational phases: univocal identification of the flow and execution of the flow.
The first phase allows the user to interactively specify the flow, given the mother methodology pictorial representation. From the system point of view, this representation consists of the identification of the mother methodology plus a sequence of node numbers plus the name of the design files which activate the flow. This representation solves the problem of non-sequential flows (i.e. flows which contain alternative points and/or cycles and/or feedbacks). For example, a cycle which has to be run n times is described by means of n couples (cycle input node number, cycle output node number).
The second phase, that is the flow execution, is accomplished by a remote DMM application which reads the flow internal representation, determines the ordered sequence of tools to be run and executes it.

5. CONCLUSIONS.

The introduction of state-of-the-art CAD frameworks has already faced some of the major difficulties related to data production, data storage and data communication among many CAD tools. However, a lot of problems are still unsolved, such as the organization of the overall project, in terms of work distribution to several designers and/or design teams, and the formalization and enforcement of design methodologies. These are the topics we have taken into account in the development of a set of Workspace Management (WM) and Design Methodology Management (DMM) functionalities for the Berkeley design environment. These functions, while not modifying the VEM interface, exploit VEM editing capabilities and the OCT data format, so that WM and DMM objects are treated in the same way as design objects.
A multilevel workspace hierarchy, governed by precise scope and reaching rules, has been discussed. Two kinds of methodology views have been described. The concept of file history has been introduced so as to support methodology adherence check. History descriptions are the result of a tradeoff between storage allocation space and computation time.
The automation of tool flow executions is accomplished by means of a remote application which interprets flow descriptions and makes remote calls to CAD tools straightforwardly.

References.

[HMS86] Harrison, D.S., P. Moore, R. L. Spickelmier and A. R. Newton, "Data Management and Graphics Editing in the Berkeley Design Environment", Proceeding ICCAD, Santa Clara, California, November 1986
[HNS90] Harrison, D. S., A. R. Newton, R. L. Spickelmier and T. J. Barnes, "Electronic CAD Frameworks", Proceedings of the IEEE, Vol. 78, NO. 2, February 1990
[KaC87] Katz, R. H. and E. Chang, "Managing Change in a Computer-Aided Design Database", Proceedings VLDB, Brighton, 1987
[KAC86] Katz, R. H., M. Anwarrudin and E. Chang, "A Version Server For Computer-Aided Design Data", Proceedings of the 23rd Design Automation Conference, Las Vegas, Nevada, June 1986, pp. 27-33
[KCB86] Katz, R. H., E. Chang and R. Bhateja, "Version Modeling Concepts for Computer-Aided Design Databases", Proceedings ACM-SIGMOD, Washington, DC, May 1986
[LFS88] Leung, S. S., P.D Fisher and M. A. Shanblatt, "A Conceptual Framework for ASIC Design", Proceedings of the IEEE, Vol. 76, No. 7, July 1988
[OCT] Octtools Distribution 3.5, Department of EECS - University of California, Berkeley, March 1990

VLSI 91
A. Halaas and P.B. Denyer (Eds.)
Elsevier Science Publishers B.V. (North-Holland)

Single-Level Wiring for CMOS Functional Cells

Jan Madsen

Electronics Institute
Technical University of Denmark
DK2800 Lyngby, Denmark

This paper proposes a branch-and-bound algorithm for generating interconnections in a CMOS functional cell when only one layer is available for wiring. Topological and technology constraints are used to prune the search tree. The flexible incorporation of cost functions allows for easy trade-off between area (cell height as well as cell width) and circuit performance. The proposed algorithm has been implemented as part of a cell compiler (CELLO [1]) working in an experimental silicon compiler environment.

1 Introduction

Using *cell compilers*, which automatically generates circuit layout from a transistor netlist (or boolean functions), makes it possible to replace many primitive gates, such as NAND and NOR gates, with a single *complexgate* tuned to the circuit requirements. While the area and performance is optimized by this approach, the interconnect problem of internal signals is complicated, requiring advanced wiring schemes.

One of the popular layout styles is the line-of-diffusion introduced by Uehara and van Cleemput [2]. It consists of two rows, one of PMOS transistors and one of NMOS transistors as illustrated in Figure 1. The layout generation of this layout style includes three different wiring problems:

- Transistor gate wiring using poly. Due to the ordering of transistors in the two rows, this may be established straightforward by using a river routing algorithm.
- Power wiring. Due to the layout style this task is also straightforward.
- Wiring of internal nets in each of the rows.

Figure 1a shows a layout example of a CMOS functional cell using the line-of-diffusion layout style which requires complicated internal wiring. In [3] a modified channel routing algorithm is used to produce the internal wiring in the area between the two transistor rows. Nair [4] propose a branch-and-bound algorithm to solve the more general problem of single-level wiring in cell compilers targeted toward sea-of-gates gate arrays. The wiring problem of internal signals in each of the transistor rows may be formulated as a single-row routing (SRR) problem. Several approaches to the single-row routing problem have been reported, most of them aimed at solving PCB problems for dual-in-line packages. Tsukiyama et.al. [5] gives necessary and sufficient conditions for solving the SRR problem given a maximal allowable channel congestion, and propose an algorithm to solve the problem based upon interval graphs. An other interval graph based algorithm to solve

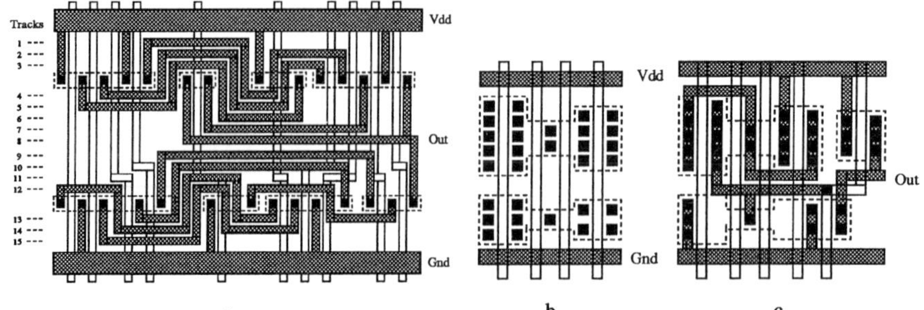

Figure 1: (a) Layout of a CMOS functional cell requiring complicated internal wiring; (b) Sample CMOS line-of-diffusion layout style; (c) the sample function ($f = xy + zw$).

the SRR problem in polynomial time is given in [6]. Algorithms based upon branch-and-bound techniques have been proposed in [7] and [8]. The algorithm in [8] has been used as backbone for solving the general multilayer wiring problem in PCBs.

Common to all of these wiring schemes is that the node (pin) positions are fixed and that optimum is the solution which minimizes the channel congestion, i.e., leading to the minimum number of horizontal routing tracks.

This paper address the problem of generating the interconnections in a circuit when only one layer is available for wiring. The proposed algorithm is dedicated to solve the wiring problem in circuits which have been laid out using the line-of-diffusion layout style. The algorithm is based upon the work of R. Nair [4], it guarantees to find a solution if one exsists and allows for easy trade-off between area (cell height as well as cell width) and circuit performance.

2 Problem Characterization

The structure of the physical layout of the line-of-diffusion layout style is illustrated in Figure 1b, while Figure 1c give an example of the metalization which implements the function $f = (xy + zw)$. The line-of-diffusion layout style for implementing complexgates may be characterized as follows:

1. Each transistor may be sized individually which in contrast to the gate-array style does not leave a fixed number of routing tracks over the transistors.

2. Power supply is routed at top and bottom of the cell leaving a routing channel between the transistor row and the power supply.

3. Horizontal routing tracks are available in the area between transistor rows. The output takes up one track.

The problem may be reduced by not allowing horizontal wiring over transistors, i.e., viewing all transistors as being of minimum width or filling the wide transistors with contact holes. From an electrical point of view this seems favorable since, when having wide transistors, placing as many contacts as possible will reduce the contact resistance, increase reliability and give an uniform field across the channel, parameters all of which are essential in high performance circuits.

Concerning the interconnect between the power track and the transistor row, this may complicate the wiring problem because it has to be connected using metal1 in order to minimize the resistance in the power path. This will result in a blockage of the routing area between the power supply and the transistor row. Also the interconnect of the output track to the transistor rows will introduce blockages in the area between the transistor rows.

From this characterization it is evident that the layout generation of a CMOS functional cell introduces two single-level wiring problems each of which may be formulated as:

> Given a sequence of contacts which has to be wired, find the connection scheme which minimize some cost function.

3 Model

The topology of the single-level wiring problem may initially be modelled as an ordered sequence of nodes representing the contacts in the transistor row as shown in Figure 2a. Formally this may be expressed as a n-tuple:

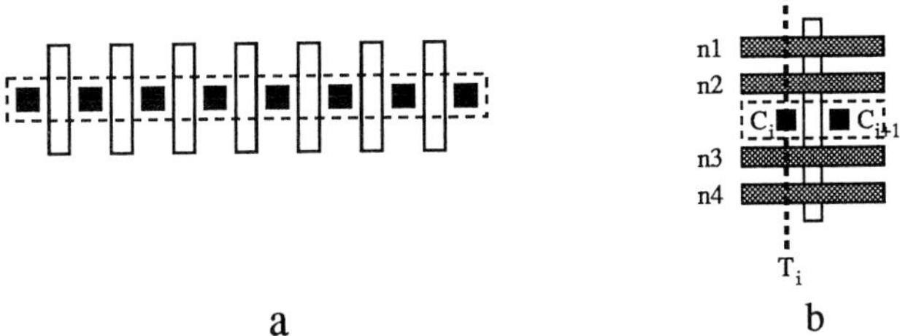

Figure 2: (a) Initial transistor row which has to be wired; (b) The 2-tuple T_i gives the wiring order in the upper- and lower-channel for node c_i.

$$T_{nodes} = < c_1, c_2, \ldots, c_i, \ldots, c_n > \tag{1}$$

where c_i denotes the i'th node in the row counted from the left. The nets which have to be wired may also be expressed as a n-tuple:

$$T_{nets} = < n_{c_1}, n_{c_2}, \ldots, n_{c_n} > \tag{2}$$

where $n_{c_i} = n_{c_j}$ if and only if c_i and c_j have to be connected. The wiring may be described as a n-tuple, where each element is a 2-tuple (T_i) representing the net positions above and below a given node, c_i as illustrated in Figure 2b. The 2-tuple describes the nets in the upper ($T_{i\uparrow}$) and lower ($T_{i\downarrow}$) channel respectively. This may be formulated as:

$$T_{wiring} \quad = \quad < T_1, T_2, \ldots, T_{n-1} > \tag{3}$$

where,

$$T_i \quad = \quad < T_{i\uparrow}, T_{i\downarrow} > \tag{4}$$

and

$$\begin{aligned} T_{i\uparrow} &= \langle n_1, n_2, \ldots, n_l \rangle \\ T_{i\downarrow} &= \langle n_{l+1}, \ldots, n_m \rangle \end{aligned}$$

where the net ordering is calculated from top to bottom, i.e., n_l from $T_{i\uparrow}$ is closet to the row in the upper channel, while n_{l+1} from $T_{i\downarrow}$ is closest to the row in the lower channel. The topology of the wiring solution can then be formulated from equation (1), (2) and (3) as:

$$T_{wiring} = \langle\langle T_{1\uparrow}, T_{1\downarrow} \rangle, \langle T_{2\uparrow}, T_{2\downarrow} \rangle, \ldots, \langle T_{n\uparrow}, T_{n\downarrow} \rangle\rangle \tag{5}$$

The final layout wiring may be recovered from the topological abstraction. As an illustrative example, lets consider the following wiring problem:

$$T_{nets} = \langle 1, 2, P, 1, 2 \rangle \tag{6}$$

where node P is a power node which blocks the upper channel. The problem leads to the topological solution:

$$T_{wiring} = \langle\langle\langle 1 \rangle, \emptyset \rangle\langle\langle 1 \rangle, \langle 2 \rangle\rangle, \langle \emptyset, \langle 1, 2 \rangle\rangle, \langle \emptyset, \langle 2 \rangle\rangle, \langle \emptyset, \emptyset \rangle\rangle$$

A graphical illutration of the topological solution is shown in Figure 3a, while Figure 3b shows the recovered layout.

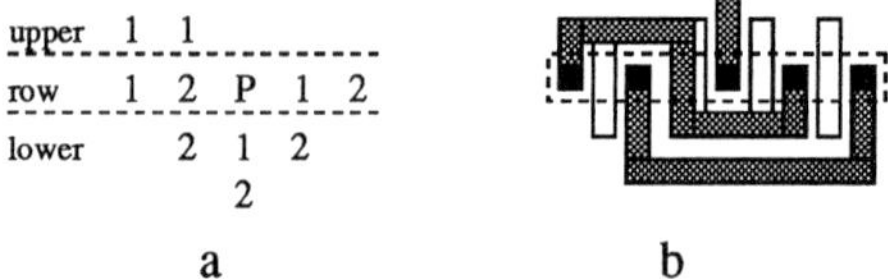

Figure 3: Wiring solution of the wiring problem $T_{nodes} = \langle 1, 2, P, 1, 2 \rangle$; (a) Topological wiring; (b) Recovered layout.

4 The Algorithm

The algorithm is a branch-and-bound algorithm based upon a systematicaly pass from left to right. The algorithm may be described as follows:

1. Proceed from left to right.
2. In each step find all possible valid permutations P_i from each of the previous 2-tuples T_{i-1} leading to a valid transition to T_i. To each valid transition associate a cost value.
3. When the rightmost column has been reached, start back-tracking in order to find the cheapest path.

The algorithm produces a permutation tree where each node correspond to a 2-tuple and each branch to a valid transition between two nodes. Figure 4 shows the permutation tree of the wiring problem of Figure 3. If all possible permutations are taken into account, the

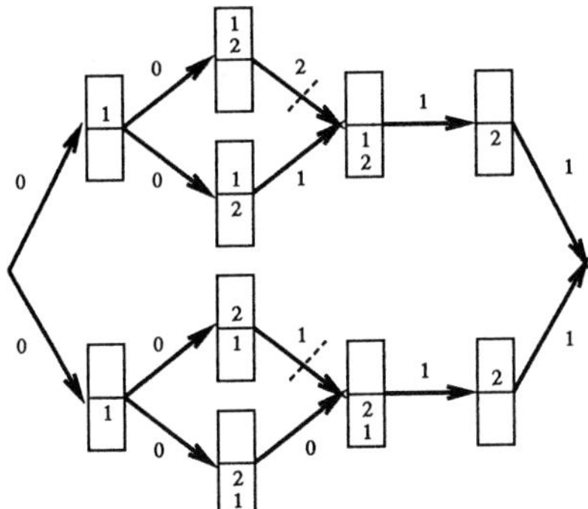

Figure 4: Pruned Permutation tree for the wiring problem presented in Figure 3.

size of the permutation tree will grow exponential even for small-size wiring problems. However, technological and layout style restrictions are used to prune the tree. Furthermore, we use a greedy approach which expands on the cheapest solution only, as well as a "delayed binding" concept where permutations are only performed when enforced by the circuit topology. This approach highly reduces the number of valid permutations and leads to suboptimal solutions which in practical cases have shown to be close to optimum and as such feasible solutions to the described wiring problem.

Before discussing the valid permutations we will define some algebraic relations which will ease the description of permutations.

Definition 1 *Let* $T = < c_1, c_2, \ldots, c_n >$ *be a n-tuple, then* $T + c_k$ *is the* $(n+1)$*-tuple given by* $< c_1, \ldots, c_n, c_k >$*, and* $c_k + T$ *the* $(n+1)$*-tuple given by* $< c_k, c_1, \ldots, c_n >$ *(concatenation).*

Definition 2 *Let* $T = < c_1, c_2, \ldots, c_n >$ *be a n-tuple, then* $T - c_k$ *is the* $(n-1)$*-tuple given by* $< c_1, \ldots, c_{n-1} >$*, i.e., removing last element, and* $c_k - T$ *the* $(n-1)$*-tuple given by* $< c_2, \ldots, c_n >$*, i.e., removing first element.*

Definition 3 *Let* $T = < c_1, \ldots, c_i, \ldots, c_j, \ldots, c_n >$ *be a n-tuple where* $i <= j$*, then* $T[i, j]$ *is the index function giving a* $(j - i + 1)$*-tuple consisting of element* i *to* j *of* T*.*

4.1 Permutations

Lets consider the vaild permutations of a given 2-tuple T_i. Let

$$P_i = \{T_{i,1}, T_{i,2}, \ldots, T_{i,m}\} \tag{7}$$

denote the set of valid permutations of T_i and $\mid P_i \mid = m$ denote the number of valid permutations of T_i. There exsists 3 basic conditions leading to a change of P_i (see Figure 5), each of these are described in the following.

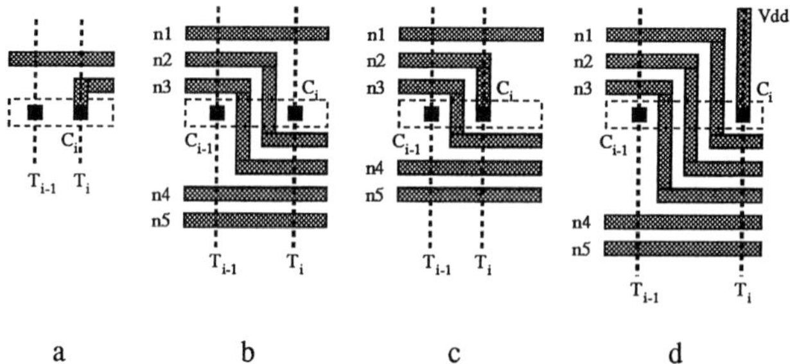

Figure 5: Valid permutations; (a) Adding a new net; (b) 2 nets crossing over; (c) Terminating net in upper-channel; (d) Power net forcing all nets in upper-channel to shift.

A New Net

When a new net is added there are 2 possible valid permutations depending on wether the net is to be routed in the upper- or lower-channel. Figure 5a illustrates a new net being routed in the upper-channel. Formally, this may be written as:

> Let c_i denote the node of the new net, then the possible valid permutations of T_i is,
>
> $$\begin{aligned} T_{i\uparrow} &= T_{i-1\uparrow} + c_i \text{ (upper-channel)}, T_{i\downarrow} = T_{i-1\downarrow} \\ \textit{and}, & \\ T_{i\downarrow} &= c_i + T_{i-1\downarrow} \text{ (lower-channel)} T_{i\uparrow} = T_{i-1\uparrow} \end{aligned}$$

Crossing Over

When k nets have to shift from one channel to the other channel, k cross-overs have to be introduced. This may be formulated as:

$$\begin{aligned} T_i &= < T_{i-1\uparrow} - T_{i-1\uparrow}[n-k+1, n], \\ & T_{i-1\uparrow}[n-k+1, n] + T_{i-1\downarrow} > \end{aligned} \tag{8}$$

when the nets have to cross from upper to lower channel. The expression for crossing from lower to upper channel is similar. Figure 5b illustrates the principle of cross-overs. In general there are 2 possible conditions leading to cross-over, net termination and special nets.

Net Termination: When a net is terminated, i.e., when the last contact has been reached the net in question has to be moved to the midle, i.e., next to the row. This cause the nets in between to be shifted as shown in Figure 5c. Formally this may be written as:

$$\begin{aligned} T_i &= < T_{i-1\uparrow} - T_{i-1\uparrow}[n-k+1, n], \\ & T_{i-1\uparrow}[n-k+2, n] + T_{i-1\downarrow} > \end{aligned} \tag{9}$$

when the net is terminating from the upper-channel. A net terminating from the lower-channel gives a similar expression.

In case of multi-point nets, internal contacts are treated as a net termination followed by a net insertion (i.e., a new net) introducing two new possible permutations. This allows for a cost-free cross-over.

Special Net: The output and power net introduce blockages in the routing channels, forcing *all* nets in that channel to be shifted to the other channel. This may be formulated as follow:

$$T_i = < \emptyset, T_{i-1\uparrow} + T_{i-1\downarrow} > \tag{10}$$

when the blockage is in the upper-channel. A similar expression for a blockage in the lower-channel is straight forward. Figure 5d illustrates the effect of special nets.

Merging Nets

When two or more permutations leads to the same 2-tuple they are merged into one. In the permutation tree this means that several branches are leading to the same 2-tuple. Only the branch with the lowest cost value is kept.

From the description of the 3 basic conditions leading to permutations of the 2-tuple it is seen, that only the addition of a new net leads to an increase of the permutation tree. Thus, an upper-bound on the size of the permutation tree may be given as:

$$\sum_{i=1}^{m} 2^i + (n-m-1) \cdot 2^m \tag{11}$$

where n is the number of nodes and m is the number of nets. The first part accounts for the breadth of the tree while the second part accounts for the depth of the tree.

4.2 The Cost Function

The cost function is used to decide among several possible valid permutations. Thus, it has to reflect physical properties, e.g., layout area and electrical performance considerations. From an area point of view two things have an important influence:

- Number of tracks used in the routing channels. This will influence the cell height and depend upon the widest part of the channel. Formally, this may be expressed as:

$$C_{height} = \max_{T_i \epsilon T_{wiring}} (\mid T_{i\uparrow} \mid + \mid T_{i\downarrow} \mid) \tag{12}$$

- Number of nets that have to cross-over. This will influence the cell width and may be formulated as:

$$C_{width} = \sum_{i=1}^{n-1} C(T_i, T_{i+1}) \tag{13}$$

where,

$$C(T_i, T_{i+1}) = min(\mid T_{i\uparrow} - T_{i+1\uparrow} \mid, \mid T_{i\downarrow} - T_{i+1\downarrow} \mid) \tag{14}$$

which corresponds to the number of cross-overs between i and $i+1$.

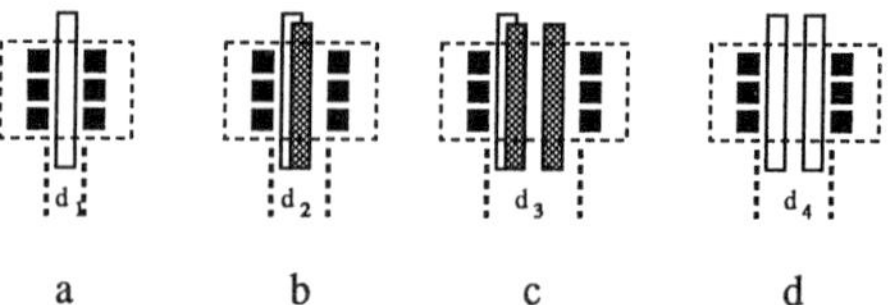

Figure 6: Realistic cost values; (a) no nets crossing; (b) one net crossing; (c) two nets crossing; (d) non-connected node.

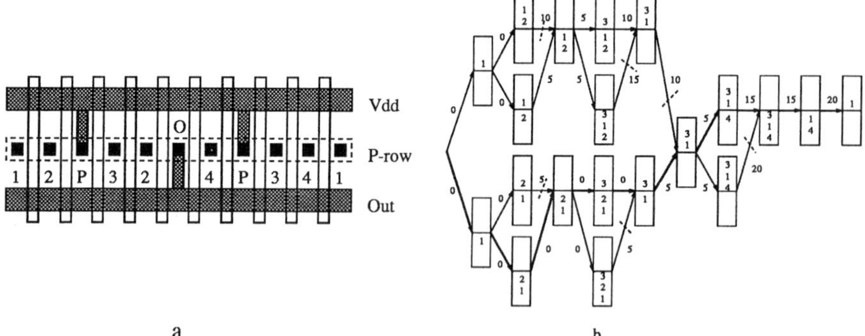

Figure 7: (a) Sample PMOS transistor row; (b) Permutation tree for the routing problem.

This flexible cost scheme allows for the calculation of realistic cost values. Consider the three examples of cross-overs shown in Figure 6, Figure 6a shows the space available for cross-overs when minimum cell width have to be obtained. In Figure 6b the space needed for crossing one net (d_2) is given, thus the cost of the first net crossing is $d_2 - d_1$. Figure 6c shows the situation of two nets crossing leading to the cost $d_3 (> d_2 - d_1)$. Finally, Figure 6d illustrates the case where a node *only* connects the two neighbouring transistors, as no contact is needed the space left for cross-overs (d_4) is increased making the first net crossing free, i.e., the cost is 0.

From a performance point of view, stretching the diffusion area between neighbouring transistors, in order to allow cross-overs, also increases the parasitic capacitance, thus slowing down the circuit. A weight factor may be used to reflect the effect on performance by stretching this area, i.e., stretching an area close to the output node gives a high cost, while stretching one close to the power supply gives a low cost. The cost function corresponding to the cell width may then be formulated as:

$$C_{width} = \sum_{i=1}^{n-1} w_i \cdot C(T_i, T_{i+1}) \tag{15}$$

5 An Example

In order to exemplify the algorithm lets consider the transistor row in Figure 7. The transistor row is modelled as a ordered sequence of nets:

$$T_{nets} = < 1, 2, P, 3, 2, O, 4, P, 3, 4, 1 > \tag{16}$$

where the power net (P) and the output net (O) are blockaging the routing areas as shown in Figure 7a. Running the algorithm from left to right results in the permutation tree shown in Figure 7b. The number associated with each edge is the total cost from the leftmost net to the edge. Only the simple cross-over cost has been taken into account, i.e., a penalty of 5 for each cross-over. It is seen that the size of the permutation tree is 20 while the upper-bound on the size is 126, i.e., the tree is highly pruned. Back-tracing from right to left in the permutation tree, adding the wiring information from each step to the topology model, yields the topological description shown in Figure 8a. The final layout

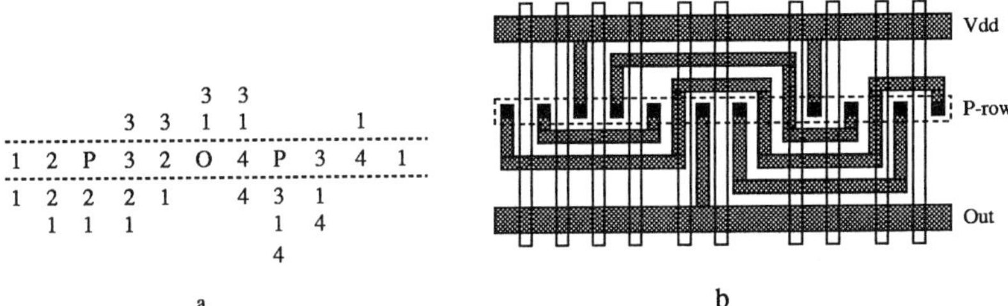

Figure 8: (a) Topological description after back-tracing; (b) Final layout after single-level routing.

may be recovered from the topological abstraction resulting in the layout in Figure 8b.

6 Implementation and Results

The branch-and-bound algorithm has been implemented in C++ under UNIX as part of the cell compiler CELLO [1]. The implementation has been applied to a large set of benchmarks, including realistic circuit examples as well as academia examples. Tabel 1 shows some of the results from the benchmarks, the table lists the number of nodes and nets for each example together with the upper-bound, the actual size and maximum width of the corresponding permutation tree, and the CPU time.

7 Conclusion

A wiring scheme for the interconnection of internal signals in the line-of-diffusion layout style has been presented. Further, a branch-and-bound algorithm to solve the interconnect problem has been proposed. The algorithm highly reduces the search space, and allows for easy trade-off between cell height and width as well as between area and circuit performance, by providing a flexible way of incorporating cost functions.

The algorithm has been implemented as part of a cell compiler CELLO, where it solves the problem of wiring internal nets.

Acknowledgements

The author would like to express special thanks to Martin Arnbjerg-Nielsen who was involved in the initial code implementation of the algorithm.

circuit name	*# nodes*	*# nets*	*upper-bound of perm. tree*	*Size of perm. tree*	*Max. width of perm. tree*	*CPU in sec.*
F1p	12	2	42	29	8	0.38
F1n	9	2	30	11	2	0.18
F2p	10	2	34	30	8	0.43
F2n	7	2	22	7	2	0.19
F5p	5	1	8	6	2	0.14
F5n	6	1	10	6	2	0.19
F7p	9	1	16	11	2	0.18
F7n	9	0	0	8	1	0.09
FAp	9	2	30	11	2	0.19
FAn	12	2	42	29	8	0.44
FBp	17	6	766	44	8	0.63
FBn	20	5	510	38	4	0.59
c1	6	2	18	10	4	0.18
c2	11	4	126	19	4	0.28
c3	11	5	222	107	32	1.32
c4	12	6	446	187	64	2.23
c5	9	3	54	14	4	0.24
c6	9	4	94	39	8	0.57
c7	22	10	13310	169	16	2.08
c8	14	4	174	50	8	0.61
c9	14	5	318	42	8	0.59

Table 1: Benchmarks; The CPU time is meassured on an Apollo DN2500.

References

[1] Jan Madsen, *"A New Approach to Optimal Cell Synthesis,"* Proceedings of IEEE International Conference on Computer-Aided Design, Santa Clara, California, 1989, pp.336–339.

[2] T. Uehara and W.M. vanCleemput, *"Optimal Layout of CMOS Functional Arrays,"* IEEE Trans. on Computers, vol. C-30, May 1981, pp. 305–312.

[3] S. Wimer, R.Y. Pinter and J.A. Feldman, *"Optimal Chaining of CMOS Transistors in a Functional Cell,"* IEEE Trans. on Computer-Aided Design, vol. CAD-6, no.5, September 1987, pp. 795–801.

[4] Ravi Nair, *"Single-Level Wiring for Cell Compilers,"* Proceedings of VLSI 89, August 16–18, 1989, pp. 13–22.

[5] Shuji Tsukiyama, Ernest S. Kuh and Isao Shirakawa, *"An Algorithm for Single-Row Routing with Prescribed Street Congestions,"* IEEE Trans. on Circuits and Systems, vol. CAS-27, no.9, September 1980, pp. 361–368.

[6] Tom Tsan-Kuo Tarng, Malgorzata Marek-Sadowska and Ernest S. Kuh, *"An Efficient Single-Row Routing Algorithm,"* IEEE Trans. on Computer-Aided Design, vol. CAD-3, no.3, July 1984, pp.178–183.

[7] Michel T. Doreau and Luther C. Abel, *"A Topologically Based Non-Minimum Distance Routing Algorithm,"* Proceedings of the 15th ACM/IEEE Design Automation Conference, 1978, pp.92–99.

[8] Raghunath Raghavan and Sartaj Sahni, *"Optimal Single Row Routing,"* Proceedings of the 19th ACM/IEEE Design Automation Conference, 1982, pp.38–45.

VLSI 91
A. Halaas and P.B. Denyer (Eds.)
Elsevier Science Publishers B.V. (North-Holland)

An Over-the-cell Channel Router

Ravi R. Pai[a] and S.S.S.P. Rao[b]

[a]VLSI Design Center, Indian Institute of Technology, Powai, Bombay 400076, India

[b]Department of Computer Science and Engineering, Indian Institute of Technology, Powai, Bombay 400076, India

Abstract

A router which formulates an *over-the-cell* channel routing problem as a 0-1 integer linear programming problem(ILP) is presented. The ILP is solved using Interior Point Methods. The router can handle limited *street capacities* over the cells. This router outperforms all the other known over-the-cell routers on all benchmark problems with significantly less number of tracks in the channel. *Deutsch Difficult Example* was solved with 15 tracks in the channel.

1 Introduction

With the two and a half layer technology (one layer of polysilicon and two layers of metal) becoming more and more popular, some more routing space can be added to the traditional *channel* by routing some of the net segments over the cells both on the upper and lower sides of the channel. Several *over-the-cell routers* [1], [3], [6], [8] have been proposed exploiting the availability of one routing layer over the cell areas.

Over-the-cell channel problems have been proved to be NP-hard[1]. Common approach is to divide the *over-the-cell* channel routing problems into 3 steps as follows:

1. Connecting pins of the same net by net segments over the cells such that no two net segments belonging to different nets cross each other *over the cells.*
2. Choosing net segments to be connected *within the channel* such that the channel density is reduced. This problem of *net selection* has been proved to be NP-hard[1]. So, some heuristics are generally used.
3. Routing the *resulting channel routing problem* using some conventional channel router.

The main emphasis of this paper is on the first two steps. The third step can be handled by any conventional 2-layer channel router.

The router in [1] finds a maximal set of the net segments which can be routed over the cells and assumes that by doing so, the channel density of the problem is minimized. A simple example in Fig. 1, shows that maximizing the number of segments routed over the cells may not necessarily lead to the minimization of channel density.

In this paper the *over-the-cell routing problem* is formulated as a 0-1 Integer Linear Programming Problem(ILP). The ILP is solved using Interior Point Methods[4],[5]. This procedure is followed by an optional '*greedy heuristic*' for improving the solution obtained through solving the ILP. A conventional 2-layer channel router, viz. EHCR[7], is used to solve the resulting channel routing problem. The solutions were obtained for all benchmark problems with much better results compared to the results in [1]. In the case of the famous Deutsch Difficult Example, the solution was obtained with 15 tracks within the channel.

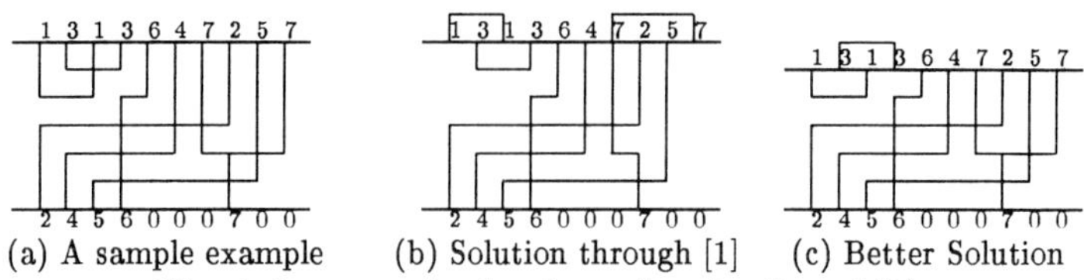

(a) A sample example (b) Solution through [1] (c) Better Solution

Fig. 1 An example showing suboptimality of [1]

2 Formulation of the Problem

A channel routing problem can be specified by two arrays UPPER and LOWER, of dimension equaling the number of columns in the channel. The ith element of array UPPER (LOWER) contains the net number of the pin in the upper (lower) side of the channel in column i.

Consider the routing problem given in Fig.2(a). For this problem,

UPPER = [0, 2, 1, 7, 3, 3, 2, 4, 5, 5, 4, 2, 5]
LOWER = [2, 1, 6, 4, 7, 2, 1, 4, 6, 5, 4, 1, 4]

Let t = (n, c, T) denote a '*terminal*' of net n in column c, where T $\in$ {U, L, B} indicates the '*type*' of terminal t. The convention is that the *type*,

$$T = \begin{cases} U & \text{if ((UPPER[c] = n) and (LOWER[c]} \neq \text{n))} \\ L & \text{if ((UPPER[c]} \neq \text{n) and (LOWER[c] = n))} \\ B & \text{if ((UPPER[c] = n) and (LOWER[c] = n))} \end{cases}$$

Let the list $l_n \equiv (t_1, t_2, ..., t_k)$ define the entire net n where each t_i is a *terminal* of net n and k is the number of *terminals* of net n. (Note: In this context, the number of *terminals* of a net equals the number of distinct columns containing pins of the same net). The *terminals* t_1, t_2, ..., t_k are so arranged in the list l_n, that the corresponding columns of the terminals increase monotonically from left to right.

For any k-terminal net, a minimum of $k-1$ two-terminal net segments are required to completely connect all the k terminals. We follow *minimum horizontal distance chain* to connect the k terminals. Each net segment $s_{n,j}$, (j = 1, ..., k-1) of net n, connects two terminals t_j and t_{j+1} of net n having the list of terminals $l_n \equiv (t_1, t_2, ..., t_k)$.

The objective of our '*over-the-cell*' routing scheme is to choose the net segments to be routed over the cells so that the *maximum channel density* of the problem is minimized. Then a conventional 2-layer channel router is used to solve the *resulting channel routing problem* within the channel.

3 0-1 Integer Programming Problem Formulation

This section deals with the way an '*over-the-cell*' routing problem is formulated as 0-1 integer programming problem.

3.1 Local-Density Constraints

Suppose W is the maximum channel density of a channel with C columns. Let the local density of a column c be w_c. Then

$$w_c \leq W \text{ for c = 1,..., C} \tag{1}$$

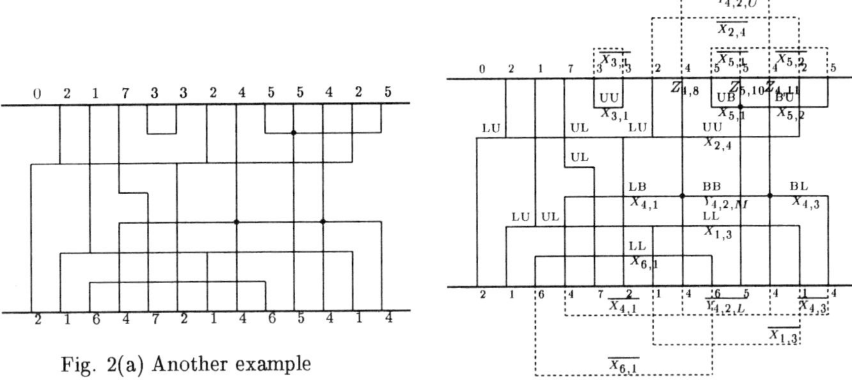

Fig. 2(a) Another example

Fig. 2(b) Possible paths for net segments and their corresponding variables in ILP

Local density w_c of a column c is defined as the number of distinct nets crossing column c *within the channel.* The segment density s_c of column c is defined as the number of net segments allotted *within the channel* crossing column c. s_c may be different from w_c because even if more than one net segment of a net are crossing the column c, the local density w_c is increased by unity, where as the *segment density* s_c is increased by the number of segments of the net. Suppose s_c distinct net segments allotted to *within the channel* cross column c. We define *horizontal detour* to be the condition when more than one segment of a net is allowed to run between two columns inside the channel or over the cells. If no *horizontal detours* are allowed,

$$s_c - 2 \leq w_c \leq s_c \tag{2}$$

The *type* of a segment depends upon the *types* of its two end terminals. If the end terminals of a segment are 'U' and 'L' respectively, the *type* of the net segment is 'UL'. Similarly net segments of *types* 'LU','LL','UU','LB','BL','UB','BU' and 'BB' are defined. Depending upon their *types* the segments are categorized into three categories:

3.1.1 CATEGORY F: Types (UL, LU):

These net segments can not be routed over the cells as their end terminals are on different sides of the channel. So, they have to be necessarily routed within the channel. Let F_c be the number of net segments belonging to type UL or LU, crossing column c. Such net segments need not be considered for routing *over-the-cells.*

3.1.2 CATEGORY X: Types (LL, UU, LB, BL, UB, BU):

The net segments belonging to this category can be routed either '*within the channel*' or '*over-the-cells*'. Thus each of the net segments of this category can have two possible paths. The types and possible paths for each of the net segments of the problem of Fig.2(a) are shown in Fig.2(b).

With each such net segment $s_{n,j}, (j = 1, ..., k-1)$ for a k-terminal net n, belonging to X category, a variable $X_{n,j}$ is associated with the condition

$$X_{n,j} \in \{0, 1\} \tag{3}$$

The convention is that if $X_{n,j} = 0$, the corresponding net segment $s_{n,j}$ is routed *over-the-cell*. If $X_{n,j} = 1$, the net segment $s_{n,j}$ is routed *within the channel*. Both the possibilities for $X_{n,j}$ variables are shown in Fig.2(b) by path $X_{n,j}$ *within the channel* and by path $\overline{X_{n,j}}$ outside the channel *over-the-cell*. $\overline{X_{n,j}}$ is a complement of $X_{n,j}$.

Suppose l and r are the left and right columns of net segment $s_{n,j}$. If the contribution of segment $s_{n,j}$ to the *segment density* s_c at column c is $X_{c,n,j}$, then

$$X_{c,n,j} = \begin{cases} 0 & \text{if } ((c < l) \text{ or } (c > r)) \\ X_{n,j} & \text{if } (l \leq c \leq r) \end{cases} \tag{4}$$

If X_c is the contribution to the *segment density* of nets belonging to X category at column c,

$$X_c = \sum_{n=1}^{N} X_{c,n,j} \tag{5}$$

3.1.3 CATEGORY Y: Type (BB):

The net segments belonging to this category can be routed *within the channel*, or *over-the-cell* on the upper or on the lower side of the channel.

With each such net segment $s_{n,j}, (j = 1, ..., k-1)$ for a k-terminal net, we associate three variables $Y_{n,j,M}, Y_{n,j,U}, and Y_{n,j,L}$ with the conditions,

$$\begin{aligned} Y_{n,j,M}, Y_{n,j,L}, Y_{n,j,U} &\in \{0,1\} && (6) \\ Y_{n,j,M} + Y_{n,j,U} + Y_{n,j,L} &= 1 && (7) \end{aligned}$$

Eqns. (6) & (7) ensure that only one variable will have an integer value of one. The convention is that if $Y_{n,j,M} = 1$, the segment $s_{n,j}$ is routed within the channel. Otherwise, if $Y_{n,j,U} = 1 (Y_{n,j,L} = 1)$, the segment $s_{n,j}$ is routed on the upper (lower) side of the channel.

Suppose l and r are the left and the right columns of net segments $s_{n,j}$ of type BB. If contribution of segment $s_{n,j}$ on the *segment density* s_c at column c is $Y_{c,n,j}$, then

$$Y_{c,n,j} = \begin{cases} 0 & \text{if } ((c < l) \text{ or } (c > r)) \\ 0 & \text{if } ((l \leq c \leq r) \text{ and } ((y_{n,j,L} = 1) \text{ or } (y_{n,j,U} = 1))) \\ 1 & \text{if } ((l \leq c \leq r) \text{ and } (y_{n,j,M} = 1)) \end{cases}$$

Suppose Y_c is the contribution to the *segment density* at column c because of nets belonging to Y category. Then,

$$Y_c = \sum_{n=1}^{N} Y_{c,n,j} \tag{8}$$

Now considering the contributions of all the net segments belonging to F, X and Y categories, *segment density* s_c at column c, becomes

$$s_c = F_c + X_c + Y_c \tag{9}$$

3.1.4 Z-variables

Our primary aim is to minimize the *local density* w_c, rather than segment density s_c. The segment density s_c at column c may be different from the local density w_c. As an example, consider net segments $s_{5,1}$ and $s_{5,2}$ denoted by variables $X_{5,1}$ and $X_{5,2}$

respectively in Fig.2(b). Each of these segments have a common pin at 10th column. If both these net segments are routed within the channel, the contribution of the above mentioned net segments to the segment density s_{10} is two. But still then the contribution to the local density w_{10} due to both the net segments together is only one.

Associated with each net segment $s_{n,j}, (j = 1, ..., k-1)$ for k-terminal net, if $s_{n,j+1}$ exists, and if both $s_{n,j}$ and $s_{n,j+1}$ do not belong to category F, and c is the column number of $j+1^{th}$ terminal, a variable $Z_{n,c}$ is introduced with the conditions

$$
\begin{aligned}
Z_{n,c} &\in \{0,1\} \qquad (10)\\
Z_{n,c} &= \begin{cases} 0 & \text{if at least one of the net segments } s_{n,j} \text{ and } s_{n,j+1} \text{ is routed } \textit{over the cell} \\ 1 & \text{if both the net segments } s_{n,j} \text{ and } s_{n,j+1} \text{ are routed } \textit{within the channel} \end{cases}
\end{aligned}
$$

If n_1 and n_2, are the two nets having pins at bottom and top sides of the column c,

$$Z_c = \begin{cases} Z_{n_1,c} + Z_{n_2,c} & \text{if } n_1 \neq n_2 \\ Z_{n,c} & \text{if } n_1 = n_2 = n \end{cases} \qquad (11)$$

Equality (12) gives the complete expression for *local density* w_c.

$$w_c = F_c + X_c + Y_c - Z_c \qquad (12)$$

Substituting in eqn.(1) we have the *local-density constraints* as

$$X_c + Y_c - Z_c - W \leq -F_c \text{ where c=1,...,C} \qquad (13)$$

3.2 Planarity Constraints

Two net segments s_i and s_j belonging to different nets can not be routed *over the cell* on the same side of the channel if the span of s_i partially overlaps the span of s_j or vice versa. Consider two net segments s_m and s_n belonging to nets m and n $(m \neq n)$. Let $l_m(l_n)$ and $r_m(r_n)$ be the left most and the right most columns of net segment $s_m(s_n)$. Let $X^U_{m,i}$ and $X^U_{n,j}$ be the corresponding net segment variables on the upper side of the channel. We define two sets M and N associated with s_m and s_n respectively.

$$
\begin{aligned}
M &\equiv [l_m, l_m+1, ..., r_m] \\
N &\equiv [l_n, l_n+1, ..., r_n]
\end{aligned} \qquad (14)
$$

If $M \cap N \neq \oslash$ and $((M \cup N \neq M)$ and $(M \cup N \neq N))$, then both the net segments $s_{m,i}$ and $s_{n,j}$ can not be routed together on one side of the channel. i.e.,

$$X^U_{m,i} + X^U_{n,j} \leq 1 \; \forall \; m, n, i, j \qquad (15)$$

where $X^U_{m,i}$ and $X^U_{n,j}$, $(m \neq n)$ are the variables associated with the upper side of the channel for net segments $s_{m,i}$ and $s_{n,j}$ with the span of each net segment partially overlapping the span of the other. Equation (15) ensures that at the most one of the two net segments $s_{m,i}$ and $s_{n,j}$ can be routed *over the cells*.

3.3 Street Capacity Constraints

So far, it is assumed that unlimited number of tracks are available on both the sides of the channel. But in practice, the number of tracks available on any side of the channel is limited. The *street capacity* of a column c *over the cell* on any side of the channel is defined as the number of independent net segments that can cross column c. Let $U_c(L_c)$ be the *street capacity* on upper(lower) side of the channel at column c at that side. Let $X_i^U, \ldots\ldots, X_{i'}^U$ be the X variables and $Y_i^U, \ldots\ldots, Y_{i'}^U$ be the Y variables on the upper side of the channel associated with net segments passing through column c. The *street-capacity constraint* at column c specifies that

$$\begin{aligned} \overline{X_i^U} + \ldots + \overline{X_{i'}^U} + \overline{Y_i^U} + \ldots + \overline{Y_{i'}^U} &\leq U_c \\ (1 - X_i^U) + \ldots + (1 - X_{i'}^U) + (1 - Y_i^U) + \ldots + (1 - Y_{i'}^U) &\leq U_c \end{aligned} \tag{16}$$

Inequalities (13),(15),(16) and (7), specify all the constraints in the ILP.

4 Solving the 0-1 ILP Problem

Interior Point Method for Integer Programming is a new algorithm for 0-1 Integer Programming. This algorithm follows an *interior point* path solving a concave quadratic programming problem. For more details [5] can be referred.

In the earlier sections, the exact formulations of 0-1 ILP are discussed in detail. First of all the ILP so formed is solved with relaxing the integer constraints. The solution for this *relaxed* LP problem gives a fractional value W_f as the *lower bound* for the the *channel density*. As a fractional value for *channel density* is unfeasible, $W_R = \lceil W_f \rceil$ is the *lower bound* on the *channel density*. At this stage, the integer constraints are reimposed and the ILP is solved for a feasible 0-1 solution with the *channel capacity* equaling the *lower bound* value of *channel density*.

Interior Point Method continues with a series of iterations following an interior point path using a quadratic objective function which prefers integral solutions over the fractional ones. Depending upon the computational resources available, the number of iterations allowable is decided. If a feasible solution is not reached by then, the *channel capacity* is increased by unity and the iterations are continued. This cycle continues until an integer solution is reached.

In brief, following steps are involved in the solution for the *over-the-cell* routing problem using Interior Point Methods:

1. Initially start with a large value W_0 for *channel density*.
2. Minimize W by relaxing the integer constraints of the ILP and solve the relaxed LP problem to get the fractional value of lower bound on the channel density, W_f
3. Round off W_f to its immediate higher integer $W_R = \lceil W_f \rceil$ and try to obtain a feasible integer solution with *channel capacity* W_R.
4. If a feasible solution is not obtained within a reasonable time, increment W_R by unity. Loop back to step 4 until a feasible solution is obtained.

We are at present trying to use other 0-1 Integer Programming solution methods like *branch and bound* methods. This will eliminate the necessity of finding an interior point initially and the integer solution can be directly obtained.

5 Physically Routing the Segments Over-the-cells

After the net segments to be routed *over-the-cell* are identified, they have to be physically routed *over-the-cells*. This problem can be solved in two steps:

1. Sort out all the net segments allotted to *over-the-cell* area, on lower(upper) side of the channel according to their *horizontal span lengths*.

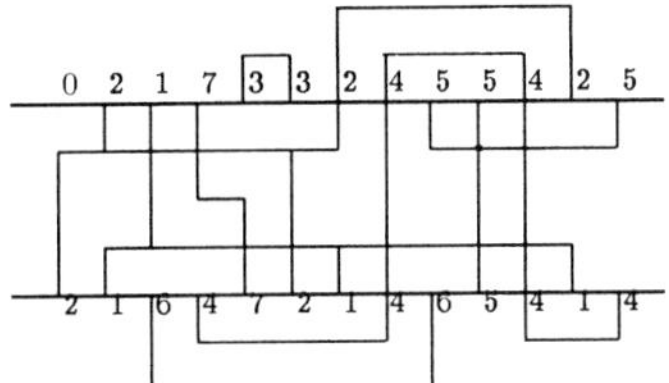

Fig. 3 Solution to the problem in Fig.2(a)

2. Route all the net segments over-the-cell on lower(upper) side of the channel with the shortest nets routed first. For any net segment, try to allot a track closest to the channel boundary but which does not have any other net segments allotted to it within the *horizontal span* of the segment under consideration. Any net segment is always allotted only to a single track.

To maintain the *planarity* of the net segments routed over the cells, the horizontal span of the net segments in tracks farther from channel boundary should cover the *horizontal span* of the net segments in the nearer tracks. Hence, the net segments allocable to the inner(nearer to channel boundary) tracks are always shorter than the net segments in the outer tracks. Hence, the shorter nets are routed first. Above mentioned method always routes all the net segments within the minimum number of tracks over the cells on lower(upper) side of the channel. Fig.3 shows the solution for the problem of Fig.2(a).

6 *Greedy* Heuristic for Improvement

Not allowing *horizontal detouring* may lead to sub-optimal results in some cases. A 'greedy' heuristics which tries to further reduce the *channel density* by introducing some *horizontal detours* (only if it helps in reducing the channel density) is discussed.

In Fig. 4(a), the solution of Fig. 3 is reproduced with *street capacity* of 2 tracks for all columns *over the cells* on both sides of the channel. The entire routing area including the *channel* and the *over the cell* areas are bounded by a *'routing box'* as shown. Another set of lines, called *street exhaustion lines* are introduced on both sides of a column, where the *street capacity* of the column is exhausted by the already routed net segments. These lines are drawn vertically from the channel boundary of the particular column up to the boundary of the *'routing box'*. These vertical lines shown in the figure indicate that no more net segments can be routed over the cell, across these columns.

All net segments routed *over the cell* and the vertical *street exhaustion lines* divide the total over-the-cell routing area, into a number of non-overlapping *zones*. Two such *zones* are shown in the figure by shading them. The physical significance of a *zone* is that the pins sharing a common *zone* can be interconnected *over the cell*, using a net segment. Each pin can belong to a maximum of two *zones*. The pins which do not share a common *zone* can not be connected *over the cell*.

The *'greedy heuristic'* employed mainly involves the identification of net segments in the most dense areas *within the channel* and try to route them *over the cell*. Each such step results into an introduction of a *horizontal detour*. At any stage, any k-terminal net is connected by $k-1$ net segments forming a tree of k nodes. Removal of any net segment results into two *subtrees*. These *subtrees* can be re-connected by connecting any two nodes, one from each *subtree*. The idea is to identify the net segments allotted

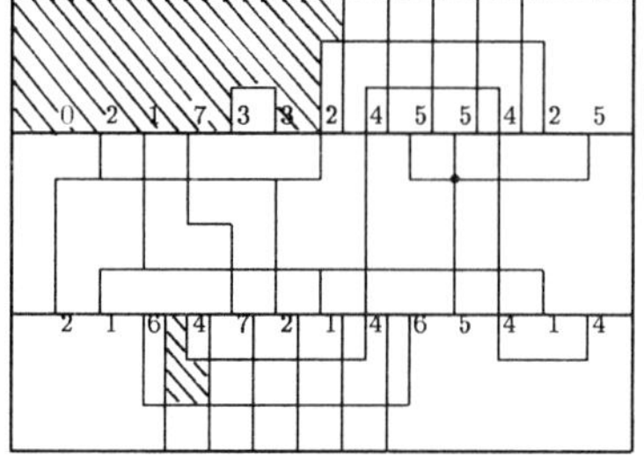

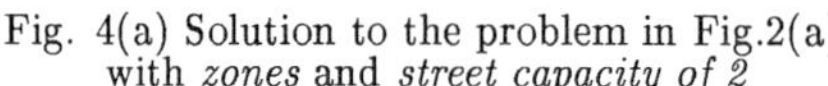

Fig. 4(a) Solution to the problem in Fig.2(a) with *zones* and *street capacity of 2*

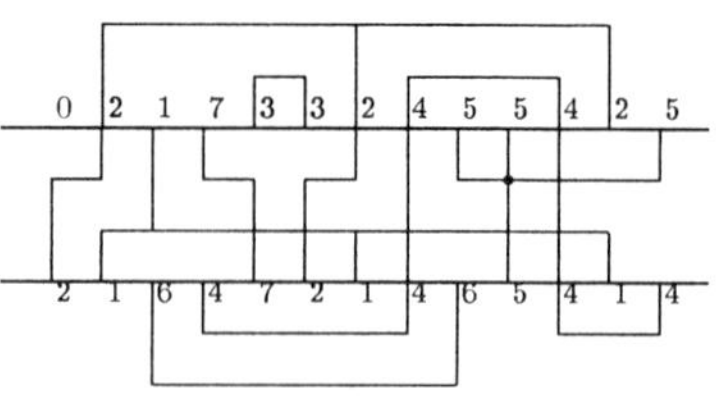

Fig. 4(b) *Improved* Solution to the problem in Fig.2(a)

to *within the channel* which contribute to the *maximum channel density* of the problem, and then try to interconnect the two *subtrees* formed, by a segment *over the cell.*

The procedure adopted involves the following steps:

1. Divide the *routing area* into *zones* as described earlier.
2. Find out all the columns having *maximum channel density.*
3. Identify all the net segments allotted to *within the channel,* crossing the columns of *maximum channel density.* Sort all such net segments into a list L, according to the number of *maximum channel density columns* they cross.
4. If the list L is not empty, remove a net segment s on the top of the list L from the *within the channel.* Otherwise, exit from the *improvement heuristic.*
5. Find out whether the two *subtrees* formed because of the *removal of net segment* s can be connected by a net segment *over the cell.* If it is not possible, re-allot net segment s back to *within the channel,* and remove the segment s from the top of the list L, and go to step 4.
6. Route the segment s *over the cell,* update the *zones* and go to step 2.

7 Experimental Results

The present *over-the-cell router* was tried on all the benchmark problems available from [9]. The results before and after the *improvement heuristics* are tabulated in Table I. It can be noticed that the *lower bound* on the *channel density* is quite tight as in most of the cases the ILP solution was obtained with the *channel capacity* equaling the *lower bound* value. The *channel router* employed, viz. EHCR[7], is found to be quite capable as all the problems were solved with *channel widths* equal to the *channel densities* of the *resulting channel routing problems.* However, the effectiveness of the formulation of *over-the-cell* routing as an Integer Programming problem does not depend upon the efficiency of the channel router.

Table II shows computing times required by the routers on the benchmark problems on SUN 3/260 workstation. The number of variables in ILP formed for each problem indicate that the number of variables for each problem is quite manageable even for reasonably large problems like Deutsch Difficult Example. The computing time required by our *over-the-cell router* is large when compared to [1]. But the increase in the computational time is more than justified by the substantially better results as shown in the Table III.

Ex.	Density	Without Improvement Heuristics					After Improvement Heuristics			
		Lower bound (LP)	Channel Density (ILP)	upper tracks	lower tracks	Channel solution (EHCR)	Channel Density	upper tracks	lower tracks	Channel solution (EHCR)
YK1	12	8	8	2	3	8	7	3	4	8
YK3a	15	11	11	3	4	11	11	3	4	11
YK3b	17	10	11	4	4	11	11	4	4	11
YK3c	18	12	13	3	3	13	13	3	3	13
YK4b	20	13	13	3	4	13	12	4	5	12
YK5	20	14	14	5	2	14	13	6	2	13
deut	19	15	16	4	4	16	15	5	4	15

TABLE I Results on some benchmark problems

Ex.	# variables	Computational time required			
		Relaxed LP Soln	ILP soln	*greedy* heuristic	EHCR
YK1	45	1.6	40.6	0.7	1.1
YK3a	67	3.1	61.2	0.9	2.2
YK3b	60	2.9	53.1	1.0	1.9
YK3c	73	6.1	55.3	1.1	2.0
YK4b	117	10.6	98.9	1.6	3.6
YK5	84	9.0	70.3	1.3	3.7
deut	153	21.0	165.8	2.3	5.9

TABLE II Computational details on Benchmark Problems

Ex.	Density	Solutions in [1]		Our Solutions		% improvement		
		#tracks in chan.	#tracks (total)	#tracks in chan.	#tracks (total)	over 2L solution	over [1] #tracks (chan)	over [1] #tracks (total)
YK1	12	10	17	7	14	33.3%	30.0%	17.6%
YK3a	15	12	21	11	18	26.7%	8.33%	14.3%
YK3b	17	13	20	11	19	35.3%	15.4%	5.0%
YK3c	18	15	22	13	19	27.8%	13.3%	13.6%
YK4b	20	16	25	12	21	40.0%	25.0%	16.0%
YK5	20	14	21	13	21	35.0%	7.1%	0.0%
deut	19	17	32	15	24	21.5%	11.8%	25.0%
Avg.						31.4%	15.8%	13.1%

TABLE III Comparison of the results with [1]

The performance of our router is consistently better than those of [1]. On an average, our router resulted in a savings of 31.4% over optimal 2-layer routing and 15.8% over the solutions of [1]. If the tracks *over the cell* and also *within the channel* are considered, on an average, our router resulted in a savings of 13.1% over [1]. It is significant that the number of net segments routed over the cell by our router is significantly less than that of [1] but still then the results are consistently better.

In the case of Deutsch Difficult Example[2], the *lower bound* was found to be 15. But ILP solution through Interior Point Method could only give a solution with *channel density* of 16. However, this result was *improved* by the *improvement heuristics* used and the solution was obtained with a *channel capacity* of 15. The best known result[1] reported so far, on this problem required 17 tracks. This solution resulted in a savings of 21.5% over optimal 2-layer routing and 12.8% over the solution of [1]. Moreover, it resulted in a savings of 25% over [1], if the total number of tracks *over-the-cells* as well as *within the channel* are considered. Our solution for Deutsch Difficult Example is shown in Fig. 5 on the next page.

References

[1] Cong J., Liu C.L. *'Over-the-cell Channel Routing'*, IEEE Trans. on CAD, vol. CAD-9., no.4, Apr 1990, pp 408-418.

[2] Deutsch D.N. *'A Dogleg Channel Router'*, Proc. 13th Design Automation Conference, 1976, pp 425-433.

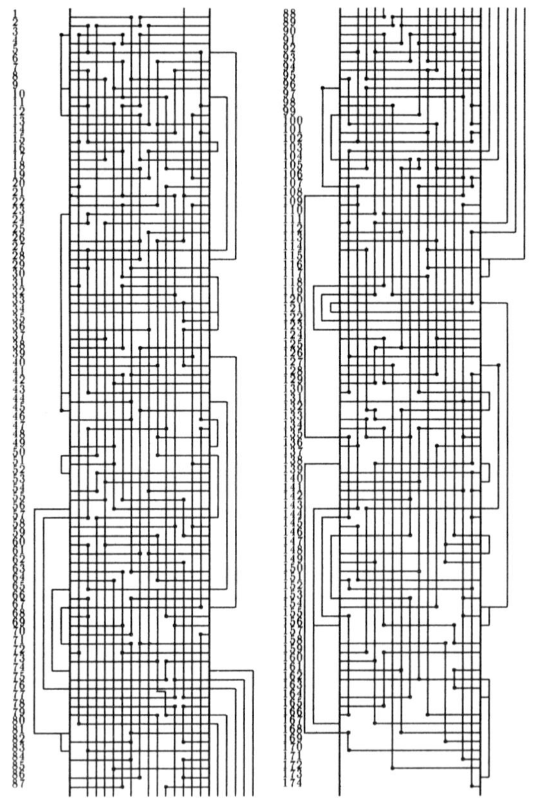

Fig. 5 Solution for Deutsch Difficult Example

[3] Deutsch D.N., Glick P. *'An Over-the-cell Router'*, Proc. 17th Design Automation Conference, 1980, pp 32-39.

[4] Karmarkar N.K. *'A New Polynomial-time Algorithm for Linear Programming'*, Combinatorica, 4: 1984, pp 373-395.

[5] Karmarkar N.K., Ramakrishnan K.G., Resende M.G.C. *'An Interior Point Algorithm for 0-1 Integer Programming'*, Internal report, AT&T Bell Laboratories, Murray Hill, New Jersey.

[6] Krohn H.E. *'An Over-the-cell Gate-Array Channel Router'*, Proc. 20th Design Automation Conference, 1983, pp 665-670.

[7] Pai R.R., Rao S.S.S.P.R. *'A Channel Router Based on Efficient Heuristics'*, Proc. 2nd International Workshop on VLSI Design, Bangalore, Dec 1988, pp 28-42.

[8] Shiraishi Y., Sakemi Y. *'A Permeation Router'*, IEEE Trans. on CAD, vol. CAD 6., no.5, May 1987, pp 462-471.

[9] Yoshimura T., Kuh E.S. *'Efficient Algorithms for Channel Routing'*, IEEE Trans on CAD, vol. CAD-1., no.1, Jan 1982, pp 25-35.

VLSI 91
A. Halaas and P.B. Denyer (Eds.)
Elsevier Science Publishers B.V. (North-Holland)

Switchbox routing by pattern matching

M. Starkey[a] and T. M. Carter[a]

[a]Department of Computer Science, University of Utah, Salt Lake City, UT 84112, U.S.A

Abstract

Many good algorithms exist that provide good solutions to the wire routing problem in VLSI. Unfortunately, many of these consider only a small subset of different parameters such as number of layers, routability of layers and technology. We believe that most routing algorithms can be applied generally using a set of parameters and a set of standard routing tools to implement them. We note that many routing algorithms use patterns directly or can easily be converted to use patterns. We present a powerful formalism for describing these patterns.

1. INTRODUCTION

The wire routing problem (finding the best, least expensive path from one point to another) has been around for many years in a number of different forms. In computer science this problem is known as the Traveling Salesperson Problem. Wire routing adds another twist to this problem. Instead of having just one tour of the nodes, there are a number of tours which must use the same set of available paths. No path can be shared among tours and generally a lower total cost of all tours is more important than the lowest cost of any individual tour. The problem is NP-Complete. However, many good algorithms have been developed to provide good solutions with reasonable computational complexity.

A typical method used to find solutions quickly is to use the constraints imposed by the environment to reduce the size of the problem. For example, assumptions are often made about the number of available layers and the directions routing occurs on these layers. Unfortunately, these assumptions make applying an algorithm to a different set of parameters almost impossible. By providing a flexible description language, many routing algorithms can easily be used in many different environments. This description must be powerful enough to allow the algorithm to be implemented easily without an undue amount of computational overhead. The method we believe to be the most flexible in this regard involves patterns. Patterns have been used in a number of different algorithms and are found in some of the best existing algorithms.

The most closely related work was presented by Soukup and Fournier[1] in which the regularity of a circuit design and a pattern router are used. They developed a simple pattern description language and provide results for routing printed circuit boards. WEAVER[2] has provided some of the best results in switchbox routing. This knowledge-based routing system has a number of "experts" that implement different phases of wire routing. One of these experts is the pattern router expert which attempts to find and route certain patterns. The BEAVER[3] router is also targeted towards routing switchboxes. It attempts to interconnect pins using certain patterns, some of which are given priority over others. It provides good results on the classic examples. In addition to these direct implementations of patterns, Burstein and Pelavin use small patterns that describe all possible 2xN Steiner trees in their Hierarchical Routing algorithm[4].

These approaches to using pattern matching to route wires show that patterns have value in finding good solutions in reasonable amounts of time. We expand on this by proposing that many routing algorithms can be implemented in terms of patterns. Patterns increase the flexibility and power of routing algorithms by allowing them to operate in many different environments. We present a formal pattern grammar and some powerful properties of the chosen description method.

2. ENVIRONMENT INDEPENDENCE

A number of good routing algorithms exist that have been applied to problems in environments that restrict the number of routing layers available and the directions used on these layers. There are many different permutations of these constraints. The goal is to utilize good routing algorithms in any environment. Some of the additional flexibility required involves varying the numbers of layers, the technology cost factors (e.g., for GaAs) and routing directions (including non-Manhattan routing).

Most algorithms are designed around a process which permits two routing layers. There are then two ways to allocate the tracks and columns for routing. The simplest of these constrain one layer to be horizontal and another to be vertical. Jogs in wires can only be added by changing layers twice. The extra vias thus introduced are often undesirable; this leads to preferred directions for a given layer with exceptions made for small jogs or doglegs.

Introducing a routing algorithm designed for one set of assumptions in another environment with different assumptions may be detrimental to routing a circuit. If an algorithm expecting to make same-layer jogs is placed in an environment where doing so is impossible, it could conceivably fail when it tried to avoid an obstacle. Similarly, if an algorithm was introduced to an environment where the directions were only preferred then the result may not be as compact as possible and may contain an unacceptable number of vias.

These differences require rewriting a routing algorithm to be flexible in the number of layers and the directions allowed on each layer and possibly the relative cost of using certain layers over others. This is facilitated by using a powerful pattern description representation.

In addition to having a pattern description method, there must be a method of interacting with the underlying design system. Four functions are defined to provide this interface. The $owner$ function '$owner(x_1, y_1, z_1, x_2, y_2, z_2)$' returns the net to which the straight segment between these points has been allocated or 0 if it is not allocated. In traditional routing terminology, the x values represent the columns, the y values are the tracks and the z values correspond to the layer number. Wire segments can be unavailable if they are eliminated due to wires or other obstacles already placed. These obstacles must be routed around, $owner$ will return an invalid net indicator in such cases.

Three additional functions are also required. '$allocate(net, x_1, y_1, z_1, x_2, y_2, z_2)$' allocates the segment between (x_1, y_1, z_1) and (x_2, y_2, z_2) to net, '$free(x_1, y_1, z_1, x_2, y_2, z_2)$' will deallocate the segment and '$cost(x_1, y_1, z_1, x_2, y_2, z_2)$' returns the cost of the wire segment.

The $cost$ function is very important in reducing the search space of the problem and in ensuring that a lowest cost route is obtained. In wire routing, there are a number of costs which are associated with different solutions to interconnecting an entire circuit. These costs are dependent on physical properties of the routing media. Capacitance and resistance play a major role in circuits. Limiting the number of vias used is also important as is minimizing the number of jogs in a wire.

3. PATTERN DESCRIPTIONS

Patterns offer a very powerful method with which to describe algorithms. Patterns may describe everything between a single segment of a wire and a complex path.

3.1. Describing Patterns as Regular Expressions

Regular expressions are a powerful, natural way of describing patterns. For example,if a string of symbols is desired which starts with a **1** followed by one or more **0**s then any number of **0**s or **1**s finally terminated by a **1**, then it can be simply described as $\mathbf{10^+(0+1)^*1}$. The pattern **10001** satisfies this regular expression but **110** does not. Although the acceptance of strings for this regular expression can be analyzed by inspection, regular expressions can become complex and acceptance may require the use of large state machines.

Regular expressions utilize some operations that facilitate describing complex expressions: the Kleene star (*) which permits zero or more replications of the symbol, the plus ($^+$) which signifies one or more replications of the symbol and the or symbol ($+$) which provides alternation among symbols. In addition, parentheses identify sub-expressions.

In wire routing, regular expressions are not only used to accept patterns but also define how the pattern should be generated. Therefore, information must be encoded in the characters used for the above operations. The semantics of the logical or and the parentheses remain, the changes affect the Kleene star and the plus. When running wires in certain directions, we usually wish to maximize or minimize the length of the wire in that direction. For example, vias are usually minimized. Other directions may be preferred and therefore maximized. This information is encoded in the symbols $^<$ and $^>$. The $^<$ symbol replaces the Kleene star and is defined to allow zero or more replications of the symbol, trying to minimize its use. The $^>$ is similar to the plus, there must be one or more replications of the symbol and its use should be maximized. Naturally, the * and $^+$ characters could be used but the new symbols clarify the goals of the regular expression in terms of its application to routing.

Traditionally, only six directions are used for wires: horizontal tracks (East/West), vertical columns (North/South) and vias (Up/Down). This set is enough for most routing environments, however, other symbols can be easily added to the alphabet for any additional directions or any other distinctions between directions. Pattern regular expressions have the alphabet $\Sigma = \{\mathbf{N}, \mathbf{S}, \mathbf{E}, \mathbf{W}, \mathbf{U}, \mathbf{D}\}$.

An example of a regular expression is one that represents a step of the "Greedy" channel router[5]. As this algorithm proceeds from left to right column by column, each point on the top and bottom of the channel in the current column must be brought into the channel only as far as required to continue rightward on a track. The description which can be converted to a regular expression is "minimize the wire segments in the north or south direction (depending on whether the connection is on the bottom or top) where there is an available track which allows the connection to extend to the right". As a regular expression this becomes $\mathbf{(N+S)^<(U+D)^<E}$.

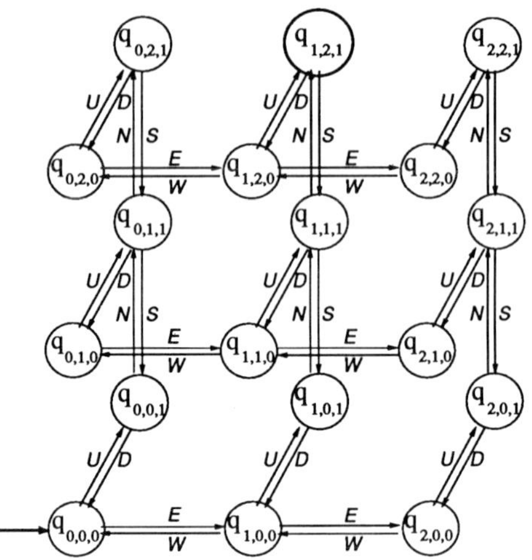

Figure 1: The DFA corresponding to the bounds $0 \leq x \leq 2$, $0 \leq y \leq 2$, $0 \leq z \leq 1$ where only column wires are allowed on layer 1 and track wires on layer 0. The arrow indicates the start state for a string starting at $q_{0,0,0}$ and proceeding to $q_{1,2,1}$.

The $(\mathrm{U} + \mathrm{D})^{<}$ is required since a layer change may be necessary between directions. This direction change will be minimized and will generate no symbols if no layer changes are required.

Using regular expressions provides a very simple, concise and understandable method to describe patterns. Regular expressions also have some other desirable properties that can be exploited. Regular expressions are closed under certain operations. These operations include replicating expressions, extracting sub-expressions, concatenating multiple expressions and combining expressions into more complex expressions. These closures ensure that the resulting regular expression is also regular and therefore can be accepted by a Deterministic Finite Automaton. These operations are extremely useful for regular expressions used to represent wire routing directions. Hopcroft and Ullman provide an excellent description of regular expressions and Deterministic Finite Automata in [6].

3.2. Accepting Patterns With Deterministic Finite Automata

The acceptance of strings of symbols matching certain regular expressions is performed by a Deterministic Finite Automaton (DFA). A DFA is defined by a 5-tuple, $(Q, \Sigma, \delta, q_0, F)$. The state space, Q, is a set of all states in the DFA. Σ is the alphabet whose symbols form the string to be accepted. The δ function describes the transition from one state to another beginning with the start state, q_0, until one of the final states in the set F is reached.

For routing wires, the obvious way to build the DFA is to have the arcs represent wires in the circuit. Therefore, if a transition is taken from one state to another, the arc used represents the wire which is allocated to the net being routed. Figure 1 shows the DFA for a traditional routing environment. This environment has wires running horizontally on one layer and vertically on a second layer. Layers can be connected by running vias between them. To clarify the correspondence between position in the circuit and the wires being allocated, a state is identified as $q_{x,y,z}$ where x is the column, y is the track and z represents the layer.

Each of the components of the DFA $M = (Q, \Sigma, \delta, q_0, F)$ can now be described. To increase the power of the DFA, a bounding box of the area defined as $[x_{min}, x_{max}]$, $[y_{min}, y_{max}]$, $[z_{min}, z_{max}]$ must be provided. Also the node where the route is to start must be specified. Initially, the DFA will be described for finding point to point connections. This constraint will be relaxed to allow routed wires to stop when a segment of the same net is reached. Also, the initial definition will only provide for a pattern using unallocated segments, this constraint will also be relaxed.

The state space consists of all of the states which are nodes in the routing area. More formally, $Q = \{q_{x,y,z} | x_{min} \leq x \leq x_{max}, y_{min} \leq y \leq y_{max}, z_{min} \leq z \leq z_{max})\}$. The alphabet is the same alphabet as for the regular expressions, $\sum = \{\mathbf{N}, \mathbf{S}, \mathbf{E}, \mathbf{W}, \mathbf{U}, \mathbf{D}\}$. The δ function describes which arcs are present for routing the current net. The *owner* function, described previously, is required to find the owner of an arc. The δ function then becomes,

$$\begin{array}{lclcl}
\delta(q_{x,y,z}, \mathbf{E}) & = & \{q_{x+1,y,z} & | & owner(x,y,z,x+1,y,z) = 0\} \\
\delta(q_{x,y,z}, \mathbf{W}) & = & \{q_{x-1,y,z} & | & owner(x,y,z,x-1,y,z) = 0\} \\
\delta(q_{x,y,z}, \mathbf{N}) & = & \{q_{x,y+1,z} & | & owner(x,y,z,x,y+1,z) = 0\} \\
\delta(q_{x,y,z}, \mathbf{S}) & = & \{q_{x,y-1,z} & | & owner(x,y,z,x,y-1,z) = 0\} \\
\delta(q_{x,y,z}, \mathbf{U}) & = & \{q_{x,y,z+1} & | & owner(x,y,z,x,y,z+1) = 0\} \\
\delta(q_{x,y,z}, \mathbf{D}) & = & \{q_{x,y,z-1} & | & owner(x,y,z,x,y,z-1) = 0\}
\end{array}$$

The start state, q_0, is $q_{x,y,z}$ where the node at (x, y, z) is where the route is to start. The final state at node (x_f, y_f, z_f) will be the only member of the set $F = \{q_{x_f,y_f,z_f}\}$.

The DFA defined above demonstrates a route which will only follow unused segments of wire. Usually, a connection is made if the current net being routed, *current_net*, connects to a wire segment already allocated to this net. Then the test in each δ function can be relaxed to be $owner(x, y, z, x', y', z') = 0$ or $current\_net$. The set of acceptable final states becomes $F = \{q_{x_f,y_f,z_f}\} \cup \{q_{x,y,z} | owner(x, y, z, x, y, z) = current\_net\}$.

This DFA is very powerful for accepting patterns in the routing environment. It is very general and can be applied to any physical constraints. The *owner* function provides the information on the directions which can be used on certain layers. The use of the DFA aids in reducing the search space required for pattern string generation from regular expressions. It also allows a string of symbols to be accepted or not accepted for allocating or deallocating wire segments as well as calculating a total cost for all segments allocated.

3.3. Pattern String Generation

Deterministic Finite Automata are useful as accepters of strings of symbols ($\sum^*$). Therefore, some method is required to generate these strings. The string generation is performed by the traversal of a tree corresponding to the regular expression. The DFA helps to limit the search of this tree by providing information which prunes the tree.

Pruning is essential for efficiency; the best pruning method is to make the regular expression as descriptive as possible. Occasionally this is not desirable if a number of patterns or variations on a pattern are to be compared to find a best solution.

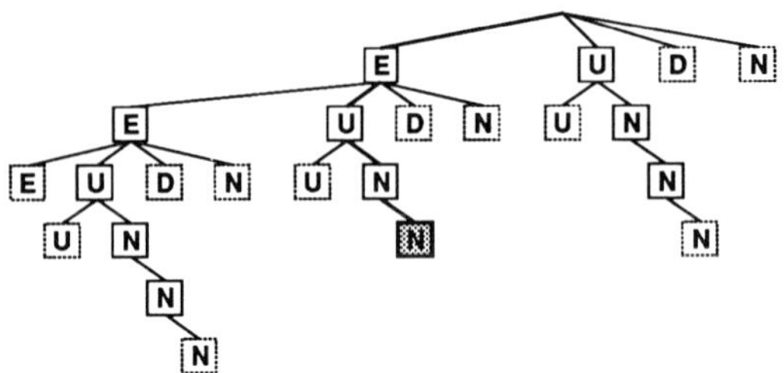

Figure 2: Search tree for the pattern $\mathbf{E}^{<}(\mathbf{U}+\mathbf{D})^{<}\mathbf{N}^{<}$ starting at $q_{0,0,0}$ and ending at $q_{1,2,1}$. Dashed boxes indicate pruned nodes, the shaded box is the solution node.

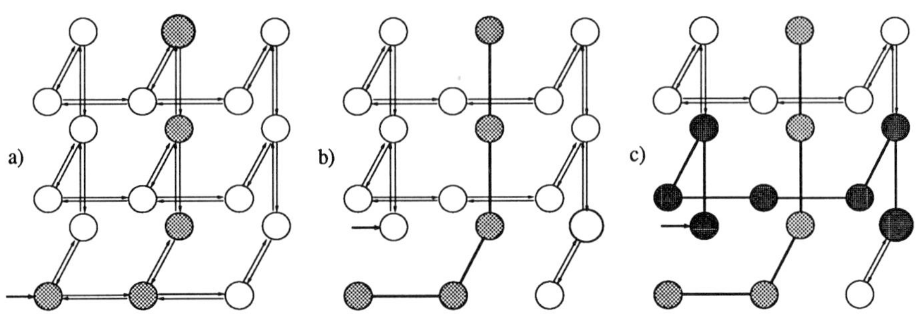

Figure 3: The DFA showing the path from figure 2 and the subsequent paths.

Another method of pruning is the branch and bound method[7]. Finding a method to calculate a reasonable lower bound for each node of the current configuration is difficult, the actual cost can be easily calculated using the *cost* function defined previously. The lower bound on the cost can be obtained by finding the smallest distance possible along the available routing directions if none of the wire segments along that path were allocated. This can provide a method of deciding which of a number of paths should be attempted.

In addition to these methods, the states in the DFA are also used to prune the tree. As the tree is traversed, each node visited will correspond to a state in the DFA. The tree can be pruned if the DFA does not have an arc that permits the corresponding traversal.

Finally, there is one additional method of pruning where a number of solutions exist and a global physical constraint is known. As the traversal of the tree proceeds, the cost of the wires required to implement the symbols is accumulated. If this cost exceeds some desired maximum along any branch, it can be pruned since any further allocation along that path will only increase the cost.

As an example of this tree, assume that the first net must start at $q_{0,0,0}$ and reach $q_{1,2,0}$ with the pattern specified by $\mathbf{E}^{<}(\mathbf{U}+\mathbf{D})^{<}\mathbf{N}^{<}$. There is only one solution to this pattern in this environment demonstrated by figure 2. This is allocated in the routing area (figure 3a) and forms an obstacle to the second net. The obstacle is handled by the DFA described above. This

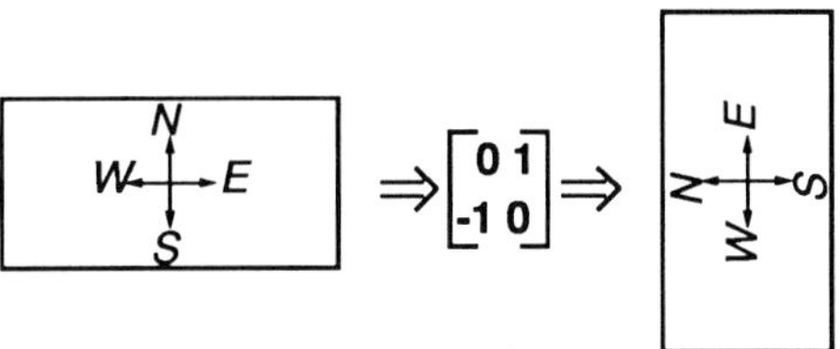

Figure 4: Transforming the desired routing area to the real area.

obstacle essentially eliminates the arcs which have been implemented as wires to make the first connection as well as arcs which are directed away from or towards any nodes along the path of the first route.

The second net now starts at $q_{0,0,1}$ and tries to connect to $q_{2,0,1}$ with the pattern $\mathrm{N}^{<}(\mathrm{U}+\mathrm{D})^{<}\mathrm{E}^{<}(\mathrm{U}+\mathrm{D})^{<}\mathrm{S}^{<}$. If the first net had not already been allocated as shown in figure 3b then the solution would be **DEEU** however, this is not possible since the arc from $q_{0,0,0}$ to $q_{1,0,0}$ is not available. Therefore, the resulting pattern is **NDEEUS** shown in figure 3c.

4. ABSTRACTION

Using patterns and regular expressions has another benefit: the abstraction of physical information, permitting mathematical manipulations of the real routing area. Seemingly different routes may be achieved in an identical manner simply by transforming the routing area into that which is understood by a given algorithm and then back to the physical domain.

There are two ways of providing this transformation. One method essentially transforms the points that are to be connected, transforming the DFA and the *owner* function, performing the route and then transforming the wire segments to their proper position. This method is efficient unless obstacles exist in the routing area. If there are obstacles, then the DFA is difficult to transform. This difficulty is caused by the *owner* function since it must transform all of the obstacles. The shapes of these obstacles in terms of the wire segments that they invalidate is unknown to the router, and can only be discovered by trying to allocate wire segments that cross the obstacle.

A better method is to incorporate the transformation in the symbols. For example, if N is defined as a unit vector in the direction $(0,1,0)$ then a 90 degree rotation of this would produce a unit vector $(-1,0,0)$. Figure 4 shows the transformation of the entire set of directions. Since this is only a two-dimensional rotation, the symbols for changing layers are not transformed. The identical patterns and regular expressions using them can then be used.

Another useful feature which the patterns abstract is that each symbol represents a unit length segment. The units may be different for different layers or different directions. For example, the N symbol on one layer may have a different length than a N on another layer. In addition the N symbol may represent a unit length on each layer in a different direction. There is also no requirement that the opposite direction on the same layer must have the same unit length.

5. EXAMPLES IN ALGORITHMS

Three algorithms are chosen here to illustrate how simply patterns can be used. These algorithms provide a small cross-section of different types of algorithms.

5.1. "Greedy" Channel Router[5]

This algorithm is designed to route a channel - a routing area with all of the connections on the top and bottom of the bounding area. The algorithm is not described with explicit patterns, but it can be easily converted to use patterns. Starting from the left for each column in turn until the rightmost column is reached, do the following:

1. Bring the connections on the top and the bottom of the channel into the routing area. The connection should be brought to the closest track which ensures that the track which the new point is on can be extended to the right. If an available track cannot be found then insert a track. The pattern used here is simple. The directions up or down, depending on whether the connection is on the top or the bottom, should be minimized. The pattern $(\mathbf{N}+\mathbf{S})^{<}(\mathbf{U}+\mathbf{D})^{<}\mathbf{E}$ also checks that a segment can be placed in a track. This pattern can then be intersected with the pattern $(\mathbf{N}+\mathbf{S})^{<}(\mathbf{U}+\mathbf{D})^{<}$ to remove the symbol which verifies that the track is available, from the initial string.
2. Collapse any nets which may be occupying two different rows in the same column by connecting them if possible. The two points which are to be collapsed will be provided to the pattern matcher as it tries to maximize the number of symbols between these two points. The pattern used is $(\mathbf{U}+\mathbf{D})^{<}(\mathbf{N}+\mathbf{S})^{>}(\mathbf{U}+\mathbf{D})^{<}$. The layer changing symbols are required in case the two points are on different layers.
3. Shrink the number of rows between connections on the same net in the same column which could not be collapsed. The idea is to take the connections as close to each other as possible since they could not be connected. The pattern $(\mathbf{N}+\mathbf{S})^{>}(\mathbf{U}+\mathbf{D})^{<}\mathbf{E}$ with the correct bounding box will perform this.
4. Make preference jogs which attempt to get the wire to a row closer to the next point on the net. The pattern required for preference jogs is the same as for shrinking nets since it requires essentially the same functionality.
5. Extend each wire from this column into the next one to the right. This requires simply adding a single segment. The pattern $(\mathbf{U}+\mathbf{D})^{<}\mathbf{E}$ will change layers if the one it is on does not permit wires in this direction.

5.2. BEAVER[3]

The BEAVER router uses five patterns to interconnect points on the same net in a switchbox. Each connection on the boundary of the switchbox is given some length of the row or column perpendicular to its side. This is unusable by other patterns until a connection is made thereby freeing up some of this segment. The amount of the row or column wire which is reserved for

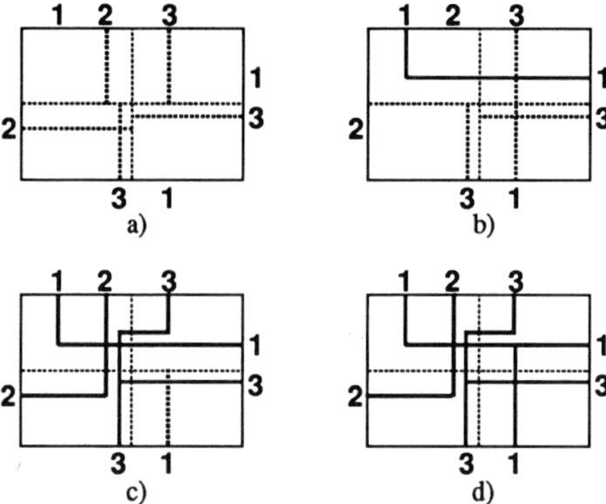

Figure 5: BEAVER algorithm example. a) allocate half of each row or column from any point not in net 1. b) allocate corner connection between two net 1 connections, net 3 precludes the other point being connected. c) allocate nets 2 and 3, with a corner then a stairstep. d) complete net 1 since net 3 has now been routed.

a connection starts at 50%. This number is decreased until the entire route is complete. Figure 5 demonstrates this algorithm.

Start the factor at 50% and keep decreasing until finished. For each value of the factor, perform the following steps for each net:

1. Allocate wire segments for all nets except for the one being routed. The bounding box used for pattern generation is based on the current factor. The patterns are just single direction connections and therefore can be described by $(\mathbf{N}^{>} + \mathbf{S}^{>} + \mathbf{E}^{>} + \mathbf{W}^{>})$. Save these patterns to permit them to be freed later.

2. For the current net, try to connect with corner patterns as many points as possible to other points or wire segments on the net. The description for any corner is $((\mathbf{N}+\mathbf{S})^{<}(\mathbf{U}+\mathbf{D})^{<}(\mathbf{E}+\mathbf{W})^{<}) + ((\mathbf{N}+\mathbf{S})^{<}(\mathbf{U}+\mathbf{D})^{<}(\mathbf{E}+\mathbf{W})^{<})$.

3. For the current net, try to connect any points not connected with corners using straight, dogleg, stairstep or horseshoe patterns which are all described by the regular expression: $((\mathbf{N}+\mathbf{S})^{<}(\mathbf{U}+\mathbf{D})^{<}(\mathbf{E}+\mathbf{W})^{<}(\mathbf{U}+\mathbf{D})^{<}(\mathbf{N}+\mathbf{S})^{<})$
$+((\mathbf{E}+\mathbf{W})^{<}(\mathbf{U}+\mathbf{D})^{<}(\mathbf{N}+\mathbf{S})^{<}(\mathbf{U}+\mathbf{D})^{<}(\mathbf{E}+\mathbf{W})^{<})$.

4. Free all wire segments used to reserve columns and tracks for unconnected points.

5.3. The Lee Algorithm[8]

This algorithm is one of the oldest and also one of the most widely used. It is usually incorporated inside more complex algorithms to provide optimal point-to-point connections. The algorithm starts at a point and propagates outward one unit in each direction if possible. A cost is associated with each of these new locations and is the cost of the initial location plus the additional cost required to get to the new location. This is continued until the destination point is reached. By backtracking along the path, always taking the lowest cost direction, the optimal path is assured.

This algorithm is defined by the simple pattern $(\mathbf{N}^< + \mathbf{S}^< + \mathbf{E}^< + \mathbf{W}^< + \mathbf{U}^< + \mathbf{D}^<)^<$. This demonstrates the need for good pattern tree pruning, otherwise the search space is very large.

6. CONCLUSIONS

This powerful pattern description and matching system can be used for many existing algorithms. These algorithms are not restricted to methods that already are described with patterns. The use of these patterns will produce more flexible algorithms that can take advantage of parameters for the layout being performed.

In addition, once these algorithms are implemented with a common basis in patterns, some of the similarities and differences will become apparent. These comparisons will enable algorithms to be analyzed to possibly discover why some methods perform better than others in certain situations.

Apart from comparing different algorithms, the effects of using different patterns within the same algorithm framework can be studied. By keeping the control and strategy mechanisms constant and changing the patterns, perhaps better results in some situations can be achieved. The patterns can then be specified by the user for a specific route while maintaining the goals and functionality of the original algorithm. This will permit algorithms to be easily tuned to different examples or different environments.

7. REFERENCES

[1] J. Soukup and S. Fournier, “Pattern router,” in *ISCAS*, pp. 486–489, 1979.

[2] R. Joobbani and D. P. Siewiorek, “Weaver: A knowledge-based routing expert,” *IEEE Design and Test of Computers*, vol. 3, pp. 12–23, February 1986.

[3] J. P. Cohoon and P. L. Heck, “Beaver: A computational-geometry-based tool for switchbox routing,” *IEEE Trans. on Computer-Aided Design*, vol. 7, pp. 684–697, June 1988.

[4] M. Burstein and R. Pelavin, “Hierarchical channel router,” in *20th Design Automation Conference*, pp. 591–597, 1983.

[5] R. L. Rivest and C. M. Fiduccia, “A “greedy” channel router,” in *19th Design Automation Conference*, pp. 418–424, 1982.

[6] J. E. Hopcroft and J. D. Ullman, *Introduction to Automata Theory, Languages, and Computation.* Addison-Wesley Publishing Company, 1979.

[7] E. Horowitz and S. Sahni, *Fundamentals of Computer Algorithms.* CS Press, 1978.

[8] C. Y. Lee, “An algorithm for path connections and its applications,” *IRE Transactions on Electronic Computers*, vol. 10, pp. 346–365, September 1961.

VLSI 91
A. Halaas and P.B. Denyer (Eds.)
Elsevier Science Publishers B.V. (North-Holland)

GPFP: A SIMD PE for Higher VLSI Densities

Don Beal and Costas Lambrinoudakis

Department of Computer Science, Queen Mary and Westfield College, University of London, Mile End Road, London E1 4NS, United Kingdom

Abstract

VLSI advances continue to pack more and more circuitry onto individual chips, but pinout technology fails to keep pace. SIMD array processors such as the DAP, MPP and Connection Machine are examples of architectures that become seriously pinout limited if existing machine designs are simply scaled to higher circuit densities on chip.

The essential difficulties of scaling simple SIMD machines are discussed. Existing architectural trends are presented briefly and the ***General Purpose with Floating Point*** support (GPFP) processing element, which uses the expansion of circuitry from VLSI advances to provide on-chip memory, cost-effective extra functionality and a suitable degree of local autonomy, is discussed in detail. The GPFP design accelerates floating point arithmetic, while retaining the architectural aims to optimise cost-effectiveness and general purpose capability.

Performance is dependent on technology, but readily available current processes (1.5 micron CMOS) give over 2.5 GigaFlops from a 32x32 array.

1. INTRODUCTION

Simple minded 1-bit processors used in SIMD (Single Instruction Multiple Data) architectures, like the AMT DAP [1-2], the Goodyear MPP [3] and the General Electric GRID [4], can be conveniently fabricated as an array of PEs on a single VLSI chip, connected to external memory build from conventional memory chips. With this arrangement, it is essential that each PE has its own data line to memory. As VLSI technology advances more components can be integrated into the same chip area [5-6], effectively enlarging the usable area. Although the natural reaction would be to exploit this extra circuitry by fabricating thousands of simple 1-bit PEs on a single chip — which is feasible nowadays (1991) — this does not take account of trends in current IC technology and in particular the divergence between gate and pin count. Moreover, simple geometric reasoning argues against packaging technology ever keeping pace with increasing PE density. If the chip area allows an array of nxn PEs, then the package has to provide $4n + n^2$ pins for PE interconnection: n^2 memory connections and n along each edge. However, the space available for pin connection (basically around the edges of the chip) will at best increase linearly with n, not as n^2.

Alongside the pin problem there are other known limitations/deficiencies associated with the fairly elementary PEs, currently in use, which suggest the necessity for seeking ways other than increasing the number of PEs for utilising the extra circuitry available. More specifically problems like the dominant effect of the long external memory access times on the cycle times that the PEs are clocked, or the inability of elementary 1-bit PEs to execute efficiently several complex operations (like multiplication, or normalisation during floating point operations) [7], are currently imposing severe performance constraints to array machines and therefore require close attention.

Using the higher VLSI densities, architectural progressions like integration of on-chip memory for improving the memory bandwidth or provision of hardware components, like shift registers, for accelerating the execution of more complex operations or even multi-bit PEs, are now feasible. Indeed, examination of recent PE designs like the BLITZEN PE [8] as an improved version of the MPP PE, the COPE design [9] as a coprocessor for the DAP PE, the new DAP CP/8 [10] range and the microprocessor array (μPA) [7] architecture, illustrate the tendency towards more advanced processing units.

The GPFP PE follows this trend abandoning the single-bit nature and also extends the PEs to include the minimum number of support components necessary for the efficient and fast execution of floating point operations. However, since the multi-bit nature introduced runs the risk of sacrificing eminent advantages enjoyed by single-bit PEs [11-12], like low manufacturing cost, precision flexibility and freedom in data organisation, the GPFP design pays careful attention to preserving as much as possible of the general purpose character of simpler machines and also to retaining all the positive points that make SIMD architectures so attractive.

The GPFP design includes an optimised amount of on-chip memory and a powerful local address modification mechanism that is an essential contribution to an appropriate degree of local autonomy for the PEs. This paper discusses some of the design requirements and solutions found. Finally the structure of an array processor which merges the advantages contributed by the multi-bit nature of the processing elements, with those obtained by the conserved ability to perform bit serial operations, is discussed.

All the architectural concerns are important to the continuing evolution of SIMD machines and require effective solutions for the next array processor generation. Some of the architectural issues are introduced to the reader through the discussion of design decisions and implementation details encountered during the development of the GPFP PE. A prototype chip is being fabricated in 1.5 micron custom CMOS which, even with the conservative design standards used, is expected to allow the PEs to execute instructions at more than 50MHz. As processor arrays of 32x32 or larger are readily feasible, this implies number crunching performance of 2.5 GigaFlops upwards.

2. THE GPFP PROCESSING ELEMENT

2.1. On-chip Local Memory

Experimental results produced on a DAP-510 array processor [13], employing a selected variety of real applications, indicate that over 70% of the executed instructions access data in the array memory. Since the memory access times are usually very high, compared with the cycle times that the PEs are capable of operating at, the processor-memory bandwidth becomes a dominant factor for the overall performance of the machine.

The major problem associated with the external array memory is that the data has to travel long distances in order to reach its final destination, which inevitably introduces considerable propagation delays. Even if extremely fast memory is used, the overall memory cycle still suffers the overheads of the lengthy data routing. The preceding problem can be eliminated if memory, local to each processing element, is integrated on the chip used to accommodate the processors. Unfortunately, although the use of on-chip memory can improve considerably the processor-memory bandwidth, it must always be of limited capacity and hence it can not replace the external storage. Consequently new questions concerning the productive use of the on-chip memory are raised.

A possibility is to use this memory as a transparent cache between the processing elements and the array memory. However the complexity introduced into the design, due to the address decoding logic, compares to the complexity of the PE itself and moreover it may even dominate the cycle time of the machine. We therefore feel that as far as array architectures are concerned, an efficient solution is to implement the memory as blocks of directly addressed registers. This high speed storage can be used to accommodate a number of frequently required operands under some static allocation scheme and hence contribute towards an improved overall system performance.

In an attempt to utilise the processor capabilities thoroughly, the on-chip memory for the GPFP PE was designed to be *triple ported*. That enables a PE to fetch two operands from the memory, operate on them and store the result back to the memory on the same cycle. For an existing architecture with no on-chip memory, this would normally take two or three cycles.

Having simulated the operation of the on-chip memory on a DAP-510 array processor, we performed a number of experiments, based on real applications, aiming to highlight the variations of the overall system performance as the capacity of the simulated memory was altered. By analysing the results obtained it was deduced that a suitable choice for the size of the on-chip memory is 512 bits [14]. A PE featuring the above amount of memory will be able to handle efficiently programs that use operands of large precision (e.g. 64-bit floating point numbers). The memory integrated into the prototype GPFP PE is 256 bits. This amount was thought to be sufficient for the demonstration of the design concepts and is able to hold seven 32-bit operands (plus 32 bits available as workspace).

2.2. Multi-bit Nature of the GPFP Processing Element

Since array processors are designed to exploit the fine grain data parallelism inherent in various problems, they usually feature an extremely large number of elementary processing elements. Although it is true that the bit-oriented PEs offer almost unlimited flexibility of precision and data representation, it was also the case that the integration technology available in the previous decade could not cope, at least within a sensible cost, with the required number of processors if the PEs were more powerful. Nowadays these serious fabrication constraints have been eliminated, making it feasible to consider array architectures featuring much more complex PEs.

The GPFP processing element, shown in Figure 1, has been designed as a multi-bit unit, with additional architectural properties that enhance existing PEs. The multi-bit character introduced enables the PE to support local address modifications and higher level instructions for efficient execution of several numeric operations. However, the multi-bit width should be increased with caution, as extremely low propagation delays can be achieved with single-bit PEs encouraging the use of really fast system clocks. Also it should be stressed once more that the flexibility of (vertical) data word length and data organisation offered by the single-bit PEs are important and they should not lightly be given up. An array system organisation that retains all the operations of single-bit PEs alongside the multi-bit operations is discussed in section 3.

A crucial decision during the design stages of the GPFP PE was related to the amount of complexity that should be introduced into the design, as well as to the processor width that should be adopted. The approach that we finally followed was to combine a 4-bit ALU with the minimum number of support components required, for essentially optimal execution of floating point arithmetic. A further improvement of the performance figures obtained by the GPFP PE (Section 4) , can only be achieved by increasing the width of the processor's word. However, this higher performance can only vary linearly as the width is increased, and it is approximately cancelled by the reduced number of processors that the same amount of circuitry can build and also offset by the fact the PEs gradually turn into specialised processors not talented to operate competently on short precision numbers. We therefore believe that the 4-bit choice, made for the GPFP PE, achieves a cost-effective exploitation of the available resources and furthermore it can demonstrate outstanding performance / functionality on arbitrary precision operations.

The computing power of the processing element is provided from a 4-bit ALU, a 4-bit multiplier, local address modifications and several 4-bit wide data paths. Some important operations that can be executed in one cycle are listed below:

- Add two 4-bit words, stored into the local memory, with a single-bit carry input, generating a 4-bit sum and a carry output.
- Move a 4-bit word to a different location inside the local memory.
- Perform the basic boolean operations, AND, OR, XOR as well as the complement of those, on either two operands producing a 4-bit result, or on one operand producing a single bit result (e.g. the logical AND of the data bits consisting a word) which subsequently is extended to form a 4-bit word.
- Shift a data word by one, two or maximum three bits in either direction. The bits that are shifted out can either propagate into the next memory location or disregarded, while the bits that are shifted in can either be zeros or data bits from the next/previous (depending on the

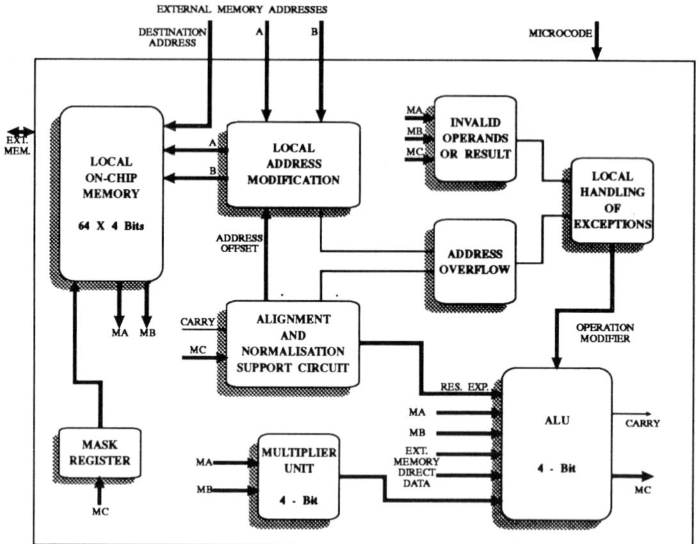

Figure 1. GPFP Processing Element

direction of shifting) memory location. This function enables relatively fast shifting of an operand stored into the local memory, forming a 4-bit wide column, by amounts which are not multiples of 4-bits. If the shifting required is a multiple of 4-bits then the address modification mechanism, explained later, is used instead. It is also worth mentioning that the direction and amount of shifting is provided to the ALU by a *local* register, empowering each PE to shift its associated data in a unique way although simultaneously with the rest of the PEs.

Multiply two 4-bit words, stored into the local memory, and produce an eight-bit result.

The multiplier unit is specially designed to multiply either unsigned (if the numbers are stored in sign and magnitude form then the magnitude of the operand is fed to the multiplier) or 2's complement operands [15] (in order to provide flexibility concerning the representation scheme used for floating point operands) of any precision, producing in both cases a result of analogous type. The operation of the multiplier is tuned to the particularities of the two different formats through explicit microcode control without introducing any extra operational costs for that additional functionality. The interconnection of that unit with the rest of the PE components was very carefully designed in order to accomplish a productive collaboration with the ALU and hence minimise the machine cycles required for the multiplication of two operands that are longer than four bits.

Each GPFP PE includes a 4-bit register, called the *mask register*, which is used to execute conditional write operations to the on-chip memory. More specifically, each PE may be directed to update just the memory bits whose *corresponding mask register bits* are set. Hence, by setting up an appropriate memory mask (by either loading data directly into the mask register or by masking, logical AND, its contents with the ALU output) it is feasible to alter only specific bits of the local memory.

The rest of the components, shown in Figure 1, mainly designed to support the local address modification scheme and the fast execution of floating point operations, are examined throughout the succeeding sections.

2.3. Local Address Modification and Floating Point Support

The choice of the hardware components integrated into the GPFP PE for achieving the desired speed-up of floating point operations, was based on an extended analysis of the distribution of machine cycles during the execution of a 32-bit floating point addition/subtraction on a DAP-510 array processor. The results of this study are illustrated in Figure 2 and can be safely considered as representative for array machines based on single-bit processing elements.

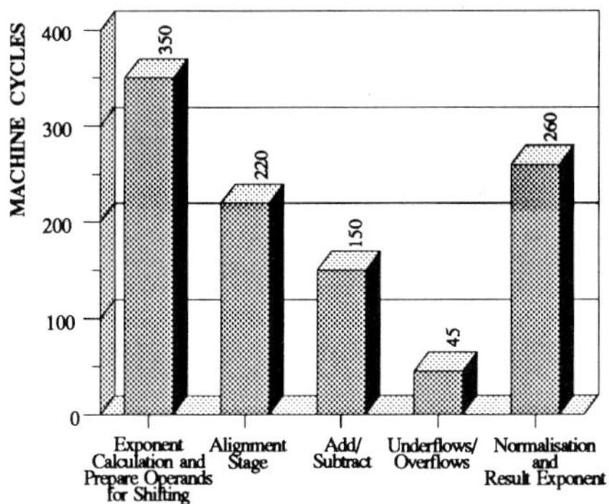

Figure 2. Machine Cycle Distribution During a 32-bit Floating Point Addition/Subtraction on a DAP-510 Array Processor

It is revealed from Figure 2 that the majority of machine cycles are dissipated for reorganising the mantissas inside the array store, in order to facilitate the concurrent execution of shift operations for aligning the two operands (some of the PEs may need to shift the mantissa of the first operand, while the rest of the PEs must shift the mantissa of the second operand), and subsequently for accomplishing all the shift operations during the alignment and normalisation stages (the number of required cycles is dominated by the longer shift in the array, even if that applies to a single processing element).

The GPFP processing element is enriched with several support components which, in conjunction with the 4-bit ALU, the 4-bit multiplier and the on-chip memory, make a substantial contribution towards an improved computational throughput during the execution of "time consuming" operations, like multiplication, data shifts and others. Moreover a sophisticated form of local address modification (much more powerful than elementary address offset mechanisms proposed in other PE designs, like the BLITZEN PE [8]), illustrated in Figure 3, is implemented. This mechanism enables each PE to locally modify one of the two globally broadcasted addresses —the address selection depends on a local state— in a way that explicit shift operations during a floating point operation are eliminated. The address offset, used for the modification, is formed by maneuvering values, stored into local registers, that reflect the particularities of the data stream allocated to each PE. An extended description of the hardware support components functionality, as well as of the operational principles of the local address modification scheme, can be found in [16].

The GPFP PE has been designed to achieve the highest possible execution speed for IEEE/ANSI floating point operations while maintaining generality and simplicity. During the design stages of the processing element we noted that some deviations from the IEEE/ANSI floating point standard [17] could result in significantly faster execution times. In order to obtain cost-effective use, in terms of performance, design complexity and generality, of the invested silicon we have designed enough flexibility into the microcode control [18] to allow, under user control, operation in two distinct modes, namely the *IEEE mode* and the *Maximum Speed mode*.

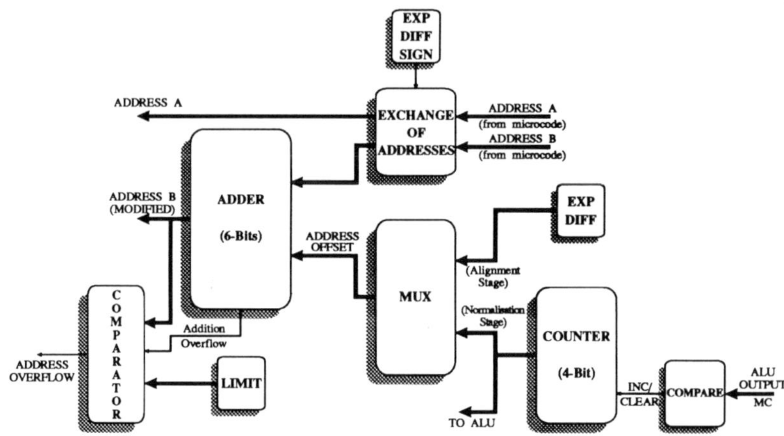

Figure 3. Local Address Modification Mechanism and
F.P. Support Hardware of the GPFP Processing Element

In IEEE mode full compatibility with the IEEE/ANSI standard is maintained. Problems that do not rely on the strict implementation of the standard can take advantage of the Maximum Speed mode of operation, which can result to a potential speed up of approximately 30%. In this mode we have settled to abandon some strict rules of the IEEE/ANSI standard concerning the representation of floating point numbers, error handling and rounding matters.

Specifically, during the Maximum Speed mode of operation, both the mantissa and the exponent of a floating point number are represented in the twos complement format with a hexadecimal exponent radix. Also a simplified method for handling overflows and underflows is enabled by using the two most significant bits of the exponent field to signal those exceptions (*Invalid Operands or Result* unit in Figure 1) and subsequently to locally modify the ALU operation as appropriate (*Local Handling of Exceptions* unit in Figure 1). Finally in this mode results are truncated to the precision used, instead of rounded.

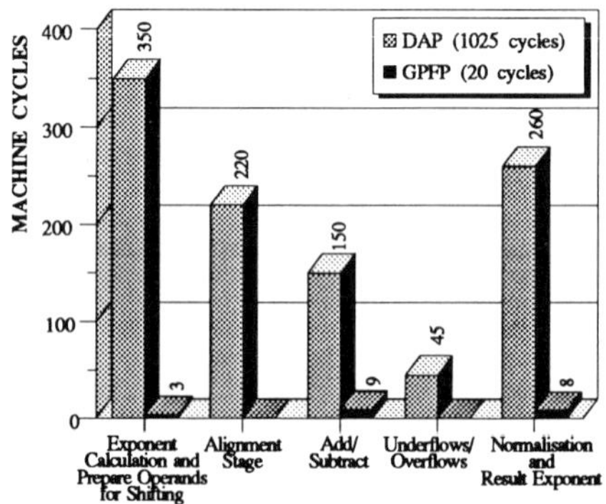

Figure 4. Machine Cycle Distribution During a 32-bit
Floating Point Addition/Subtraction

The implication of the extra hardware supplied is a substantial improvement in the performance that can be exhibited by an array processor while executing floating point operations. From Figure 4, which compares the DAP PE with the GPFP PE, it can be observed that the number of cycles required to complete a 32-bit floating point addition has been reduced from 1025 to 20. This is almost down to the absolute minimum (17) that is theoretically possible with 4-bit operations that rewrite into memory in each cycle.

3. AN ARRAY PROCESSOR USING THE GPFP PE

The goal of our research work was to develop an array processor which combines *high performance* and *general purpose capability* in the most cost-effective way. Section 2 emphasised the high performance issue, by presenting the architectural and operational principles of the GPFP processing element. In addition, the GPFP processor was carefully designed to avoid over-specialisation, for example loss of word-length flexibility, and we also sought to retain as much as possible of the benefits of the single bit organisation, namely data organisation and manipulation flexibility.

To this end, the GPFP array processor is designed to retain all the manipulation capability available in an array of simple 1-bit PEs, where four 1-bit PEs are regarded as mapping onto a single GPFP PE. The programmer's view of the situation is as Figure 5. The machine can be regarded as consisting of simple-style processors implementing all Boolean operations and planar shifts in the *Bitmap Section* of the machine, while a new processor plane implements the *Number Crunching* section performing vertical arithmetic and shift operations. The on-chip memory designed to increase the processor-memory bandwidth is accessible to all operations.

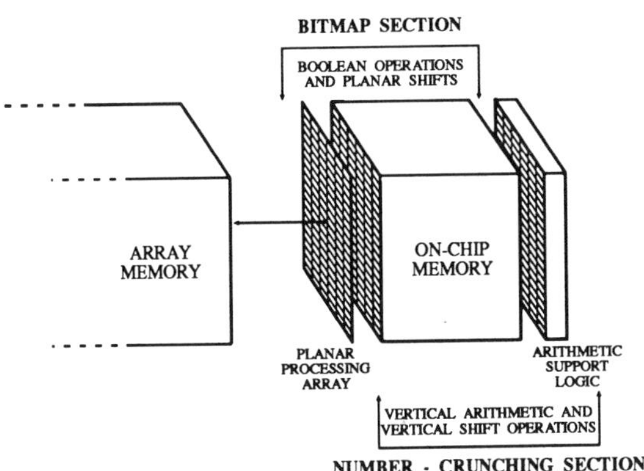

Figure 5. Organisation of an Array Processor Using both Single-bit and Multi-bit Processor Planes

Each processing element, in either processor plane, can communicate with its four nearest neighbours through a 2-Dimensional mesh interconnection network, as implied by Figure 5. Considering the bandwidth between the on-chip and array memory, it was found that a single-bit link per 4-bit GPFP PE, with 512 bits of on-chip memory, provides sufficient memory bandwidth to fully exploit the fast arithmetic operations dominant in the numerical case study programs examined during the architectural design (data transfers from the external memory to the on-chip memory can be overlapped with the operation of the processing elements). Smaller models of machines using the GPFP PEs could have 4-bit links between each PE and array memory. Larger models may have to have 1-bit links, to reduce chip pinout, and so will require more than 512 bits of on-chip memory per PE or the acceptance of slower operation for data movement intensive applications.

Having already invested a significant amount of circuitry for implementing the powerful GPFP processing element, a major concern was to analyse cost and design complexity issues in order to provide *architectural flexibility* to a variety of distinct application requirements. This is achieved by using microcode and local control signals to alter the operational characteristics and behaviour of the numerous support components, in a way suitable to the every time requirements. Research into the architectural details of providing sufficient microcode flexibility is still continuing and is the subject of another paper [18]. The effects of this flexibility are highly reflected on the general purpose capability of the machine and they also ensure that the majority of the PE hardware is used in the best possible way.

4. PERFORMANCE OF THE GPFP PE

The operation of the GPFP processing element has been fully simulated in software. The performance figures, presented in Table 1, have been verified by feeding the necessary microcode sequence into the simulator. The performance of the GPFP PE, the DAP PE and the MPP PE, is presented in terms of machine cycles, in an attempt to enable the reader to do a comparison of the PE's power, independently from the clock period and the number of PEs used in a system.

Table 1
Machine Cycles Required for Representative Integer and F.P. Operations

Operation	DAP-510	MPP	GPFP
8-bit Integer Add / Subtract	48	25	2
8-bit Integer Multiplication	250	88	11
32-bit Floating Point Add / Subtract	1025	384	20
32-bit Floating Point Multiplication	1443	769	61

Figure 6 compares the performance of the GPFP PE with the performance of some other well known PEs. The figures are given in MFLOPS/PE assuming that all the processors are driven by a 10MHz clock.

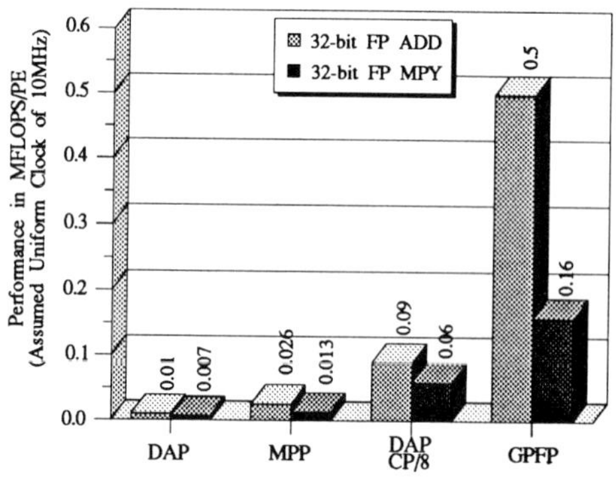

Figure 6. Comparing the Performance of Different PEs

Table 2 gives the performance that would be obtained from a typical array of size 32x32, using a clock rate of 50MHz.

Table 2
Execution Speed of Typical Operations for a 32x32 Array of GPFP PEs

Operation	GPFP
8-bit Integer Add / Subtract	25600 MIPS
8-bit Integer Multiplication	4654 MIPS
32-bit Floating Point Add / Subtract	2560 MFLOPS
32-bit Floating Point Multiplication	839 MFLOPS

Application programs for array processors can be upgraded to the GPFP by supplying alternative library routines for array arithmetic. Although some applications will need program modifications (mainly concerning the organisation and movement of data in the array store) to exploit the GPFP architecture to the full, simply replacing a few of the high level language library arithmetic routines with ones that support the new features of the enhanced PEs is sufficient to produce more than an order of magnitude improvement to the execution time. This is based on an analysis of the execution cycles for two case-study application programs. The first program was an image processing application, implementing the Canny Operator (Edge Detector) on a 512x512 image, involving a large amount of floating point and integer arithmetic operations over a wide range of precisions, while the other one, called Nolen, solves Poisson's equations on a 96x96 grid with 32-bit floating point operations.

Linking the GPFP library routines into the application programs at compile time reduces the cycle count (applicable to the arithmetic part of the programs) from 41178768 to 1377677 for the Canny Operator program and from 3813153 to 135945 for the Nolen program, factors of 30 and 28 respectively. This is close to the theoretical maximum speed-up of 50 for individual floating point additions given in Table 1.

5. CONCLUSIONS

In this paper we have highlighted the major features that we identified as desirable in order to build a successful next generation array processor, by presenting the architectural and operational principles of the GPFP processing element. The highly numerical nature of several modern scientific and industrial applications and the extra availability of circuitry, due to the advances of fabrication techniques, provided the major initiatives for the development of the new processing element.

The validity of most architectural enhancements were verified, prior to their implementation, by simulating them on a DAP-510 array processor. One of the early decisions made was to provide 256 or 512 bits of on-chip memory which according to experimental results [14] can increase the overall system performance, under some static allocation scheme, by approximately 50% to 65%.

A further investment in silicon utilisation was made by enhancing the computational capability of the processor. The functionality of the support components provided were determined after extensive study of the distribution of machine cycles required for the completion of a floating point operation on the DAP-510 array processor. The 4-bit width for PEs was the preferred choice between design complexity and computational throughput.

The resulting PE is approximately 50 times faster for floating point operations than the simplest 1-bit PEs, while retaining nearly all the data flexibility and high speed of clocking.

Acknowledgements

We would like to thank David Hunt from Active Memory Technology Ltd. (AMT) for supplying technical information concerning the DAP array processor and also for his useful comments throughout this work. Finally, but by no means least, we would like to thank Peter Flanders and Stewart Reddaway, also from AMT, for their valuable critique of our work.

6. REFERENCES

1 D. PARKINSON, *The Distributed Array Processor (DAP)*, Computer Physics Communications, North-Holland Publishing Company, Vol 28, 1983, pp. 325-336.

2 S. F. REDDAWAY, *DAP-A Distributed Array Processor*, Proc. of the first Annual Symposium on Computer Architecture, 1973, pp. 61-65.

3 K. E. BATCHER, *Architecture of a Massively Parallel Processor*, IEEE Transactions on Computers, Vol C-29, No 9, Sept. 1980, pp. 836-840.

4 I. N. ROBINSON, W. R. MOORE, *A Parallel Processor Array Architecture and its Implementation in Silicon*, Proc. of IEEE Custom Integrated Circuits Conference, Rochester N.Y., 1982, pp. 41-45.

5 R. M. LEA, *An Overview of the Influence of Technology on Parallelism*, Major Advances in Parallel Processing, Ed: C. Jesshope, Technical Press, 1987.

6 J. D. MEINDL, *Chips for Advanced Computing*, Scientific American, Trends in Computing, Special issue, Vol 1, 1989, pp. 99-107.

7 C. R. JESSHOPE, R. O'GORMAN, J. M. STEWART, *Design of SIMD microprocessor array*, IEE Proceedings, Pt.E, Vol 136, No 3, May 1989, pp. 197-204.

8 E. W. DAVIS, J. H. REIF, *Architecture and Operation of the BLITZEN Processing Element*, Proc. of the 3rd International Conference on Supercomputing, Boston MA, May 1988, pp. 128-137.

9 R. E. MORLEY, G. E. CHRISTENSEN, T. J. SULLIVAN, O. KAMIN, *The Design of a Bit-Serial Coprocessor to Perform Multiplication and Division on a Massively Parallel Architecture*, Proc. of 2nd Symposium on Frontiers of Massively Parallel Computation, IEEE Computer Society, 1988, pp. 419-422.

10 ACTIVE MEMORY TECHNOLOGY LTD (AMT), *DAP Series Technical Overview, Introducing the DAP/CP8 range*, Sales Support Note 7, Apr. 1990.

11 K. E. BATCHER, *Bit-Serial Parallel Processing Systems*, IEEE Transactions on Computers, Vol C-31, No 5, May 1982, pp. 377-384.

12 S. F. REDDAWAY, *The DAP Approach*, in Supercomputers Vol 2, Infotech State of the Art Report, Infotech International Ltd, Maidenhead Berks, 1979, pp. 311-329.

13 D. BEAL, C. LAMBRINOUDAKIS, *Workspace Usage and Memory Hierarchies on the AMT DAP*, Internal Report 487, Department of Computer Science, Queen Mary and Westfield College, University of London, 1988.

14 D. BEAL, C. LAMBRINOUDAKIS, *Performance Improvements to Array Processors from Fast (on-chip) Memory*, Internal Report 475, Department of Computer Science, Queen Mary and Westfield College, University of London, 1988.

15 K. HWANG, *Global and Modular Two's Complement Cellular Array Multipliers*, IEEE Transactions on Computers, Vol C-28, No 4, Apr. 1979, pp. 300-306.

16 D. BEAL, C. LAMBRINOUDAKIS, *Floating Point Support for SIMD Array Processors*, Internal Report 511, Department of Computer Science, Queen Mary and Westfield College, University of London, 1990.

17 IEEE, *IEEE Standard for binary floating Point Arithmetic*, ANSI/IEEE std. 754-1985, 1985.

18 D. BEAL, C. LAMBRINOUDAKIS, *RLC: An Intelligent Technique for Enhancing Microcode Flexibility in Array Processors*, Internal Report 512, Department of Computer Science, Queen Mary and Westfield College, University of London, 1990.

VLSI 91
A. Halaas and P.B. Denyer (Eds.)
Elsevier Science Publishers B.V. (North-Holland)

Input/Output Design for VLSI Array Architectures

Wayne P. Burleson[a] and Louis L. Scharf [b]

[a]Department of Electrical and Computer Engineering, University of Massachusetts, Amherst, MA, 01003, USA

[b]Department of Electrical and Computer Engineering, University of Colorado, Boulder, CO 80309, USA

Abstract

VLSI array architectures which exploit vector-, word- and bit-level parallelism impose constraints on the input and output of data. These input/output constraints are frequently neglected in the complexity analysis of VLSI array architectures. The design of highly-parallel VLSI systems can require cumbersome conversion between different space-time formats of data. A simple example of this is serial-parallel and parallel-serial conversion. We explore two approaches to the generalized I/O design problem.

First, we formalize the specification of a particular data format in terms of a space-time transformation of the index set. We can then determine the conversion between two different formats. From this formalism we systematically generate conversion hardware and obtain an upper bound on the area complexity of a VLSI implementation for an arbitrary conversion. We illustrate our design method with a systolic array for orthogonal transforms based on Given's rotations. We compare our approach of I/O hardware design with previous work in several diverse fields including the theory of systolic arrays, interconnection networks, systolic system design, and optoelectronics.

A second approach to the format conversion problem is to incorporate the I/O constraints in the design of the array. Space-time maps of bit-level algorithms may be chosen to satisfy constraints on I/O format. We show how the proper choice of map for a bit-level rotator cell results in trivial conversion hardware when the cell is used in the word-level array from the previous example.

1. INTRODUCTION

Analysis of numerical computations at the bit level has yielded novel architectures with unique input/output requirements. Numerically intensive computations can be accelerated with these architectures; however, the processors must interface with host computers, and with A/D and D/A converters that have word-level, bus-oriented I/O. In addition, heterogeneous systems must be designed such that several different arrays may be interfaced with minimum conversion hardware and performance degradation. We suggest that when conversion hardware is included in an overall system complexity analysis, arrays which are optimum when viewed alone may no longer be optimum in the context of the surrounding system.

Unlike conventional word-level computers, bit-serial [16], online(MSB-first) [6], bit-level systolic [9] and distributed arithmetic [1,4] architectures require I/O hardware consisting of word and bit-level registers with serial-parallel and parallel-serial conversion capability. The complexity of this conversion hardware is rarely included in the analysis of these architectures.

Space-time transformations of regular iterative algorithms [8] have been used to map algorithms to parallel architectures. The VLSI model of Thompson [17] makes some simplifying assumptions about computation in VLSI and offers a formalism for deriving asymptotic bounds for the area, latency and period of a particular architecture. We formalize the specification of a particular data format and calculate the necessary space-time transformation between two formats. The transformation formalism allows us to calculate bounds on the VLSI complexity of the conversion hardware. For certain conversions, these bounds are tight and the complexity of the conversion hardware can be included in comparisons and optimizations of different bit-level architectures. A data format may be transformed in space and time with the same map as the corresponding regular iterative algorithm for the computation. Once a conversion is designed, the hardware may be put on the VLSI chip, potentially reducing the package count for a particular system. Likewise, pin count may be modified to fit packaging limitations by altering the interface data format.

Section 2 introduces the notation for describing data formats and establishes criteria for the existence and solution of a transformation between formats. Section 3 presents a systematic method for determining architectures for implementing a particular transformation. We suggest a general factoring of the transformation matrix which maps to well-known VLSI structures. We show an example, solve for the transformation and design conversion hardware.

In Section 4 we show how to design VLSI arrays to satisfy I/O constraints. Previous work in the theory of systolic arrays [8,12] addresses the problem of projecting the dependency graph such that all I/O occurs at boundary processors. We extend this work to the design of two-level arrays, where a bit-level array is designed to match I/O constraints imposed by the word-level array. A design example is shown in which a bit-pipelined implementation of the Givens rotation is used in a word-level array for orthogonal transformations [2].

2. NOTATION FOR DATA FORMATS

Digital signal processing algorithms may often be decomposed into a matrix notation where the input and output are blocks of fixed point data. When viewed at the bit-level, s-dimensional data becomes $(s+1)$-dimensional. We use a regular iterative description of the bits in a data block. This description consists of a set of variables $[x_1, x_2, \dots x_v]$ which take on binary values. These variables are indexed by a set of integer variables $[i_1, i_2, \dots i_s]$, each of which has a finite domain, forming a finite s-dimensional lattice.

Without much loss of generality and for ease of notation, we will study input bit blocks of dimension 2. The input data for a single instance of a computation consists of a vector of length n whose elements are 2's complement words each of length p. In the binary data matrix A= $\{a_{ij}\}$, words are indexed by i and bits by j.

$$\mathrm{A} = \begin{bmatrix} a_{11} & a_{12} & \dots & a_{1p} \\ a_{21} & a_{22} & \dots & a_{2p} \\ \vdots & \vdots & \ddots & \vdots \\ a_{n1} & a_{n2} & \dots & a_{np} \end{bmatrix}$$

A data format is a mapping of the bits of A onto space and time coordinates. The map U of the index set $[i,j]$ maps the elements of A onto B, which is characterized by a new index lattice $[s_1, s_2, t]$.

$$a_{i,j} = b_{U[i,j]}$$
$$b_{s_1,s_2,t} = a_{U^{-1}[s_1,s_2,t]}$$

We now present three types of data formats and show that a conversion between two formats exists only if both formats are *1-1* maps of the index set. We also show that the *data rate* must be preserved across the two formats.

2.1 Linear Formats

A *linear format* is specified by a linear transformation U from one set of indices *onto* a new set of indices. Both the domain and range of this mapping are finite, and are bounded by the values assumable by each index. For U with full column rank, the mapping is *1-1* and *onto* and therefore is invertible.

$$\begin{bmatrix} s_1 \\ \vdots \\ s_m \\ t \end{bmatrix} = U \begin{bmatrix} i \\ j \end{bmatrix}$$

Examples of space-time transformations U include:

$$U_1 = \begin{matrix} s \\ t \end{matrix} \begin{bmatrix} 1 & 0 \\ 0 & 1 \end{bmatrix} \quad \text{bit-serial word-parallel}$$

$$U_2 = \begin{matrix} s \\ t \end{matrix} \begin{bmatrix} 0 & 1 \\ 1 & 0 \end{bmatrix} \quad \text{bit-parallel word-serial}$$

$$U_3 = \begin{matrix} s_1 \\ s_2 \\ t \end{matrix} \begin{bmatrix} 1 & 0 \\ 0 & 1 \\ 1 & 1 \end{bmatrix} \quad \text{pipelined bit-parallel word-parallel}$$

$$U_4 = \begin{matrix} s \\ t \end{matrix} \begin{bmatrix} 1 & 0 \\ 0 & -1 \end{bmatrix} \quad \text{bit-serial(MSB-first) word-parallel}$$

$$U_5 = t \quad [p \quad 1] \quad \text{bit-serial word-serial}$$

The transformations U are frequently binary matrices; however, the only formal restriction is that the resulting indices be integers. The resulting index lattice may no longer be a parallelepiped due to linear transforms of the indices. (An example is U_3, where the t index is bound by sums of the upper and lower bounds on i and j.)

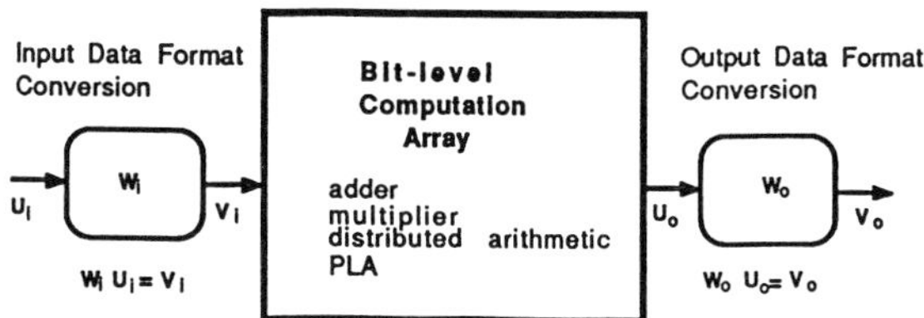

Figure 1. Data Format Conversion for VLSI Arrays

The design of conversion hardware between data formats can be specified in terms of the corresponding space-time transforms. Figure 1 shows the general problem of data format conversion for bit-level processor arrays. If we assume that input data arrives in a format U_i and a processor array requires the format U_j, we must solve for the conversion W_{ji}.

$$W_{ji} U_i = U_j$$

$$W_{ji} = U_j U_i^+$$

where

$$U_i^+ = (U_i^T U_i)^{-1} U_i^T$$

A simple example with input format U_2 and processor format U_1 results in

$$W_{12} = U_1 U_2^{-1} = \begin{matrix} s \\ t \end{matrix} \begin{bmatrix} 0 & 1 \\ 1 & 0 \end{bmatrix}$$

The matrix for the opposite conversion W_{21} is identical indicating that the hardware for the two complementary conversions is identical.

2.2 Permuted Formats

If we look at a more general class of formats, there will be cases where the format U can not be expressed as a linear transformation on the index set. In [3] we show two types of permuted formats, *index permutation format* and *bit permutation format.* Unlike the linear transformation representation of formats, the index permutation representation allows arbitrary permutations within an index, and permutations of entire indices. Unfortunately, this representation does not permit "mixing" of indices. For completely general permutations, bit permutation format allows the arbitrary permutation of all bits in an s-dimensional data block. This very general permutation operates on a vector which contains an element for every bit in the data block.

A necessary condition for the existence of a conversion W_{ij}, is that both formats U_i and U_j be *1-1*, and therefore invertible, maps of the original index set. In [3] we have demonstrated this property for linear, index permutation and bit permutation formats.

2.3 Data Rates

For every format U, there is an associated *data rate*, $D_U(t)$. The *cycle time* (t_{cycle}) of a format is the difference between adjacent indices in the time dimension. The data rate is measured in bits per unit time and may be time-varying. We count the number of bits in a hyperplane which is orthogonal to the time axis and divide by the cycle time of the format.

$$D_U(t) = \frac{|a_t|}{t_{cycle}}$$

A necessary condition for the existence of a transform W_{ij} is the equality of the input and output data rates when summed over their respective time domains.

$$\sum_{t \in T_i} D_{U_i}(t) = \sum_{t \in T_j} D_{U_j}(t)$$

3. CONVERSION HARDWARE

For a transform W, we need to measure the area A, latency L and period P of the hardware implemented in a VLSI model. To determine upper bounds on these complexity measures, we suggest a systematic method for designing hardware structures for arbitrary conversions. While for certain conversions the resulting hardware may not be minimal in terms of A, L and P, it does give an upper bound on the complexity for an arbitrary conversion. We use the VLSI model of Thompson [17] with some minor modifications.

Definition: A (area) is the area of the convex hull of the space sublattice. We use a_{node} to account for the size of local memory within a node.

Definition: L (latency) is the total time (*cycles x* t_{cycle}) from when the first bit for a particular data block arrives at a conversion apparatus until the first bit leaves (not necessarily the first to arrive).

Definition: P (period) is the time (*cycles x* t_{cycle}) from when a bit of a particular data block arrives at a conversion apparatus and the corresponding bit from the next block arrives. This is the inverse of the throughput of the conversion apparatus.

Note that our definitions of L and P differ from those of Thompson [17] in that we allow $L < P$ while they have $L = dP$, where d is the pipelining depth. This is due to our definition of L which is peculiar to the format conversion problem.

3.1 Factoring the Conversion Matrix

The matrix W may be factored

$$W = RQS$$

where the factors map between the following domains,

$$S : (S_i, t_i) \mapsto (S_s)$$
$$Q : (S_s) \mapsto (S_r)$$
$$R : (S_r) \mapsto (S_j, t_j)$$

S maps the entire data block into space coordinates, Q performs an arbitrary permutation on the block in the space domain and R maps the bits into the output space-time format. S and R are shift register structures. Bounds on the VLSI complexity of Q are known for some classes of permutations [19]. The VLSI complexity of the conversion W is:

$$A(W) = A(R) + A(Q) + A(S)$$
$$L(W) = L(R) + L(Q) + L(S)$$
$$P(W) = \max(P(R), P(Q), P(S))$$

We can develop an upper bound on the area required by an arbitrary transformation W by computing the area upper bounds of the factors, R, Q and S. In [3], we show that for a conversion W where $|a|$ is the number of bits in the input data block , $A(W) \leq 3|a|$. This bound can be quite loose for specific cases. Two reasons for this are the complexity of the general Q transform and the memory required to store the entire block of data in R and S.

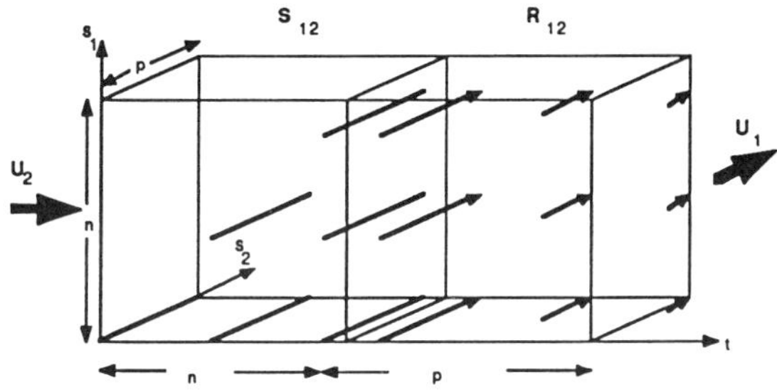

Figure 2. W_{12} Conversion Hardware

We now show an example for which the RQS factoring produces a minimum area structure. The conversion W_{12} transforms a block of data from word-serial bit-parallel to word-parallel bit-serial format. In Figure 2 the 3-space of s_1, s_2 and t illustrates the conversion apparatus. The input data in format U_2 arrives on the left while output data in format U_1 emerges from the back. Time slices indicate the state of the converter array at each time. The area, latency and period of the array are described by:

$$A = A(R) + A(Q) + A(S) = np + 0 + np = 2np$$
$$L = L(R) + L(Q) + L(S) = 0 + 0 + n = n$$
$$P = \max(P(R), P(Q), P(S)) = \max(p, 0, n)$$

Alternate factorings for special cases of the W matrix have been proposed which map directly to regular VLSI hardware. Wah, Aboelaze and Shang [18] propose a factoring, $W = T_1T_2T_3T_4T_5T_6$ where each T_i corresponds to a particular aspect of the transformation such as reversal, multiplexing and skewing. Products of these sub-transformations cover many of the data-format conversions needed by systolic arrays. Rajopadhye and Lisper [13] suggest a group theoretic solution for a certain class of permutations, where all elements of the group may be generated by rotation and reflection.

3.2 Design Example: Orthogonal Transforms with Givens Decomposition

Using the systematic method shown above, we will specify conversion hardware for interfacing a word-oriented bus with an orthogonal transform computer using the Givens decomposition [2]. The word-oriented bus could be the connection to a host computer, or to A/D and D/A converters in a real-time application.

Orthogonal transforms may be decomposed into a product of Given's rotations. This decomposition provides a convenient mapping to a regular, mesh-connected processor array [2]. The processor array operates on a vector of length n which is an $n\ x\ p$ binary data block when viewed at the bit level. The computation is pipelined at the word level, and packaging constraints force us to transfer the data for a particular word in a bit-serial fashion. These constraints result in the processor data format

$$U_j = \begin{matrix} s \\ t \end{matrix} \overset{\begin{matrix} i & j \end{matrix}}{\begin{bmatrix} 1 & 0 \\ p & 1 \end{bmatrix}}$$

with data rate

$$D(U_j) = \frac{|a|}{t_{cycle}(U_j)} = \frac{n}{t_{cycle}(U_j)}$$

The data arrives in format $U_i = U_2$ with required data rate

$$D(U_i) = D(U_j) = \frac{n}{t_{cycle}(U_j)}$$

resulting in an input cycle time $t_{cycle}(U_i) = pn^{-1}t_{cycle}(U_j)$

The necessary format conversion is then

$$W_{ij} = U_jU_i^{-1} = \begin{matrix} s \\ t \end{matrix} \overset{\begin{matrix} i & j \end{matrix}}{\begin{bmatrix} 1 & 0 \\ p & 1 \end{bmatrix}} \overset{\begin{matrix} s & t \end{matrix}}{\begin{bmatrix} 0 & 1 \\ 1 & 0 \end{bmatrix}} = \begin{matrix} s \\ t \end{matrix} \overset{\begin{matrix} s & t \end{matrix}}{\begin{bmatrix} 0 & 1 \\ 1 & p \end{bmatrix}}$$

Factoring W_i,

$$W_i = RQS == \begin{matrix} s \\ t \end{matrix} \overset{\begin{matrix} s_1 & s_2 \end{matrix}}{\begin{bmatrix} 0 & 1 \\ 1 & p \end{bmatrix}} \overset{\begin{matrix} s_1 & s_2 \end{matrix}}{\begin{bmatrix} 1 & 0 \\ 0 & 1 \end{bmatrix}} \overset{\begin{matrix} s & t \end{matrix}}{\begin{bmatrix} 1 & 0 \\ 0 & 1 \end{bmatrix}}$$

The complexity of the conversion is

$$A(W_i) = np + n(n+1)p/2$$
$$L(W_i) = t_{cycle}(U_j)$$
$$P(W_i) = pt_{cycle}(U_j) = nt_{cycle}(U_i)$$

The inverse conversion occurs for the output of the processor array. The rotator array with conversion hardware is shown in Figure 3.

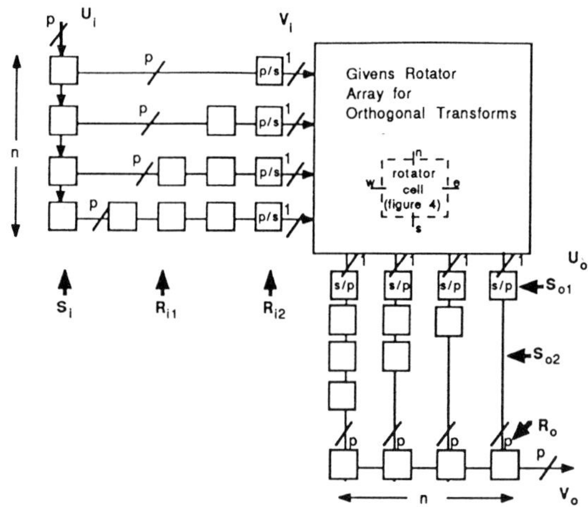

Figure 3. Rotator Array with Conversion Hardware

4. DESIGNING ARRAYS TO SATISFY I/O CONSTRAINTS

Kung [8] and Rajopadhye [12] explore the problem of I/O for systolic arrays. The main idea is to expand the iteration lattice such that all I/O occurs on boundary processors. The expanded portion of the lattice consists only of registers, thus making the overall algorithm heterogeneous. Note that the lattice expansion is a formal method for deriving R and S factors which can be implemented within the computational array for a special class of conversions.

We now show an example in which we choose a space-time map of a bit-level algorithm to match the I/O constraints specified by a higher-level array. Unfortunately, the map which satisfies the I/O constraints will not necessarily be the optimal map for implementing the actual array.

4.1 Design Example: Givens Rotator using Distributed Arithmetic

We use distributed arithmetic for the bit-level implementation of the Given's rotator from Section 3.2. Distributed arithmetic [1,4] is a method for computing inner products with one stored vector through the use of tables and adders instead of multipliers. The bit-level algorithm may be expressed as a regular iterative algorithm which is a lattice indexed by (a, b, c, d). In this RIA the input and output data are sub-lattices indexed by a, b and a, d respectively. The rotation coefficient bits are indexed by c.

The Givens algorithm has the property that the south and east outputs are the north and west inputs to adjacent cells. We now look for a space-time map of the regular iterative algorithm such that the input and output formats U_i and U_j are identical, meaning that there is no conversion hardware!

Kung [8] suggests the enumeration of all maps as a brute force approach to performance optimization of VLSI array architectures. Although the search space of all maps results in a relatively small optimization problem, a non-exhaustive search method is still an interesting open research problem.

The following map T,

$$T = \begin{matrix} s_1 \\ s_2 \\ t \end{matrix} \overset{\begin{matrix} a & b & c & d \end{matrix}}{\begin{bmatrix} 0 & 1 & 0 & 1 \\ 0 & 0 & 1 & 0 \\ 1 & 0 & 1 & 0 \end{bmatrix}}$$

produces a rotator architecture in which the input and output formats are identical. In addition, this map results in circulating coefficients which allows simple serial updating of the rotation angles.

$$U_i = \begin{matrix} a \\ b \end{matrix} \overset{\begin{matrix} s & t \end{matrix}}{\begin{bmatrix} 0 & 1 \\ 1 & 0 \end{bmatrix}} = U_o = \begin{matrix} a \\ d \end{matrix} \overset{\begin{matrix} s & t \end{matrix}}{\begin{bmatrix} 0 & 1 \\ 1 & 0 \end{bmatrix}}$$

This solution is not unique, and all maps which result in $U_i = U_o$ must be analyzed in terms of the resulting area, latency and period.

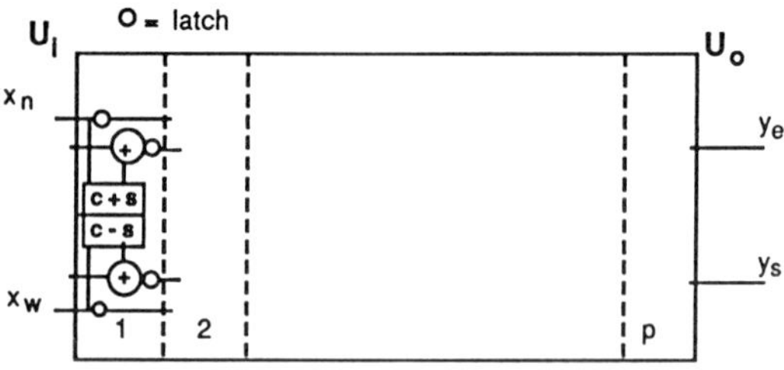

Figure 4. Bit-serial Rotator Cell

Figure 4 illustrates the bit-level Givens rotator cell. This structure is similar to a bit-level multiplier of complex numbers reported by Smith and Denyer [15]. Although this example may seem rather obvious by producing purely bit-serial structures, more complex RIAs result in I/O formats which are much less intuitive.

5. DISCUSSION

We must evaluate the results generated by our systematic method of deriving conversion hardware and determine classes of conversions for which we can produce optimal architectures. The architectures for data format conversion suggested in this paper are for arbitrary permutations, therefore there may exist architectures for special case transformations which are tighter than our upper bounds.

The class of *corner-turning* conversions between linear data formats such as U_1,U_2,U_3,U_4 and U_5 from Section 2 permute or sign switch entire indices. Corner-turning includes serial-parallel, parallel-serial, transposition, time reversal, and space reversal in addition to literal spatial corner turning. Table 1 lists the size, area, latency, and period of RQS-derived hardware for a variety of corner-turning problems. We know of no spatially- and temporally-local VLSI architectures for implementing these conversions that are smaller(A) or faster(L) than those derived by the RQS factoring. We should note that this is the same class addressed in [18] and [13].

More general permutations, such as those requiring index and bit permutation format can be implemented with shuffle-exchange networks [19], cellular permutation networks

Description	U_i	$\mapsto$	U_j	size	Area	Latency	Period
Corner-turn	s_1	$\mapsto$	s_2	n	n^2	n	n
Pipelined Corner-turn	s_1, t	$\mapsto$	s_2, t	n^2	n^2	n	n
Par-Ser	s_1	$\mapsto$	t	n	n	1	n
Ser-Par	t	$\mapsto$	s_1	n	n	n	n
Pipelined Par-Ser	s_1, s_2	$\mapsto$	s_1, t	n^2	n^2	1	n
Pipelined Ser-Par	s_1, t	$\mapsto$	s_1, s_2	n^2	n^2	n	n
ParPar-Ser	s_1, s_2	$\mapsto$	t	n^2	n^2	1	n^2
Ser-ParPar	t	$\mapsto$	s_1, s_2	n^2	n^2	n^2	n^2
Transpose	s_1, s_2	$\mapsto$	s_2, s_1	n^2	n^2	n	n
Space-time transpose	s_1, t	$\mapsto$	t, s_2	n^2	n^2	n	n
Time reversal	t	$\mapsto$	$-t$	n	n	n	n
Space reversal	s	$\mapsto$	$-s$	n	n	n	n
Pipelined space reversal	s	$\mapsto$	$-s$	n	n^2	n	1
Space and time reversal	s, t	$\mapsto$	$-s, -t$	n^2	n^2	n	n

Table 1. Corner-turning conversions

[11], and address sorting networks [referred to in 13]. The generality offered by these methods is very expensive in terms of VLSI area and latency, therefore, in specific applications the design of dedicated hardware is preferable.

An important distinction of our work is that we address formats which are expressed in space and time, as opposed to space only [7,10,19] or time only [14]. We also present a method for factoring transformations which maps directly to VLSI hardware. Using this method we can re-derive a number of structures which were previously reported without derivation [10,13,14].

The methods and analysis presented in this paper are useful as a formalism for expressing input/output formats. For this work to be of more practical significance, we must find ways to convert the derived architectures to those known for special cases. One method we are pursuing is an alternate factoring of the W transformation. The current method of RQS factoring is rather simplistic and alternative factorings which are dependent on the particular transformation may yield tighter upper bound architectures. An example is the class of *time permutations* from [14]. Another possibility is the further factoring of R, Q and S to distribute the permutations of Q in the unpacking and repacking hardware of R and S, thus saving area and latency by reducing the total number of tridiagonal permutation matrices. The problem of determining an optimal factoring of the conversion matrix may become one of developing a taxonomy of space-time permutations and factors which map to efficient hardware. Obviously for different technologies, the optimal factoring will vary.

Different conversion hardware schemes may be appropriate for different notions of optimality. For example, we may wish to minimize area while satisfying a constraint on period. The optimal architecture in this case may reuse hardware such as a shuffle-exchange network, or it might require several instantiations of that network to fully pipeline the conversion. We can formulate a constrained optimization problem in A, L and P to find the appropriate architecture. We have investigated an optimizing compiler for bit-level distributed arithmetic arrays that includes format conversion hardware as part of the design problem [4]. We are currently investigating a particularly useful application of this work, the derivation of I/O hardware for partitioned problems on fixed size arrays [5].

Further work in this area includes investigation of special permutation hardware (i.e.[10,13,14]), more general-purpose conversion hardware (i.e.[18,19]), and theoretical VLSI lower bounds (i.e.[17]) on the conversion problem.

6. CONCLUSION

We have developed a formalism for describing the space-time format of a block of data at the bit level. We use this formalism to specify a conversion between two data formats and suggest a measure of VLSI complexity associated with a particular transformation. An alternate method of dealing with I/O issues is designing the array to satisfy I/O constraints. The methods discussed in this paper allow the complexity associated with the I/O of array processor data to be included with the computational complexity in a VLSI cost model. This is consistent with the current philosophy that VLSI computation is bound not by operations, but by internal and external communication.

REFERENCES

[1] Burleson,W. and Scharf,L. "VLSI Design of Inner-Product Computers Using Distributed Arithmetic", Proceedings of Intl. Conf. on Circ. Sys., May, 1989.
[2] Burleson,W., Scharf,L., Endsley, N. and Gabriel, A. "A Systolic VLSI Chip for Implementing Orthogonal Transforms", Jou. of Solid State Circ., April 1989.
[3] Burleson,W. and Scharf, L., "Input/Output Complexity of Bit-level VLSI Array Architectures", Proc. 24th Asilomar Conf, Oct., 1989.
[4] Burleson,W. "Efficient Computation in VLSI with Distributed Arithmetic", PhD dissertation, Dept. of ECE , Univ. of Colorado-Boulder, 1989.
[5] Burleson,W. "The Partitioning Problem on VLSI Arrays: I/O and Local Memory Complexity", Proceedings of ICASSP'91.
[6] Ercegovac, M.D. "An Online Arithmetic: An Overview", SPIE Vol. 495, *Real Time Signal Processing VII*, 1984.
[7] Flanders, P.M., "A unified approach to a class of data movement on an array processor", IEEE Trans. on Comp., C-31(9):809-819, Sept. 1982.
[8] Kung, S.Y., *VLSI Array Processors*, Prentice-Hall, 1988.
[9] McCanny,J.V., McWhirter,J.G. "Bit-level Systolic Array Circuit for Matrix-vector Multiplication", IEE Proc. 130, pt. G., 4(Aug 1983), 125-130.
[10] O'Leary, D.P., "Systolic arrays for matrix transpose and other reorderings", IEEE Trans. on Computers, C-36(1):117-122, Jan., 1987.
[11] Oruc, A.Y. and Oruc, M.Y., "Programming cellular permutation networds through decomposition of symmetric groups", IEEE Trans. on Computers, C-36(7):802-809, July 1987.
[12] Rajopadhye, S. "I/O Behavior of Systolic Arrays", Technical Report CIS-TR-87-16A, Dept. of Computer and Information Science, University of Oregon, August, 1988.
[13] Rajopadhye, S. and Lisper, B. "Matrix Permutations on Mesh-Connected Arrays", Proc. Intl. Conf. Circ. Sys., 1990, p. 2626-2629.
[14] Ramanan, S. and Jordan, H. "Photonic Architectures for Performing Perfect Shuffle on a Time-Division Multiplexed Signal", OCS Technical Report 89-03, University of Colorado, January, 1989.
[15] Smith, S.G. and Denyer, P.B., "Efficient Bit-Serial Complex Multiplication and Sum-of-Products Computation Using Distributed Arithmetic," Proc. ICASSP'86.
[16] Smith, S.G., Denyer,P.B. , *Serial Data Computation*, Kluwer, 1988.
[17] Thompson,C.D., *A Complexity Theory for VLSI*, PhD dissertation, Dept. of Computer Science, Carnegie-Mellon University, 1980.
[18] Wah, B.W., Aboelaze, M., and Shang, W., "Systematic Designs of Buffers in Macropipelines of Systolic Arrays", Journal of Parallel and Distributed Computing, 5:1 25, 1988.
[19] Wu, C.L. and Feng, T.Y. "The Universality of the Shuffle-Exchange Network", IEEE Trans. on Computers, C-30(5):324-332, May, 1981.

VLSI 91
A. Halaas and P.B. Denyer (Eds.)
Elsevier Science Publishers B.V. (North-Holland)
1992 IFIP.

Comparing Transformation Schemes for VLSI Array Processor Design — A Case Study

Anders Færgemand Nielsen, Poul Martin Rands Jensen[1], Kallol Kumar Bagchi and Ole Olsen

Department of Communication Technology, Institute of Electronic Systems, Aalborg University, Fr. Bajers Vej 7, DK-9220 Aalborg, Denmark

Abstract

During the past ten years a multitude of transformation techniques for automating the design of systolic arrays have been developed. Fortes et al have tabulated 19 different transformation schemes, [1], and compared two methods and found them to be equivalent, [2]. Since then many new transformations have been proposed and designers have a hard time selecting an appropriate one for a particular application, algorithm and architecture. In this paper we propose a new scheme for evaluation and comparison of different transformations. The scheme is based on forming an evaluation matrix of design parameters and assigning various weights to the parameters for a specific application. We demonstrate the use of such a scheme with a case study from the Digital Signal Processing (DSP) domain, an ARMA filter design. Based on this evaluation, similarities and differences as well as strengths and weaknesses of various transformations can be noted and a transformation can be selected. Side results of the evaluation are: 1) a guide line for the amount of pre- and post-processing required in using a transformation in an overall design scheme and 2) a scheme for making design tradeoffs from high- to medium-performance applications. Finally we relate a standard cell prototype implementation of one of the solutions to conventional von Neumann and DSP Signal Processor implementations.

1. Introduction

In applications like real-time DSP, the requirement of high-performance, reliable and optimal parallel systems is getting increasingly felt. The development of VLSI technology has produced architectures like systolic and wavefront arrays [3], which have been used for implementing various such algorithms. However, the very initial problem that a designer faces in such an array design is to select and apply a formal transformation that could be relevant for the application. A multitude of transformation schemes exist and even for simple algorithms two transformations may yield relatively different solutions and therefore some kind of characterization of the transformations is needed.

In this paper we propose a systematic evaluation scheme for different such transformations. As a case study we will use the DSP ARMA algorithm and illustrate how a meaningful comparison can be done and propose an evaluation matrix of design parameters which can be used ef-

[1]Supported by a grant from The Danish Technical Research Council, (STVF, No. 16-4895-105)

fectively for the selection of a method. Four often-referred transformations from literature have been selected for the case-study: Moldovan and Fortes' method (MoFo) [4], S.Y. Kung's method (SYKu) [3], Jagadish-Rao-Kailath's method (JaRaKA)[5] and Rosseel, Swaaij, Catthoor and De Man's method (RoSwCaDe) [6].

First the parameters are selected carefully so that simplification is not made in difficult cases and effectiveness is maintained when simpler comparisons are needed. Selecting various such parameters that characterize the application and the target implementation, we next try to quantify these. The selection of a method is finally based on a careful study of the weighted parameters.

Most of these schemes require the algorithm to be put in a particular input format, for which an algorithm re-hashing exercise is usually required. The transformations mostly aim at arriving at a logical architecture, from which additional work is usually required to arrive at a chip solution. So these transformations form only the middle-part of the overall design. The steps mentioned above form an integral part of our overall proposed design method for VLSI arrays, from an application to a chip, as described in [7].

2. The Evaluation Matrix and its Parameters

The evaluation matrix consists of: 1) parameters describing the working of the transformations, 2) various physical parameters like area, time, utilization, input/output etc. and 3) parameters for comparison of different solutions. This matrix provides a convenient framework for comparing various attributes of such transformations and has several advantages:

- It helps a designer to gain insight in studying a particular transformation scheme, based on the subset of performance parameters selected. Effects of individual parameters on the overall design could also be studied. In particular, studies can be made to observe the amount of tradeoff needed, if possible, for arriving at pre-selected values of a given set of parameters (the specification),
- Two or more transformations can be compared by contrasting values of selective elements from the evaluation matrix where similarities and differences in various schemes can thus be noted; such comparisons are almost non-existent at present,
- Transformations can be classified more succinctly by observing the values of different parameters.

This is not an easy task to accomplish, given so many parameters of interest in such a design space. Many of these are important, and difficult to quantify at the same time. Note that some of these parameters are known to the designers; however, our aim is to make a formal base for deriving objective observations. The most important parameters that can be used for different evaluations are listed below:

Parameters Related to the Input to a Transformation: Amount of pre-processing: Algorithm restructuring; Use of zero or non-zero data dependencies; Specifying hardware constraints;

Parameters Related to the Behaviour of a Transformation: Formal or heuristics-based; General or specialized-form solver; Search and optimization strategies; Verification needed or not, verification built-in or outside; Ease of use;

Parameters Related to the Output of a Transformation: Number of different solutions in terms of area, time, utilization of the resources, input/output, data movement; Amount of post-processing: Post-scheduling and post-mapping for physical implementations; Regular/irregular system (different PEs); Granularity (levels of pipelining); Directly realisable (schedule, structure, geometry);

Physical Parameters for a VLSI implementation: Area (computation, memory, control, interconnection); Time (latency, throughput, completion); Input/output (Data Flow (DF)i/o, single and/or multiple fan-in and/or fan-out, # of i/o, PE i/o); A and T for different technologies and styles (full and semi custom); Pins/chip (computation, control, power/-ground); Power supply; Power consumption; Utilization of silicon;

Parameters for Comparing Solutions from other Computational Models: Parameters involving robust symbolic or logical predictions of area, time and other hardware measures using different computational models like: von Neumann, PRAM etc. It must be noted that not all transformations can yield solutions whose parameters can be compared to this class of parameters.

3. A Case Study: The ARMA Algorithm

We evaluate the transformations with three goals in mind: 1) the solution should have *minimum pre- and post processing*, 2) the solution should be expressed at a *physical rather than at a logical level* and 3) the solution should be *optimal with respect to area, time and utilization of the individual PEs*, and it should compare favorably with the solutions derived from other computational models.

We discuss below how the various transformations perform with the ARMA algorithm.

3.1. Input to the Transformations

The ARMA algorithm in its original form is not suitable for any of these transformation schemes. The re-hashing of it has to be done obeying the DSP constraints. For example, an input $x(n)$ at time n should arrive before $x(n+1)$; the original data flow needs to be changed carefully, preserving the structure of the original algorithm, etc. Note that the restructuring may result in many different equivalent algorithms or structures of the original description, and therefore this pre-processing of the algorithm plays a very important role on the behaviour of the algorithm as also on the performance of the final chip. So the algorithm has to be converted to a form where it has a regular pipelined structure. Such a solution fulfills nearest-neighbour constraint, and i/o takes place at boundary PEs. These are basically some of the VLSI-array architectural constraints.

The forms used by the different transformations are: Nested Loop (NL) [4], Regular Iterative Algorithm (RIA) [5], Dependence Graph (DG) [3] or Linear Recurrence (LR) [6] like Conditional Uniform Recurrence Equation (CURE). We have developed a pre-processing procedure which ends up in a newly defined form called Controlled CURE (CCURE). It encapsulates all the features from the above mentioned forms like, for example uniform data dependencies allowing conditions in the index space and a single operator form with an optimal pre-scheduled atomic computation. Not all these features are used by every transformation, however, they can all extract the necessary information from the CCURE.

When converted in terms of algorithm re-hashing, such re-hashing may consist of several steps. The re-hashing of the ARMA algorithm starts from the original DSP description, Direct Form I, and ends at the required CCURE form.The ARMA algorithm can be written as follows:

$$y(i) = \sum_{j=0}^{M} b_j x(i-j) + \sum_{j=1}^{M} a_j y(i-j), \text{ for } i = 1...N \tag{1}$$

where M is the order of the filter and N is the number of samples, a, b are coefficients and x, y are input and output variables.

The pre-processing is a non-trivial exercise and our systematic pre-processing procedure which uses known techniques from existing methods ,consists of the following steps:

- Step 1: Deduce the iterative sequential description from equation 1,
- Step 2: Obtain a single (index) assignment form, which ensures that every variable in the algorithm is assigned only one value during the execution of the program,
- Step 3: Derive an Affine Recurrence Equation (ARE) form, which is an index matching single assignment form and here a number of different possibilities can emerge,
- Step 4: Obtain a (Conditional) Uniform Recurrence Equation ((C)URE) form, where the index matching and uniformization (steps 3 and 4) are tied together and the effect of the process is more obvious having a graphical representation of the same problem. In these steps a forward recurrence form has been selected because it yields a regular data dependency.
- Step 5: Derive a Controlled CURE (CCURE) form, which is an attempt to help the transformation in finding an optimal solution. First new variables are needed to implement the single operator form (for example, for the atomic computation: $y = b * x + a * y \Rightarrow bx = b * x, ay = a * y, y = bx + ay$). Then an optimal pre-scheduling for the atomic computation is selected using primitive pre-scheduling techniques.

We have defined step 5, a redefinition of step 4, to clarify what the design tradeoffs are like, for example, the single operator form which enables total control on allocation and scheduling of resources.

Some of these steps can be automated, as, for example, in [8] some sequential iterative algorithms are converted to URE (steps 1 or 2 to 4) and in [9, 10] algebraic transformations are used for converting an ARE to an URE (steps 3 to 4). In both cases a user interaction is needed and the conversion is not easy. We conclude that human interaction is still needed to select the optimal solutions. Our pre-processing procedure is performed in close interaction with the designer and some of the steps could be computer-assisted.

Comparing the transformations with respect to the pre-processing it is observed that all the methods need the CURE form as input. However, information like zero dependencies and hardware constraints are not used when applying MoFo's and SYKu's methods, whereas JaRaKa's and RoSwCaDe's methods utilize information for all dependence variables and associate hardware constraints.

3.2. The Transformations

The details of the methods are not discussed here see references, [4, 3, 5, 6]. The transformations basically use the dependency relations between variables to produce the scheduled hardware resources in terms of regular-connected processing units. A common characteristic for the various transformations is that they all exploit the inherent parallelism by observing the data dependencies. Primarily it is the approach in doing this that makes the transformations different.

S.Y. Kung uses a graphical or an algebraic approach and a linear projection of the DGs or dependence vectors onto Signal Flow Graphs (SFGs) or systolic arrays, resulting in 1- or 2-dimensional structures. Although graphical forms always help greatly in visualizing the data-dependencies and the working of the method, for large applications it may not always be possible to handle the problem easily. An optimization criteria is not built-in and an effective search strategy is also needed in the method and the obtained solutions need post-processing and simulation for further VLSI implementations.

The MoFo, JaRaKa's and RoSwCaDe's methods use data-dependence vectors and interconnection constraints. This allows a mapping by linear projections using algebraic methods. A linear scheduling function is also used in all the cases. However, MoFo's method specifies a schedule ($\underline{q}$) common to all variables at index point $\underline{I}$, $t = \underline{q}\,\underline{I}$, whereas in JaRaKa and RoSwCaDe's methods, a schedule is made for each variable, $t_a = \underline{q}\,\underline{I} + \gamma_a$, which includes detailed hardware constraints back annotated from the physical level, resulting in γ-values. The salient features of MoFo's method are that it is comparatively easier to understand, space mappings are generated by the method and scalability is also ensured. The JaRaKa's method ensures implementation in 1- or 2-dimensional arrays based on user-selected space mappings. RoSwCaDe's method is an extension of JaRaKa's method which enables optimal utilization of the hardware resources. Further it is the only one that has a kind of search and optimization strategy to obtain an efficient solution. It is implemented by solving small subproblems guided by the designer like making local hardware multiplexing and finding local and global schedules separately.

An algebraic approach as in MoFo's, JaRaKa's and RoSwCaDe's methods includes some formalism in the design process. Further JaRaKa's and RoSwCaDe's methods are capable of generating general and specialized solutions depending on the need which reduces the amount of post-verification. All the methods require a great deal of work to learn and use effectively, but the algebraic ones have built-in formalism and provide accurate results. For most of these transformations, additional verification workout may be needed.

3.3. Output from the Transformations

Most of the transformations need additional post-processing. In MoFo's and SYKu's methods post-scheduling is required to tailor the logical structures to physical VLSI implementations. All the methods need additional mappings for some of the solutions like elimination of multiple fan-in and fan-out, broadcasting and mapping from infinite to finite sized arrays, etc.

Even for this simple algorithm, one can arrive at entirely different solutions based on the re-hashed algorithm: different number of PEs, data flow and i/o requirements. In this case 12 different logical structures can be derived from all transformations. These are constrained out of a total of $3^3 = 27$, resulting from three interconnection directions and three different data flow dependencies. Out of these 12, only six are topologically different and the others have just reversed data flow. However, they may differ in various parameters, which results in different

Statistics for Regular Array Solutions						
Parameters	MoFo/SYKu			JaRaKa/RoSwCaDe		
	Case I	Case II	Case III	Case I	Case II	Case III
Area (#PEs)	N	$N+M$	M	N	$N+M$	M
comput.	2 add, 2mult			2 add, 2 mult		
memory	28{7}			14		13
control	3	2	1	3	2	1
Time						
throughput	8{2}			4		
latency	4{1}			6		
completion	$8(N-1)+4M\{2(N-1)+M\}$			$4(N-1)+2M$		
Utilization						
multiplier	$\frac{1}{2}\frac{M}{N}$	$\frac{1}{2}\frac{M}{N}$	$\frac{1}{4}$	$1\frac{M}{N}$		$\frac{1}{2}$
adder	$\frac{1}{4}\frac{M}{N}$	$\frac{1}{4}\frac{M}{N}$	$\frac{1}{8}$	$\frac{1}{2}\frac{M}{N}$		$\frac{1}{4}$
Input/Output						
DF i/o	multi	multi	single	multi	multi	single
# i/o	$2(N+1)$	$2(N+M)$	2	$2(N+1)$	$2(N+M)$	2
PE i/o	5	4	3	5	4	3

Table 1: Parameters for various ARMA filter regular array solutions obtained by using different transformations. Cases I, II and III MoFo/SYKu are based on a scheduling function $\underline{q} = [8\ -4]\ \{[2\ -1]\}$ and for JaRaKa/RoSwCaDe it is $\underline{q} = [4\ -2]$.The numbers in "{}" are based on the original schedule $\underline{q} = \{[2\ -1]\}$ found by MoFo's method. "multi" means broadcast of i/o and "single" means i/o taking place at boundary PEs.

solutions. An optimal application specific solution can be selected from these solutions.

To make it easier to compare the results generated by the transformations, all the methods should have equivalent assumptions. Therefore the assumed underlying hardware, for all solutions, has latency times like: 1 time unit for addition and delay, and 2 time units for multiplication. These hardware constraints correspond to word-level or bit-parallel arithmetic, and three out of the six logical structures are shown as Cases I, II and III in tables 1 and 2.

In table 1 MoFo and SYKu are placed in one group and JaRaKa and RoSwCaDe in another, as the methods in a group yield the same type of solutions. Table 2 lists two other types of solutions, obtained by refining RoSwCaDe's method further by including some special hardware sharing and multiplexing constraints in the transformation, denoted by "local mux" and "global mux". The obtained logical structures can be quite different and the values of some salient physical parameters are shown in the tables.

From the data presented in table 1 it appears that the controlled and detailed schedule as used in JaRaKa/RoSwCaDe yields more optimal solutions in terms of utilization, hardware requirements and timing statistics in this specific case. In table 2 the design tradeoffs are illustrated and here it is seen that further improvements in utilization and hardware cost can be achieved by doing local multiplexing within the PEs denoted by "(RoSwCaDe (local mux))", in contrast to the "JaRaKa/RoSwCaDe" solutions, without losing drastically in timing performance. The last example denoted by "RoSwCaDe (global mux)" shows an extreme case of area and time tradeoff, where the full array has been mapped onto a single PE.

The scheduling function found for MoFo's method needs some post-processing to include realistic hardware constraints, because in this case only 1 time unit has been allocated to perform the atomic computation in the PEs. The global symbolic like scheduling function obtained by

STATISTICS FOR THE LOCAL AND GLOBAL MULTIPLEXED ARRAY SOLUTIONS					
Parameters	RoSwCaDe (local mux)				RoSwCaDe (global mux)
	Case I	Case II	Case III	Case III(bit-serial)	Case III($M = 8$, 1PE)
Area (#PEs)	N	$N+M$	M	M	1
comput.	1 add, 1 mult				1 add, 1 mult
memory	16	26	13	$18b+40$	154
control	7	6	5	5	8
Time					
throughput	5	6	4	$4b$	32
latency	6	9	7	$5b+1$	33
completion	$q_i(N-1) - q_j M$				$32(N-1)+4M$
Utilization					
multiplier	$1\ \frac{M}{N}$	$1\ \frac{M}{N}$	1	1	1
adder	$\frac{1}{2}\ \frac{M}{N}$	$\frac{1}{2}\ \frac{M}{N}$	$\frac{1}{2}$	$\frac{1}{2}$	$\frac{1}{2}$
Input/Output					
DF i/o	multi	multi	single	single	single
# i/o	$2(N+1)$	$2(N+M)$	2	2	2
PE i/o	5	4	3	3	3

Table 2: Parameters for the local and global multiplexed array ARMA filter solutions obtained by using some of RoSwCaDe and our constraints in the mapping. For the "local mux" Case I: $\underline{q} = [5\ -4]$, Case II: $\underline{q} = [6\ -2]$, Case III: $\underline{q} = [4\ -2]$, Case III(bit-serial): $\underline{q} = [4b\ -(2b+1)]$ and for "global mux" single PE it is Case III($M = 8, 1\text{PE}$): $\underline{q} = [32\ -4]$. Note that the "completion"-time, due to lack of space, is specified symbolically as a function of $\underline{q} = [q_i\ q_j]$.

the method, $\underline{q} = [2\ -1]$, has therefore been multiplied by the latency time for the critical path in the PE, $(2(*) + 1(+) + 1(+) =)\ 4$, in order to obtain a valid detailed schedule. Further the found schedule only allows construction of 6 out of a total 12 solutions, or rather 3 topologically different ones: Cases I, II and III. To obtain the others a new and less time-optimal scheduling function has to be found, according to the optimality measure proposed in [4]. Constraints could be added but, then a new method would evolve and eventually it will get closer to JaRaKa's and RoSwCaDe's methods.

In JaRaKa's and RoSwCaDe's methods the hardware constraints are used as input for finding a schedule. In the "JaRaKa/RoSwCaDe" cases all the solutions have the same scheduling function, provided that there are no multiplexing constraints. But in the local and global multiplexing cases new constraints are added and different scheduling functions characterize the different solutions. But no post-scheduling is needed in either of these cases.

Seen from the VLSI designer's point of view, the RoSwCaDe's method is the one generating the output, which for this particular example, is closest to the final chip solution. During the evaluation of the transformations new constraints were defined and applied successfully in Cases III($M = 8, 1$ PE and bit-serial). The new defined pipelining and local/global multiplexing constraints, provide the possibility of folding a full array to smaller ones, and it provided the capability of modeling the hardware both on latency and throughput, in a formal manner.

3.4. Physical Implementation Details

Table 2 contains a global multiplexing case, which is a mapping of an array of size M onto a single PE for an ARMA filter of order $M = 8$, denoted by Case III ($M = 8, 1PE$). The global multiplexing is performed for Case III because it has a finite array size M, boundary i/o communication and therefore no ad hoc post-processing is needed.

The local multiplexing Case III is the optimal solution based on the selected case study parameters. It has been re-transformed using new hardware constraints, Case III(bit-serial). The statistics for the new solution are timing numbers derived from the physical bit-serial Least Significant Digit First (LSDF) arithmetic. The hardware constraints here also refer to the latency time of the resource in bit clock periods. For two's complement b-bits arithmetic the latency constraints are: 1 time unit for addition and delay and $b + 1$ time units for multiplication. Throughput time is b time units for addition and delay and $2b$ for multiplication. The area parameter "memory" describes the number of one bit registers.

The selected optimal solution , Case III(bit-serial), has been implemented in a semicustom system generating standard cell layout, SOLO 1400, [11]. The main reason for selecting a standard cell solution is that it is fast, simple, user-friendly and a number of industrial problems have been implemented using such systems. The interface to the tool is either a hierarchically schematic capture (DRAFT) or a procedure-oriented Hardware Description Language (MODEL) and we opted for MODEL [11].

The process of converting the output from the RoSwCaDe's method to an HDL consists of the following steps:

- design of the computational building blocks like delay, memory, adder and multiplier at the gate level,
- design of the data path for the atomic computation, similar to the implementation of the data flow graph selected in the pre-processing (step 5 in the re-hashing procedure),
- synthesis and implementation of the control signals, using the obtained scheduling functions for the corresponding variables as inputs to the control structures.

The physical layout implements a 16-bits fix point two's complement bit-serial arithmetic and a filter of order $M = 25$. The layout showed that the area is equal to the number of PEs multiplied by the area of one PE. The total area is $25 \cdot 6mm^2$/PE $\sim 150mm^2$ for a 1.5 μ CMOS standard cell layout and control logic occupies approximately 15% of that. With the standard clocking scheme (two non-overlapping clocks) and an automatic clock distribution, the clock rate is $16MHz$ which gives a sampling rate of $16MHz/(4 \cdot 16bits) \sim 250kHz$. Further the regularity factor is 24/25 $\sim$ 96%, because we have only designed one PE out of 25. The design of the PE includes computation and control. A comparison of the transformation output to the simulation results shows an almost one-to-one correspondence, with clock-ticks used in one and nano-seconds in another.

3.5. Results from other Computational Models

The following summarizes the results derived from the regular VLSI structures and estimates from theoretical complexity models.

Complexity is used here to estimate measures obtained from various theoretical computational models. For the von Neumann solution the Random Access Memory (RAM) computational model has been used, where estimates for operations are 1 time unit for addition, read, write, memory move and $3b$ time units for multiplication [$b \sim$ *#bits*]. The signal processor is based on the Harvard architecture of the general ADSP 2101, $12.5MHz$ (80ns), 16 bits, processor from Analog Devices Inc. It has special instructions that in our case can be utilized 100%. It is capable of doing multiplication, accumulation, fetch new data and coefficients and update pointers and memory simultaneously. Given these two models, the number of cycles to perform an Mth order ARMA filter is $114M + 51$ for the von Neumann and $2M + 1$ for the DSP. Using the same sampling rate as for the systolic array solution, i.e. $250kHz$, the DSP is just capable of implementing a 25-order ARMA filter while being utilized 100% ($12.5MHz/250kHz \sim 50[cycles] = 2M + 1 \Rightarrow M \sim 25$). For general and practical usage, it is not possible to utilize the processor 100% as in this case but only 80%, because 20% is used for administration of data, loops, control structures etc. The von Neumann machine needs a clock frequency of approximately $(114 \cdot 25 + 51)250kHz \sim 725MHz$ to implement an ARMA filter of order $M = 25$. If it is assumed that the clock of the von Neumann is $20MHz$ then the sampling rate, $1/t_{throughput}$, is approximately $20MHz/(114 \cdot 25 + 51) \sim 7kHz$.

The above measures give an idea about where the obtained Regular VLSI solutions are placed in area and time. It is observed that the bit-serial based architectural solution, Case III(bit-serial), using a semicustom 1.5μ CMOS process is equivalent to a state-of-the-art signal processor implemented in 1.0μ CMOS. If the computational power of the signal processor is too high for the application the local and the global multiplexed array solutions can be used to make the needed tradeoffs to fit the specification. A better solution can be obtained by using a full custom design style where a realistic gain could be a factor of 3-4 in area and time and another factor of 2 going from 1.5 to 1.0 μ CMOS. So the RoSwCaDe's method with our pre-processing and additional constraints is capable of generating various design solutions ranging from high- to low-speed applications.

3.6. Final Remarks

We have observed that JaRaKa's and RoSwCaDe's methods are those requiring the least amount of post-processing. Compared to the other schemes the stated solutions of RoSwCaDe are at a lower level of detail and most optimal in terms of area, time and utilization. Further it requires the least amount of post-processing and the performance of the solution obtained is comparable to von Neumann and state-of-the-art DSP signal processors.

By applying global multiplexing constraints and using area, time and utilization parameters as well as the amount of pre- and post-processing as the main parameters for weighted evaluation, one can select RoSwCaDe's method as the optimal one suitable for implementing the algorithm. *Thus this is the only method that meets the stated goals of section 3*. Initial experiments with other adaptive filter algorithms, like QRD, Lattice, Kalman and Transversal, also confirm the above conclusion.

It should be noted that the standard cell implementation completes our design trajectory. The main purpose of the semicustom case was to illustrate the interaction and processing needed to obtain a fast VLSI chip solution.

4. Conclusion

In this paper we have discussed how to compare various formal transformation schemes using an evaluation matrix of design parameters. We have reported application of this scheme to a non-trivial example from the DSP domain. A full-blown evaluation matrix may contain many more parameters of interest. However, for many applications a subset of the selective parameters may yield the needed oversight for selecting a given scheme for optimal and correct implementation.

The use of transformations on more advanced DSP algorithm structures, such as SVD will be studied next. The semi- or full-automation of the pre- and post-processing steps is the short term goal, whereas automating the generation of the full-blown evaluation matrix is the long term goal of our research. It should be mentioned, however, that for a given application, the selection of an appropriate transformation technique is not an easy task, because all the transformations need to be tried exhaustively and the tools for these are still immature [12]. No precise method can be provided at this stage for selection beforehand, without trying all these techniques. So, our limited experience with this case study from the DSP domain indicates that such a comparison of transformations with the help of an evaluation matrix, should be a necessary step in selecting a given method for an application.

References

[1] J.A.B Fortes, K.S. Fu, and B.W. Wah. Systematic approaches to the design of algorithmically specified systolic arrays. IEEE Confrence on ICASSP 85, vol.1, pp. 8.9.1-4, 1985.

[2] M.T. O'Keefe and J.A.B. Fortes. A comparative study of two systematic design methodologies for systolic arrays. IEEE International Conf. on Parallel processing, pp. 672-675, 1986.

[3] S.Y. Kung. VLSI array processors. Prentice Hall Information and System Sciences Series, Prentice Hall,Prentice Hall, Englewood Cliffs, 1988.

[4] D.I. Moldovan. Towards a computerized optimal design of VLSI systolic arrays. Design Methodologies, S. Goto (Editor), Elsevier Science Publishers B.V. (North-Holland), pp. 215-234, 1986.

[5] H.V. Jagadish, S.K. Rao, and T. Kailath. Array architectures for iterative algorithms. Proceedings of the IEEE, vol. 75, no. 9, pp. 1304-1321, 1987.

[6] I. Rosseel, F. Catthoor, and H. De Man. Extensions to linear mapping for regular arrays with complex processing elements. Conf. on Application Specific Array Processing, ASAP90, New York, 1990.

[7] A. Færgemand, K.K. Bagchi, P.M.R. Jensen, and O. Olsen. A VLSI design scheme for a class of algorithms. Int. Workshop on Algorithms and Parallel VLSI Architectures, Part B, pp. 78-82, Pont-a-Mousson, France, 1990.

[8] J. Bu and E.F. Deprettere. Converting sequential iterative algorithms to recurrent equations for automatic design of systolic arrays. IEEE Confrence on ICASSP 88, vol. IV, V2.7, pp. 2025-2028, 1988.

[9] Y. Wong and J.M. Delosme. Broadcast removal in systolic algorithms. Proceedings of the International Workeshop on Systolic Arrays, pp. 403-413, 1988.

[10] S.V. Rajopadhye. Synthesizing systolic arrays with control signals from recurrent equations. Distributed Computing, 3, pp. 88-105, Springer-Verlag, 1989.

[11] ES2. Solo 1400 reference manual. ES2 Publications Unit, 1990.

[12] E.S. Manolakos, H.M. Stellakis, and D.H. Brooks. Parallel processing for biomedical signal processing. IEEE Computer, pp. 33-43, March 1991.

VLSI 91
A. Halaas and P.B. Denyer (Eds.)
Elsevier Science Publishers B.V. (North-Holland)

A New Chip Architecture for VLSIs - Optical Coupled 3D Common Memory and Optical Interconnections

Mitsumasa Koyanagi

Hiroshima University, Research Center for Integrated Systems
1-4-2 Kagamiyama, Higashi-Hiroshima, Japan 724

ABSTRACT

A new intelligent memory with the optical interconnection and parallel processing function is proposed to overcome the problems of the increased signal delay and the cross talk in the electrical wiring. This new memory is called an optically coupled three-dimensional common memory (3D-OCC memory). A very fast parallel processing computer system can be achieved by using the 3D-OCC memory. The 3D-OCC memory consists of the multilayered structure of two-dimensional memory with LEDs and photoconductors. Memory layers are optically coupled with each other through the LEDs and the photoconductors which compose the optical interconnection. Data are transferred in the vertical direction by optical coupling, while the conventional memory operations are performed in the horizontal planes. Very fast optical data transfer of 128 Gbits / s is confirmed in the simulation.

1. INTRODUCTION

As the capacity and the packing density of VLSIs are increased, the signal delay on the electrical wiring becomes the most significant problems. The total wiring length is increased with increasing the capacity and the packing density of VLSIs because the chip size is increased although the device size is shrunk. The increased wiring length causes the wiring capacitance and resistance increased and the operation speed decreased. The decrease of the operation speed is accelerated when the supply voltage is reduced according to the scaling rule. Consequently, the performance of VLSIs comes to be determined by the wiring delay rather than the device performance. Accordingly, it is a key issue in the future VLSI how to overcome the problem of the chip performance deterioration caused by the electrical wiring. It is considered for solving these electrical wiring problems to decrease the wiring delay itself by employing the optical interconnection and to increase the effective operation speed by adopting the parallel processing method. Then, we propose a new intelligent memory as an example of new chip architectures with the optical interconnection and the parallel processing function. This new intelligent memory is called an optically coupled three-dimensional common memory which is abbreviated as 3D-OCC memory [1-4].

The 3D-OCC memory can be used for achieving the very fast parallel processing computer system. Rapid transfer of memory data and simultaneous use by many CPU's are essential for performing fast parallel processing in such computer system. However, the performance of the bus connection type parallel processing computer system with multiprocessors becomes to be limited by the data communication efficiency of the signal bus if the conventional electrical signal bus is used. This is because it is difficult to simultaneously transfer a large volume of data among many processors and memories by using the conventional electrical signal bus. The 3D-OCC memory can be used instead of the conventional signal bus in the parallel processing computer system because it has a new function of simultaneously implementing the parallel data transfer and the data latch for a block of data. The 3D-OCC memory is considered as a multi-bus with memory function. A large block of data can be simultaneously transferred with very high speed by optical coupling and these transferred data can be shared among many CPU's without conflict in the 3D-OCC memory.

2. Interconnection in Future VLSI

According to the scaling rule, the CR time constant of electrical wiring remains constant even after the device size is scaled-down because the wiring capacitance is decreased although the wiring resistance is increased. However, in fact the wiring capacitance is rather increased. This is because the fringing and coupling capacitances are increased as a result that the wiring spacing is narrowed without reducing the film thickness of metal wiring. Consequently, the signal delay time on the electrical wiring is dramatically increased as the wiring line width is decreased as shown in Fig.1 where the wiring spacing is taken to be identical with the wiring line width and the wiring length is 1 cm. The wiring delay can be reduced by employing the copper wiring in place of the aluminum wiring as shown in Fig.1. However, the wiring delay is still quite large for the narrow line wiring. Another serious problem in the electrical wiring when the device is scaled-down is the cross talk. The coupling capacitances between adjacent wirings are significantly increased when the wiring spacing is reduced. As a result, a very large coupling noise is easily superimposed on the adjacent wirings as a cross talk in the deep submicron wiring. Thus, the wiring delay and the cross talk become significant problem in the future VLSI.

These problems can be overcome by employing the optical interconnection. Two kinds of optical interconnections are applicable for VLSI chip or module. One is the horizontal optical interconnection and the other is the vertical optical interconnection as shown in Fig.2. The horizontal optical interconnection is suitable for the long distance wiring and hence can be used in place of the electrical main bus. This optical main bus can connect the several large circuit blocks with very small wiring delay. The advantage of the horizontal optical interconnection becomes more pronounced as the wiring length is increased. The wiring flexibility is very high in the optical interconnection because the two light beams can easily pass across each other without disturbing. The vertical optical interconnection is

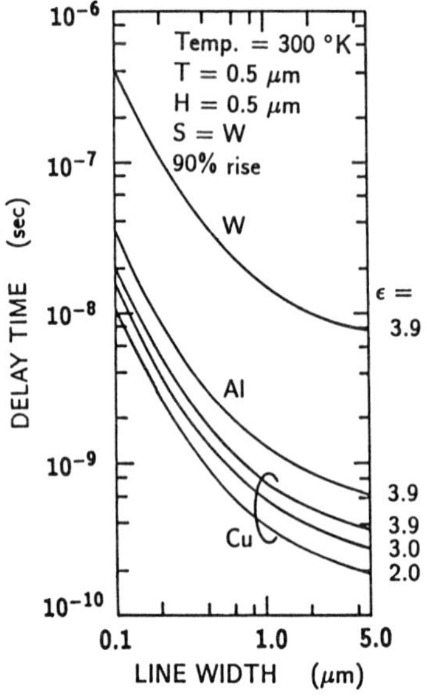

Fig.1. Signal delay time as a function of wiring line width (= wiring spacing). metal film t_M = dielectrics t_{ox}

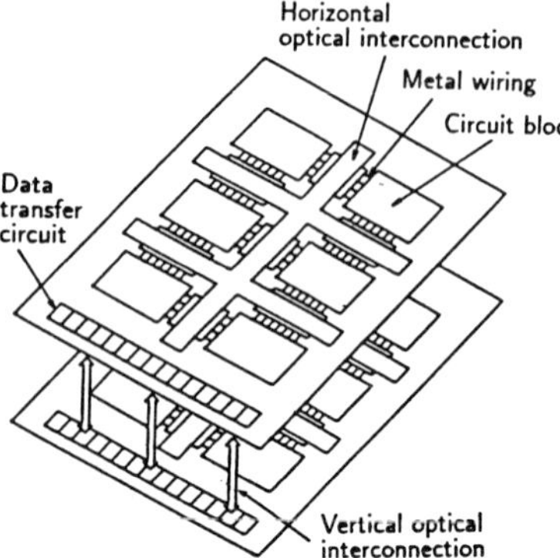

Fig.2. VLSI chip or module with optical interconnection.

very attractive for parallel processing because it becomes possible to transfer a large block of data simultaneously. Furthermore, the vertical optical interconnection makes it easy to construct the three-dimensional structure which is also preferable for the parallel processing. The three-dimensional structure is useful for reducing the long distance wiring as well. Thus, the combination of the optical interconnection and the parallel processing is the most promising approach for the future VLSI chip or module. The vertical optical interconnection and the parallel processing method are utilized in the 3D-OCC memory.

3. Optically Coupled 3D Common Memory

3.1. Parallel processor system with 3D-OCC Memory

An example of a parallel processor system with 3D-OCC memory is shown in Fig.3. The 3D-OCC memory consists of a multilayered structure of 2D memory with LED's and photoconductors. Each memory layer in the 3D-OCC memory is connected to the respective CPU. In addition, a memory layer is optically coupled with the upper and lower memory layers by LED's and photoconductors which compose the vertical optical interconnection. In this system, a large block of data can be simultaneously transferred with very high speed by optical coupling in the vertical direction, while the conventional memory operations are carried out in the horizontal planes. Memory cells with identical addresses in all memory layers have the same data after the optical data transfer. The stored data in the memory can be simultaneously read by many CPU's without conflict. Consequently, very fast real-time parallel processing can be achieved in this system.

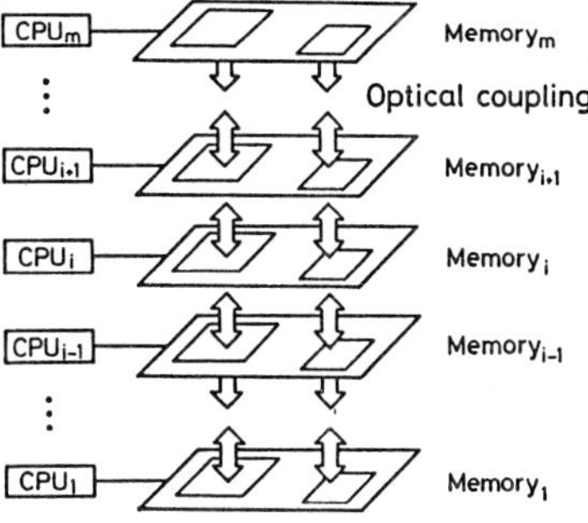

Fig.3. Parallel processing system with 3D-OCC memory.

The optical coupling in the 3D-OCC memory is carried out by using the optically coupling flip-flop with two pairs of LED's and photoconductors as shown in Fig.4. This flip-flop consists of two parts for data store and data transfer. The data store part is similar to the conventional high resistive load SRAM memory cell. In the optically coupling flip-flop, however, the high resistive loads, PC1 and PC2, act as the photoconductors to detect the signal light as well. The data transfer part, which contains two LED's, LED1 and LED2, and three NMOS transistors, Tr5, Tr6 and Tr7, is connected to the nodes of the data store flip-flop. In this flip-flop, the electrical writing and reading are performed in the similar manner to the conventional SRAM. Meanwhile, the optical writing is performed by utilizing the phenomenon that the photoconductor resistance is decreased by absorbing the signal light.

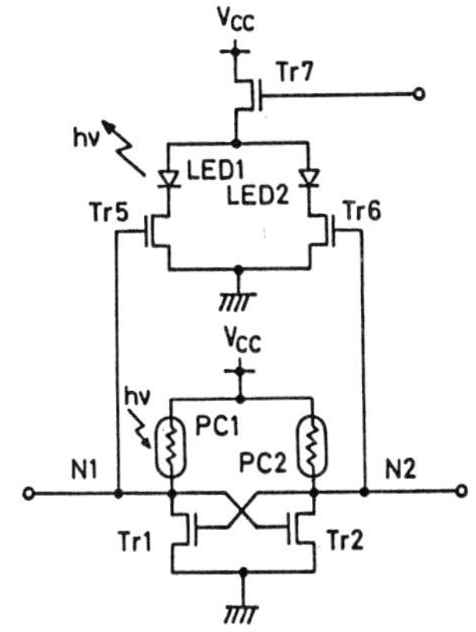

Fig.4. Optically coupling flip-flop circuit.

When either of the two photoconductors receives the signal light, the voltage of the flip-flop node where the illuminated photoconductor is connected is set to "high" irrespective of the previous node condition. The content of written data is determined by selecting the photoconductor to receive the signal light. The optical data transfer operation starts by selecting the LED control line which is connected to the pull-up transistor Tr7. The LED connected to "high" node of the flip-flop emits the signal light to the upper and lower memory layers. This emitted light impinges onto the corresponding photoconductors of the upper and lower memory layers. Therefore, each of LED's is placed face to face with the corresponding photoconductor between every two memory layers. Thus, data are transferred vertically by optical coupling through LED's and photoconductors.

A cross-sectional view of the 3D-OCC memory is shown in Fig.5 where only two memory layers are illustrated. As is clear in the figure, LED's and photoconductors are fabricated by using GaAs-on-Si technology [5,6] after completing the CMOS wafer process. Silicon wafers with LED's and photoconductors are thinned and glued to each other by means of the wafer-to-wafer bonding technique to form the 3D structure [7].

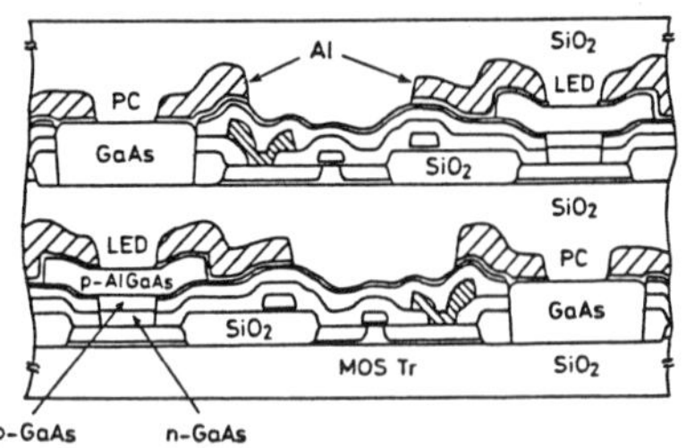

Fig.5. Cross section of the 3D-OCC memory.

3.2. Memory-cell-coupled 3D-OCC Memory

The optically coupling flip-flop can be used as a memory cell in 3D-OCC memory. In such 3D-OCC memory, very-high-speed data transfer can be achieved since the memory cells are directly coupled by light. We call this type of 3D-OCC memory, a memory-cell-coupled 3D-OCC memory. In the memory-cell-coupled 3D-OCC memory, two LED's and two photoconductors are contained in the memory cell and are laid out face to face between the upper and lower memory layers. Therefore, in this memory, the horizontal spacing between LED's or photoconductors should be minimized for decreasing the memory cell size and increasing the packing density. However, decreasing the horizontal spacing results in the increased stray light effect, because the light emitted from an LED impinges not only on the corresponding photoconductor as a signal light but also on the other photoconductors as a stray light. Then, the influence of such stray light should be precisely evaluated based on the optical coupling efficiency between LED and photoconductor so as to optimally design the memory cell. The optical coupling efficiency is determined by the light-emission efficiency of LED, the transmission efficiency through an insulating layer between the LED and the photoconductor, and the

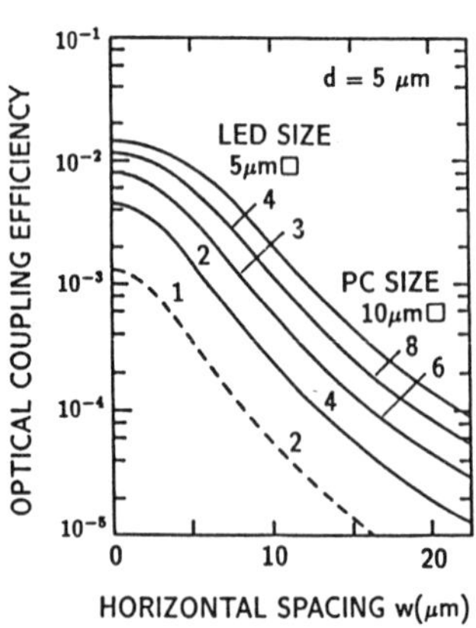

Fig.6. Optical coupling efficiency vs. horizontal spacing for different sizes of LED and photoconductor (PC) pairs.

light-collection efficiency of the photoconductor.

The optical coupling efficiency calculated is shown as a function of the horizontal spacing between the LED and the photoconductor w in Fig.6. The LED and photoconductor sizes are varied as parameters. As is obvious in the figure, the optical coupling efficiency rapidly decreases with increasing the horizontal spacing w. The optical coupling efficiency also decreases with reducing the sizes of LED and photoconductor. The optical coupling efficiency for the signal light is obtained at $w = 0$ while that for the stray light is given for $w > 0$. Therefore, the influence of stray light can be considerably reduced by increasing the horizontal spacing w. However, increasing the horizontal spacing results in the increase of the memory cell size. Therefore, the horizontal spacing should be optimized considering the optical coupling action. Thus, the optical coupling efficiency for the stray light can be reduced to less than 0.01% when the horizontal spacing w is 20 μm and the sizes of LED and photoconductor are $5 \times 5\mu m^2$ and $10 \times 10\mu m^2$, respectively. Meanwhile, the optical coupling efficiency of 1.4% can be obtained for the signal light at the same condition.

The optical data transfer speed in the optically coupling flip-flop strongly depends on the optical coupling efficiency between the LED and photoconductor. It is required to increase the optical coupling efficiency for the signal light, decreasing that for the stray lights in order to increase the optical data transfer speed. Figure 7 shows the data transfer time of the optically coupling flip-flop as a function of the optical coupling efficiency for the signal light. The LED current is changed as a parameter. The data transfer time is significantly increased as the optical coupling efficiency is decreased. On the other hand, the data transfer time is decreased with increasing the LED current because the signal light intensity is increased. Therefore, it is obvious from Fig.7 that the LED current should be $275\mu A$ or more in order to reduce the optical data transfer time to less than 10 ns when the optical coupling efficiency is 1.4%.

Based on the results discussed above, 4-kbit × 4-layer memory-cell-coupled 3D-OCC memory are designed with $2\mu m$ CMOS technology. For performing the optical/electrical circuit simulation in the chip design, optoelectronic device models were newly developed and incorporated into the SPICE simulator. A circuit block diagram for one

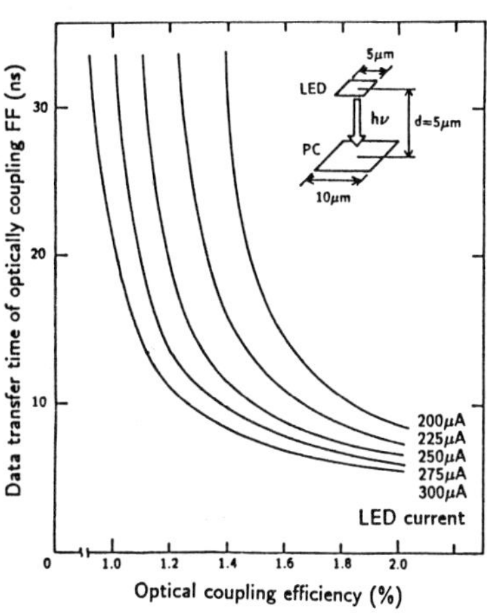

Fig.7. Data transfer time vs. optical coupling efficiency of signal light.

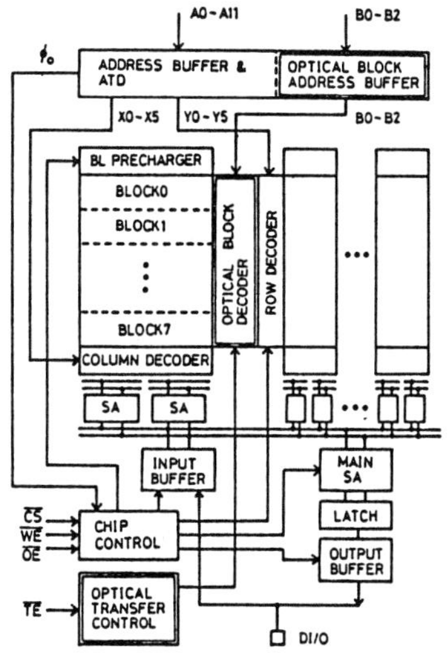

Fig.8. Circuit block diagram for one memory layer of the 4-kbit × 4-layer 3D-OCC memory.

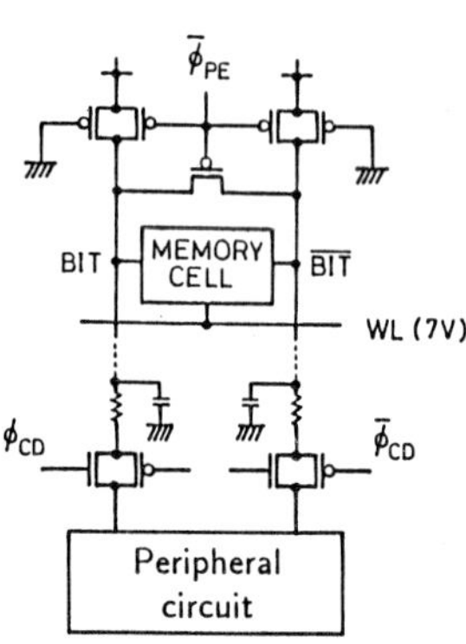

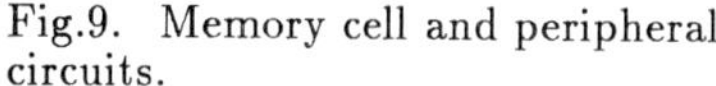

Fig.9. Memory cell and peripheral circuits.

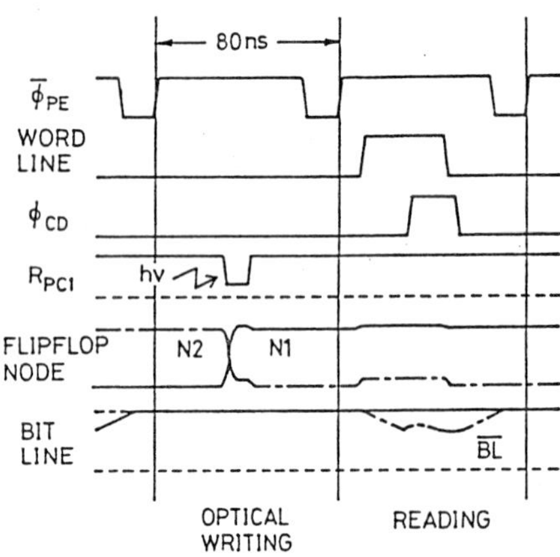

Fig.10. Waveforms for the optical writing and electrical reading operation in the memory-cell-coupled 3D-OCC memory.

memory layer of the memory-cell-coupled 3D-OCC memory designed by using this optical/electrical simulator is shown in Fig.8. In order to reduce the power consumption, the memory cell arrays are divided into eight blocks and a data block of 512 bits is simultaneously transferred. In this condition, one LED consumes the active power of about $23\mu W$ with $300\mu A$ drive current. The optical data transfer characteristics of this memory are simulated in the memory cell and peripheral circuits shown in Fig.9. The waveforms in the optical writing and electrical reading operation in one memory layer are shown in Fig.10. As is obvious in the figure, the photoconductor resistance is decreased and the node voltages of memory cell flip-flop is changed after receiving signal light, and thus the data are optically written. The optically written data do not appear on the bit lines during the optical writing operation because the data are written directly into the memory cell flip-flop. The written data are correctly read into bit lines in the following electrical reading operation as shown in the figure. Figure 11 depicts the waveforms for the multilayer optical data transfer operation in the 3D-OCC memory. The data is electrically written into the memory cell flip- flop of the first memory layer by applying the pulse WL to the memory cell word line and the pulse ϕ_{CD} to the transfer gate after the bit lines are precharged to the same voltage by applying pulse ϕ_{PE}.

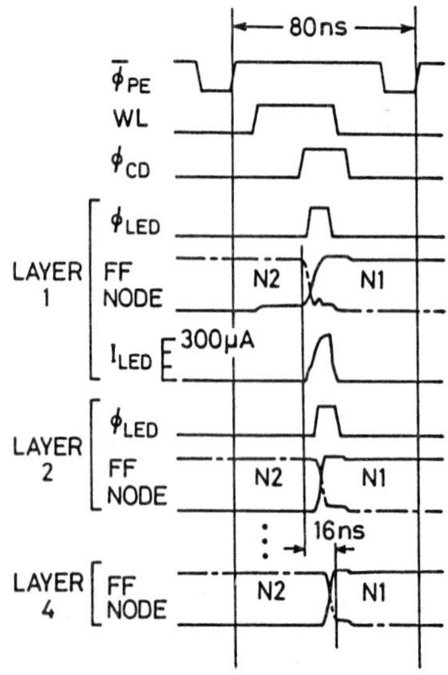

Fig.11. Waveforms for the multilayer optical data transfer operation in the memory-cell-coupled 3D-OCC memory.

Then, if the pulse ϕ_{LED} is applied to the LED control line, the current I_{LED} flows through the LED connected to the "high" node of the memory cell flip-flop to emit the signal light toward the upper and lower memory layers. This emitted light changes the flip-flop node voltage in the upper and lower memory layers. Thus, the data electrically written into the first layer memory cell flip-flop is sequentially transferred through the second and third layers to the fourth layer memory cell by optical coupling. A block of 512 bit data is transferred through four memory layers within 16 ns. This is equivalent to the data transfer speed of 128 Gbits/s. The waveforms for the data transfer operation during electrical reading are plotted in Fig.12. As is clear in the figure, the data are correctly transferred by optical coupling even during electrical reading without disturbing the electrical reading operation. Thus, it is confirmed by simulation that a large number of data are successfully transferred with a very high speed in the memory-cell-coupled 3D-OCC memory.

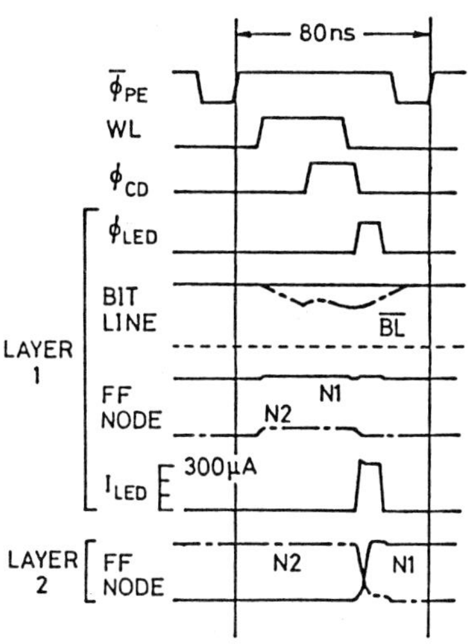

Fig.12. Waveforms for the optical data transfer operation during electrical reading in the memory-cell-coupled 3D-OCC memory.

3.3. Sense-amplifier-coupled 3D-OCC Memory

As is obvious in Fig.6, the optical coupling efficiency for the signal light rapidly decreases with decrease in the sizes of LED and photoconductor. Therefore, it seems difficult to reduce the sizes of LED and photoconductor to less than $1\mu m \times 1\mu m$. This means that the packing density and the memory capacity are not sufficiently increased even if CMOS device sizes are drastically scaled-down. Accordingly, the memory-cell-coupled 3D-OCC memory is not suited for a large capacity of memory application. In order to reduce the memory cell size for increasing the packing density, LED's and photoconductors should be removed from the memory cell.

In the 3D-OCC memory, all LED's do not always simultaneously emit the signal lights during the vertical data transfer because the total power consumption for LED's is limited. This means that each of the memory cells does not need LED's, and hence the optical data transfer part can be separated from the memory cell array part. As a result, we can use conventional memory cells such as SRAM cells or DRAM cells and reduce the 3D-OCC memory chip size. Figure 13 shows an example of this type of 3D-OCC memory. In this memory, optically coupling sense amplifiers are used in the optical data transfer parts. Data transfer between the upper and lower memory cells is implemented through the optically coupling sense amplifier, and all data stored in the memory cells which are connected to the one word line are simultaneously transferred. We call this 3D-OCC memory, a sense-amplifier-coupled 3D-OCC memory [8,9]. In Fig.13, an SRAM cell is

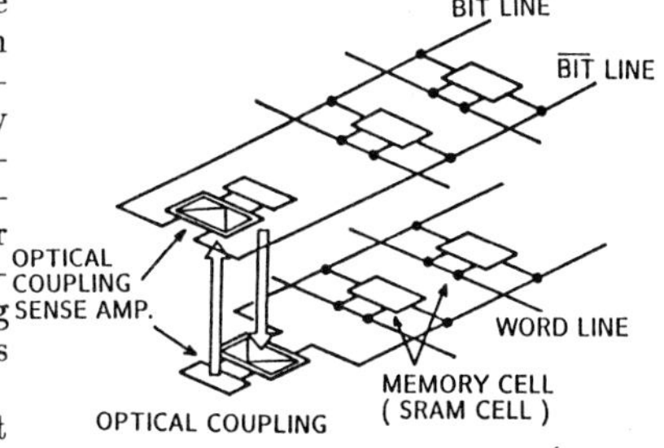

Fig.13. Sense-amplifier-coupled 3D-OCC memory circuit.

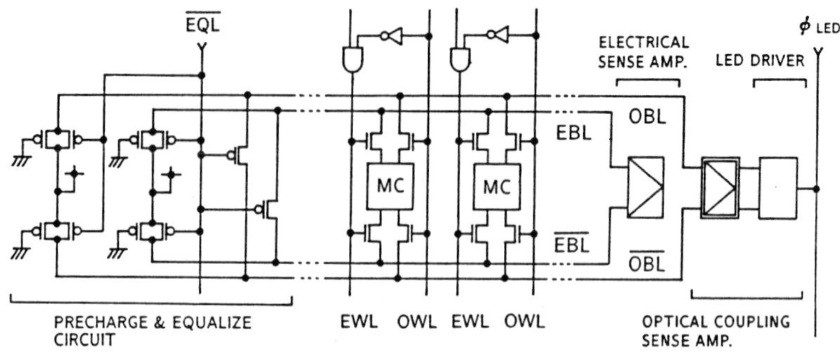

Fig.14. Sense-amplifier-coupled 3D-OCC memory circuit with dual-port SRAM cell.

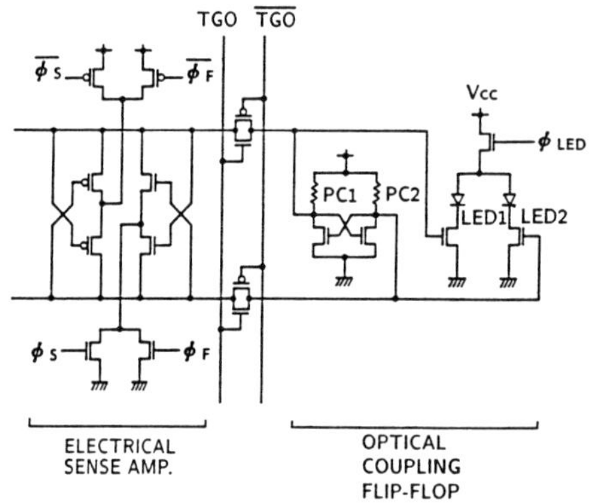

Fig.15. Optically coupling sense-amplifier circuit.

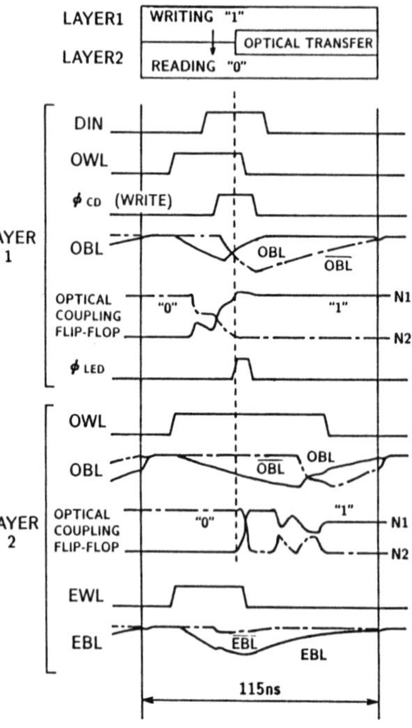

Fig.16. Waveforms in the sense-amplifier-coupled 3D-OCC memory with dual-port SRAM cell.

employed as a memory cell. Suppose that data are transferred from the upper memory cell to the lower memory cell in Fig.13. Data being read from the upper memory cell are amplified by the upper optically coupling sense amplifier and transferred to the lower optically coupling sense amplifier. The transferred data are again amplified by the lower optically coupling sense amplifier and written into the lower memory cell. Thus, this sense-amplifier-coupled 3D-OCC memory needs two operation steps per cycle and hence its operation speed becomes slower, while the memory-cell-coupled 3D-OCC memory requires only one step. However, if the word line and bit line are duplicated and a two-port memory cell is employed as shown in Fig.14, one operation step per cycle is also possible in the sense-amplifier-coupled 3D-OCC memory.

The optically coupling sense-amplifier circuit is shown in Fig.15. It consists of the optically coupling flip-flop and the electrical dynamic CMOS sense amplifier. The CMOS sense amplifier is used for amplifying not only electrically read-out signals but also optically transferred data. Data transferred to the optically coupling flip-flop are written into the memory cell after they are amplified by a CMOS sense amplifier.

The waveforms for electrical writing/reading and optical data transfer operation in the sense-amplifier-coupled 3D-OCC memory with dual-port SRAM cell are shown in Fig.16, where the data transfer operation between two memory layers are described. As is clear in Fig.16, the electrical writing operation of data "1" in the layer 1, the electrical reading operation of data "0" in the layer 2, and the optical data transfer of data "1" from the layer 1 to the layer 2 are performed in one cycle, although the cycle time is slightly larger than that of the memory-cell-coupled 3D-OCC memory. The data "1" written into the layer-1 memory cell are transferred to the layer-2 memory cell through optically coupling sense amplifiers by applying clock pulses ϕ_{LED} to the LED control line and by activating the electrical sense amplifier part of the optically coupling sense amplifier in the layer-2. It takes 16 ns to transfer data continuously through optically coupling flip-flops from layer-1 to layer-4. Thus, it is confirmed that the data are successfully transferred by sense amplifier coupling.

Accordingly, the packing density and the memory capacity can be increased to the degree comparable to those of the conventional SRAM or DRAM by using the sense-amplifier-coupled 3D-OCC memory, although the data transfer speed is slightly slower than that of the memory-cell-coupled 3D-OCC memory.

4. CONCLUSION

The optically coupled three-dimensional common memory (3D-OCC memory) has been proposed as a new intelligent LSI to overcome the wiring problem which becomes very serious in the future VLSI. The 3D-OCC memory has the optical interconnection and the parallel processing function together with three-dimensional structure. It is confirmed in simulation that the 3D-OCC memory is very suitable for simultaneously transferring a large number of data with a very high speed and sharing these data among many CPU's. A block data of 512 bits can be transferred within 16 ns in the memory-cell-coupled 3D-OCC memory which is designed by using 2 μm design rule. This is equivalent to the data transfer speed of 128 Gbits/s. The packing density and the memory capacity comparable to these of SRAM or DRAM are achieved in the sense-amplifier-coupled 3D-OCC memory where the optically coupling parts for the data transfer are placed in the periphery of the memory cell array area, although the data transfer speed is slightly slower than that of the memory-cell-coupled 3D-OCC memory. A parallel processing system with very high data transfer speed can be achieved by utilizing such 3D-OCC memory.

5. ACKNOWLEDGMENTS

The author would like to thank Prof. M. Hirose, Prof. M. Ae., Prof. M. Yamanishi, Prof. S. Yokoyama, Prof. R. Aibara, Mr. H. Takata and Mr. H. Okano for useful discussions of device characteristics and memory architecture.

6. REFERENCES

1. T. Ae, Dig. 5th Int. Workshop Future Electron Devices, pp. 55-60, 1988.
2. M. Koyanagi, H. Takata, T. Maemoto and M. Hirose, Optoelectron., vol. 3, no. 1, pp. 83-98, June 1988.
3. M. Koyanagi, H. Takata, H. Mori and M. Hirose, Symp. VLSI Circuit Dig. Tech. Papers, pp. 81-82, 1989.
4. M. Koyanagi, H. Takata, H. Mori and J. Iba, IEEE J. Solid-State Circuits, Vol. 25, no. 1, pp. 109-116, Feb. 1990.
5. H. Shichijo and R. J. Matyi, IEDM Tech. Dig., pp. 88-91, 1987.
6. I. Suemune, Y. Kunitsugu, Y. Tanaka, Y. Kan and M. Yamanishi, Appl. Phys. Lett., vol. 53, no. 22, pp. 2173-2175, Nov. 1988.
7. T. Hamaguchi, N. Endo, M. Kimura and M. Nakamae, IEDM Tech. Dig., pp. 688-691, 1985.
8. H. Takata, H. Mori, J. Iba and M. Koyanagi, Extended Abstr. 21st Conf. Solid State Devices and Materials, pp. 441-444, 1989.
9. H. Takata, H. Mori, J. Iba and M. Koyanagi, Japan. J. Appl. Phys., Vol. 28, no. 12, pp. L2305-L2308, 1989.

VLSI 91
A. Halaas and P.B. Denyer (Eds.)
Elsevier Science Publishers B.V. (North-Holland)

Pass-Transistor Self-Clocked Asynchronous Sequential Circuits

Farhad Aghdasi[1]

Department of Electrical and Electronic Engineering
University of Bristol, Bristol BS8 1TR, UK

Abstract

The contribution of this work is to present a new, systematic and easy to apply method of design for VLSI implementation of asynchronous sequential circuits without having to consider critical races and hazards. This is achieved by generating templates in the CAD library of the layout which include a locally generated clock and master-slave flip-flops. The state assignment is arbitrary and the number of state variables can be the absolute minimum. Having called the appropriate template from the library, depending on the number of state variables, the design simply requires the addition of inputs which can be derived directly from the state diagram or the excitation table of the sequential circuit to be designed. The hardware realization of the templates incorporates pass-transistors, has a high degree of regularity, and in most cases the outputs associated with the various states are also generated.

1 Introduction

The vast majority of digital hardware designs are presently based on the global communication and central control of the synchronous approach where all the activities of the system are synchronized by a common clock. The design of synchronous finite state machines are well established and many CAD tools have been developed to assist the designers in the design of large synchronous systems. However there are many applications where insisting on a synchronous methodology proves a limiting factor in the speed of operation and the reliability of the system. The same considerations of managing the design of very large integrated systems that provide the motivation for dividing a system into modular parts argue that the parts be independently timed. If the parts are each synchronous systems with independent clocks, information communicated from one part to another must be synchronized to the receiver's clock. Trying to synchronize inherently asynchronous signals faces the problem of *metastability*. Synchronizers can not avoid metastability and there is no bound for the time the bistable element may remain

[1]This work was supported by the Association of Commonwealth Universities through a commonwealth academic staff scholarship.

in this metastable condition. There are many methods to reduce the probability that such a fault would crash a system, but they all cost time and so would reduce efficiency. As the integration density of VLSI circuits increase, synchronous circuits face clock skew and power distribution problems. The speed of operation of synchronous circuits is always limited by the slowest element. The limitations imposed by the synchronous discipline in certain applications suggest that other disciplines be also considered.

Asynchronous sequential circuits offer improved speed of operation when compared with their synchronous counterparts since they respond directly to input changes and do not have to wait for the arrival of the system clock. The modules within an asynchronous circuit switch over a period of time instead of at a clock edge thereby reducing instantaneous current requirements and the complexity of the power distributio n problem in a VLSI circuit. Moreover, they do not suffer from the problem of metastability associated with synchronous systems. The major factor preventing the widespread use of asynchronous sequential circuits is that they involve a complex design procedure which must account for critical races and hazards. For VLSI applications, the relative lack of CAD tools for such designs compounds the problem. It has long been established that self-clocked circuits which generate a clock locally and use edge-triggered flip-flops to hold state variables enable asynchronous sequential circuits to be designed without having to eliminate critical races [4,10,7]. However, the use of edge-triggered flip-flops and delay elements needed in these methods do not lend themselves to easy implementation in VLSI designs. Other methods are particularly suitable for special Programmable Logic Devices [1,3] or Logic Cell Arrays [2].

It has been shown that pass transistor logic has advantages over conventional CMOS gate logic structures in factors like DC power requirements and compactness [9]. Two recent works suggest regular structures for the design of asynchronous sequential circuits using pass-transistors [12,6]. However, both these methods require a critical race free state assignment.

the self-clocked design method presented in this paper enables asynchronous state machines to be implemented in VLSI circuits using pass transistors without having to consider critical races and hazards which allows the state assignment to be arbitrary and the number of state variables to be the absolute minimum. The template for any number of state variables can be included in the CAD library of the layout and be called upon for the relevant designs.

2 The design method

To enable the widespread use of asynchronous sequential circuits in VLSI applications it is necessary to reduce the design complexity and layout time of such circuits. At the same time economy of the silicon area and the speed of operation should be preserved. There are, however, strong arguments that in many VLSI designs shortening of the design and fabrication time is more important than achieving the absolute minimum transistor count. Therefore, there are two main aims in this method:

1. To allow the design of asynchronous state machines without special state assignments that are usually necessary for the elimination of critical races and hazards. The achievement of this objective simplifies the design process and allows arbitrary state assignments with the minimum of state variables.

2. To include all the feedback connections in a template of the design. This template can then be added to the CAD library of the layout and be called upon during the layout of any asynchronous state machine. The achievement of this objective enables the design and layout time to be reduced and focused on the input circuitry and enhance the testability of the design.

It is assumed that the sequential machine to be designed operates in the fundamental mode in which any state transition moves the machine from a stable state to another stable state via a single unstable state. It is also assumed that the circuit inputs are of the level type and that they do not change until the circuit is stable internally.

To achieve the first aim outlined above, it is necessary to use the principle of self-synchronization in which a clock is generated locally and a master-slave configuration isolates the changing state variables from the feedback. To achieve the second aim outlined above, a decoded version of the state variables can be used to control the pass-transistor gating of the input variables that move the machine from every present state to the next state according to the state diagram or the excitation table that defines the machine. The template model of the realization for state machines of up to 4 states (2 state variables) is shown in figure 1. This template can be expanded for more state variables. The outputs of the two master-slave flip-flops, y_1 and y_2, are decoded by a 2 to 4 decoder producing $m_0, .., m_3$ which control the gates of the pass transistors whose outputs are wire ORed and provide D_1 and D_2 of the flip-flops. Since at any time the state machine is in one and only one internal state, only one of the pass transistors for each of the D inputs is conducting. I_1^0 through I_1^3 and I_2^0 through I_2^3, therefore, should be the combination of inputs which takes the machine to the next stable state and should pass the value of the next state. The clock pulse is generated whenever a change of input requires a change of state i.e. during unstable states. The pass transistors of each branch representing each state of the machine are in series to ensure that when the clock goes high the D inputs have already established the next state values. The gating of the pass transistors of the clock use the outputs of the master section of the flip-flops to ensure that when a change of input provides the rising edge of the clock, once the outputs of the master section have assumed the values of the next stable states, the clock falls and transfers the values of the stable state to the outputs of the slave section. The D inputs should have the correct values of the next states even during stable states to ensure that possible spurious pulses, due to races and hazards, do not take the machine to a wrong state. The fact that the feedback loop is broken by the master-slave action enables the disregard of possible critical races. With this template the design of the sequential machine would require the derivation of I_1^0 through I_1^3 and I_2^0 through I_2^3 for the next state values and C_0 through C_3 for the clock to generate a pulse during each unstable total states. All the variables are functions of the inputs to the state machine only and can be easily derived from the relevant state diagram or the

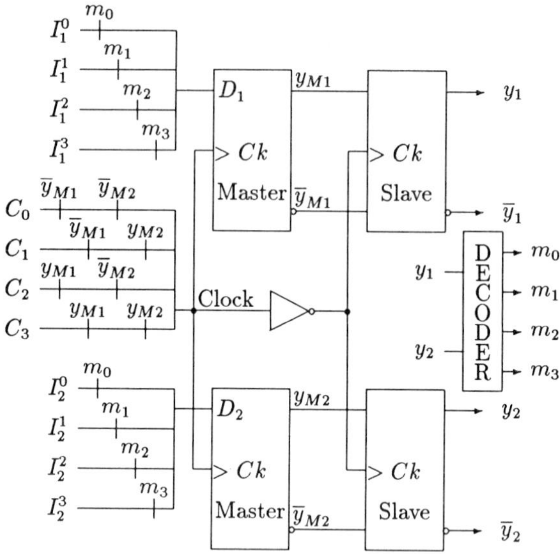

Figure 1: Template for 2 state Variable Self-Clocked State Machines.

excitation table. To expand the template for n state variables (2^n states) there will be n master-slave flip-flops, the decoder would have the corresponding size, there would be 2^n parallel pass transistors for each D input and the pass transistor circuit for the clock would have 2^n branches (each branch being active for only one state).

3 Design example

The method presented in this paper can be illustrated by the design of a VME bus requester. Bus requesters are used in common bus systems that support multiple processors controlling bus transfers. The function of a bus requester is to request permission to control data bus transfers and to indicate to the master when control has been granted. The VME Bus is a common, high performance asynchronous bus that supports multiple bus masters. A self-synchronized design approach for a VME bus requester is appropriate because the VME Bus is asynchronous and high-performance. The bus request function is asynchronously initiated and sequential. A synchronous approach requires an external clock to synchronize and time the sequence. The VME Bus provides a 16 MHz system clock. Our proposed self-synchronized design is much higher performance than a synchronous design using the system clock.

The arbitration bus consists of six bused VMEbus lines and four daisy-chained lines. These daisy-chained lines require special signal names. The signals entering Requesters are identified as Bus Grant In lines (BGxIN), while the signals leaving the requesters are identified as the Bus Grant Out lines (BGxOUT). Therefore, the lines which leave slot N as BGxOUT enter slot N+1 as BGxIN. In the VME bus arbitration system, a Requester will drive a bus request line (one of BR0 through BR3), a bus grant out line (one of BG0OUT through BG3OUT) and the bus busy line (BBSY). If a VME board does not use some bus request levels, it must jumper t he respective bus grant in lines to their respective bus grant out lines. The arbiter will drive the bus grant in lines (BG0IN through BG3IN).

The state diagram for a simple VME Bus Requester is shown in figure 2. A simplified VME arbitration timing is shown in figure 3. The requester initiates a request when an on-board request (OBR-) has been detected. The requester then d rives the bus request (BR-) line active and waits for the bus grant in (BGIN-) line to become active. Once the requester detects BGIN- active, bus busy (BBSY-) and on board grant (OBG-) are driven active and BR- is released to inactive. OBG- indicates to the master that it has the bus and may perform a data transfer once the previous transfer has complete d. Transfer completion is indicated when the address strobe (AS-) is inactive. The requester releases the bus by releasing BBSY- and OBG- when BGIN- and OBR- have become inactive. If BGIN- becomes active, but OBR- is not, the requester passes the grant down the daisy-chain by making bus grant out (BGOUT-) active. All inputs and outputs are active low indicated by a dash following each signal name. Notice that since in an asynchronous state diagram there is no system clock to activate the path each stable state has a path to itself which sustains the system in that state.

To implement this VME bus requester using our template of figure 1 (the state machine has 4 states, therefore the template does not need to be expanded), the inputs to the pass transistor networks for the next states and the clock can be derived directly from the state diagram of figure 2. For example when the machine is in state 01, I_1^1, I_2^1 and C_1 will be passed to D_1, D_2 and the Clock respectively. The machine should remain in that state as long as BGIN- is not asserted and should move to state 10 when BGIN- is asserted. Therefore, $I_1^1 = \overline{BGIN-}, I_2^1 = BGIN-$ and $C_1 = \overline{BGIN-}$. I_1^1 and I_2^1 hold the correct next state values even during stable states to ensure any spurious pulses on the clock, due to races and hazards, do not take the machine to a wrong state. The list of the inputs to the template is as follows:

$I_1^0 = OBR-.\overline{BGIN-}$ $\quad$ $I_2^0 = \overline{OBR-} + \overline{BGIN-}$

$I_1^1 = \overline{BGIN-}$ $\quad$ $I_2^1 = BGIN-$

$I_1^2 = \overline{OBR-.BGIN-.AS-}$ $\quad$ $I_2^2 = 0$

$I_1^3 = \overline{BGIN-}$ $\quad$ $I_2^3 = \overline{BGIN-}$

$$
\begin{aligned}
C_0 &= \overline{OBR-} + \overline{BGIN-} \\
C_1 &= \overline{BGIN-} \\
C_2 &= OBR-.BGIN-.AS- \\
C_3 &= BGIN-
\end{aligned}
$$

These inputs can be added to the template either by gate or pass transistor logic. The design must respond to the system reset signal, SYSRESET-, taking the state machine to state 00. Therefore, the asynchronous reset input of the flip-flops should be connected to SYSRESET-. A CMOS master slave D type flip-flop with reset capability is shown in figure 4. If such flip flops are used, it is necessary to ensure that each of the three pass networks of the inputs of the template drives a load such as an invertor. The addition of the invertors before D_1, D_2 and the Clock requires that the inputs to the template be inverted accordingly. The layout of the example design was carried out using CHIPWISE on an Appolo workstation and was simulated with satisfactory results. The design of the outputs of the machine is standard and is not elaborated here. However, it is clear that the outputs associated with each state (e.g. BR- is only asserted during state 01 and BGOUT- during state 11) are already available at the outputs of the decoder. Any further restrictions on the outputs, such as open collector outputs for connection on the bus, should be catered for separately.

4 Conclusions

A systematic and easy to implement design method and templates for the synthesis of asynchronous sequential machines have been presented to facilitate such designs and to enable their widespread use in VLSI circuits. Self-synchronization of the circuit has made the design methodology comparable with synchronous systems while retaining the basic advantages of asynchronous circuits. The proposed templates containing the skeleton of the designs which can be added to the CAD library speeds up the design process and the layout of the circuit. The circuitry for the inputs can directly be derived from the state diagram or the excitation table of the sequential machine to be synthesized. The use of pass-transistor logic and the regularity of the circuit contribute towards its compactness. The main advantage of this method over existing pass-transistor asynchronous sequential circuits [12,6] is that it does not require a critical race free state assignment, providing simple and automatable design and layout procedures. To maintain regularity the transistor count is not the absolute minimum but it ensures a speedy and reliable design and layout process comparable with synchronous systems. The design can directly be converted to a synchronous machine by eliminating the locally generated clock and connecting it to the global clock of the system.

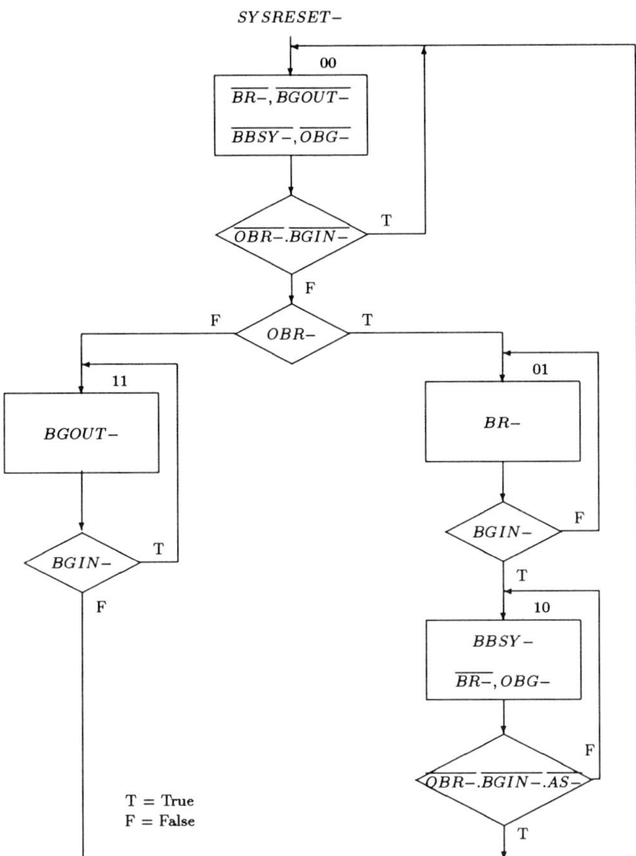

Figure 2: VME Bus Requester State Diagram.

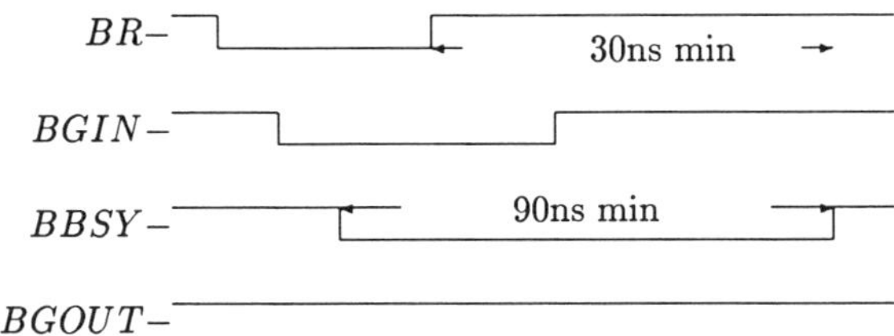

Figure 3: VME Arbitration Timing.

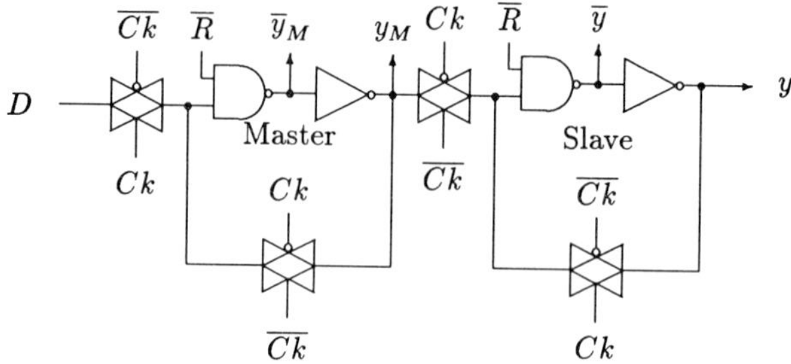

Figure 4: CMOS Master-Slave D Flip-Flop With Reset.

References

[1] F. Aghdasi, "Design of asynchronous sequential circuits using interface protocol asynchronous cell (IPAC) PAL device," Proceedings of International Symposium on Computer Architecture and Digital Signal Processing, CA-DSP '89, Hong-Kong, pp 426-430, Oct. 1989.

[2] F.Aghdasi, "Application of Logic Cell Arrays in Design of Self-Clocked Sequential Circuits," Proceedings of IEEE TENCON'90, Computer and Communication systems, Hong-Kong, 24-27 Sept. 1990, pp 519-523.

[3] F.Aghdasi, M.Bolton, "Self-clocked asynchronous state machine design with PAL22IP6 device," Microprocessors and Microsystems, Feb. 1991,

[4] H. Y. H. Chuang, S. Das (Washington Univ.), "Synthesis of multiple input change asynchronous machines using controlled excitation and flip-flops," IEEE Transactions on Computers, Vol. C-22, No. 12, pp 1103-1109, Dec. 1973.

[5] Cypress Semiconductor, "CY7C331 application example: asynchronous, self-timed VME bus requester," 1988.

[6] S.K. Gopalakrishnan, G.K. Maki, "VLSI asynchronous sequential circuit design", Proceedings of IEEE International Conference on Computer Design: VLSI in Computers and Processors, Cambridge, Massachusetts, Sept 1990, pp 238-242.

[7] J. L. Huertas, and J. I. Acha, "Self-synchronization of asynchronous sequential circuits employing a general clock function," IEEE Transactions on Computers, vol. C-25, No. 3, pp 297-300, March 1976.

[8] Motorola, "VMEbus Specification Manual," MVMEBS/D1, Aug 1982.

[9] D. Radhakrishnan, S. Witaker and G. Maki, "Formal design procedures for pass transistor switching circuits", IEEE JSSC, Vol. SC-20, pp 531-536, April 1985.

[10] C. A. Ray, J. Vaucher (Univ. de Montreal), "Self-synchronized sequential machines," IEEE Transactions on Computers, vol. C-23, No. 12, pp 1306-1311, Dec. 1974.

[11] S. H. Unger, "Asynchronous sequential switching circuits," New York: Willey, 1969.

[12] S. Whitaker, G. Maki, "Pass-transistor asynchronous sequential circuits", IEEE JSSC, Vol. 24, No. 1, Feb. 1989, pp 71-78.

VLSI 91
A. Halaas and P.B. Denyer (Eds.)
Elsevier Science Publishers B.V. (North-Holland)

Theoretical and Practical Issues in CMOS Wave Pipelining

C. Thomas Gray, Thomas Hughes, Sanjay Arora, Wentai Liu, Ralph Cavin, III

Department of Electrical and Computer Engineering, North Carolina State University, Raleigh, NC 27695-7911

Abstract

Wave pipelining (also known as maximal rate pipelining) is a timing methodology used in digital systems to increase the number of effective pipelined stages without increasing the number of physical registers in the pipeline. Using this technique, new data are applied to the inputs of a combinational block before the previous outputs are available thus effectively pipelining the combinational logic. In this paper, we investigate issues involved in wave pipelining for CMOS systems from both a theoretical approach and a practical approach. We consider the effect of manufacturing process variations, operating environment variations, and clock distribution methodologies. We conclude by describing a CMOS adder circuit demonstrating wave pipelined operation.

1 Introduction

As system performance requirements continue to push the limits of current CMOS processes, it is clear that new techniques and design methodologies must be developed for high speed applications. While these requirements could be met by moving to faster technologies such as GaAs or ECL, our research indicates that significant performance increases can be gained in CMOS systems by using the technique of wave pipelining.

Wave pipelining (or maximal rate pipelining), as first identified by Cotten[1], can be used in digital systems to increase the number of effective pipeline stages in a block of combinational logic without increasing the number of physical registers. With conventional logic design, combinational logic between registers is allowed to fully stabilize before new input data is applied. In contrast, wave pipelined systems apply new inputs to combinational logic before the previous results are available. In this manner, the combinational logic is effectively pipelined with multiple coherent "waves" of data present at various stages of the logic block. The minimal clock period at which new inputs can be applied to the combinational logic is, in the ideal case, limited only by the difference in the delay times of the maximal and minimal paths through the logic. Thus, clock frequencies can be maximized when logic delay paths are equalized.

Wave pipelined systems also have a major advantage in the area of clock distribution. Conventional high performance pipelined systems divide combinational blocks into many pipeline segments and separate these segments with registers. High speed clock lines must be routed to each of these registers and skew must be controlled to insure proper clocking. In order to gain higher pipeline performance, it is necessary to increase the number of these pipeline segments making this clock distribution even more difficult. Many times, high performance pipelined systems are limited in speed by the requirement of skew free clock signals. However, in wave pipelined systems, since combinational logic is being used as storage elements for the waves of data, the number of physical registers can be minimized thus reducing potential clock distribution problems.

This method of design is not without its disadvantages however. First, wave pipelining requires careful design to keep the path delay differences small. This may require adding delays at various points in the circuit. Second, wave pipelined systems are difficult to debug. In conventional designs, the clock can be stopped to observe the state of the machine. In wave pipelined systems, however, the machine must be run and debugged at its operational speed. Lastly, unlike conventional pipelined systems, clock signals must be synchronized with the data "waves", which implies that local skew on clock and data lines must be controlled such that data emerging from the combinational logic is latched at the correct time.

Previous work in the area of wave pipelining has included theoretical work on the timing constraints of pipelined systems [1, 2, 3, 4], work on techniques for path equalization [5, 6], and the design of systems using the techniques of wave pipelining [7, 8, 9]. Most of the previous experimental work in this area, including the work described in [7], [8], and [5], was based on bipolar ECL technologies. In contrast, our work concentrates on the application of wave pipelining methodologies to CMOS systems and the associated design considerations.

In this paper, we first identify general timing constraints that limit the degree of wave pipelining that can be incorporated into logic systems and develop general wave pipelining concepts. Next, we describe practical considerations including choice of circuit family, the effects of process and environmental variations, clock distribution, and physical layout. Finally, we describe an implementation of a wave pipelined 16-bit CMOS adder circuit. We conclude with the current status of our project and identify areas of future work.

2 Wave Pipelining Concepts and Timing Constraints

2.1 Notation

The following notation will be used throughout this paper in describing the various constraints on the timing of wave pipelined systems:

T_{clk}	clock period
t_{max}	maximum delay from input to output
t_{min}	minimum delay from input to output
$t_{max}(i)$	maximum delay input to internal node i
$t_{min}(i)$	minimum delay input to internal node i
t_d	delay time of latch

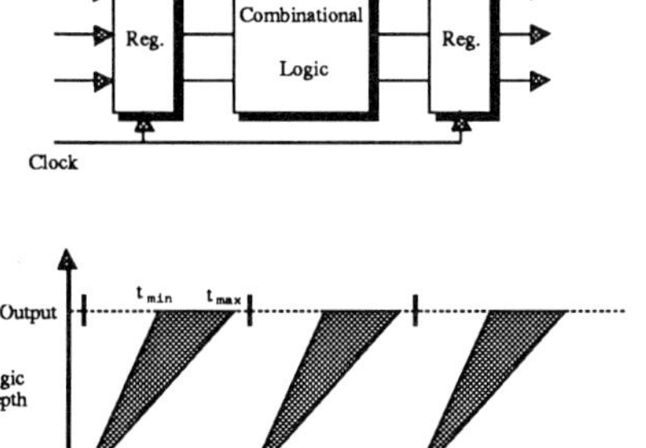

Figure 1: Combinational Logic with No Pipelining

t_{setup}	set up time for the latch
t_{hold}	hold time for the latch
Δt_e^n	earliest possible clock skew at input register
Δt_l^n	latest possible clock skew at input register
Δt_e^o	earliest possible clock skew at output reg.
Δt_l^o	latest possible clock skew at output register
$t_{node}(i)$	response time of internal node i due to driving gate and nodal capacitance
Δ	intentional delay added to clock line

2.2 Comparison with Conventional Pipelining

Figure 1 pictures a block of combinational logic with no pipelining. As is shown in the block diagram, data enters the combinational logic through an input register and leaves through an output register activated by a common clock line. The logic depth diagram [4] shows the flow of data through the combinational logic block versus time. After new data is latched into the input register, the data begins to flow through the combinational logic block. There is some variation in the time at which data will emerge from the combinational logic due to path delay differences, process/temperature variations, etc. The minimal and maximal delay times due to these variations are represented by t_{min} and t_{max} and are shown in the figure. The shaded areas between the t_{max} and t_{min} lines depict regions of uncertainty in the data value. As the diagram indicates, outputs must be strobed outside these cones of uncertainty to allow time for all data to be stable, for set-up/hold time of the latches, and for worst case clock skew. In this diagram, new data is clocked into the input registers only after data in the combinational logic is completely stable. Thus, the minimal clock period $T_{clk}$is limited by t_{max} in addition to setup/hold times and clock.

With conventional pipelining the combinational logic block is divided into multiple segments The clock period can thus be shortened by reducing t_{max} of the pipeline segments. Now, separate data cones are present at each stage of the pipeline. Again, as in the non-pipelined case, new inputs are applied only after the combinational logic is allowed to fully stabilize.

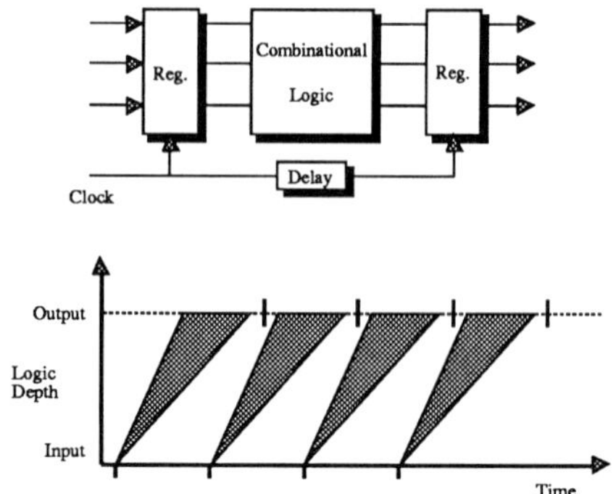

Figure 2: Combinational Logic with Wave Pipelining

In Figure 2, the wave pipelining concept is illustrated. The input data is now clocked into the combinational logic before the previous data is available at the output register. In this way, the clock period can decreased as long as the data uncertainty cones do not overlap. This insures that new data will not arrive at a particular point before old data has fully passed. This decrease in clock period can be observed graphically by "squeezing" the cones together in time. If after the clock period has been minimized, the input and output clock are not in phase, a delay element must be added on the clock line or in the data paths to effect this phase shift.

2.3 Constraints on Minimal Clock Period – Edge-Triggered Circuits

For this section, we assume the clocking model shown in Figure 2 in which the output clock is derived from the input clock delayed by an amount Δ. In order to achieve proper latching of data, relatively stringent double sided timing requirements must be met. The clock period must be such that the output data is latched *after* the latest data has arrived at the outputs and *before* the earliest data from the next set of inputs has arrived. This can be seen in Figure 3.

To ensure the latest data is latched correctly, the following constraint must be true.

$$kT_{clk} + \Delta - \Delta t_e^o \geq \Delta t_l^n + t_d + t_{max} + t_{setup}, \qquad \textit{for integer } k, \tag{1}$$

where kT_{clk} is the amount of clock delay between input and output latches derived from the cyclic nature of the clock and Δ is the amount of delay introduced by the physical delay unit. Note that in general (when $\Delta > T_{clk}$), k does *not* correspond to the number of waves in the circuit. This constraint formulation is developed in more detail in [10].

To ensure the earliest data is latched correctly, the following constraint must be also true.

$$T_{clk} + t_d + t_{min} - t_{hold} - \Delta t_e^n \geq kT_{clk} + \Delta + \Delta t_l^o, \qquad \textit{for integer } k \tag{2}$$

Consequently, by combining constraints (1) and (2), the following condition on the number

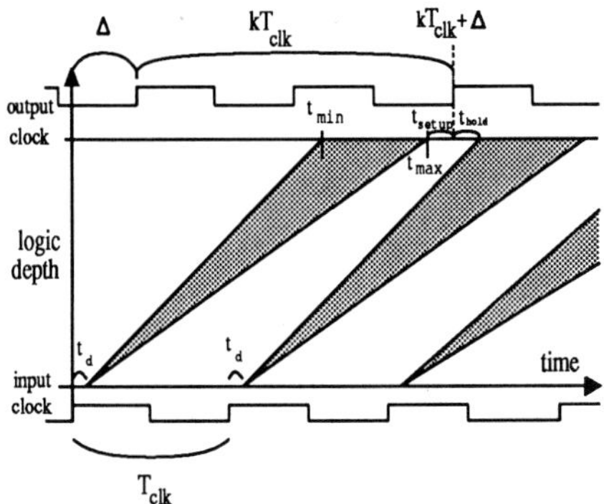

Figure 3: Constraints on Clock Period

of waves is necessary for proper latching.

$$T_{clk} + t_{min} - t_{hold} - \Delta t_e^n - \Delta t_l^o \geq kT_{clk} + \Delta - t_d \geq t_{max} + t_{setup} + \Delta t_l^n + \Delta t_e^o, \quad \text{for integer } k \tag{3}$$

Constraint (3) then implies constraint (4).

$$T_{clk} \geq (t_{max} - t_{min}) + t_{setup} + t_{hold} + \Delta t_e^n + \Delta t_e^o + \Delta t_l^n + \Delta t_l^o \tag{4}$$

Thus, the maximal clock rate is bounded by the difference of maximal and minimum delays between two consecutive pipelining stages.

Physical limits on switching delay of individual gates impose additional constraints on the minimal clock period. Constraint (4) indicates the clock period is bounded only by setup/hold time and clock skew when logic path are equalized ($t_{max} = t_{min}$). However, factors, such as non-zero rising/falling time, signal dispersion, technology, manufacturing and temperature variation, certainly affect the clock speed. Accordingly, additional constraints due to physical properties of internal nodes must be considered such that data waves do not overlap at any of the internal nodes. In other words, the next earliest possible wave at any particular node i should not arrive earlier than the latest possible wave. Mathematically these constraints are expressed by constraint (5).

$$T_{clk} + t_{min}(i) - \Delta t_e^n \geq t_{max}(i) + \Delta t_l^n + t_{node}(i) \qquad \text{for all internal nodes } i \tag{5}$$

2.4 Minimizing the Clock Period

Given the constraints identified in the previous section, we can now proceed to minimize the clock period (T_{clk}). Given variable $T_{clk}$and Δ and all other parameters fixed, this can be formulated as a linear programming problem:

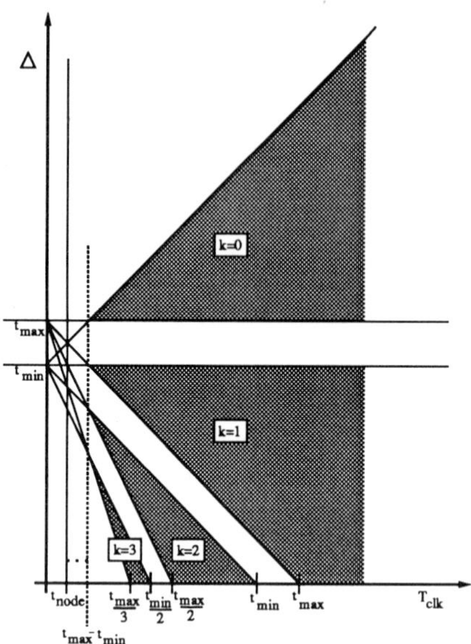

Figure 4: Timing Constraints in Terms of Circuit Parameters

Min: T_{clk}

Subject to:

$$kT_{clk} + \Delta - \Delta t_e^o \geq \Delta t_l^n + t_d + t_{max} + t_{setup}$$

$$T_{clk} + t_d + t_{min} - t_{hold} - \Delta t_e^n \geq kT_{clk} + \Delta + \Delta t_l^o$$

$$T_{clk} \geq t_{max}(i) - t_{min}(i) + \Delta t_l^n + t_{node}(i) \qquad \forall i$$

Figure 4 illustrates the timing constraints of this problem by setting setup time, hold time, latch delay, and unintentional clock skew to zero. The feasible solution space is a convex region (shaded in Figure 4) bounded by $T_{clk} + t_{min} = kT_{clk} + \Delta$, $kT_{clk} + \Delta = t_{max}$, and $\max_{\forall i}(t_{max}(i) - t_{min}(i) + t_{node}(i)\ )$. For fixed values of k, distinct feasible regions of operation are available. For example, when $k = 2$, the feasible region is bounded by $T_{clk} + t_{min} = 2T_{clk} + \Delta$, $2T_{clk} + \Delta = t_{max}$, and $\Delta = 0$. Note that as k is increased, the feasible region is reduced. From the figure is easy to see that the minimal clock period is indeed bounded below by $t_{max} - t_{min}$ for all values of k and that the feasible region for a particular value of k is shrunk as $t_{max} - t_{min}$ is increased. These facts emphasize the necessity of delay balancing.

2.5 Multiple Stage Pipelines

Although wave pipelining can eliminate many of the internal latches in pipelined systems, some latches may still be required. For example, in long combinational logic blocks, the retiming effect of internal latches may be necessary to curb data signal dispersion. Also, external timing constraints may dictate the inclusion of internal latches. Thus, it is important to consider the constraints of wave pipelining in multi-stage pipelines.

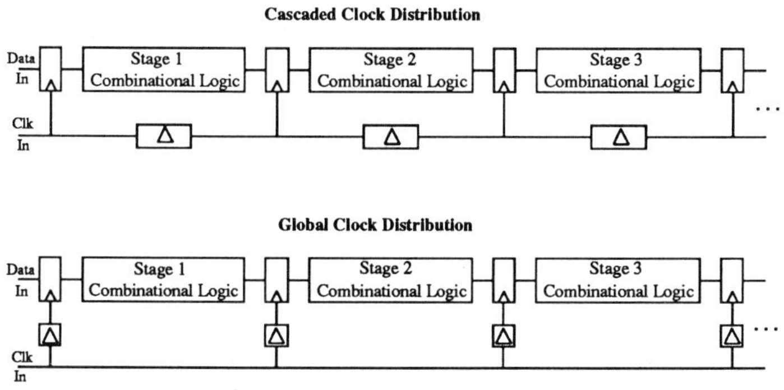

Figure 5: Clock Distribution Strategies

Wave pipelined circuits can be clocked using a variety of different clock distribution schemes. Two of which are shown in Figure 5. Design tradeoffs between these strategies will be discussed in section 3.4. For this section, consider the *cascaded clock distribution* scheme.

Extending the formulation of the previous section and setting setup time, hold time, and unintentional clock skew to zero, the constraints on $T_{clk}$for the multi-stage problem can be expressed as follows:

$$k_i T_{clk} + \Delta_i \geq t_d + t_{max\ i} + t_{setup}, \quad \textit{for i=1...n}, \tag{6}$$

$$T_{clk} + t_{min\ i} - t_{hold} \geq k_i T_{clk} + \Delta_i, \quad \textit{for i=1...n}, \tag{7}$$

where i represents the pipeline stage number and n is the number of stages. As can be seen, these constraints along with constraint that arises from non-zero physical node delay (constraint 5 in Section 2.3) give rise to a set of independent constraint spaces such as shown in Figure 4.

3 Design Considerations for Practical Circuits

3.1 Circuit Choice

The choice of circuit family for use in a wave pipelined system can have a significant impact on the performance the system though the effect of unequal t_{max} and t_{min} at the gate level. This will occur when gate delay differences arise for different input patterns or different signal timing patterns. This is of particular importance in CMOS systems and implies that a gate should be chosen that minimizes this effect. Parallel transistor paths, such as in a static CMOS NAND gate, can give a factor of two difference in t_{max} and t_{min} depending on input data. However, if the parallel paths in the NAND gate are replaced by a single pullup transistor (as in pseudo NMOS), a large component of the data dependency can be eliminated. The serial path can still contribute to differences in t_{max} and t_{min} depending on arrival times of input signals, however this difference is much less than that due to parallel paths. The serial path imbalance can be minimized by the used of a cross-coupled NAND (Figure 6) although at a cost of more area[11].

Minimization of $\max_{\forall i}(t_{node}(i))$, the response time of the slowest node (constraint 5), is another consideration in choosing gate level circuitry. This constraint implies that excessively slow gates or heavily loaded nodes due to long interconnect or high fanout should be avoided. Thus, combinational logic should be implemented using multiple level logic with smaller (faster) gates rather than in two-level logic using larger (slower) gates.

3.2 Parametric Variations due to Manufacturing and Environmental Factors

With the requirements of wave pipelined systems, the control of both early and late skew in clock and data signals is a major concern. For proper operation, it is necessary to keep proper phase alignment between the clock at each latch and the data arriving at the latch.

From Figure 4, we can make some interesting observations concerning operation in different k regions. Consider an operational frequency (T_{clk}) between $t_{max}/2$ and t_{min}. Depending on the length of the delay unit (Δ) the circuit can operate in either $k = 1$ mode or $k = 2$ mode. Note, however that there is a major difference between the two cases in terms of the amount of variation in clock frequency that can be tolerated. In the $k = 2$ mode with $\Delta = 0$ the frequency must be between $t_{max}/2$ and t_{min} for correct operation. However, in the $k = 1$ case with $\Delta = t_{min}$, the frequency range is only bounded below by $t_{max} - t_{min}$. This implies that lower k modes of operation will be able to tolerate more variation in frequency and will this have a greater range of operation. This, however, is at the expense of larger delay units.

This same effect can also be observed when process and environmental variations on combinational logic path delays are considered. Assuming a constant scaling factor c (per chip) representing the amount of delay variation from nominal values due to process and environmental effect, the delay parameters t_{min}, t_{max}, and Δ become ct_{min}, ct_{max}, and $c\Delta$. This will modify the feasible constraint space of Figure 4. An upper bound (assuming $t_{max} = t_{min}$ and $\Delta = 0$) on the maximal amount of variation, c, that can be tolerated for a given k such that correct operation is maintained is $\frac{1}{2k-1}$ (for $k > 1$)[10]. Thus, if $k = 10$, under ideal conditions only approximately 5% variation in c can be tolerated. This implies that wave pipelined operation in high k modes will be very difficult. Note also that, in contrast to previous work, our formulation does not assume $\Delta < T_{clk}$. This allows operation in the lower k regions up to the maximum frequency.

Unfortunately, there are many sources for these delay variations that must be considered – especially considering the high frequencies operation that the wave pipelining technique makes possible. These sources can be categorized into two major groups: *static variation* and *dynamic operational variation.*

The static component refers to the inevitable processing variations that occur between chips of different fabrication runs and between chips at different locations on the same wafer. Slight variations on channel lengths of fabricated devices can produce variations in nominal delay of the logic gates which could cause phase misalignment at a latch. If the delay units are made programmable, this static component can be compensated for after fabrication.

The dynamic component is caused by the variation in gate delay induced by changes in

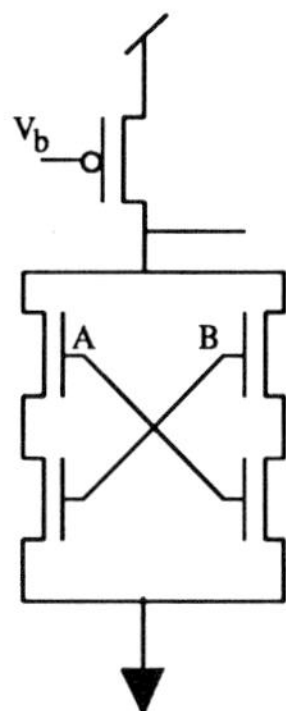

Figure 6: Cross Coupled Pseudo NMOS NAND

the operating environment of the device. For example, the amount of power will vary widely depending on whether the data causes many devices to switch or few devices to switch. This will, in turn, cause variations in temperature and thus the delay of the device. This dynamic variation must be considered in t_{max} and t_{min} or compensated for after fabrication using structures such as phase locked loops and dynamically programmable delays. This type of compensation has been considered in the design of a high speed wave pipelined FIFO[12].

The considerations outlined in this section imply that, for reliable wave pipelined operation, the lower k modes should be used along with careful consideration of environmental and process variations.

3.3 Balancing t_{max} and t_{min}

Due to the circuit family used, static and dynamic variations present, and circuit topology, a given combinational logic block will have an associated t_{max} and t_{min}. From constraints 4 and 5, it is evident that $T_{clk}$is minimized when $t_{max} - t_{min}$ and $t_{max}(i) - t_{min}(i)\forall(i)$ is minimized. This implies that optimal tuning at every internal node and at latches must be done.

Several group including Wong [5, 6] and Ekroot [4] have presented methods for padding combinational logic paths with delay units to equalize path lengths and thus minimize T_{clk}. These methods of path delay equalization can be applied to CMOS circuits with the understanding that they do not fully consider the effects of data dependent and process/environment dependent delay variations.

3.4 Clock Distribution and Physical Layout Schemes

As shown in Figure 5, the clock can be distributed to data latches in two basic ways. The clock can either be cascaded from stage to stage with each stage deriving its local clock from the previous or the clock can be routed globally with each stage deriving its clock from the global clock. The major differences are: 1) with the cascade distribution, the timing constraints at a particular stage depend only on that stage while in the global distribution, the timing constraints at a stage depend on the amount of clock delay at that stage and the previous stage, 2) cascade scheme introduces more clock signal dispersion (dispersion through all delay units), 3) cascade scheme has less potential clock skew problems since clock is routed point to point, 4) cascade scheme loads clock line less that global scheme, and 5) phase drift between the input clock and the output data due to dynamic variations in delay unit is more severe with the cascade scheme since the drift is through delays of all stages as opposed to the delay unit of only the last stage in the global scheme.

Care must also be taken in the physical layout of CMOS wave pipelined systems. For example, differences in routing of power and ground to different circuit elements of a combination logic block can cause differences in effective t_{max} and t_{min} thus limiting the minimal clock period. Also, the routing of the clock lines will have an effect on the amount of skew between stages.

4 Wave Pipelined Adder

4.1 Adder Design

In order to validate our theoretical formulation, we have designed a 16-bit wave pipelined CMOS adder circuit[11]. A conventional carry-ripple adder is not well suited for wave pipelining since the ith bit of the sum is nominally available before the $(i+1)$th bit. This is because the carry must ripple through all the bits. This staggering of the sum outputs means the difference between $t_{max}$and $t_{min}$is large which, as described earlier, limits the maximum operating frequency. Thus, we chose to implement a carry lookahead scheme.

Since full single stage lookahead for all 16 bits would be very expensive in terms of circuit area and circuit delay, many lookahead schemes are implemented by combining four bits into one lookahead stage and letting the carry ripple through these four bit stages. This scheme works if the objective is to speed up the circuit by decreasing the maximum delay. However, this scheme is not acceptable for wave pipelining since it does not eliminate carry ripple completely.

The adder shown in (Figure 7) and described below uses a two level carry lookahead circuitry. This scheme eliminates ripple completely without excessively complicating the logic. The following notation will be used. a, b, and c_{in} are the inputs to the adder with a_i, b_i being the ith bits of the inputs. c_i is the carry out from the ith bit (c_{15} is the carry out from the adder). $P_i = a_i + b_i$ (logical OR) is the ith carry propagate signal and $G_i = a_i \cdot b_i$ (logical AND) is the ith carry generate signal. Thus,

$$c_i = G_i + P_i \cdot c_{i-1}.$$

The first level of lookahead circuitry combines P and G signals of four adjacent bits. The second level of lookahead circuitry combines the first level outputs so that the carry out of every bit is a function only of the carry in to the adder. After the carry out of every bit has been computed, all that is required is to compute the exclusive OR (XOR) of a_i, b_i and c_i to get S_i, the ith bit of the sum. Note that the carry out of the adder has already been computed (c_{15}).

The outputs are then latched by an edge triggered latch. The latch is clocked by a clock signal which has been appropriately skewed to arrive at the latch when the sum outputs have had enough time to set up.

The circuit was constructed of cross-coupled pseudo-NMOS NAND gates (Figure 6), and static CMOS inverters. Pseudo-NMOS NAND gates were chosen over static CMOS because of their data independent behavior and the static CMOS gates were appropriately sized to reduce differences in rise time and fall time delays. The circuit was simulated using the CAzM circuit level simulator developed at MCNC[13] with nominal MOSIS 2 micron parameters.

4.2 Maximum and Minimum Delays

The circuit was rough tuned by hand to decrease the difference between the maximum and minimum delays. Short paths were identified and padded with delay buffers to raise the minimum delay closer to the maximum.

Observing the signals before they are latched gives an estimate of $t_{max}$and t_{min}. $t_{max}$was found to be about 14.4 ns, and t_{min}about 10.5 ns. This difference between $t_{max}$and $t_{min}$is due mainly to suboptimal delay path tuning and the fact that data dependencies could not be completely eliminated from the gates. With more careful fine tuning of delay paths, t_{max}-t_{min}could possibly be further reduced.

4.3 Simulation Results

The adder was simulated at different data speeds and various intentional clock skews (Δ) to identify regions of valid circuit operation. The simulation results are summarized in the constraint space plot of Figure 8 (compare with Figure 4). A circle represents correct operation of the adder and a cross shows incorrect operation. Note that correct operation was observed in the k =0, 1, 2, and 3 feasible regions and incorrect operation was observed in the infeasible regions as predicted by the theoretical formulation developed in Sections 2.3 and 2.4.

5 Conclusions

In this paper we have investigated theoretical and practical issues in the design of CMOS wave pipelined circuits and have verified our results through the design of a wave pipelined 16-bit CMOS adder circuit.

The major contributions of this paper are the application of wave pipelining techniques to CMOS circuits, general theoretical formulation of timing constraints on wave pipelined systems, consideration of process and environmental variations and their effect on practical wave pipelined circuits, and the design of a wave pipelined CMOS adder circuit.

We plan to continue our work with wave pipelining for general logic designs by considering the application of wave pipelining techniques to existing systems for optimization of timing parameters and by considering the design of new systems that explicitly take advantage of wave pipelining techniques. Many high performance systems that already rely heavily on conventional

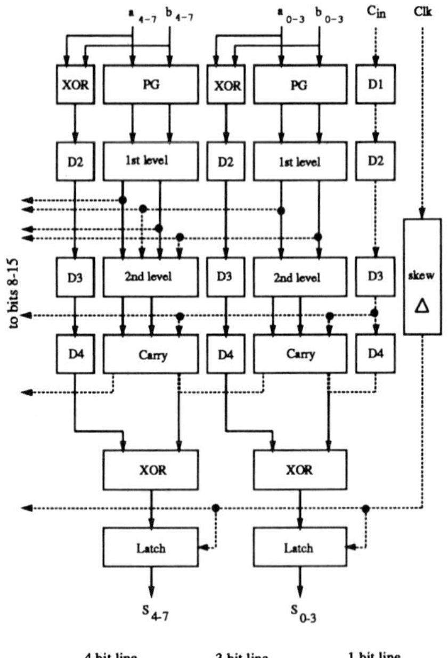

Figure 7: Wave Pipelined CMOS Adder

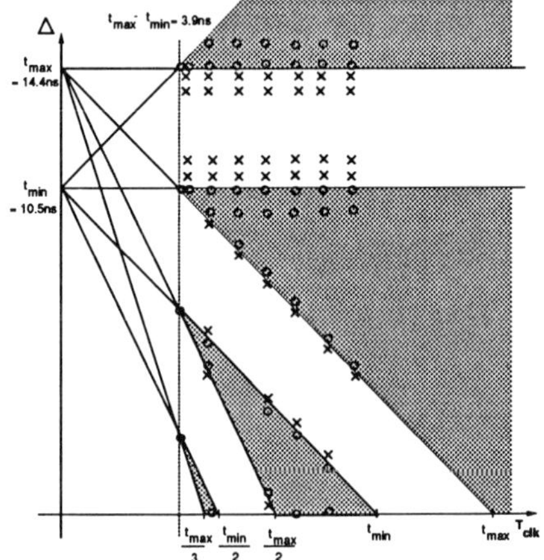

Figure 8: Constraint Space of Adder

pipelining can be modified to take advantage of wave pipelining. In addition, we plan to investigate the development of CAD tools to support design and analysis of CMOS wave pipelined systems.

Acknowledgements

The authors thank David Fan, Gary Moyer, and William Farlow for their suggestions and discussions.Thomas Gray and Thomas Hughes were supported by Semiconductor Research Corp. fellowships. Prof. Wentai Liu was supported by NSF under NSF-MIP-882142.

References

[1] L. W. Cotten. Maximum-rate pipeline systems. In *1969 AFIPS Proc. Spring Joint Computer Conf.*, pages 581–586, 1969.

[2] P. Kogge. *The Architecture of Pipelined Computers.* McGraw-Hill, 1981.

[3] B. Fawcett. Maximal clocking rates for pipelined digital systems. Master's thesis, University of Illinois, 1976.

[4] B. Ekroot. *Optimization of Pipelined Processors by Insertion of Combinational Logic Delay.* PhD thesis, Stanford University, 1987.

[5] D. Wong, G. De Micheli, and M. Flynn. Designing high-performance digital circuits using wave pipelining. In *Proc. of VLSI '89*, Munich, W. Germany, August 1989.

[6] D. Wong, G. DeMicheli, and M. Flynn. Inserting active delay elements to achieve wave pipelining. Technical Report CSL-TR-89-386, Computer System Laboratory, Stanford University, 1989.

[7] S. Anderson, J. Earle, R. Goldschmidt, and D. Powers. The IBM system/360 model 91 floating point execution unit. *IBM Journal of Research and Development*, pages 34–53, January 1967.

[8] L. Qi and X. Peisu. The design and implementation of a very fast experimental pipelining computer. *J. of Computer Science and Technology*, 3(1):1–6, 1988.

[9] F. Klass and J. M. Mulder. CMOS implementation of wave pipelining. Technical Report 1-68340-44(1990)02, Delft University of Technology, December 1990.

[10] T. Hughes, T. Gray, W. Liu, and R. Cavin. A general theory for wave pipelined systems. Technical Report NCSU-VLSI-91-02, North Carolina State University, 1991. In preparation.

[11] Sanjay Arora. Design of a pseudo NMOS wave pipelined adder. Master's thesis, North Carolina State University, 1991.

[12] C.T. Gray, T. Hughes, D. Fan, G. Moyer, W. Liu, and R. Cavin. A high speed CMOS FIFO using wave pipelining. Technical Report NCSU-VLSI-91-01, North Carolina State University, 1991.

[13] D. Rose D. Erdman and G. Nifong. CAzM: Circuit analyzer with macromodeling user's guide. Technical report, MCNC, June 1990.

VLSI 91
A. Halaas and P.B. Denyer (Eds.)
Elsevier Science Publishers B.V. (North-Holland)

Automatic Interfacing of Synchronous Modules to an Asynchronous Environment

Naser Awad[a] and David R. Smith[b]

[a]Elec. Engrg. Dept, [b]Comp. Sci. Dept, SUNY, Stony Brook, NY, 11794, USA

Abstract

Locally-synchronous globally-asynchronous hardware modules employ a locally generated clock to control interactions among their components and a handshaking protocol to govern interactions with the outside environment. This paper describes a new automatic method to convert a purely synchronous hardware structure to a locally-synchronous globally-asynchronous one. Interface hardware is first allocated from a set of specially designed templates. The controller of the synchronous module, which is produced by an existing automatic synthesis system, is then modified to relay data production and data consumption events to the communication hardware. The interface circuits implement the details of the handshaking protocol in a way that is transparent to the synchronous module. This leads to *concurrent* execution of the computation and communication tasks. The results of applying the method to several examples are reported.

1 Introduction

In synchronous systems, a single clock is employed to coordinate the operation of all system components. This clock is distributed to all locations in the system. As chip area increases with the advancement of technology, *clock skew* becomes a major concern. This is also a problem in board design and limits attainable clock speeds.

As a means of overcoming this there has been a recent interest in asynchronous circuits and their synthesis [Mar89] [MBM89] [Sut89] [Chu87]. Here, all components operate at their own speed exchanging data through handshake signals. Employing such a handshaking scheme at all levels results in a significant area overhead as compared to a purely synchronous design.

Locally synchronous globally asynchronous hardware modules employ a local clock to control the interactions between their components and a handshaking protocol to communicate with their environment. Such modules have the advantage of the simpler synchronous design without suffering from the clock skew problem because of their limited size. At the same time they are complex enough to justify the area overhead of the handshaking circuits.

Our goal is to provide an *automatic* method to transform a synchronous hardware structure into a locally synchronous globally asynchronous one. Our motivation is to isolate the designer from specifying the details of data transactions with the asynchronous environment. In this way, design efforts would concentrate on the synchronous part and the rest left to automation tools. In this paper, we describe such a method.

2 Related Work

Automatic synthesis of purely asynchronous circuits has received increasing attention in recent years. Starting from a signal transition graph which specifies the triggering relationships between inputs and outputs of an asynchronous circuit, [CG86] describes a method to automatically synthesize the circuit. Adopting a standard handshaking protocol, [MBM89] starts from a *high level* description of an asynchronous interconnection network and derives the corresponding signal transition graph and hazard-free implementation of the network. A purely asynchronous microprocessor has also been built [Mar89].

Interface synthesis has previously been addressed in connection with the high level synthesis of *synchronous* circuits. [NT86] describes a method to incorporate interface timing constraints into a high level synthesis system. Input/output interface hardware which interacts with a controller is chosen from a set of templates. [BK87] synthesizes *transducer logic* to bridge two interface behaviors specified as timing diagrams. Interface hardware is first chosen from a set of templates, then it is modified to satisfy minimum and maximum timing constraints and eliminate races.

Unlike purely synchronous and purely asynchronous hardware however, locally synchronous globally asynchronous modules have received little attention. [Cha84] describes two classes of such modules. Both employ a *stretchable* local clock: i.e. the clock phase can be extended *indefinitely* depending on an external signal. One important characteristic of both classes is that the synchronous module is *tightly* coupled with the asynchronous environment; i.e. the computation performed by the synchronous core is highly dependent on events in the asynchronous environment. [Ros88] describes the design of a locally-clocked finite state machine operating in an asynchronous environment. Here, too, a stretchable local clock is employed. In neither of these works is there any design automation of the interaction with the synchronous module.

Our goal is different. It is as follows: Given a specification of a synchronous module, then convert it *automatically* to a locally synchronous globally asynchronous one. We are not aware of any such tool in the literature. The benefits (over explicit specification with wait statements) are simplicity of functional specification and efficiency in time and space, since the area overhead is smaller and concurrency of communication and computation is automatically exploited.

Finally we mention two more comparisons with previous work: Unlike [Cha84] and [Ros88], we do not employ stretchable local clocks since our synchronous hardware is usually implemented in dynamic logic with a maximum time limit before charge restoration. We do not think this will be a significant problem in practice. Also although the techniques developed in [CG86] and [BK87] can be used to *synthesize* handshaking circuits, a detailed specification must be provided by the designer. Of course, these authors do not address the question of the modification of an existing synthesized *synchronous controller*.

3 Synchronous Design Environment

We first introduce our design environment for synchronous modules and some terminology.

We use SBL (Structural and Behavioral Language) as our specification environment for synchronous modules [GSS85]. In this, each hardware module is viewed functionally as an algebraic datatype with explicit state and encapsulated *external operations*. The *abstract specification* views the module purely as a black box with each operation described functionally in terms only of the data ports. The *realization specification* separately describes the internal structure of the module, with each module operation specified as a composition of operations on its component *submodules* in the same functional style. Finally, the *time specification* describes the control and timing of the module. The instant at which an operation is invoked is termed the *activation* time. All timing parameters associated with that operation are measured relative to this instant. Input port timing is specified as the interval for which its value is required to be stable. Similarly, output port timing is specified as the interval during which its value is guaranteed valid.

The design language is supported by a suite of tools. These include a *compiler* and a *high-level simulator* for specification development, a *timing validator* to verify the consistency of timing constraints, a *controller synthesis* package, a *mixed-level simulator* to check the high-level simulation output against that of a mask-level implementation, and a *netlist generator* that infers the interconnections between module components. Using these with a place and route tool, synchronous hardware modules are designed and implemented from high-level specifications to chip layout.

4 Extension for Global Asynchronism

Our **asynchronous environment** employs the *bundled data convention* [Sut89]. Here each data port (input/output) has associated with it two handshaking wires; *request* and *acknowledge*. All data transactions are governed by the 4-phase handshaking protocol [Sei80] as illustrated in fig. 1.

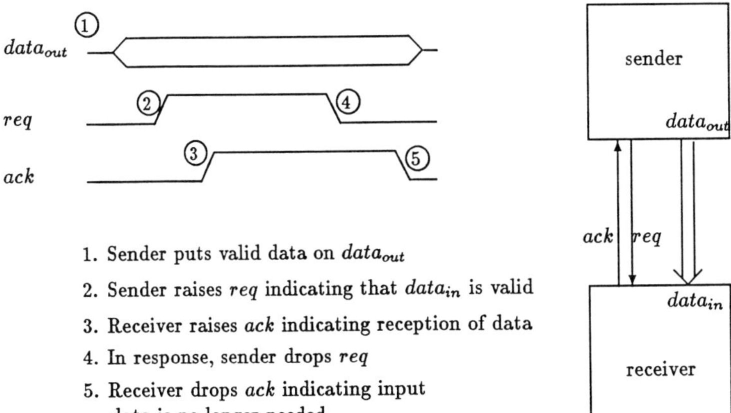

1
$data_{out}$
req
2
4
ack
3
5
sender
$data_{out}$
ack
req
$data_{in}$
receiver
1. Sender puts valid data on $data_{out}$
2. Sender raises *req* indicating that $data_{in}$ is valid
3. Receiver raises *ack* indicating reception of data
4. In response, sender drops *req*
5. Receiver drops *ack* indicating input data is no longer needed

Figure 1: **The 4-phase bundled data convention**

To introduce the method and facilitate the presentation, we initially restrict the discussion to modules having a single external operation referencing nonempty sets of input and output ports. This will be generalized later.

In the design environment of the previous section, once an operation of a synchronous module is activated, we bind the times at which input values are required to be stable and at which output values are produced. In an asynchronous environment however, such timing relationships are inherently nondeterministic. To meet the input timing requirements of the synchronous circuitry, an operation cannot be activated until all its input arguments are valid. Otherwise, it could have to wait for a nondeterministic period of time and violate an input time constraint. In addition, all output values produced in its most recent execution must have been already consumed, otherwise they could be overwritten with newly produced values.

During execution of an operation, the asynchronous environment should be *notified* as soon as a given argument is no longer needed or a result is produced. The times at which these events occur are determined by the timing behavior of the synchronous circuitry.

The interfacing method is performed in two steps. First, the hardware to perform the communication tasks is allocated using a set of templates. Second, the controller of the synchronous circuitry is modified to determine when an operation can be activated and to communicate data consumption/production events to the asynchronous environment.

4.1 The Communication Interface

To convert a synchronous module into a locally synchronous globally asynchronous one, an *envelope* of communication interface circuitry is created to encapsulate the *synchronous core* as illustrated in fig. 2.

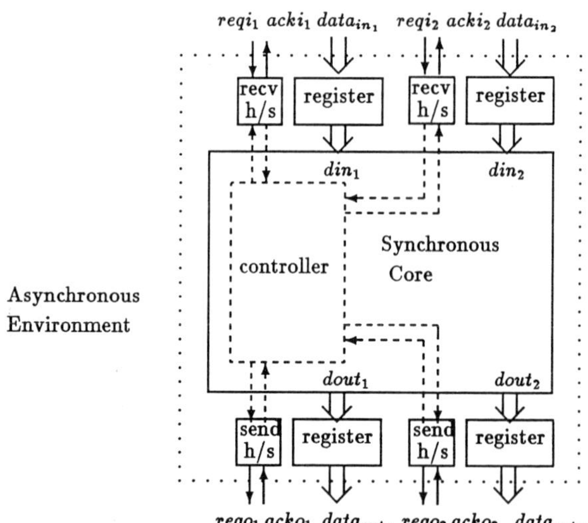

Figure 2: **The Communication Envelope**

The envelope is composed from three templates. Those labelled *register* and *h/s* respectively are parameterized registers and predesigned templates to carry out the handshake protocol. Following the bundled data convention each port is allocated one *register* of corresponding dimension and one *handshake* template of corresponding type (recv or send).

The handshake templates contain both synchronous and asynchronous elements since they interact asynchronously with the outside environment and synchronously with the controller of the synchronous core. They have been designed to ensure that the communication with the external environment can be performed in parallel with local computation whenever possible. Their operation is central to our approach and therefore we explain it in detail. The logic diagram of the input handshake circuit is shown in Fig. 3 together with signal transition sequences.

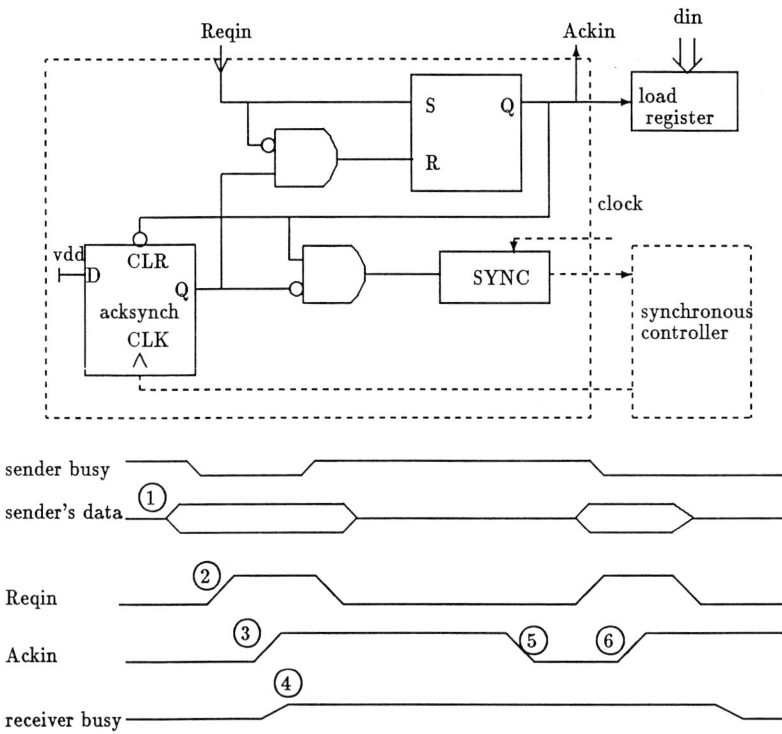

1. sender puts valid data on its outputs then raises Reqin in 2
3. receiver handshake module latches data (sender can change value)
4. receiver core starts computing
5. receiver core releases input port (sender can overwrite register)
6. receiver latches next data while its core is computing

Figure 3: **Receiver Handshake Circuit**

We have deliberately inserted registers at *both* the input and the output to accommodate and exploit the general input signal stability requirements and output guarantees of complex synchronous modules as supported by SBL. We were able to modify the four phase protocol of figure 1 accordingly. This is also the reason we selected it over the two phase protocol [Sut89]. Note in particular the following points: Firstly, the sender can invalidate the data immediately after it is latched by *Ackin*, and therefore can proceed to update the data in its own output register. This means that both the sender and the receiver can be executing at the same time. Secondly, the synchronous core controller indicates that it no longer needs the contents of the input register by setting the *acksync* flip-flop at the earliest time possible. In this way the data source can overwrite the input register while the receiver is completing the execution of the current operation. The time at which the data is released is determined automatically as will be explained shortly.

The logic diagram of the output handshake circuit is shown in fig. 4. Assuming all signals are initially zero, the controller latches output data in the output register *as soon as* it is produced during the execution of an operation. It then clocks the *reqsynch* flip-flop indicating the presence of valid data in the output register. The remainder of the transaction is straightforward.

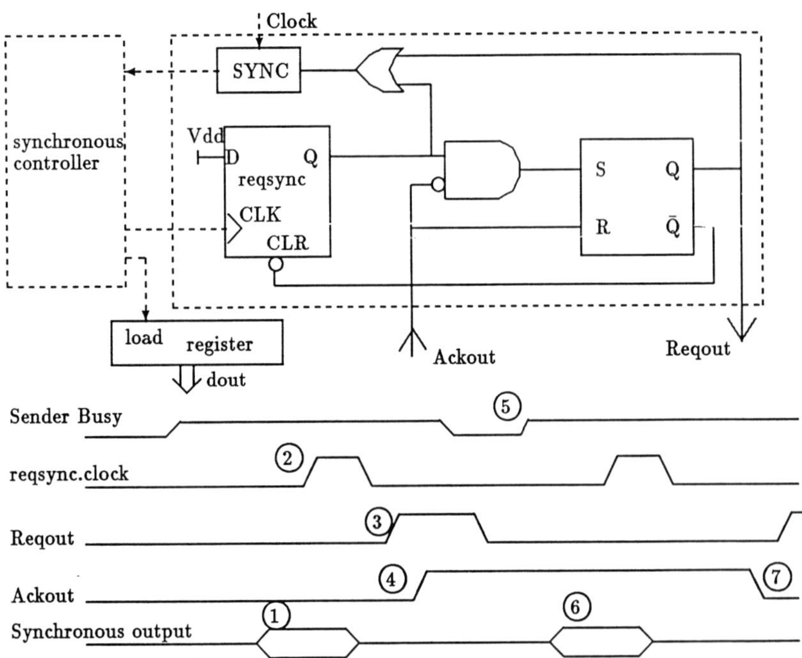

1. Synchronous data is clocked into output register
2. Controller sets the request flip-flop (reqsync)
3. Reqout is set
4. Receiver latches data
5. The sender starts the next computation
6. Next data value is clocked into output register
7. Reqout is delayed until receiver releases its input port (Ackout is low)

Figure 4: **Sender Handshake Circuit**

It is important to note that the synchronous core can proceed to compute the next data value and latch it into the output register as soon as the acknowledgement of the previous data is received. This is the case even if the consumer has not completed the previous transaction; i.e. it has not reset its *Ackin* line. Again, the time at which output data is written is determined automatically from the specification.

Since the handshake circuits are predesigned templates, they have been carefully optimized in terms of both area and speed, and to ensure hazard free operation. Two considerations have affected our designs of the handshake circuits. First, the execution of an external operation is required to be atomic. By this we mean that an external operation, once activated, will execute in a manner identical to that if the core were operating in a purely synchronous environment. For example, the times at which output values are produced and the time it takes an operation to complete, all measured relative to its activation time, should be identical in both cases. This should be the case irrespective of any transactions with the asynchronous environment. Secondly, efficiency is not sacrificed. Since interactions with the external environment are independent from the *busy* state of the core, they can be carried out in parallel with the execution of an operation. Finally, resources (e.g. input registers) are released as soon as they are no longer needed while results are promptly made available to the environment.

4.2 Modification of the Controller

In previous work [ALSS91], we demonstrated how the controller of the synchronous core can be derived from its structural specification and the timing specifications of its components as a byproduct of a timing validation procedure. Also, we have described how the timing parameters of the input and output ports of the core can be derived from the timing parameters of its components and the results of the timing validation phase. Note that the input/output ports of the synchronous core are compositions of the input/output ports of its component submodules.

The derived controller needs to be modified to carry out the interactions with the communication envelope described in the previous section. Two aspects are involved here, the consumption and production of data and the start of the external operation. To relay a data consumption event, a pulsed output to reset the *Ackin* signal is added to all state transitions corresponding to the release time of an input port (the time at which it is no longer needed). Similarly, a pulsed output to load an output register is added to all state transitions corresponding to the time at which an output port becomes valid. A delayed version of this pulse is used to set the *reqsynch* flip-flop of this port, conveying a data-production event.

The external operation of the synchronous core cannot be started until all the input registers have valid data and until the data in all the output registers has been acknowledged. This status information is available to the controller from the handshake circuits and can be incorporated in its start state. For modules with several external operations, the procedure is more involved as will be explained in the next section.

4.3 Modules with multiple operations

In previous sections we restricted the discussion to hardware modules with a single external operation. More generally the synchronous core has several external operations, each referencing a different subset of its input and output ports. The value of a control port (i.e. an opcode) specifies which operation to execute.

To handle such modules, the control port is treated as another input port and itself is allocated communication hardware. The controller of the synchronous core is modified to wait until the *instruction register* contains a valid operation code. The input and output ports referenced by the selected operation are checked to determine if the operation can be activated. If activation conditions are not met then the controller goes into a wait state until they are satisfied. The selected operation is then executed and the acknowledge of the control port is reset indicating that the module is ready to execute the next operation.

A large class of asynchronous circuits do not require a control port since they employ a data flow computation model. This means simply that each computation starts as soon as all input arguments become valid. A subclass of synchronous modules follow this model, once they are interfaced to operate in an asynchronous environment. They have the following characteristics. First, module operations reference disjoint sets of data ports (either input or output). Second, operations that only reference output ports execute depending on the internal state of the module. And finally, the result of executing an operation is independent of the order in which operations are executed. An example of such a module is a first-in first-out (FIFO) store. It has two operations *insert* and *remove.* The *insert* operation executes when data is available at the input ports provided that the buffer is not already full. The *remove* operation executes when the buffer is not empty. The order in which operations are executed is immaterial.

For this class of modules, the *external* control port of the synchronous core can be eliminated. This is possible since we can independently determine which operations are *ready* to execute from the status of the ports they reference and the internal state of the module. Of course if more than one operation can become ready at the same time and if two operations both utilize some submodule hardware which cannot be operated concurrently, then an arbitration may be needed to choose which operation is to execute first. Standard templates for either a fixed or alternating priority scheme could be added and would fit within the philosophy of our system.

4.4 Optimizations

In the design method described so far, Every port of the synchronous core is allocated a handshake circuit and a register to hold the data. Each handshake circuit in turn requires one controller input (with its associated synchronizer) and one controller output.

A number of optimizations are possible. First, the controller inputs associated with all input ports referenced by the same external operation can be combined into a single *data available* signal using an *AND* gate. Similarly, controller inputs associated with all output ports referenced by the same external operation can be combined through an *OR* gate before being synchronized to the controller's clock. In this way, the number of synchronizers needed is dependent on the number of external operations rather than the on number of input or output ports. Reducing the number of inputs to the controller results in a simpler finite state machine and hence a smaller PLA implementation. Second, registers that the *synchronous* module might already have at the inputs or at the outputs can be *automatically* identified and eliminated. They are replaced by those of the communication envelope. Finally, if the concurrent operation of the data source and the synchronous module is not of primary concern, then the input port *registers* could be eliminated. Clearly, modified handshake templates must be used in this case.

5 Results and Current Status

The method described above has been implemented in C on SUN/UNIX. Finite State Machine specifications of the modified controller are derived in Berkeley's MEG format. The handshaking templates have been extensively simulated and their operation verified.

Table 1 summarizes the results of applying the method to three circuits. The increase in the PLA size of the synthesized (synchronous) controller due to the modification process is shown. For this purpose estimates of the increase in the overall chip area (including handshaking circuits and input/output registers) were computed assuming that the controller's PLA occupies $\frac{1}{5}$ of the overall chip area. The portion of the total area overhead due to the increase in the controller's size is also shown.

The self-testing RAM employs a testing scheme similar to that reported in [Dek88]. It has the highest area overhead of the three circuits. This is expected since it has a total of 5 ports: 4 data ports (*datain, dataout, address and testfail*) and one control port which results in a total increase of 10 PLA inputs and outputs.

The FIFO is an 8-bit first-in-first-out store. Here the increase of the PLA area is negligible. This is due to dispensing with the external control port as described in section 4.3.

The final circuit is a 16-bit shift-and-add multiplier. The increase in the PLA size appears to be high due to the small size of the original controller. This fact also leads to a small chip area estimate which in turn contributes to the high area overhead. Input and output registers are the major contributors to the increase in the chip area in this case. With the optimizations described in section 4.4 these could be eliminated and replaced by the existing registers of the multiplier.

Circuit	*total # of ports*	*Increase in PLA area*	*Estimated area overhead*	*Control portion of overhead*
		$\frac{NewPLA-OldPLA}{OldPLA}$	$\frac{NewChip-OldChip}{OldChip}$	$\frac{NewPLA-OldPLA}{NewChip-OldChip}$
Self-testing RAM	5	27%	9.6%	57%
FIFO buffer	2	1.7%	8.8%	4%
16-bit multiplier	3	21%	21%	20%

Table 1: **Area Overhead of Example Circuits**

These experiments are the first step in determining the size range of the synchronous modules for which the method is applicable. Clock skew imposes an upper limit on module sizes while area overhead imposes a lower limit. We envision that the method will best be applicable to complex synchronous structures having their own controllers. One such module we plan to target is a novel associative buffer chip. It is intended to serve as a cache for associative memories and is currently in the final stages of specification and design using the SBL design system. We are also investigating ways to improve the synchronous SBL design system in order to permit than one operation to be executing concurrently by controlling access to common resources. The interface method could then be extended to handle this case.

Acknowledgements: this research has benefited from numerous discussions with Jing Lin and Ramesh Sathianathan. Their comments are greatly appreciated.

References

[ALSS91] Awad, N., Lin, J., Sathianathan, R., and Smith, D. Timing Validation of Hardware Description. In *Proc. International Workshop on Formal Methods in VLSI*, Jan. 1991.

[BK87] Borriello, G. and Katz, R. Synthesis and optimization of Interface Transducer Logic. In *Proc. IEEE International Conference on CAD*, Nov. 1987.

[CG86] Chu, T.A. and Glasser. Synthesis of self-timed circuits from graphs: an example. In *Proc. IEEE International Conference on Computer Design*, pages 565–571, oct. 1986.

[Cha84] Chapiro, D. *Globally asynchronous locally synchronous systems.* PhD thesis, Stanford University, Oct. 1984.

[Chu87] Chu, T.A. *Synthesis of Self-timed VLSI circuits from Graph Theoretic Specifications.* PhD thesis, Massachussets Institute of Technology, Sept. 1987.

[Dek88] Dekker, R., et al. A Realistic Self-Test Machine for Static Random Access Memories. In *Proc 1988 International Test Conference*, pages 353–361, 1988.

[GSS85] Gopalakrishnan, G., M. K. Srivas, D. R. Smith, "An Algebraic Approach to the Specification and Realization of VLSI Designs", *Proc. Seventh International Symposium on Computer Hardware Description Languages*, Tokyo August 1985.

[Mar89] Martin, et. al. The Design of an Asynchronous Microprocessor. In *Advanced Research in VLSI*, pages 351–373, March. 1989.

[MBM89] Meng, T., Brodersen , R., and Messerschmitt, D. Automatic Synthesis of Asynchronous Circuits from High Level Specicification. *IEEE transactions on CAD/ICAS*, 8(11):1185–1205, Nov 1989.

[NT86] Nestor, J. and Thomas, D. Behavioural Synthesis With Interfaces. In *Proc. IEEE International Conference on CAD*, Nov. 1986.

[Ros88] Rosenberger et. al. Q-Modules: Internally Clocked Delay-Insensitive Modules. *IEEE transactions on Computers*, 37(9):1005–1018, Sept. 1988.

[Sei80] Seitz. System Timing. In *Introduction to VLSI systems, Chap. 7, Mead and Conway*, Addison-Wesely, 1980.

[Sut89] Sutherland, I. Micropipelines. *CACM*, 32(6):720–738, June 1989.

VLSI 91
A. Halaas and P.B. Denyer (Eds.)
Elsevier Science Publishers B.V. (North-Holland)

How to Compare Analog Results

Bernhard Klaassen

Gesellschaft für Mathematik und Datenverarbeitung, Postfach 1240,
D-5205 St. Augustin, Germany, Tel.: xx49-2241-142851, FAX: xx49-2241-142889

Abstract
Users (as well as program designers) of CAD tools often need to compare analog results of different tools or versions. Also on-chip measurements have to be checked against simulator outputs. An algorithm is presented to compare two waveforms. Displacements both in vertical and horizontal direction are checked in a reliable manner. Although this task looks simple at first glance it is performed incorrectly by several existing tools. The method discussed in this paper has proven to be accurate and efficient.

1 Problem definition

In a wide variety of CAD applications either the user or an algorithm has to check if two given waveforms are 'equal' or 'different' within certain tolerances. Some common applications are:

- results of two different versions of a simulator (e.g. SPICE [Qua89])
- two simulations with different process parameters
- comparison between logic and analog simulation
- check of simulated results against e-beam measurements
- convergence check of an iterative method (e.g. Waveform Relaxation [Kla89])

All these comparisons may allow deviation not only in vertical (function value) direction but as well in horizontal (e.g. time value) direction. This means that the user could consider two voltage curves as 'equal' if the distance between them is e.g. either 0.05V in vertical or 0.1ns in horizontal direction (or less) for all points on both curves.

Now, if you take a look at the literature or source code of tools, you may find that such routines exist in many cases. But they often rely, and this is a startling fact, on a 'lemma' which is apparently wrong in this context:

> **Lemma 1** (true or false?)
> Given two piecewise linear functions f, g defined by a set of data points out of the same interval.
> If all data points of both curves lie in close distance $d < \varepsilon$ to the other curve, then the same holds for all intermediate points on the line segments.

We need such a lemma, since we do not want to inspect more points on both curves than necessary.

If we define the term 'distance' in lemma 1 only in one direction (let's say as usual $d := |f(x) - g(x)|$) then lemma 1 is obviously true. But in the interesting case with $d := \min(\ \textit{horizontal\_distance},\ \textit{vertical\_distance}\ )$ we have to take a closer look:

To keep our examples simple we assume in this section that horizontal and vertical tolerances have the same length on the graph (otherwise we could scale our t-axis). Figure 1 is a typical example from which one might be convinced that lemma 1 is true.

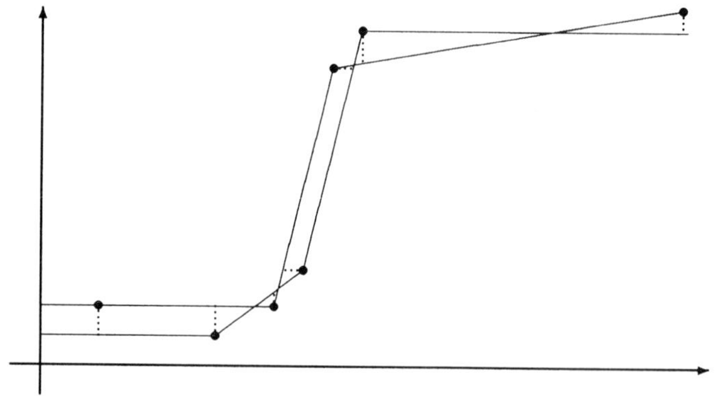

Fig. 1

But Figure 2 presents a counterexample: Although all data points lie close to the other curve there are intermediate points (indicated by the arrow) which violate our rule.

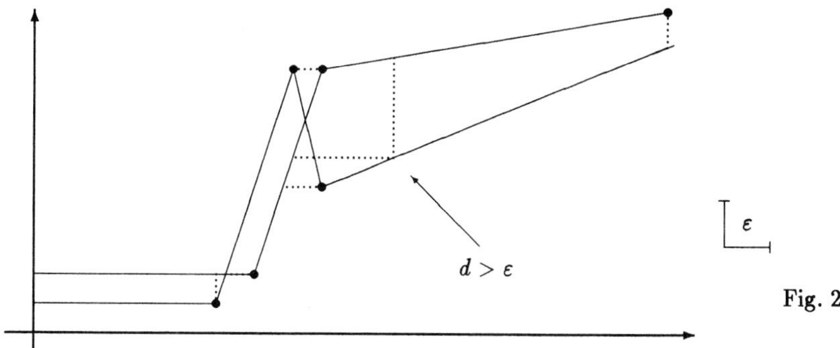

Fig. 2

Since 'lemma 1' is wrong, we have to look for a safer method. In the following three sections all functions are piecewise linear (PWL) and defined by a finite number of data points over the same compact interval I.

2 Mathematical analysis

First we explain the term 'neighbourhood' for points and for curves in an appropriate manner.

Consider a neighbourhood of a point $p = (x_0, y_0)$ which is defined by the tolerances δ in direction x and ε in direction y. The tightest mathematical neighbourhood with these parameters is

$$N(p, \delta, \varepsilon) = \{x, y | \; \varepsilon |x - x_0| + \delta |y - y_0| \leq \varepsilon\delta\}$$

In graphical terms this is a diamond around p with horizontal diameter 2δ and vertical diameter 2ε:

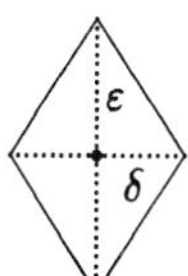

Fig. 3

Now we can transfer the term 'neighbourhood' to curves as follows:

Definition: The neighbourhood $N(f,\delta,\varepsilon)$ of the graph of $f : I \to \mathbb{R}$

$$\text{is defined as} \quad N(f,\delta,\varepsilon) := \bigcup_{t \in I} N((t, f(t)), \delta, \varepsilon)$$

In the following sections this term will sometimes be abbreviated to $N(f)$, when δ and ε are fixed. The upper and lower border of $N(f)$ shall be denoted as $\overline{n}(f)$ and $\underline{n}(f)$ resp.

As a graphical representation this definition is much more suggestive: Observe that $\overline{n}(f)$ and $\underline{n}(f)$ are again PWL functions. And the horizontal respectively vertical "width" of $N(f)$ corresponds to δ and ε. That is the special feature of this type of neighbourhood (Fig. 3).

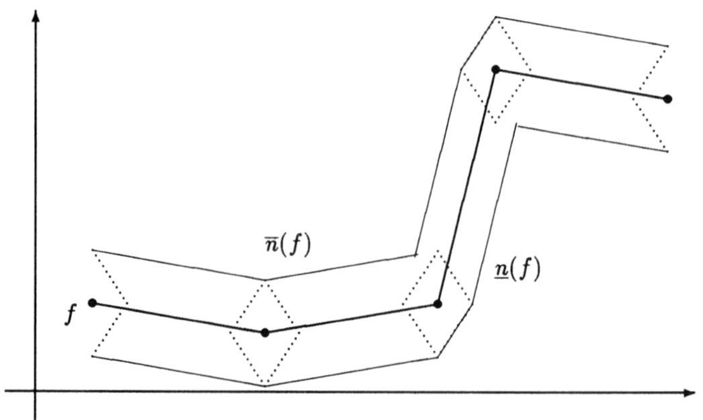

Fig. 4

This leads us to a practical definition of the term 'close distance':

Definition We say f and g lie in **close distance** to each other w.r.t. δ and ε if $g \subset N(f,\delta,\varepsilon)$ and $f \subset N(g,\delta,\varepsilon)$

(where $g \subset \ldots$ means 'graph of g is subset of ...')

The second term with '$f \subset \ldots$' might look redundant to the reader, but one can construct many examples for f and g where g lies within $N(f)$ but not vice versa. (Think of a nearly constant function f with one steep spike in it compared to a constant g.)

In the mathematical literature this topic is treated as "metrics of onedimensional subsets in $\mathbb{R}^2$ ".

3 Algorithms

We can reduce our original problem to the following task, to check if f and g lie in close distance:

ALGORITHM CHECK_CLOSE:
Given two curves f, g and the tolerances δ and ε.

- Build $N(f, \delta, \varepsilon)$

- Check if g lies under $\overline{n}(f)$ and over $\underline{n}(f)$

- Build $N(g, \delta, \varepsilon)$

- Check if f lies under $\overline{n}(g)$ and over $\underline{n}(g)$

Here we can make use of the fact that the upper and lower borders of $N(f)$ are again PWL functions over I. Hence we need two subroutines to perform the whole task:

UPPER_BORDER to build $\overline{n}(f)$, and
CHECK_OVER to check if the graph of a function lies totally over another one.

(to build $\underline{n}(f)$ we use UPPER_BORDER with $-f$ and invert the result again.)

ALGORITHM UPPER_BORDER:
Given curve f and the tolerances δ and ε.

- Build $f_+ \ := \ f$ shifted to the right by δ

- Build $f_- \ := \ f$ shifted to the left by δ

- Build m with $m(t) := max(f_+(t), f_-(t))$ for all $t \epsilon I$
  (with constant extrapolation at the right and left borders of I)

- Build $\hat{f} \ := \ f$ shifted up by ε

- Build $\hat{m}$ with $\hat{m}(t) := max(\hat{f}(t), m(t))$ for all $t \epsilon I$

- If $\hat{m}$ runs inside an $N(p)$ (for any data point $p = (x, y)$), (see Fig. 5)
  replace $\hat{m}$ locally by $max(\hat{m}(t), \Delta(t))$, $t \epsilon [x - \delta, x + \delta]$ w. $\Delta :=$ upper border of $N(p)$

- Return $\hat{m}$ as $\overline{n}(f)$.

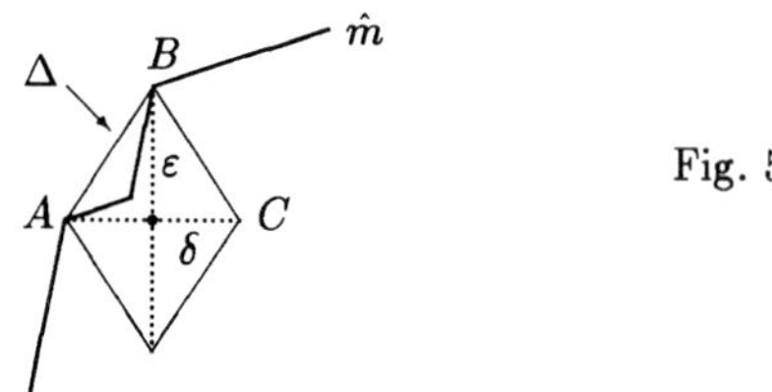

Fig. 5

One should keep in mind that $\hat{m}$ must lie (by construction) above or on the points A, B, C. Hence the last correction step of algorithm UPPER_BORDER (Fig. 5) must always result in a continous PWL function.

To perform the CHECK_OVER algorithm efficiently we need a simple lemma which guarantees that we are now allowed to inspect data points only:

Lemma 2
Given two piecewise linear functions f, g defined over I
Let $D(f), D(g)$ be the set of data points for f and g

If $f(t) \geq g(t)$ for all $t \epsilon D(f) \cup D(g)$
then $f(t) \geq g(t)$ for all $t \epsilon I$

If the reader likes short proofs, he will find one in the Appendix of this paper.

With these results the CHECK_OVER routine can easily be implemented using linear interpolation at each data point. But using the scalar product is much cheaper:

CHECK_DATA_POINT
Given a data point (X, Y) and a line segment l from (x_a, y_a) to (x_b, y_b)
with $x_a \leq X \leq x_b$

- compute $s := < \begin{pmatrix} y_a - y_b \\ x_b - x_a \end{pmatrix}, \begin{pmatrix} X - x_a \\ Y - y_a \end{pmatrix} >$ (scalar product)

(observe that the first vector is a normal vector relative to l.)

- if s $\begin{matrix} > \\ = \\ < \end{matrix}$ 0 then (X, Y) lies $\begin{matrix} \text{over} \\ \text{on} \\ \text{under} \end{matrix}$ l

Hence the CHECK_OVER routine consists of a loop over all data points in $D(f)$ and $D(g)$, with a call to CHECK_DATA_POINT. Observe that no division is needed here, since the scalar product is done by two multiplications only.

4 Performance and Conclusion

The described method has been implemented as a C-routine, and was tested with 'real-life' waveforms (up to 700 data points) on a VAX, a SUN4, and a PC. Although performance is not the most critical point for such a task, it should be mentioned that the computational effort is $O(N)$. (N is the number of data points in both PWL functions to be compared.)

The runtimes on all three machines (e.g. 0.07 sec for 2×700 pts. on a SUN4) were significantly smaller than those for reading the waveforms from their input files, which should be expected for an efficient compare routine.

If a deviation larger than the tolerances is detected, the compare routine returns the critical time interval. An additional test can be performed by checking the curve lengths of both function graphs. E. g. a slightly oscillating behaviour within the tolerances would be detected if it leads to different curve lengths (differing more than a given percentage).

The concept of neighbourhoods can be easily transferred also to piecewise polynomial curves (like splines, Hermitian or Akima interpolation [Aki70], etc.). There is no principal obstacle against using CHECK_CLOSE for the general case. Of course additional effort has to be made for UPPER_BORDER and CHECK_OVER. But for polynomials up to order 3 the problem of intersection points is easily solved by elementary calculus (e. g. Cardani's formula).

Summarizing we can say that the described method has proven to be both accurate and efficient and can be used for a wide range of applications.

References

[Aki70] H. Akima, "A New Method of Interpolation and Smooth Curve Fitting Based on Local Procedures", Journ. of the ACM, **17**, 4, Oct. 1970

[Kla89] B. Klaassen, K.L. Paap, P.G. Ploeger, "Improved Waveform-Relaxation-Newton Method", Proc. of the 1989 IEEE ISCAS, p.856-859, Portland OR., May 1989

[Qua89] T.L. Quarles, "Analysis of performance and convergence issues for circuit simulation", Memo UCB/ERL M89/42, Berkeley CA, 1989.

Appendix

Proof of Lemma 2:
Function $p := f - g$ is PWL; apply Lemma 3. □

> **Lemma 3** If a piecewise linear function $p : I \to \mathbb{R}$ is nonnegative at all its data points $\epsilon D(p)$, it is nonnegative in the whole interval I

Proof of Lemma 3:

$$p(t) \geq 0 \ \forall \, t \epsilon D(p) \ \Rightarrow \ \min_{t \epsilon D(p)} p(t) \geq 0 \ \overset{PWL}{\Rightarrow} \ \min_{t \epsilon I} p(t) \geq 0 \ \Rightarrow \ p(t) \geq 0 \ \forall \, t \epsilon I \quad \Box$$

VLSI 91
A. Halaas and P.B. Denyer (Eds.)
Elsevier Science Publishers B.V. (North-Holland)

Application of scan-based design-for-test methodology for detecting static and timing failures in VLSI components

Bulent I. Dervisoglu [a] and Gayvin E. Stong [b]

[a] Hewlett Packard, Apollo Systems Division, Chelmsford, MA, 01824

[b] Hewlett Packard, Integrated Circuits Division, Ft. Collins, CO, 80525

Abstract

This paper describes using scanpath techniques for testing both static and timing failures inside VLSI components. In particular, a new flip-flop design is presented which can be used to simplify test pattern generation and the application of test patterns for test and measurement of internal path-delays. Design and application of a boundary-scan cell that enables input/output delays to be measured is also described.

1. INTRODUCTION

The ever increasing density and functional complexity of digital integrated circuits requires more efficient and narrowly focused methodologies for testing them. Whereas a batch of *broadside* test patterns (i.e. test patterns that are applied and test results that are observed directly on the input/output pins of the device under test) might have been sufficient to test most LSI and some VLSI components of recent vintage this is no longer the case for present day VLSI components. There are two major components to the problem of testing modern VLSI components. First, there is the all too familiar problem of testing for *stuck-at* failures which manifest themselves as circuits permanently (and undesirably!) tied to constant logic levels (logic 0 or logic 1). Furthermore, there are other *static* failures which, though their failure mechanisms are different than the *stuck-at* failures, can be detected using the same models and test methods used for the *stuck-at* faults. Examples of such faults include the *CMOS stuck-open* as well as signal *short-circuit* and open-circuit conditions. The common characteristic of all of these *static* failures is that they are *permanent* and hence repeatable, regardless of the operating environment (i.e. operating voltage or temperature). The second component of the problem is testing for *timing* failures whereby nodes of the circuit undergo all the desired signal transitions but do so only at slower (or faster!) than expected speeds. As a result, the component might be able to perform its intended function only at slower operating frequencies. Indeed, if certain signal changes occur much faster than expected this might create *hold-time* violations which prevent the component from performing its intended functions correctly at any frequency. In many cases the effects of such faults may become more or less pronounced if the operating environment is changed. Therefore, testing for these failures takes on the form of measuring the acceptable operating frequency at a given operating environment and comparing that with the expected frequency.

It is highly desirable to use a unified approach for testing both the *static* and *timing* failures in a digital integrated circuit. Scanpath technology, which originated more than 15 years ago[1], is finally becoming the method of choice for implementing Design-For-Testability (DFT) features in complex integrated circuits and systems[2],[3]. This paper describes how scanpath techniques and other *ad-hoc* DFT techniques have been used in a state-of-the-art VLSI component for detection of *static* faults. Furthermore, section 2.2.2 describes a significant improvement to the traditional scanpath methodology whereby it has become possible to measure internal path-delays with great accuracy. With this, scanpath methods have been made applicable for detecting both *static* and *timing* errors in VLSI components.

2. USING SCANPATH TECHNIQUES FOR TESTING VLSI COMPONENTS

During the design of the original APOLLO DN10000 workstation it was decided that the complexity of the product coupled with the time-to-market pressures necessitated implementing scanpath techniques for testing the individual VLSI components as well as the Printed Circuit Board (PCB) assemblies[4]. Since then a board upgrade to the original DN10000 Central Processing Unit (CPU) has been developed (dubbed the DN10000TX) [5]) to perform at 36.36 MHz, which is twice the clock frequency of its predecessor. The two versions of the CPU represent different implementations of the same architecture varying in their operating frequencies as well as the level of integration in their VLSI components. The DN10000 CPU is implemented using a 23000 gate gate-array for the Integer Processor (IP) and separate gate-arrays to implement the Memory Management Unit (MMU), the Floating-Point Controller (FPC) as well as other VLSI components to implement the floating-point operations and the interface to the system bus. In the DN10000TX the functions of the IP, MMU and the FPC are integrated into a single custom VLSI component (dubbed the IPU) whose clock frequency is double that of the DN10000 IP. It is evident that circuit complexity and operating speed of state-of-the-art VLSI components are increasing at a very fast pace. It is not reasonable to expect that with every such increase new test methodologies can be developed. Therefore, it was felt necessary to refine the scanpath techniques used in the design of the DN10000 and utilize them again in the DN10000TX.

2.1. Testing for static faults

The DN10000TX CPU has been designed using full-scale implementation of scan as a means of providing design-for-testability. To achieve this all Application Specific Integrated Circuits (ASICs) have been designed such that all internal flip-flops are accessible via scan to provide controllability and observability of all internal state variables. The only exception is the on-chip General Purpose Register (GPR) arrays that are present on some of the ASICs; the GPR array on the IPU chip is not directly scannable though the GPR array on another chip (the floating-point Register File) is directly scannable. In the case of the IPU, since all remaining flip-flops on the chip are directly scannable, it is possible to access the GPR as a memory array by providing address/data values and read/write controls through the regular scanpath of the chip. In addition, all ASICs implement boundary-scan which is used for testing the PCB interconnect by allowing output values to be determined by the contents of the boundary-scan registers. Thus, from the tester (or the system diagnostics processor) perspective all internal state variables as well as all I/O pins of the ASICs are controllable and observable using standard scanpath access.

Testing of individual ASICs with scanpath access as described above is achieved using a straight forward process. First, an Automatic Test Pattern Generation (ATPG) program is used to generate test patterns for the combinational logic on the ASIC, treating all internal flip-flops as primary inputs or outputs. Fault simulation is used to determine coverage achieved by the generated test patterns and test generation is stopped when adequate coverage is reached. In most cases better than 99.5% fault coverage is achieved using between 500 and 700 test patterns. Furthermore, certain *ad-hoc* techniques are used whenever necessary to improve fault coverage as well as to improve the efficiency of testing with pseudo-random test patterns [5]. For ASICs that include embedded RAM structures in them design-rules were established to assure that the RAM contents are preserved despite the fact that control and data signals may be changing during internal scan operations. Furthermore, special registers were placed at the RAM output ports to capture their values before these are passed through combinational logic on the chip. This allows writing into individual RAM locations and then reading the same data back and makes it possible to develop software to test the embedded RAMs using standard memory test algorithms. Other design-rules were established to assist in test generation and pseudo-random test coverage around comparator circuits such that comparators having more than 5 bit operands were provided with a separate input, controlled directly from a scannable flip-flop, to force the comparator output to the particular logic state that indicates *equal operands.* By counteracting the property of the original comparator that there are many more input combinations that correspond to the *non-equal operands* case than the *equal operands* case, search algorithms in the ATPG program can find desired controlling conditions for the comparator outputs with greater ease. Yet other design-rules were established to prevent generating *illegal* internal-state conditions (e.g. internal bus fights) that may produce indeterminate results and/or cause damage to the component when performing regular or pseudo-random testing.

As stated above, all ASICs implement boundary-scan in order to test the board-level interconnect as well as the input and output drivers of the chips. Figure 1. shows a somewhat simplified logic diagram for a boundary-scan cell which can be associated with a *bi-directional* I/O pin. It can be seen that the output driver can be turned-off by asserting the *DRV_OFF* signal which is generated by the on-chip test control logic. This feature is used during the power-on sequence of the DN10000TX CPU such that all bi-directional outputs are turned-off to prevent possible bus clashes that may result before the CPU is properly initialized. Furthermore the logic value which has been scanned into flip-flop #1 can be driven onto the I/O pin by asserting the DRV_ON signal, regardless of the state of the *system logic*. Also, the current value on the I/O pin can be captured into the boundary-scan cell (flip-flop #2). This allows testing the *output-driver* and the *input-receiver* as well as testing the integrity of the board-level interconnections among the components. Finally, figure 1. shows that the boundary-scan cell can also supply the data values received by the *system logic* (using flip-flop #3) so that the chip may be tested in total isolation from its environment. For example, this feature may be used to perform the *INTEST* function that is supported by the IEEE 1149.1 Boundary-Scan Standard [6].

All ASICs used in the DN10000TX use the same test-control logic implementation with the exception of the IPU chip whose test-control logic has two different modes of operation one of which is compatible with the other ASICs and the other mode is designed for use on a proprietary tester. Thus, DN10000TX CPU printed circuit board assemblies can be tested using a common diagnostics processor interface [7].

2.2. Testing for timing faults

Testing of the internal path-delays of an IC component requires that, no matter how the test patterns are generated or applied to the chip, the response of the chip should

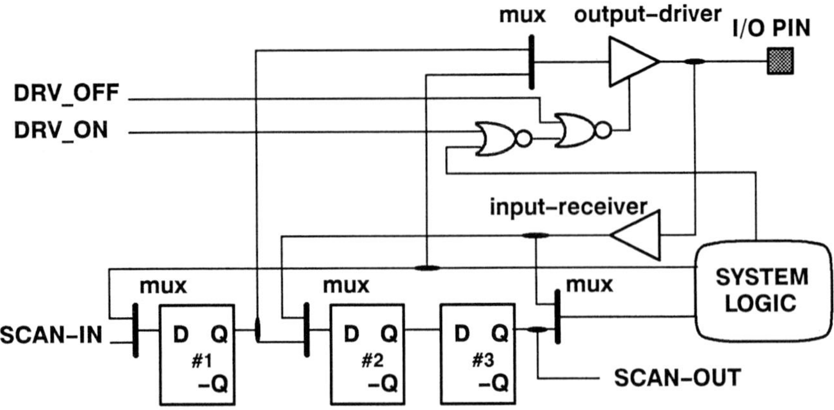

Figure 1. Boundary-cell design.

be captured using the chip's own functional clock signal. There are two reasons for this. First, internal clock delay and skew as well as the data set-up times are considered to be part of the internal path-delay and hence their effects must be included in any test or measurement of the path-delays. Second, in most (if not all) cases the functional clock signal is the only accurate mechanism which exists on the IC component to enable its internal flip-flops to operate. Whereas special test and/or scan clocks may also be routed to each internal flip-flop these signals exhibit greater delay/skew due to the often reduced frequency of scan operations compared to the actual system operation. Since test results are to be captured using the system clock it follows that signal transitions which are part of the path-delay pattern must also be generated using the system clock. However, if scanpath techniques are to be used in loading the internal flip-flops of the chip with the desired test pattern it follows that after the loading of the *initial* values of the internal path-delay test pattern at leas one system clock must be applied to create signal transitions at the outputs of the internal flip-flops. An optimistic approach to achieving the above described behavior is to perform scan-in to load the internal flip-flops with a pattern other than the desired test pattern, having performed prior circuit analysis to determine that the response of the circuit to the first system clock pulse shall transform that pattern into the intended test pattern. However, this is a very difficult approach to implement since it requires performing complicated circuit analysis in order to determine what state the device under test should be placed into (using scan) so that its next state corresponds to the desired test pattern and the state transitions between scanned-in and final states correspond to the precise signal transitions which are required as part of the path-delay test pattern.

2.2.1. Test and measurement of I/O delays

Referring to the boundary-scan cell shown in figure 1. it is seen that a total of three scannable flip-flops are used for applying the test vectors and capturing test results. During regular system operation the output-driver drives values from the chip logic to the I/O pin and the input-receiver drives values from the I/O pin to the chip logic. Tri-state control (i.e. output active or inactive) of the output-driver is determined jointly by the system logic and the test-control logic of the chip. This is necessary so that any arbitrary output value (logic 1, logic 0 or high_impedance) can be produced on the I/O pin using the boundary–scan cell, regardless of the internal state of the system logic. To perform path–delay testing between the I/O pin and the chip logic an initial value is loaded into flip–flop #3 and the transition value is loaded into flip–flop #2. When the first system clock is applied the value in flip–flop #2 replaces the value in flip–flop #3, thereby creating a signal transition through the receiver into the chip logic. The next system clock causes the chip flip–flops to update along with flip–flop #1 and flip–flop #2 which capture the output of the system logic and the I/O pin value, respectively. By repeating the tests using different clock frequencies it is possible to measure the *clock–to–output–pin* delay or the *input–pin–to–clock* setup times. Flip–flop #1 also contains circuitry that allows an *initial* value to be scanned in and held there while a different *transition* value is being scanned in. A system clock causes the output of flip–flop #1 to update with the transition value which causes the same transition to appear at the output of the driver circuit and the I/O pin. The output of the driver can be captured by the next system clock in flip–flop #2 and/or in the system flip–flops. This allows measurement of the delay through the *output–driver and the input–receiver* or the path–delay through the *output–driver, input–receiver and system logic*.

2.2.2. Test and measurement of internal path–delays

This section describes the design and operation of a scannable flip–flop which can be used in a VLSI component so that scanpath techniques can be used for testing both *static* and *timing* failures of the system logic. A key feature of the boundary–scan cell described above is that different *initial* and *final* values can be stored inside the boundary–scan cell such that the *initial* value is replaced by the *final* value when the first functional clock pulse is applied. This way a signal transition can be started at a precise point in time (i.e. on the rising edge of the functional clock pulse). A similar idea was described in [10] which discusses how a three–latch circuit might be used to store two different bit values inside the same flip–flop. Figure 2. shows the logic diagram of a different (and improved) flip–flop which implements this feature using CMOS gate–array technology[8]. During regular system operation control signals DS (double–strobe) and ML (master–load) as well as scan clocks SI_CLK (scan–in clock) and SO_CLK (scan–out clock) are held at logic 0 to allow the master–latch to follow the D input signal while CLK (system clock) is low and transfer the master–latch contents into the slave–latch on the rising edge of CLK. During regular scan operations the system clock must be held low and the non–overlapping SI_CLK and SO_CLK are used to capture the SI (scan–in) input into the slave–latch and move the contents of the slave–latch into the scan–latch, respectively. Multiple flip–flops are chained together by connecting the SO (scan–out) output from the current flip–flop to the SI input of the next one. Whereas this arrangement results in inversion of data polarity between the slave–latch output and the SO output as well as between the SI input and the slave–latch output, both inversions can be compensated for by inserting an inverter before the SI input to the first flip–flop and inserting another inverter after the SO output from the last flip–flop along the scan chain.

Asserting the DS control signal (i.e. DS = 1) isolates the master–latch from the D input and the CLK signal. This way, whenever a scan–in operation is performed with the master–load (ML) signal in its asserted state (ML = 1) the master–latch and the slave–latch are set to the same value whereas performing scan–in with ML = 0 preserves the master–latch contents while loading the slave–latch from the SI input. This allows different values to be scanned into the master– and the slave–latch sections of the flip–flops. When the scan–in operations are completed the slave–latches contain *initial* flip–flop values whereas the master–latches contain the complement of the *final* values that comprise the test pattern for testing internal path–delays. Internal path–delays can be measured by applying two system clock (CLK) pulses such that

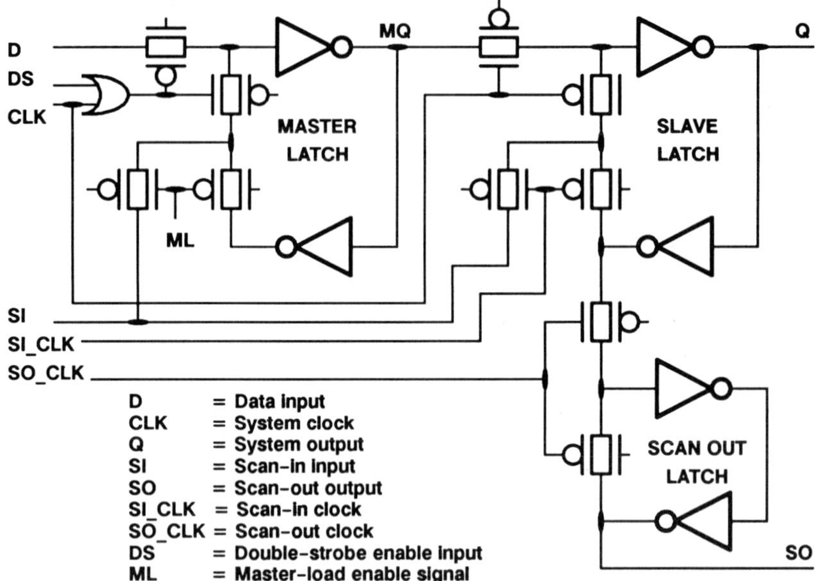

Figure 2. Scannable flip-flop with Double-strobe and Master-load features.
(Note: secondary connections to transmission gates not shown for space reasons)

the first CLK pulse moves the inverse of the contents of the master-latch into the slave-latch, thereby creating signal transitions at selected flip-flop outputs. Test control logic for the double-strobe (DS) signal is designed so that the first CLK pulse also causes the DS signal to be de-asserted. This is necessary so that the master-latch sections of the flip-flops become enabled (i.e. regular master/slave operation is enabled) before the second CLK pulse which is used for capturing the test results. By repeating the same test over and over again while systematically reducing the time distance between the two CLK pulses it is possible to arrive at the minimum separation of the two CLK pulses required for proper operation. This value represents a very accurate measurement of the internal path-delay.

2.2.3. Generating delay-test patterns

There are two dimensions to the problem of generating test patterns for measuring (or at least determining the pass/fail status) of internal path-delays. First, it is necessary to determine which internal path-delays need to be measured/tested. Given the very large number of different signal paths inside a complex digital IC it is obvious that the selection of a subset from among all possible internal paths is necessary. One way to select such a subset is to use the results of timing analysis whereby some number of the longest (i.e. slowest) paths can be selected and their delays measured. Another approach might be to include sufficient paths in the subset in order to cover all different types of cells (gates or other macro blocks) used as building-blocks so that errors due to wrong characterization of these cells may be identified.. It is not uncommon to find out that a timing *failure* may have been caused due to errors in the characterization of the building-blocks in the timing library. This is a very typical cause of timing discrepancies between the circuit model and the actual component, especially when dealing with state-of-the-art technologies where timing libraries may not yet have become stable. Other criteria for selecting the internal paths might be to include some number of paths that pass through circuit nodes which are heavily loaded.

The second dimension to the path-delay test generation problem is the identification of the actual test pattern itself. A delay-test pattern consists of a *static* and a *transition* component. The *transition* component is responsible for generating the desired signal transition(s) and the *static* component provides the sensitization of the signal paths whereby the generated transition is sensitized through a combinational circuit to the input of a flip-flop where it will be captured when the system clock is applied. A delay-test pattern is similar to a stuck-at test pattern except that in the case of the former the specific sensitization path is also a key part of the test pattern. For example, figure 3. shows a combinational circuit and identifies a specific signal-path for which the path-delay is to be measured. To determine the appropriate path-delay test pattern first a dummy AND-gate is added to the circuit as shown. An input to the AND-gate is derived from the output of the combinational circuit through which an input signal transition is to be propagated. This signal is used in its true or complemented form depending upon whether the final value of the signal transitions is a logic 1 or logic 0, respectively. Other inputs to the dummy AND-gate come from all remaining inputs of gates through which the desired signal transitions must flow. If the desired signal transition is flowing through and AND- or NAND- gate, the remaining inputs of these gates are also fed to the inputs of the dummy AND-gate whereas if the desired signal transitions flow through OR- or NOR-gates their remaining inputs are inverted and then connected to the inputs of the dummy AND-gate. The dummy AND-gate is not actually implemented as part of the combinational logic but rather is presented to a standard Automatic Test Pattern Generation (ATPG) program [9] which is used for generating the necessary delay-test patterns.This way a test pattern which has been generated to detect the *stuck-at-0* fault at the output of the dummy AND-gate becomes usable as the *delay-test* pattern for the identified signal path. For example, in the example given in figure 3. there are two such test patterns. The first pattern requires input flip-flops B, C and D all to be set to the logic 1 value and hence can only be used to sensitize a *low-to-high* transition at the D input. The second test pattern requires A, B and C all to be set to the logic 1 value. Since the second test pattern is independent of the value of the D input this pattern can be used to sensitize transitions at the D input changing in either direction. In either case the transitions created on input D will travel through the identified signal path to reach the destination flip-flop Z.

If flip-flops shown in figure 3.have the *double-strobe* and *master-load* features described above, testing would proceed as follows. First, all flip-flops would be placed into *double-strobe/master-load* mode and *scan-in* would be performed to set both the *master-* and *slave-* latches in flip-flops A, B, C and D to the logic 0 state. Next, the *master-load* control would be de-asserted while the *double-strobe* control is kept in its asserted state and a new round of *scan-in* would be performed to set the *slave*-latches in flip-flops A, B and C to the logic 1 state and set the *slave*-latch in flip-flop D to the logic 0 state. This completes the procedure to set input flip-flops A, B, C and D to their *initial* values (which is determined by the contents of their *slave* latches) as well as storing the *complement* of their *final* values in their *master*-latches. At this point, applying the first clock pulse causes the *slave*-latches of the flip-flops to be updated from their respective *master* latches. As a result of the particular values that were stored in them,. this causes only the "D" flip-flop to change its value from logic 0 to logic 1 while the A, B and C flip-flop outputs all remain at the logic 1 value. The control circuit for the DS signal has been designed so that DS is reset to the logic 0 whenever a clock pulse is applied to the chip. Hence the first clock pulse also causes the *double-strobe* mode to be canceled and enters the flip-flops into their regular *master/slave* mode of operation. Finally, applying the second clock pulse to the chip

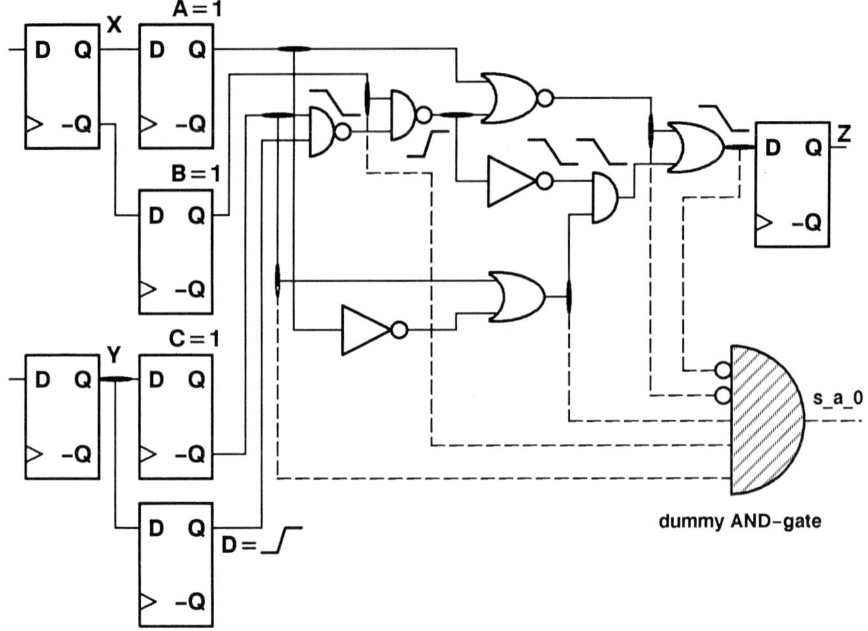

Figure 3. Circuit example to illustrate delay-path test pattern generation (all flip-flops are clocked using a common clock signal which has not been shown).

causes the destination flip-flop Z to be updated from the combinational circuit output. The value captured in the destination flip-flop will either be the *initial* or the *final* logic value at the combinational circuit output, depending upon whether or not the the two clock pulse are separated from one another by at least the propagational-delay through the identified signal path. By applying the same test over and over again while systematically bringing together the two successive clock pulses until the destination flip-flop is no longer able to capture the *final* value of the signal transition applied to its input terminal a very accurate measurement of the path-delay can be made.

A different and somewhat optimistic approach for generating test patterns for path-delay measurement is to perform scan-in to load the internal flip-flops with such a special pattern which prior circuit analysis will have determined will be transformed into the actually intended test pattern when the first functional clock pulse is applied. Using this approach, the *initial* values loaded into the flip-flops would be transformed into the *final* values which are required as part of the timing test pattern. As stated earlier, this is a very difficult approach to implement due to the difficulty of performing the required circuit analysis which, in effect, requires performing simulation in reverse time flow in order to determine what state the device under test should be placed in (using scan) so that its next state corresponds to the desired test pattern. In addition, the transition from the current-state to the next state must be achieved by allowing only the pre-determined signal transitions of the state-variables (i.e. flip-flops). Furthermore, this technique allows creating signal transitions only as the combinational logic of the chip permits, test patterns which appear correct when internal paths are considered on their own may not be applicable when the entire circuit of the IC component is put together. For example, examining figure 3. shows that to create the necessary *low-to-high* transition on the D signal and to do so by loading flip-flop D from its regular data input requires flip-flop Y to have been pre-loaded with a logic 1 value. However, this causes the C signal to be set to logic 0 at the same time as the signal transition on D is achieved. Since C = 1 is required as part of the test pattern this analysis shows that the desired test pattern can not be generated without using the *double-strobe* flip-flop described above.

2.2.4. An application case study

The technique described above has been used successfully for performing path-delay testing on the ASIC chips used in the DN10000TX project where a particular timing problem was discovered. The cell-library used for the ASIC gate-arrays contained a particular 3-to-8 decoder with a common ENABLE input, as illustrated in figure 4.a. A simplified physical layout for this cell is illustrated in figure 4.b where only portions of the poly-silicon and the first-level metal layers are shown. The CAD tool which performed routing recognizes the fact that the ENABLE input (EN) can be connected to the cell at any one of the pad points either directly on the poly-silicon layer or on the first-level metal and it chooses the one best suited for the particular instantiation of the cell. Whereas gate-array physical design rules prevented crossing through more than 3 poly-silicon segments in a single signal run this rule was inadvertently violated inside the 3-to-8 decoder cell depending upon where the ENABLE signal had been connected to the cell. Furthermore, due to a processing failure the poly-silicon layer resistance and the propagational-delay through a unit length of poly-silicon was much higher than specifications. Thus, different output ports in different instances of the 3-to-8 decoder cell exhibited grossly increased propagational-delay from the common ENABLE input to one of the cell outputs. This problem was first discovered by measuring the internal path-delays along several paths and checking these against the expected values produced by the timing analysis program. At this point our ability to take actual measurements on any desired internal path enabled us to identify the failure very quickly. It is also worth noting that such ability to deal with any chosen internal path is crucial in convincing the component manufacturer that the problem indeed lies along that path.

3. CONCLUSIONS

Aggressive use of scanpath techniques is becoming necessary to achieve proper testing of *static* and *timing* failures in state-of-the-art VLSI components. Furthermore, as circuit complexities increase and embedded RAM structures become more widely used inside VLSI components it is becoming necessary to go beyond using the basic scanpath technology just to improve controllability and observability of internal circuit nodes. Instead, specific *ad-hoc* circuits must be included inside VLSI components in order to improve testability of such embedded structures.

This paper presents a significant advancement of the basic scanpath technology and shows how the original concept can be improved and become usable for testing timing failures of VLSI components. A clear advantage of using the techniques described in section 2.2.2 for test and measurement of internal path-delays is the ability to store two independent values inside the same flip-flop. This eliminates the need to load initial values into the flip-flops and clock the system such that the system's response is captured and used as the actually intended test pattern. Ability to take actual measurements on specific internal paths quickly enables the identification of "slower than expected" paths which may be indicative of a problem. Also, since exotic

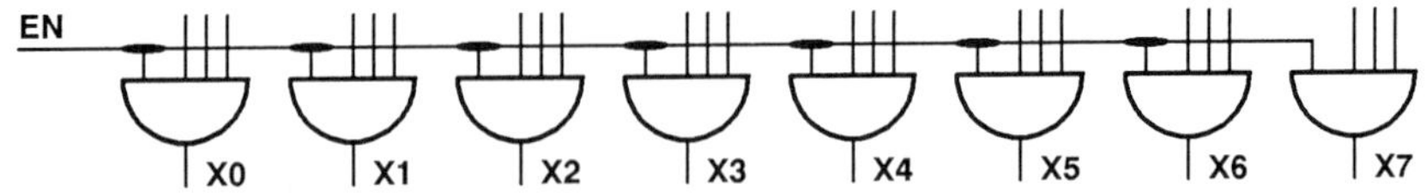

Figure 4.a. Gate-level schematic for 3-to-8 decoder.

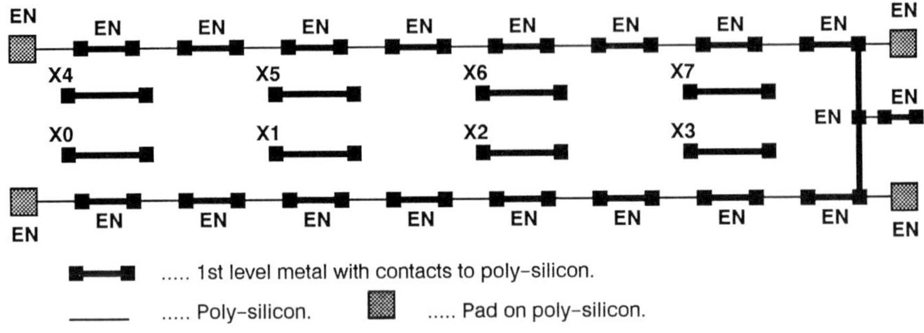

Figure 4.b. Simplified physical layout (showing 1st level metal) for the 3-to-8 decoder.

test equipment is not needed, test and measurement of internal path-delays can be done both at the manufacturer and the end-user locations and the results obtained in one location are repeatable at the other location as well. This is a major time saver since otherwise valuable time could be spent by the end-user trying to convince the manufacturer that a problem exists.

To the best of the authors' knowledge, the design of the scannable flip-flop shown in figure 2. is novel and this paper is the first one to describe an actual *implementation /application* of the techniques described here.

4. REFERENCES

1 . R. Bahr, S. Ciavaglia, B. Flahive, M. Kline, P. Mageau and D. Nickel, *The DN10000TX: A New High-Performance PRISM Processor*, COMPCON'91, San Fransisco, CA, Jan. 1991.

2 . M. J. Williams amd J. B. Angel, *Enhancing Testability of Large Scale Integrated Circuits via Test Points and Additional Logic,* IEEE TC, C-22, No. 1, Jan. 1973, pp. 46-50.

3 . E. B. Eichelberger and T. W. Williams, *A Logic Design Structure for LSI Testability,* Proc. 14th DAC, New Orleans, June 1977, pp. 462-468.

4 . B. I. Dervisoglu, *Using Scan Technology for Debug and Diagnostics in a Workstation Environment,* ITC'88, Washington D.C., September 1988, pp. 976-986.

5 . B. I. Dervisoglu, *Scan-path Architecture for Pseudorandom Testing,* IEEE Design and Test of Computers, 1989, Vol. 6, No. 4, pp. 32-48.

6 . Test Technology Technical Committee of the IEEE Computer Society, *Standard Test Access Port and Boundary-Scan Architecture,* IEEE Standard 1149.1/D6, 1989.

7 . B. I. Dervisoglu, *Towards a Standard Approach for Controlling Board-Level Test Functions,* ITC'90, Washington D.C., September, pp. 582-590.

8 . Motorola Inc., *HDC Series Design Reference Guide,* Rev. 1, 1991.

9 . AIDA Corporation, *AIDA Design System Reference Manuals,* Rev. 2.0, 1988.

10 . Y. K. Malaiya and R. Narayanaswamy, *Testing for Timing Faults in Synchronous Sequential Integrated Circuits,* ITC'83, Philadelphia, PA, October, pp. 560-571.

VLSI 91
A. Halaas and P.B. Denyer (Eds.)
Elsevier Science Publishers B.V. (North-Holland)

Identification and Resynthesis of Pipelines in Sequential Networks

Sujit Dey[1] Franc Brglez[2] Gershon Kedem[1]

[1]Department of Computer Science
Duke University, Durham, NC 27706

[2]Microelectronics Center of North Carolina
Research Triangle Park, NC 27709

Abstract

Pipelines are the only sequential circuits which can be peripherally retimed and resynthesized using combinational synthesis methods. We develop a technique to identify pipelines in general sequential networks. We extract a set of pipelines from the sequential circuit based on their resynthesis potential. We resynthesize each pipeline individually, using combinational techniques. The process of identification, retiming and resynthesis of pipeline structures have led to considerable optimization of a set of large sequential benchmarks.

1 Introduction

Compiling a typical high level design specification produces a netlist of combinational and sequential logic elements. Given a structural description of a sequential circuit, current logic synthesis systems optimize only the combinational parts of the circuit between the latches. More recently, there have been efforts to optimize sequential circuits, considering the combinational as well as the sequential elements of the circuit [1,2,3,4].

Retiming has been introduced in [5,6] as a way to reposition latches in a sequential circuit so that the clock period is minimized. In [2], a set of logic synthesis operations has been combined with retiming techniques to optimize sequential networks. The approach in [1] has introduced the notion of peripheral retiming, a technique where latches are moved temporarily to the boundaries of the circuit. Subsequently, the logic between the latches can be optimized using existing combinational techniques, and then the circuit retimed to reposition the latches. An efficient technique has been proposed in [3] to compute the invalid and equivalent states information from a sequential circuit. The corresponding sequential don't cares can be used to optimize sequential networks using combinational techniques. In [4], the concept of permissible fuctions has been extended for sequential circuits to minimize logic accross latches.

Encouraging results have been reported in [1] in the case of acyclic sequential circuits. However, most finite state machines realizations have cycles. Sequential circuits with cycles, and even some acyclic circuits, cannot be peripherally retimed. Once cycles are cut, acyclic sub-circuits need to be identified which can be peripherally retimed. Cycles can be cut in many ways and the choice may affect the subsequent identification of the sub-circuits for peripheral retiming and the ultimate synthesis results. The best way, to make the retiming and resynthesis of general sequential circuits effective, is to develop an algorithm that can determine how to break cycles as well as identify sub-circuits that can be peripherally retimed with best resynthesis results.

In this paper, we show that pipelines are the only sequential circuits which can be peripherally retimed and resynthesized using combinational methods. We extend the concepts of corolla partitioning based on an analyis of signal reconvergence [7,8] to cyclic sequential circuits. We show how to identify reconvergent regions in sequential circuits, and how to identify pipeline structures as sets of overlapping reconvergent regions. We extract a set of pipelines from the sequential circuit based upon their resynthesis potential [8]. We resynthesize each pipeline individually, using combinational techniques. The process of identification and resynthesis of pipeline structures have led to considerable optimization of a set of large sequential benchmarks.

2 Retiming and Peripheral Retiming

We consider sequential circuits consisting of combinational logic gates and latches. We model a sequential circuit with a directed graph $G(V, E)$. Each primary input, gate, fanout point and latch of the circuit is represented by a distinct node in the graph, and each wire in the circuit is represented by an edge in the graph. The circuit graph may have cycles. Figure 1 shows a sequential circuit consisting of simple gates and latches, and the corresponding graph. The rectangular nodes in the graph represent the latches of the circuit. The shaded circular nodes in the graph represent the fanout points to distinguish them from combinational gates which

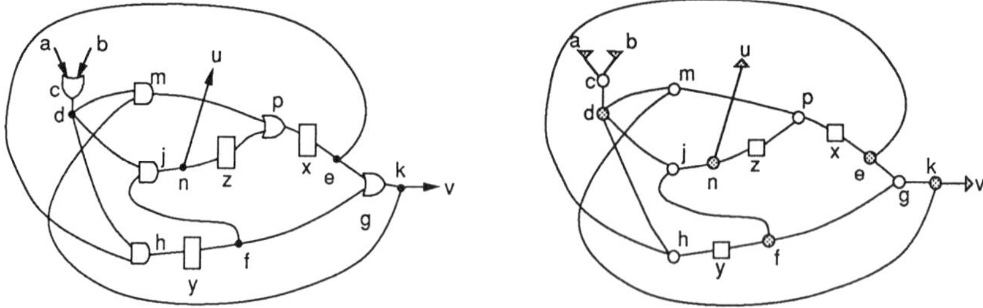

Figure 1: A sequential circuit and its corresponding graph

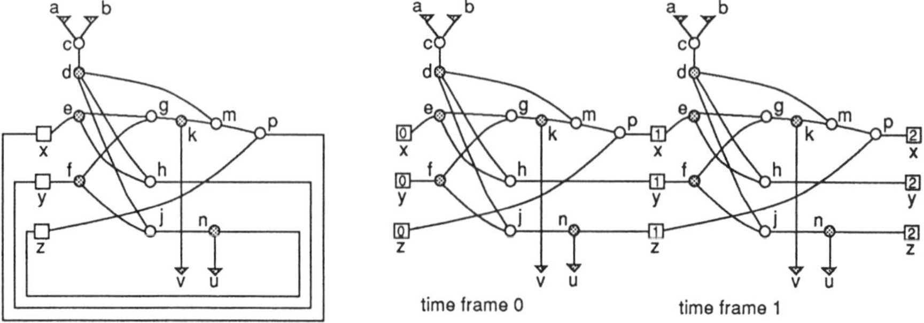

Figure 2: Time-frame graph showing two time frames

are drawn as unshaded circular nodes. The Primary Inputs and Primary Outputs are shown as triangles. Any sequential circuit can be viewed as a Finite State Machine (FSM). We represent a sequential circuit S by an equivalent graph whose nodes have been ordered to correspond to an FSM view of the circuit S. Figure 2 shows the equivalent graph which corresponds to a view of the circuit as an FSM.

A **path** p from node x to node y is a sequence of nodes and edges starting with x and ending with y. A path is *simple* if all nodes and all edges on the path, except the first and last node, are distinct. A **cycle** is a simple path which begins and ends with the same node. Throughout this paper, by a path, we refer to a simple path.

The **weight** of a path is the number of latch nodes present along a path. The weight between any two nodes x and y, $w(x, y)$, is defined if and only if: (i) There exists at least one path from node x to node y, and, (ii) All paths from node x to node y have the same weight. If the above are true, $w(x, y)$ is defined to be the weight of any path from x to y; otherwise, $w(x, y)$ is not defined.

Retiming [5,6] is a process of repositioning the latches in a synchronous sequential circuit. It is an integer assignment, $r(v)$, to each combinational block v, which represent the number of latches to be moved from each fanout of block v to each of the fanins of block v. A retiming is **legal** if all cycles in the retimed circuit have weight ≥ 1. A legally retimed circuit has been proven to be functionally equivalent to the original circuit [6].

Peripheral Retiming [1] of a synchronous sequential circuit is a retiming such that all the latches are moved to the peripheries of the circuit, leaving purely combinational logic inside. Peripheral retiming may leave negative number of latches at the inputs/outputs of the circuit, but a subsequent retiming step can always remove the negative latches from the circuit. In other words, a circuit undergoing peripheral retiming followed by a legal retiming is equivalent to the original circuit [1]. It has been proven [1] that a sequential circuit can be peripherally retimed if and only if it satisfies the following conditions:

R1. The circuit is acyclic;

R2. For each input i, output j, $w(i, j)$ is defined, that is, all paths from i to j have equal weight;

R3. $\exists$ integers α_i, β_j such that $\forall$ inputs i, outputs j, if $w(i, j)$ is defined, $w(i, j) = \alpha_i + \beta_j$.

2.1 Pipeline and Peripheral Retiming

A pipeline is an acyclic sequential circuit consisting of m stages of combinational logic, $\{0, \ldots, m-1\}$, each stage i separated from the next stage $i+1$ by a set of latches as shown in Figure 3. There are primary inputs, I_k entering stage k, and primary outputs O_k leaving stage k. We say a sequential circuit Π is a **pipeline** if and only if:

1. Π is acyclic;

2. Any path from a node x in stage i of Π to a node y in the next stage $i+1$ contains exactly one latch.

Theorem 1 *A sequential circuit can be peripherally retimed if and only if it is a pipeline.*

Proof: Please see [9,10].

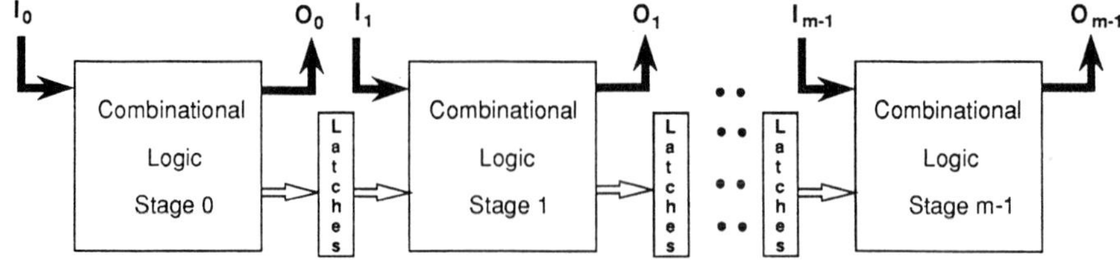

Figure 3: An m-stage pipeline circuit

3 Pipelines consisting of Reconvergent Regions

Fanout free regions of a circuit do not have any resynthesis potential [8]. It has been shown [8] that reconvergent fanout regions have good resynthesis potential. Since our goal is to extract pipeline structures which have good resynthesis potential, we identify pipelines consisting of overlapping reconvergent regions, which we term **reconvergent pipelines**.

Our approach to identify reconvergent pipelines is based on an analysis of signal reconvergence. We first introduce a time frame graph as a conceptual aid to understanding the notions of reconvergence and corollas across latch boundaries. The key ideas of primary stem region, petal and corolla were introduced in [8] for combinational circuits. We generalize them to be applicable to sequential circuits. We introduce concepts of time-frame-consistency and shifting petals over time frames. We finally define consistent corollas which are pipelines.

A **time-frame graph**, G_T, is composed of t $(t \geq 0)$ identical instances $G_i(V_i, E_i)$ of the sequential circuit graph $G(V, E)$, with each edge $(x, Y) \in E_i$ going from a non-latch node $x \in V_i$ to a latch node $Y \in V_i$ redirected in G_T to go from $x \in V_i$ to $Y \in V_{i+1}$. Each instance $G_i(V_i, E_i)$ is referred to as **time frame** i. Figure 2 shows the time frame graph corresponding to the graph G whose nodes have been organized as an FSM. A node v appears in all the time frames of G_T. We refer to a particular instance of a node by stating the time frame to which it belongs. We only label the latch nodes explicitly, marking the latches associated with time frame i by integer i.

In a time-frame graph, a path from node x in time frame i to a node y in time frame j is identical to a path from node x in time frame $(i + \delta)$ to a node y in time frame $(j + \delta)$, for any integer δ; they are not considered as different paths. That is, a path shifted by δ time frames is considered to be just an identical copy of the original path. The weight of a path from node x to node y implicitly considers x to be in time frame 0, and y in any time frame $i > 0$. If all paths from node x to node y are from time frames 0 to k, then $w(x, y) = k$, else $w(x, y)$ is not defined. A cycle in the sequential circuit graph, starting and ending with node x and having weight k, is reflected in the equivalent time-frame graph as a path starting with node x in time frame i and ending with node x in time frame $(i + k)$.

3.1 Reconvergence Across Latch Boundaries

The concept of stem regions was defined in [11], and was later modified to "primary stem regions" in [7] for combinational circuits. We redefine reconvergence and primary stem region to extend to sequential circuits.

If there are more than one disjoint path from fanout stem s to another node $r \in V$, each path having equal weight $w(s, r)$, then s is a **reconvergent fanout stem** and r is a **reconvergent**

node for s. If s has no such reconvergent node, then s is a **non-reconvergent fanout stem**. A **closing reconvergent node** of a reconvergent fanout stem s is a reconvergent node of s that does not drive any other reconvergent node of s.

Considering stem d of Figure 2, node g is a reconvergent node since two distinct paths from d, having equal weight 1, d-m-p-x-e-g and d-h-y-f-g, reconverge at g. Similarly, node p is a reconvergent node for stem d. Though two distinct paths from stem d reconverge at node j, j is not a reconvergent node because the paths have unequal weights (0 and 1).

The **primary stem region** of a reconvergent fanout stem s consist of all the nodes (and their output edges) along all the paths from s to any of its closing reconvergent nodes, such that (i) the stem region graph is acyclic, and (ii) for each closing reconvergent node r, weight $w(s,r)$ is defined. In the graph of Figure 2, both the reconvergent nodes g and p of stem d cannot be considered closing reconvergent nodes, otherwise the stem region of d will have cycles like g-k-m-p-x-e-g. If only one of them, say g, is considered the closing reconvergent node for d, the resulting stem region consists of nodes $\{d,h,m,p,x,y,e,f,g\}$ and their output edges. Also, there can be more than one set of equal-weight paths, each having different weights, from stem s to a reconvergent node r. A choice has to be made as to which set of paths to include in the stem region such that $w(s,r)$ is defined. We will later show how the time-frame graph helps us resolve the above choices in forming an acyclic stem region.

3.2 Partitioning Stem Region into Petals

Each primary stem region is partitioned into a set of disjoint petals. Each **petal** of a reconvergent stem s is a biconnected component [12] of the primary stem region of s, with directions of edges ignored. A **tip** of a petal of reconvergent stem s is a closing reconvergent node of s included in the petal. An **exit** of a petal is any edge in the petal which drives a node not included in the petal.

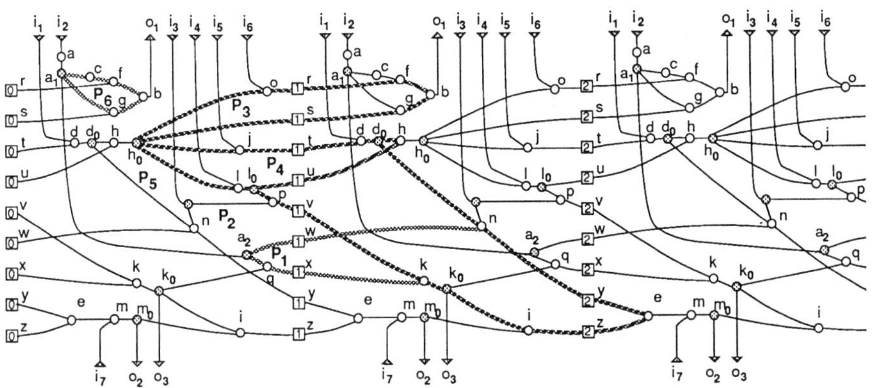

Figure 4: A time-frame graph showing petals and pipelines

Let us consider the time-frame graph of a sequential circuit in Figure 4, which shows the circuit unwrapped over three time frames. The reconvergent nodes of fanout stem h_0 are b, h

and e. The (undirected) primary stem region of h_0 has two biconnected components as shown in Figure 4. The stem node h_0 should be split to have two petals $P_3 = \{h_1, o, r, s, f, g, b\}$ and $P_4 = \{h_2, j, l, l_0, p, t, u, v, d, d_0, h, n, k, k_0, i, y, z, e\}$. The new nodes h_1 and h_2 are the new stem nodes of P_3 and P_4, respectively, and P_3 and P_4 are disjoint. Each petal is acyclic. Petal P_3 has a single tip, b, and no exits. The tips of P_4 are h and e, and it has an exit, (k_0, o_3). All paths from stem h_0 to each tip have equal weights: $w(h_0, b) = 1$, $w(h_0, h) = 1$ and $w(h_0, e) = 2$. Similarly, the stem region of a is partitioned into two petals with new stems a_1 and a_2, and tips b and e respectively. $w(a, b) = 0$ and $w(a, e) = 2$.

3.3 Overlap of Petals

We define a relation **overlap** (ρ) on the set of petals of a sequential circuit by saying that petal P_i overlaps petal P_k, denoted by $P_i \rho P_k$, if either (i) $P_i \cap P_k \neq \phi$, or, (ii) $P_i \cap P_j \neq \phi$ AND $P_j \rho P_k$. The relation, overlap, is an equivalence relation that partitions the petals of G into equivalence classes $C_1, C_2, \ldots, C_k$ such that two distinct petals are in the same class if and only if they overlap. In the circuit of Figure 4, the two equivalence classes formed by the relation overlap are: $C_1 = \{P_3, P_6\}$, and $C_2 = \{P_1, P_2, P_4, P_5\}$.

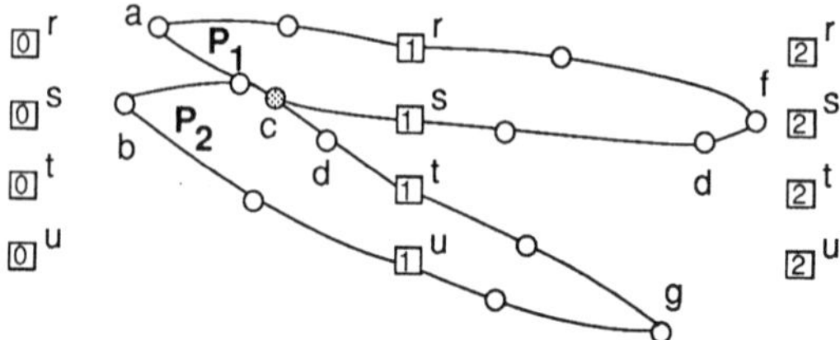

Figure 5: Overlapping petals may not create a pipeline

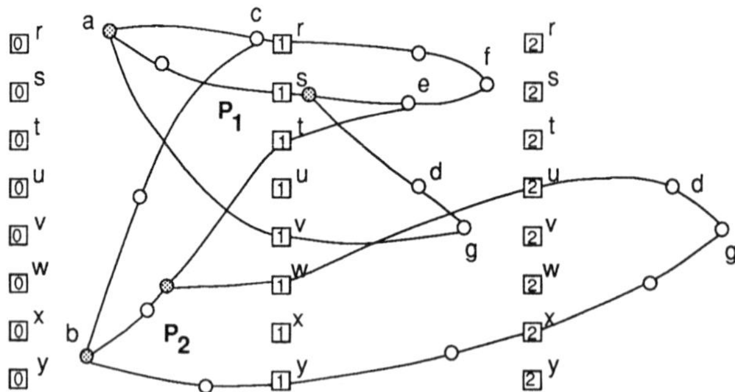

Figure 6: Overlapping petals may make input-outputs require conflicting number of latches

The equivalence classes, formed by the relation *overlap* of petals, are not necessarily pipeline circuits. As opposed to the case of combinational circuits, an arbitrary subset of overlapping petals may contain cycles. In Figure 4, the circuit formed by the overlapping petals P_4 and P_2 contains a cycle e-m-m_0-b-z-e. Even when overlapping petals form an acyclic circuit, the circuit may not be a pipeline. Let us consider the time frame graphs in Figure 5 and Figure 6, which show, for clarity, only the nodes belonging to two overlapping petals. The circuit formed by the overlapping petals P_1 and P_2 in Figure 5, though acyclic, is not a pipeline. Node c has two paths to node d, one path having no latches on it, the other path having a single latch on it. Consequently, node d cannot be assigned to any particular stage of a pipeline. Similarly, the circuit formed by the overlapping petals in Figure 6 is not a pipeline. Nodes d and g cannot be placed in any stage without violating the properties of a pipeline. To determine which subset of overlapping petals may constitute a pipeline structure, we study the time frame behaviour of overlapping petals.

3.4 Time-Frame-Consistency of Petals

The **Time Frame** of a node x with respect to a petal P, $TF_P(x)$, is the time frame in which x belongs to petal P. In Figure 4, node f belongs to Time Frame 0 when $f \in P_6$, and Time Frame 1 when $f \in P_3$. $TF_{P_6}(f) = 1, TF_{P_3}(f) = 2$.

Time Frame Shift is an operation by which a petal is shifted from one time frame to another. The Shift operation, $\Gamma(P, \delta)$, shifts the Time Frame of petal P so that each node $x \in P$ having $TF_P(x) = k$ is shifted to have $TF_P(x) = k + \delta$; we say that petal P has been shifted by δ time frames.

For example, $\Gamma(P_6, 1)$ shifts petal P_6 from time frame 0 to time frame 1, so that $TF_{P_6}(f) = TF_{P_3}(f) = 1$. That is, the petal can be now viewed as having its stem a_1 in time frame 1, and not time frame 0 as it was till now. Note that a Time Frame Shift is a virtual operation in the sense that it does not affect the timing or functional behaviour of the nodes; it just shifts our view of the petal in the time-frame graph.

A node belonging to two different petals may have two different TF associated with it. That is, the node belongs to the two petals in different time frames. Let a node $x \in P_i \cap P_j$. Let δ represent the difference in time frames for node x. Then, $\delta = |(TF_{P_i}(x) - TF_{P_j}(x))|$.

Two petals P_i and P_k are said to be **Time-Frame-Consistent** if and only if $\exists \delta$ such that shifting one of the petals by δ results in $TF_{P_i}(x) = TF_{P_j}(x), \forall x \in P_i \cap P_j$. That is, $\Gamma(P_i, \delta)$ or $\Gamma(P_j, \delta)$ results in $TF_{P_i}(x) = TF_{P_j}(x), \forall x \in P_i \cap P_j$.

In Figure 4, petals P_1 and P_3 are trivially time-frame-consistent as they have no nodes in common; petals P_1 and P_4 are time frame consistent since each of their common nodes belong to the petals in the same time frames. Though the common nodes of petals P_3 and P_6 are apparently in different time frames, petal P_3 can be shifted by 1 time frame to make all nodes common to the two petals belong to the same time frame. Consider a node f belonging to both the petals, $\delta = TF_{P_6}(f) - TF_{P_3}(f) = 1$. Applying $\Gamma(P_6, 1)$, we shift P_6 by one time frame. Now, $\forall x \in P_i \cap P_j$ (nodes $\{f, g, b\}$), $TF_{P_6}(x) = TF_{P_3}(x)$. Consequently, all nodes belonging to petals P_6, P_3 have an unique time frame associated with them, and the petals are time-frame-consistent. Petals P_2 and P_4 are not time-frame-consistent, as no δ can be found by which either of the petals can be shifted such that the common nodes of the petals belong to a single time frame.

3.5 Consistent Corollas: the Reconvergent Pipelines

A **consistent corolla** is any subset of overlapping petals which are Time-Frame-Consistent (with each other). We will show that a consistent corolla is a pipeline.

Since all petals of a consistent corolla are time-frame-consistent, every node belonging to a consistent corolla C ocurrs in only one time frame. Since the corollas are required to be disjoint, a node may belong to only one corolla. Once a node has been assigned to a consistent corolla, we term the unique time frame associated with the node as the **time stamp** of the node.

In the circuit of Figure 4, petals P_3 and P_6 overlap and are time-frame-consistent. Hence, $C_1 = \{P_3, P_6\}$ is a consistent corolla. Nodes $\{f, g, b\}$ belong to two different time frames, 0 and 1. After shifting petal P_6 from time frame 0 to 1, each node in the corolla occurs in a single time frame. Similarly, from amongst the second set of overlapping petals, only petals $\{P_1, P_4\}$ are time-frame-consistent, to form the second consistent corolla $C_2 = \{P_1, P_4\}$. The time stamp of node f is 1, as it belongs to corolla C_1 in time frame 1, while the time stamp of node q is 0, as it belongs to corolla C_2 in time frame 0. The following lemmas show that a consistent corolla satisfies the properties of a pipeline structure. For the proofs of the lemmas and theorem, please see [9].

Lemma 1 *A consistent corolla is acyclic.*

Lemma 2 *For any pair of nodes u, v that belong to a consistent corolla C, all paths from node u to node v have equal weight.*

Theorem 2 *A consistent corolla is a pipeline.*

Proof: From lemmas 1 and 2. □

Corollary 1 *A consistent corolla has a peripheral retiming.*

Consistent corollas are the only pipelines in a sequential circuit which contain reconvergent regions. Since we aim to identify pipelines for optimizing logic, we do not care to identify non-reconvergent pipelines or pipelines consisting of fanout-free regions.

4 Extracting Pipelines Using Time-Frame Graph

We outline the process of extracting reconvergent pipelines from a sequential circuit, using the time frame graph, G_T. To identify all possible pipelines, we have to consider the stems belonging to only time frame 0. However, to identify the stem region and petals of each stem s, and including petals in corollas, the maximum number of time frames that need to be traveresed is $\lfloor n/2 \rfloor + 1$, where n is the number of latches in the sequential circuit [9]. In practice, we do not require to traverse $\lfloor n/2 \rfloor + 1$ time frames, as more than two latches are used up per time frame by the stem regions and petals crossing the time frame. We first outline the process of extracting the stem region for a stem s. Then we describe how to dynamically form pipelines by including petals which are time-frame-consistent.

To identify the primary stem region of a stem s, we use a modified version of the algorithm in [11] which detects stem regions in combinational circuits. Starting at the stem node s, we do a breadth-first traversal of the nodes in the first time frame, then all the nodes in the second time frame, and so on. We mark the fanout nodes of stem s with different tags. For any further stem node that is encountered, we mark all its fanout nodes with the same tag. When we process a node r in time frame k whose inputs have different tags, r is marked a reconvergent node, with $w(s, r) = k$. During the breadth-first search, if we visit a node v a second time, there is a cycle starting and ending with v. We implicitly break the cycle by stopping the breadth-first search at v and pushing those inputs to v which are in the current time frame in an output stack S. Similarly, when a primary output node is reached, it is pushed in stack S. When there is no other node to process, a depth-first search is ensued, in the reverse direction, from every node $v \in S$. A node is detagged if all its fanouts are detagged and it is not a reconvergent node. This process ends with the closing reconvergent nodes identified as those reconvergent nodes whose fanouts have been detagged.

Processing time frame i completely before processing time frame $i+1$, and stopping along a path when a node is repeated, ensures that no two nodes are identified as reconvergent nodes whose inclusion creates a cycle in the stem region of s. Considering the time frame graph in Figure 2, both nodes g and p are reconvergent nodes for stem d, since there are two distinct paths of weight 1 to each of the nodes g and p. But a stem region based on simultaneously considering both as reconvergent nodes leads to a cycle starting and ending with g. However, our algorithm, after declaring g as a reconvergent node in time frame 1, stops at m as it is visited twice. Hence, p is not declared as a reconvergent node and the resulting stem region does not contain a cycle.

Once a stem region for stem s has been identified, it is partitioned into petals. For each petal P, we explore the possibility of including it in a new or existing corolla. If P overlaps with some petals belonging to an existent corolla C, P is a candidate for inclusion in C. We include the petal in C only if it is time frame consistent with the petals in C with which it overlaps, and its inclusion increases the resynthesis potential of C, $\psi(C)$ [8]. If P does not overlap with any petal, a new corolla is formed consisting of P. After a petal P is included in C, each node in P is assigned an unique time-stamp which is the time frame in which it belongs to C.

We check whether petal P is time frame consistent with its overlapping petals in corolla C by checking to see whether $\exists$ any node $x \in P$, such that $TF_P(x) \neq$ time-stamp(x). If no such x exists, then P is time frame consistent. Else, petal P is shifted by $\delta =$ time-stamp$(x) - TF_P(x)$ time frames, so that $TF_P(x) =$ time-stamp(x). If all nodes of P which are common to C have the same TF_P and time-stamp values after the time frame shift, P is time frame consistent, otherwise it is not. The details of the algorithm for identifying pipelines in sequential circuits can be found in [9].

5 Resynthesis of Pipelines

For each pipeline we have identified, we move all the latches to the inputs and outputs of the pipeline and resynthesize the interior combinational logic of the pipeline by combinational techniques like [13]. If resynthesis has reduced the area of the pipeline, we reposition the latches in the pipeline by legal retiming [5], removing any negative latches from the pipeline peripheries. We replace the original pipeline by the resynthesized pipeline preserving the I/O mappings of the original pipeline. The process of peripheral retiming, resynthesis and legal retiming of a pipeline preserves the functional as well as the timing behaviour of each pipeline, and hence the whole sequential circuit. We treat the problem of preserving the initial states by introducing the initializing circuitry, along with an initializing input, as an integral part of the original

Ckt	**Init**		**Comb Resynth**			**Seq Resynth**				Total
	Latch	Lits	Lits	%Rdn	CPU (sec)	Lits	%Rdn	CPU (sec)		%Rdn
								Pipe	Resyn	
s400	21	320	307	4.1	7.2	276	10.1	1.5	18.0	13.8
s526	21	445	309	30.6	11.0	285	7.8	3.3	33.6	36.0
s5378	179	4295	3198	25.5	694.1	2876	10.1	165.4	671.5	33.0
s9234	211	7971	5731	28.1	729.4	5391	6.0	123.0	546.8	32.4
s13207	638	11193	9557	14.6	1207	8598	10.0	335.0	228.6	23.2

Table 1: Optimization by Identification, Retiming and Resynthesis of Pipelines.

sequential circuit. Our resynthesis approach preserves the functionality of the initialization circuitry, thereby retaining the effect of the initializing input.

5.1 Experimental Results

We have applied our technique of extraction and resynthesis of pipelines to a subset of ISCAS89 sequential benchmarks [14]. Table 1 shows some preliminary results. The initial number of latches and literals in a circuit are shown in column `Init`. As in [8], we first resynthesize the combinational logic between the latches. Column `Comb Resynth` shows the number of literals and % reduction of literals after applying combinational techniques alone.

Next, we extract pipelines across latch boundaries and resynthesize them. The results shown in Table 1 are based on resynthesizing pipelines extracted over time frames 0 and 1 only. The entries under `Seq Resynth` represent the number of literals after applying the pipeline identification and resynthesis process to the combinationally minimized circuits, and the % reduction. The column `CPU (sec)` shows the CPU time (in seconds) on a Vax 8650 to extract pipelines (`Pipe`) and resynthesize the pipelines using mis2.2 (`Resyn`). The column `Total %Rdn` shows the aggregrate % reduction in area achieved by combinational and sequential resynthesis.

	init	mis2.2		secore	
ckt	tpairs	tpairs	secs	tpairs	secs
s400	257	220	28	217	26
s526	379	258	52	230	48
s5378	2401	2197	12275	1976	1531
s9234	3887	2938	4025	3001	1399
s13207	5610	4231	3876	3978	1771

Table 2: Comparison with mis2.2

Table 2 compares the synthesis results and running times of our technique, `secore`, with `mis2.2` standard script. For each benchmark, the initial circuit and the circuits produced by mis2.2 and secore are technology mapped to mcnc.genlib library [15] using mis technology mapper, and the resultant transistor pairs `tpairs` compared. The column `secs` shows the CPU time on Vax 8650. Our sequential optimization technique compares favorably with mis2.2, in terms of synthesis results (`tpairs`) and, in particular, the CPU times required. We expect to see even better synthesis results as we process beyond time frame 1 in G_T, identifying and resynthesizing pipeline structures consisting of several stages, at the cost of more CPU time.

6 Conclusions

We have discussed a technique to optimize sequential circuits, by identifying, retiming and resynthesizing pipeline structures. The feedbacks in the circuit are broken naturally while forming the pipelines, in a way that is best for resynthesis. The preliminary results are based on extracting and resynthesizing a subset of the possible pipelines that can be formed, those which are across a single time frame. Even then, the results are very encouraging. We expect to see even better results as we experiment with pipelines formed over several time frames. The optimization procedure has focussed on minimizing literals/transistor pairs. We are currently investigating techniques to minimize the number of latches required by a sequential circuit.

References

[1] S. Malik, E. Sentovich, R. Brayton, and A. Sangiovanni-Vincentelli, "Retiming and Resynthesis: Optimizing Sequential Networks with Combinational Techniques," *IEEE Trans. on Computer Aided Design*, vol. 10, pp. 74 – 84, January 1991.

[2] G. D. Micheli, "Synchronous Logic Synthesis: Algorithms for Cycle-Time Minimization," *IEEE Trans. on Computer Aided Design*, vol. 10, pp. 63 – 73, January 1991.

[3] B. Lin, H. Touati, and A. Newton, "Don't Care Minimization of Multi-Level Sequential Logic Networks," in *Proc. IEEE International Conference on Computer-Aided Design*, pp. 414 – 417, November 1990.

[4] Y. Matsunaga, M. Fujita, and T. Kakuda, "Multi-level Logic Minimization across latch boundaries," in *Proc. IEEE International Conference on Computer-Aided Design*, pp. 406 – 409, November 1990.

[5] C. Leiserson, F. Rose, and J. Saxe, "Optimizing synchronous circuitry by retiming," in *Proc. Third Caltech Conference on VLSI*, pp. 87 – 116, 1983.

[6] C. Leiserson and J. Saxe, "Optimizing synchronous systems," *Journal of VLSI and Computer Systems*, vol. 1, pp. 41 – 67, Spring 1983.

[7] S. Dey, F. Brglez, and G. Kedem, "Corolla Based Circuit Partitioning and Resynthesis," in *27th ACM/IEEE Design Automation Conference*, pp. 607 – 612, June 1990.

[8] S. Dey, F. Brglez, and G. Kedem, "Circuit Partitioning for Logic Synthesis," *IEEE Journal of Solid-State Circuits*, vol. 26, pp. 350 – 363, March 1991.

[9] S. Dey, F. Brglez, and G. Kedem, "Synthesis of Sequential Circuits: Minimizing Combinational Logic and Latches," tech. rep., Microelectronics Center of North Carolina, Research Triangle Park, NC, July 1991.

[10] S. Malik, *Combinational Logic Optimization Techniques in Sequential Logic Synthesis*. PhD thesis, U. C. Berkeley, Nov 1990. Available as UCB/ERL Memo M90/115, Nov. 28, 1990.

[11] F. Maamari and J. Rajski, "A Reconvergent Fanout Analysis for Efficient Exact Fault Simulation of Combinational Circuits," in *18th International Symposium on Fault Tolerant Computing*, June 1988.

[12] A. V. Aho, J. E. Hopcroft, and J. D. Ullman, *The Dèsign and Analysis of Computer Algorithms*. Addison-Wesley Publishing Company, 1974.

[13] R. Brayton, R. Rudell, A. Sangiovanni-Vincentelli, and A. Wang, "MIS: A Multiple-Level Logic Optimization System," *IEEE Transactions on Computer-Aided Design*, vol. cad-6, pp. 1062 – 1081, November 1987.

[14] F. Brglez, D. Bryan, and K. Kozminski, "Combinational Profiles of Sequential Benchmark Circuits," in *IEEE International Symposium on Circuits and Systems Proceedings*, pp. 1029–1934, May 1989.

[15] R. Lisanke, " Logic Synthesis and Optimization Benchmarks, User Guide Version 2.0," in *MCNC Proceedings of the Second International Workshop on Logic Minimization*, (Research Triangle Park), May 1989.

VLSI 91
A. Halaas and P.B. Denyer (Eds.)
Elsevier Science Publishers B.V. (North-Holland)

Hierarchical Retiming Including Pipelining

A. van der Werf, B.T. McSweeney, J.L. van Meerbergen, P.E.R. Lippens, W.F.J. Verhaegh

Philips Research Laboratories, P.O. Box 80000, 5600 JA Eindhoven, The Netherlands

Abstract

In this paper, extensions of the existing theory on retiming will be presented. New modelling techniques for synchronous circuits will be introduced. New applications of retiming, e.g. hierarchical retiming, become possible by the extended theory. It will be shown that the new modelling techniques will be applicable to operator selection in combination with retiming.

1 Introduction

At the Philips Research Laboratories, a silicon compiler for high speed algorithms, called PHIDEO, is being developed [Lip] [1]. Datapath compilation is one of the subtasks of this compiler [Wer]. To make the datapaths run at the required clock frequency, retiming is applied. Most often, the datapaths have to be pipelined to meet the timing constraint, which is a certain required clock speed. Leiserson et al. described how retiming can be used to pipeline a circuit [Le1] [Le3]. To (re)synthesize pipelined circuits, Malik et al. use peripheral retiming [Ma1] [Ma2]. This concept very closely resembles our concept of a time shape of a circuit. A time shape defines for each terminal of a circuit the time (in clock periods) at which values belonging to the same function application are consumed/produced. Within PHIDEO, retiming is allowed to change the time shape of a datapath, because within the concept of PHIDEO the time shape is a result of datapath compilation rather than a constraint for it.

Besides pipelining, another aspect of retiming high speed datapaths is that the required clock period may be smaller than the largest delay of cells in a circuit. In that case, faster cells have to be selected, which may be pipelined. Therefore, retiming must be possible for circuits containing high speed pipelined cells. The solution of this problem has led to the theory of hierarchical retiming. An abstraction of the timing behaviour of a retimed circuit can be made, such that it can be used as a cell in a higher level of hierarchy where retiming can be used again.

To use retiming, including pipelining, for high speed datapath design, four new modelling techniques will be introduced. The Linear Programming (LP) solution techniques, on which the retiming theory of Leiserson et al. is based, can still be applied. The four new modelling techniques are:

1. The explicit modelling of *nets* in a circuit. It allows the use of tri-state buffers in circuits to be retimed.
2. A different *retiming function* will be introduced. Using this function different feasible solutions may be found.
3. An extension of the *timing model* of cells in a circuit will be proposed. It allows retiming of circuits consisting of synchronous cells and it allows hierarchical retiming.

[1]This project is sponsored by the CEC as ESPRIT 2260

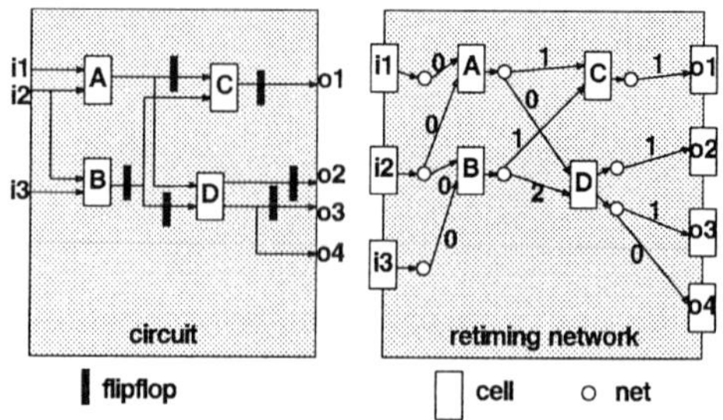

Figure 1: *An example of a circuit and its retiming network.*

4. The *time shape* of a circuit is considered flexible. As a consequence, pipelining of a circuit is possible, and the removal of pipeline stages as well. The addition or removal of pipeline stages may be different for every terminal.

The organization of this paper is as follows. In section 2 the four new modelling techniques are described. From a complete model of a circuit a mathematical formulation of the retiming problem can be derived which is described in section 3. The solution methods are based on Linear Programming (LP). How to solve an LP problem is beyond the scope of this paper, but how LP solvers can be applied is explained in section 4. Finally, in section 5, it will be shown that the new modelling techniques can also be used for operator selection.

2 Modelling

2.1 Retiming network

A synchronous circuit is modelled by a retiming network which contains two types of vertices: nets and cells. Figure 1 shows an example of a circuit and its retiming network. Every output of a cell is connected to a net, and nets are connected to inputs of cells. Therefore the retiming network is bipartite. In many circuits only one output of a cell is connected to each net. Therefore, in the model the outputs of cells could be merged with the nets to which they are connected. However, if tri-state buffers are used (busses), more than one output of a cell is connected to a net and the distinction between nets and outputs of cells is necessary. In the rest of this paper, the distinction between nets and outputs of cells will be maintained. A connection between a net, net_i, and an input, $input_i$ of a cell, $cell_j$, has a weight $w(net_i, input_i)$ representing the number of flipflops between the net and the cell. Every input and output is represented by a separate cell in the retiming network.

2.2 Retiming functions

The movement of flipflops throughout a network can be quantified by an integer $b(cell_i)$ for every cell, $cell_i$, in the network which represents the number of flipflops that are to be moved from the outgoing edges of its output nets to its incoming edges. This function will be called the backward retiming function. It is equal to the retiming function introduced by Leiserson et al.

Also a different retiming function is possible. The movement of flipflops throughout a network can be quantified by an integer $f(cell_i)$ for every cell, $cell_i$, in the network which represents the number of flipflops that are to be moved from its incoming edges to the outgoing edges of its output nets. This function will be called the forward retiming function. The distinction between the two retiming functions seems trivial, but is used to generate alternative solutions as will be shown in section 4.1.

In section 2.3 it will be shown that retiming is not only defined for cells, but also for terminals of cells. The number of flipflops, $w_r(net_a, input_i)$, on a connection from a net, net_a, to an input, $input_i$, becomes after retiming

$$w_r(net_a, input_i) = w(net_a, input_i) + b(input_i) - b(output_j) \tag{1}$$

when the backward retiming is used or

$$w_r(net_a, input_i) = w(net_a, input_i) - f(input_i) + f(output_j) \tag{2}$$

when the forward retiming function is used, where $output_j$ and $input_i$ are connected to net_a.

2.3 New timing model of cells and networks

The timing model of a cell contains for every input-output pair a delay. If no direct path through the cell from a certain input to a certain output exists, the delay belonging to that input-output pair is undefined. This part of the new timing model is equal to the complete timing model of cells used by Leiserson et al. [Le1] and Note [Not].

Leiserson et al. already mentioned that *data available times* at the input and *data ready times* at the output of the environment can be taken into account. However, these properties of a circuit are not modelled in their communication graph by the timing models of the cells. In high speed datapath circuits, the cells may be internally pipelined. Therefore the timing model of these cells should also contain these properties. Every input of a cell has a data available time which is the time before a clock period ends at which the data must be available at the input. If there is no path from an input of a cell to a flipflop inside that cell, the data available time of that input is zero. Every output of a cell has a data ready time which is the time after the beginning of the clock period at which the data is ready at the output. If there is no path from a flipflop inside a cell to an output of that cell, the data ready time of that output is zero.

If one selects an operator which has an internal pipeline stage to implement a certain operation in a circuit, a number of flipflops has been moved into the cell. This means that the inputs of that cell have been retimed differently w.r.t. the outputs of that cell. It is even possible that all inputs and outputs are retimed differently w.r.t. each other. The timing of the inputs and outputs of a cell is considered static w.r.t. each other, i.e. during the retiming of the network the internal timing of a cell is not changed. Therefore, the retiming of the inputs and outputs of one cell w.r.t. each other is represented by constants in the timing model of a cell. The definition of the retiming constant is the same as of the retiming function. Unless otherwise mentioned, the retiming constant means the backward retiming constant. The retiming constant $bc(input_i)$ of an input terminal $input_i$ quantifies the number of flipflops which are moved from its incoming edges to the inside of the cell. The retiming constant $bc(output_j)$ of an output terminal $output_j$ quantifies the number of flipflops which are moved from the inside of the cell to the outgoing edge of the net to which the output terminal is connected. The retiming constants of all terminals of a cell form together the so-called *delta time shape* of a cell. Note that adding the same constant to the retiming constants of all terminals of a cell does not alter the time shape.

The complete new timing model of a cell contains now:

1. For every input-output pair, $(input_i, output_j)$, a delay $d(input_i, output_j)$.
2. For every input, $input_i$, a data available time, $dat(input_i)$, and a retiming constant, $bc(input_i)$.
3. For every output, $output_j$, a data ready time, $drt(output_j)$, and a retiming constant, $bc(output_j)$.

Figure 2 shows an example of a circuit and its timing model. In section 3 it will be shown that $drt(output_j)$ can be replaced by a $d_{max}(output_j)$ which is the maximum of the $drt(output_j)$ and all $d(input_i, output_j)$.

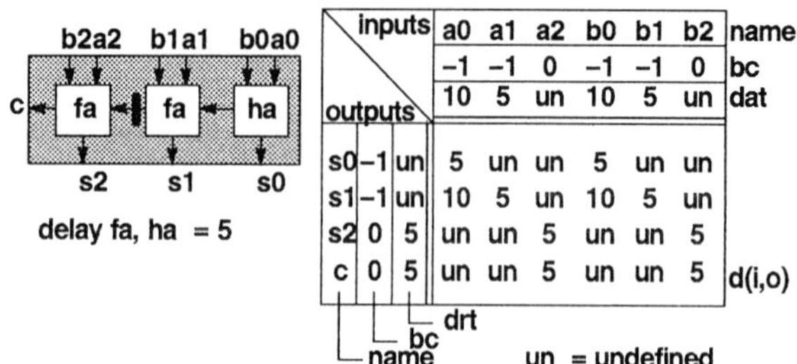

Figure 2: *An example of a circuit (3-bit adder) and its timing model.*

If the forward retiming function is used, the forward retiming constants of the terminals are denoted as $fc(input_i)$ and $fc(output_j)$. These constants are the negative of the backward retiming constants, $bc(input_i)$ and $bc(output_j)$, which relationship follows directly from their definitions.

2.4 Applications of the new timing model

The most obvious advantage of applying the new timing model is that data available and data ready times of the environment of the circuit can be modelled in the retiming network. The introduction of the new timing model allows one to select cells (often called operators) with internal flipflops to implement operations in a circuit. In section 5 it will be shown that an existing algorithm for operator selection in combination with retiming can make use of the new timing model.

Also *hierarchical retiming* can be done starting at the lowest level, bottom-up. Consider a circuit containing a multiplier. If this circuit has to be retimed and the longest path through the multiplier is longer than the clock period, the timing constraints cannot be met. A solution is to replace the multiplier in the retiming circuit by its implementation in adders. Now a retiming can be found because flipflops can be placed inside the multiplier. Suppose now that a multiplier which has an internal pipeline stage is present in a cell library. The new timing model of the pipelined multiplier allows one to use this cell in the circuit, and consequently retime the circuit. No detailed information of the internal structure of the multiplier is required here. Figure 2 shows an example of a retimed circuit and its timing model such that it can be used as a cell in higher level retiming.

The new timing model makes it possible to hierarchically retime a circuit, starting at the lowest level, bottom-up. A new timing model can be derived from a circuit which is retimed. Therefore this circuit can be a cell in a circuit at a higher level of hierarchy. This higher level circuit can be retimed recursively using the timing models of its cells which may be subcircuits.

3 Deriving an MILP description

In this section it will be described how the retiming problem can be cast into a Mixed Integer Linear Programming (MILP) problem. The MILP description will be derived in the same way as Leiserson et al. did [Le1]. The difference is that the new modelling techniques introduced in section 2 are applied. First an MILP description will be derived in which the backward retiming function is used. This derivation will be explained in detail. At the end of this section, the MILP description in which the forward retiming function is used will be given.

The longest path from a flipflop or an input of the network to a net, net_b, can be quantified by a real $s(net_b)$ for every net in the network. According to Leiserson et al. the following constraints should be met in order to obtain a legal retiming. This set of constraints can be derived from a retiming network which contains only cells without internal flipflops.

- $$s(net_b) \geq d(input_i, output_j) \tag{3}$$
  wherever $d(input_i, output_j)$ is defined and $output_j$ is connected to net_b

- $$s(net_b) \leq c \tag{4}$$
  for every net, net_b, where c is the desired clock period

- $$b(cell_u) - b(cell_v) \leq w(net_b, input_k) \tag{5}$$
  wherever $input_k$ is an input of $cell_v$ and is connected to the same net, net_b, as an output of $cell_u$

- $$s(net_b) \geq s(net_a) + d(input_i, output_j) \tag{6}$$
  wherever $d(input_i, output_j)$ is defined and $input_i$ is connected to net_a and $output_j$ is connected to net_b and where the number of flipflops between net_a and the cell of which $input_i$ and $output_j$ are terminals equals zero, i.e. $b(cell_t) - b(cell_u) = w(net_a, input_i)$, where $output_j$ and $input_i$ are terminals of $cell_u$ and an output of $cell_t$ is connected to net_a

When the new timing model is used, constraint 5 has to be changed because the retiming of inputs and outputs of a cell may differ by a constant from the retiming of the cell itself. Also two constraints (11 and 12) have to be added, because synchronous cells may be present in the retiming network. From the new retiming network, using the new timing model, the following set of constraints is derived:

- $$s(net_b) \geq d(input_i, output_j) \tag{7}$$
  wherever $d(input_i, output_j)$ is defined and $output_j$ is connected to net_b

- $$s(net_b) \leq c \tag{8}$$
  for every net, net_b, where c is the desired clock period

- $$b(output_j) - b(input_k) \leq w(net_b, input_k) \tag{9}$$
  wherever $input_k$ is connected to the same net_b as $output_j$

- $$s(net_b) \geq s(net_a) + d(input_i, output_j) \tag{10}$$
  wherever $d(input_i, output_j)$ is defined and $input_i$ is connected to net_a and $output_j$ is connected to net_b and where the number of flipflops between net_a and the cell of which $input_i$ and $output_j$ are terminals equals zero, i.e. $b(output_p) - b(input_i) = w(net_a, input_i)$, where $output_p$ is connected to net_a

- $$s(net_b) \geq drt(output_j) \tag{11}$$
  for every output, $output_j$, connected to net_b

- $$s(net_a) \leq c - dat(input_i) \tag{12}$$
  wherever $input_i$ is connected to net_a and the number of flipflops between net_a and the cell of which $input_i$ is an input equals zero, i.e. $b(output_p) - b(input_i) = w(net_a, input_i)$, where $output_p$ is connected to net_a

Constraints 7 and 11 can be merged by creating a different value in the timing model. Of every output of a cell, $output_j$, the longest path from either a flipflop or an input, $d_{max}(output_j)$, should be in the timing model. This new value does not add new information to the new timing model, but makes it possible to derive less constraints from the retiming network. The drt value for every output becomes redundant and can be dropped. The constraints 7 and 11 become now:

$$s(net_b) \geq d_{max}(output_j) \tag{13}$$

for every output, $output_j$. Constraint 12 can be dropped if the $dat(input_i)$ of an input, $input_i$, is smaller or equal than any $d(input_i, output_j)$.

To formulate a set of constraints without a clause "$b(cell_t) - b(cell_u) = w(net_a, input_i)$", a new variable B for every net is introduced. $s(net_b)$ is substituted by $c(B(net_b) - b(output_j))$ where $output_j$ is connected to net_b. This substitution makes it possible to map the constraints on a graph. After substitution, merging of constraints and some more manipulations, the set of constraints becomes:

- $$b(output_j) - B(net_b) \leq \frac{-d_{max}(output_j)}{c} \tag{14}$$
  wherever $output_j$ is connected to net_b
- $$B(net_b) - b(output_j) \leq 1 \tag{15}$$
  wherever $output_j$ is connected to net_b .
- $$b(output_j) - b(input_k) \leq w(net_b, input_k) \tag{16}$$
  wherever $input_k$ and $output_j$ are connected to the same net, net_b
- $$B(net_a) - B(net_b) \leq b(input_i) - b(output_j) + w(net_a, input_i) - \frac{d(input_i, output_j)}{c} \tag{17}$$
  wherever $d(input_i, output_j)$ is defined and $input_i$ is connected to net_a and $output_j$ is connected to net_b
- $$B(net_a) - b(input_i) \leq 1 + w(net_a, input_i) - \frac{dat(input_i)}{c} \tag{18}$$
  wherever $input_i$ is connected to net_a

Since the retiming of inputs and outputs of a cell w.r.t each other is static, the following two substitutions are applied:

- $$b(output_j) = b(cell_u) + bc(output_j) \tag{19}$$
  wherever $output_j$ is an output of $cell_u$
- $$b(input_i) = b(cell_u) + bc(input_i) \tag{20}$$
  wherever $input_i$ is an input of $cell_u$

Finally, substitution of the retiming variables of the terminals of cells results in the following set of constraints:

- $$b(cell_u) - B(net_b) \leq \frac{-d_{max}(output_j)}{c} - bc(output_j) \tag{21}$$
  wherever $output_j$ of $cell_u$ is connected to net_b
- $$B(net_b) - b(cell_u) \leq 1 + bc(output_j) \tag{22}$$
  wherever $output_j$ of $cell_u$ is connected to net_b
- $$b(cell_u) - b(cell_v) \leq w(net_b, input_k) - bc(output_j) + bc(input_k) \tag{23}$$
  wherever $input_k$ of $cell_v$ and $output_j$ of $cell_u$ are connected to net_b
- $$B(net_a) - B(net_b) \leq bc(input_i) - bc(output_j) + w(net_a, input_i) - \frac{d(input_i, output_j)}{c} \tag{24}$$
  wherever $d(input_i, output_j)$ is defined and $input_i$ is connected to net_a and $output_j$ is connected to net_b

- $$B(net_a) - b(cell_u) \leq 1 + w(net_a, input_i) - \frac{dat(input_i)}{c} + bc(input_i) \quad (25)$$

  wherever $input_i$ of $cell_u$ is connected to net_a

The other, forward retiming function can be applied as well. Therefore a new variable F for every net is introduced and $s(net_b)$ is substituted by $c(f(output_j) - F(net_b))$ where $output_j$ is connected to net_b. The forward retiming constants (fc) of the terminals in the timing model of cells are the negative of the backward retiming constants (bc). The set of constraints becomes now:

- $$F(net_b) - f(cell_u) \leq \frac{-d_{max}(output_j)}{c} + fc(output_j) \quad (26)$$

  wherever $output_j$ of $cell_u$ is connected to net_b

- $$f(cell_u) - F(net_b) \leq 1 - fc(output_j) \quad (27)$$

  wherever $output_j$ of $cell_u$ is connected to net_b

- $$f(cell_v) - f(cell_u) \leq w(net_b, input_k) + fc(output_j) - fc(input_k) \quad (28)$$

  wherever $input_k$ of $cell_v$ and $output_j$ of $cell_u$ are connected to net_b

- $$F(net_b) - F(net_a) \leq fc(output_j) - fc(input_i) + w(net_a, input_i) - \frac{d(input_i, output_j)}{c} \quad (29)$$

  wherever $d(input_i, output_j)$ is defined and $input_i$ is connected to net_a and $output_j$ is connected to net_b

- $$f(cell_u) - F(net_a) \leq 1 + w(net_a, input_i) - \frac{dat(input_i)}{c} - fc(input_i) \quad (30)$$

  wherever $input_i$ of $cell_u$ is connected to net_a

4 Solving the retiming problem

The set of constraints 21 through 25 and the set 26 through 30 both form a Mixed Integer Linear Programming (MILP) problem. The variables are placed at the left hand side of the less or equal sign, while only constants are placed at the right hand side. The variables f and b are integers and the variables F and B are reals.

Two different kinds of solutions of the MILP problem can be distinguished: *feasible solutions* and *minimal cost solutions*. A cost function is required to find a minimal cost solution.

4.1 Finding a feasible solution

A feasible solution is only useful when it does not introduce too many flipflops. An extremely pipelined circuit is a feasible solution for many required clock periods, but there may be too many pipeline stages. A feasible solution should even result in a circuit with less pipeline stages than originally present if possible. Therefore a certain strategy is required in finding a feasible solution. Leiserson et al. published an algorithm according to such a strategy for finding a feasible solution of an MILP problem [Le2]. This algorithm can be applied to the set of constraints containing the backward retiming function as well as to the set containing the forward retiming function. The two sets will lead to different solutions.

When the backward retiming function is applied, pipeline stages are shifted in from the outputs to the inputs by retiming. The timing of the outputs relative to each other will not change by the retiming. On the other hand the inputs of the circuit may have been retimed differently w.r.t. each other. When the forward retiming function is applied the timing of the inputs remains fixed w.r.t. each other, while the outputs may have been retimed differently w.r.t. each other.

The problem is that when either of the two retiming functions is applied, extra flipflops are inserted to prevent skew of either the inputs or outputs. A way to overcome this problem is to use a combination of forward and backward retiming. First a forward retiming is applied. The resulting retiming values of the outputs are used to impose additional constraints on the subsequent backward retiming. The additional constraints force the backward retiming to retime the outputs differently w.r.t each other. Therefore, skew of the outputs may occur and less flipflops may be required. The two retimings can also be applied in reverse order. First a backward retiming may be applied. Then the resulting retiming values of the inputs are used to impose additional constraints on the subsequent forward retiming.

Four alternative solutions can be generated in this way. One resulting from a forward retiming, one from a backward and two from the combination of both. The solutions differ in the resulting time shape and the number of flipflops. The best can be selected, but is not necessarily the optimal solution. Therefor a solution has to be found with minimal cost.

4.2 Finding a minimal cost solution

Finding a minimal cost solution requires a cost function. This cost function, C, expresses the number of flipflops in the retimed circuit.

$$C = \sum_{\forall \text{ net to input connections}} w_r(net, input) \tag{31}$$

In this equation, $w_r(net, input)$ can be replaced by using equation 1, resulting in

$$C = \sum_{\forall \text{ net to input connections}} [w(net, input) + b(input) - b(output)] \tag{32}$$

If only one output of a cell is connected to a net, the cost function becomes by applying further substitutions:

$$C = \sum_{\forall \text{ net to input connections}} [w(net, input) + bc(input) - bc(output)] + \sum_{\forall \text{ cells}} b(cell)[d^-(cell) - d^+(cell)] \tag{33}$$

in which $d^-(cell)$ is the number of inputs to a cell and $d^+(cell)$ is the number of outputs. Techniques to model multiway fanout and different breadth of signals can be found in [Le1]. Maybe a solution found using backward retiming may differ from a solution found using forward retiming, but since the costs are both minimal, the distinction is not important here.

Suppose that the delay can always be expressed in units of for example nanoseconds. Real constants than become integer constants. Again the set of constraints 7 through 12 containing the variable $s(net_a)$ for every net, net_a, can be transformed into a set of constraints without a qualifying clause by substituting $s(net_a)$ by $T(net_a) - c.b(output_j)$, where $output_j$ is connected to net_a and c is the required clock period. The set of constraints becomes now:

- $$c.b(cell_u) - T(net_b) \leq -d_{max}(output_j) - c.bc(output_j) \tag{34}$$
  wherever $output_j$ of $cell_u$ is connected to net_b, and c is the required clock period
- $$T(net_b) - c.b(cell_u) \leq c + c.bc(output_j) \tag{35}$$
  wherever $output_j$ of $cell_u$ is connected to net_b and c is the required clock period
- $$b(cell_u) - b(cell_v) \leq w(net_b, input_k) - bc(output_j) + bc(input_k) \tag{36}$$
  wherever $input_k$ of $cell_v$ and $output_j$ of $cell_u$ are connected to net_b

- $T(net_a) - T(net_b) \leq c.bc(input_i) - c.bc(output_j) + c.w(net_a, input_i) - d(input_i, output_j)$ (37)
  wherever $d(input_i, output_j)$ is defined, $input_i$ is connected to net_a, $output_j$ is connected to net_b and c is the required clock period

- $$T(net_a) - c.b(cell_u) \leq 1 + c.w(net_a, input_i) - dat(input_i) + c.bc(input_i) \quad (38)$$
  wherever $input_i$ of $cell_u$ is connected to net_a and c is the required clock period

A solution of this set of constraints can be found by applying Integer Linear Programming (ILP).

5 Operator selection

The selection of an operator (or cell) to implement an operation in a circuit, can be combined with retiming to find an implementation of a circuit which occupies minimal area and meets the timing constraints. The subject is treated here to show that the new timing model can be exploited there as well. Research on operator selection in combination with retiming has been reported by Note [Not]. If an operation in a circuit can be implemented by more than one operator, the timing model varies with the selected operator. The constant timing model is changed into a varying timing model; the constants become variables. The combined problem can be expressed by one (M)ILP problem. First a new variable op_{ij} is introduced. The variable op_{ij} is 1 if operation op_i is implemented by operator j, else it is 0. Since only one operator can implement an operation:

$$\sum_j op_{ij} = 1 \quad (39)$$

for every operation op_i. The variable op_{ij} is used to select the right values of the timing model:

$$d_{max}(output) = \sum_j d_{max_j}(output).op_{ij} \quad (40)$$

$$bc(output) = \sum_j bc_j(output).op_{ij} \quad (41)$$

$$dat(input) = \sum_j dat_j(input).op_{ij} \quad (42)$$

$$bc(input) = \sum_j bc_j(input).op_{ij} \quad (43)$$

The expressions can be substituted in the set of constraints, which is omitted here for briefity. The function to be minimized becomes now:

$$\begin{aligned} C \;=\; & \alpha. \sum_{\forall \text{ net to input connections}} w(net, input) + \alpha.[\sum_{\forall \text{ net to input connections}} [bc(input) - bc(output)] \\ & + \sum_{\forall \text{ cells}} b(cell)[d^-(cell) - d^+(cell)]] \\ & +[\sum_{\forall op_i} \sum_j Area_j.op_{ij}] \end{aligned} \quad (44)$$

The factor α is the area per flipflop.

6 Conclusions

To synthesize high speed datapaths, retiming is indispensable. Within the concept of PHIDEO retiming is allowed to change the time shape of datapaths. Every terminal of a circuit can now be retimed differently. Every terminal is therefore explicitly modelled as a cell in the retiming network. Should part of the time shape be preserved, additional constraints on the retiming of the terminals must be imposed. Hierarchical retiming is made possible by the introduction of the new timing model. The new timing model can also be applied to operator selection in combination with retiming. The explicit modelling of nets allows the presence of tri-state buffers in a circuit to be retimed. For large circuits, retiming in order to find a circuit occupying minimal area may take too much CPU time. Therefore it is made possible to generate different feasible solutions by either backward, forward or a combination of forward and backward retiming.

More work in the area of retiming can be done. For instance, retiming changes the places of flipflops in a circuit. Buffering of some outputs may be needed and some buffers may become superfluous. By changing the buffers in a circuit after retiming, the timing models of cells are changed. A new retiming may be necessary to meet timing constraints. The question is how buffering can be applied in combination with retiming to obtain the optimal result.

7 Acknowledgement

This work has been performed as part of the ESPRIT 2260 project (SPRITE) sponsored by the EC. The authors would like to thank the partners in this project (Siemens, Racal, Imec, Praxis, RSRE, and Inesc) for stimulating discussions.

References

[Le1] Charles E. Leiserson, Flavio M. Rose, James B. Saxe, "Optimizing Synchronous Circuitry by Retiming", Third Caltech Conference on Very Large Scale Integration, 1983.

[Le2] Charles E. Leiserson, James B. Saxe, "A Mixed-Integer Linear Programming Problem which is Efficiently Solvable", Proceedings Twenty-first Annual Allerton Conference on Communication, Control and Computing.

[Le3] Charles E. Leiserson, James B. Saxe, "Retiming Synchronous Circuitry", Algorithmica 1991 pages 5-35.

[Lip] P.E.R. Lippens, J.L. Van Meerbergen, A. van der Werf, W.F.J. Verhaegh, B.T. McSweeney, J.A. Huisken, O.P. McArdle, "PHIDEO: A Silicon Compiler for High Speed Algorithms", EDAC 1991.

[Ma1] Sharad Malik, Ellen M. Sentovich, Robert K. Brayton, Alberto Sangiovanni-Vincentelli, "Retiming and Resynthesis: Optimizing Sequential Networks with Combinational Techniques", Proceedings of the International Workshop on Logic Synthesis, Research Triangle Park, North Carolina, U.S.A., May 23-26, 1989.

[Ma2] Sharad Malik, Kanwar Jit Singh, R.K. Brayton, Alberto Sangiovanni-Vincintelli, "Performance Optimization of Pipelined Circuits", Proceedings ICCAD 1990.

[Not] Stefaan Note, "Combined Hardware Selection and Pipelining in High Performance Data-Path Design", Proceedings ICCD 1990.

[Wer] A. van der Werf, B.T. McSweeney, J.L. Van Meerbergen, P.E.R. Lippens, W.F.J. Verhaegh, "Flexible Datapath Compilation for PHIDEO", Proceedings Euro ASIC 91.

VLSI 91
A. Halaas and P.B. Denyer (Eds.)
Elsevier Science Publishers B.V. (North-Holland)

Preserving Don't Care Conditions During Retiming

Ellen M. Sentovich and Robert K. Brayton

Department of Electrical Engineering and Computer Science
University of California Berkeley

Abstract

The use of "don't care" conditions during logic optimization has been shown to be effective in producing smaller circuit implementations. Don't care conditions may be specified by the user, computed from the structure of the circuit, or, in the case of sequential circuits, extracted from a state transition graph (STG) of the circuit. These conditions may be time-consuming to recompute, or impossible to re-extract if the information is invalidated by a subsequent modification of the circuit. Therefore, it is important to preserve the don't care information during optimization. Retiming is a technique in which the cycle time or the number of registers is minimized by determining optimal register positions; the structure of the logic is unchanged while the registers are moved. In this paper, a method for preserving the **maximal** subset of the don't care conditions during retiming is proposed.

1 Introduction

Sequential circuit optimization has emerged as an important problem in logic synthesis. The techniques used in the more well-understood combinational logic synthesis domain can be applied directly to the combinational logic blocks between registers in a sequential circuit. Don't care conditions used in logic synthesis can be specified externally by the user or computed from the structure of a multi-level network which represents the circuit of interest. These conditions are used to simplify the two-level boolean functions at each node of the multi-level network [1, 6]. The don't care functions associated with a multi-level network can be large and expensive to compute [5], making it desirable to preserve the computed functions after a modification to the network has been made. The user-specified don't care conditions are impossible to recompute after a modification, and hence it is imperative that this type of don't care is preserved as much as possible. The problem of preserving don't care functions during the application of sequential optimization techniques has not been explored thus far.

Retiming [3] determines optimal positions for the registers within the circuit to minimize the cycle time or the number of registers. The logic functions at each node in the network are unchanged during retiming, but the logic blocks, which are the combinational gates between registers, are modified as the register positions change. The set of input and output signals seen by each block changes as the registers at their boundaries are shifted. As a result, preserving the don't care information in a block entails both re-expressing the don't care functions in terms of the new input signals, and computing don't care functions for the new outputs. These two operations represent the minimum amount of computation that must be done to preserve the existing don't care information.

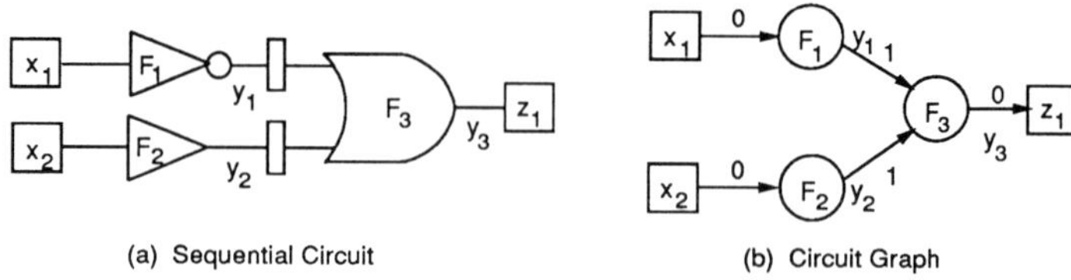

Figure 1: Sequential Circuit Representation

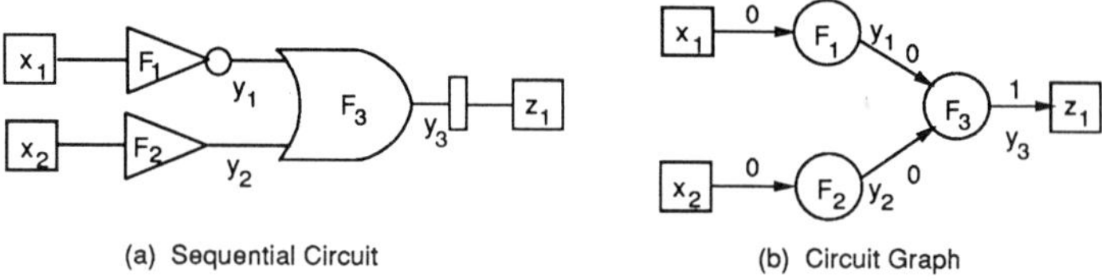

Figure 2: Retimed Sequential Circuit

2 Background

A sequential circuit is modeled by a directed graph where each vertex i represents either a) a primary input x_i $(i = 1, \ldots, p)$ b) a primary output z_i $(i = 1, \ldots, p)$ or c) a variable y_i and a representation F_i of a logic function $(i = 1, \ldots, m)$. An edge connects vertex i to vertex j if the function associated with vertex j, F_j depends explicitly on the variable y_i. Each edge e has a nonnegative integer label $w(e)$ representing the number of registers between the two logic gates it connects. Each cycle in the graph must contain at least one edge of strictly positive weight (this restriction is placed to model synchronous circuits only, thereby avoiding asynchronous problems such as race conditions). A sequential circuit is shown in Figure 1(a), with its graph in Figure 1(b). The terms circuit, network, and graph are used interchangeably whenever there is no ambiguity, as are the terms node and vertex.

2.1 Retiming

Retiming is an operation on a sequential circuit whereby registers are moved across logic gates in order to minimize the clock cycle or the number of registers while maintaining the behavior of the circuit. Retiming algorithms were first proposed by Leiserson *et al* [3]. The movement of registers can be quantified by an integer $r(v)$ for each vertex v, which represents the number of registers that are to be moved in the graph from each out-edge of vertex v to each of its in-edges. The resulting edge weight for an edge from vertex u to vertex v is $w_r(e) = w(e) + r(v) - r(u)$. The circuit shown in Figure 1 can be retimed by selecting $r(F_3) = -1$ and $r(v) = 0$ for all other vertices. The resulting retimed circuit is shown in Figure 2.

During a retiming, the registers move in a limited fashion. Lemma 1 in [3] asserts that $w_r(p) = w(p) + r(v) - r(u)$, where p is a path from u to v in the circuit, and $w(p)$ is the sum of the edge weights along that path. Since $r(v) = 0$ for all I/O pins, the weight of each path from an input pin to an output pin is unchanged during retiming. Given a particular circuit with its associated retiming, a register on a particular path can be mapped to a new position on that path in the retimed circuit because the total path weight is unchanged during retiming. As a result, each register in the circuit can be thought of as having moved backward (to some edge in its transitive fanin) or forward (to some edge in its transitive fanout). This classification of the movement of registers during retiming is essential to the method for preserving don't cares across retiming presented in Section 3.

2.2 Don't Care Sets and Boolean Operators

Three types of don't care sets have been introduced for multi-level logic optimization [1]: **External DC set** is comprised of user-specified don't care conditions, such as input combinations that never occur. In general, each output j may have a different external DC set, denoted by d_j. A property of the external don't care set, not often explicitly stated but assumed by logic optimization programs, is that each don't care for each output must remain a don't care independent of how the other don't cares at other outputs are used. This arises from the way this information is used in logic optimization. A set of don't care functions is called "compatible" if it has this independence property.
Satisfiability DC set (SDC) arises because variables at the intermediate vertices of the graph are not independent of the primary inputs, and represents all the inconsistent conditions in the network. The SDC is given by the following equation:

$$SDC = \sum_{i=1}^{m}(y_i \neq F_i) = \sum_{i=1}^{m}(\bar{y}_i F_i + y_i \bar{F}_i)$$

Observability DC set (ODC) is defined for each vertex v_i. It is derived by looking at the fanout of the variable y_i and seeing how specifically y_i is used. In [4], two sets based on observability don't care conditions were defined: MSPF's (maximum sets of permissible functions), which include for each node the on-set of the node function and the largest possible set of observability don't cares for that node, and CSPF's (compatible sets of permissible functions), which include the on-set of the node function and a maximal set of compatible observability don't cares for that node, which means that the don't cares can be used simultaneously in simplifying all the nodes without the need for recomputing the don't cares at any other nodes. These ideas were extended in [6], applied to general multi-level networks, and algorithms for computing a maximal set of CSPF's were given. Beginning at the outputs with the external don't cares and working towards the inputs, CSPF's are computed at each node and can be used in the simplification of that node simultaneously with simplifying other nodes.

The smoothing operator S (existential quantifier) for a boolean function f is defined as follows: $S_x f = f_x + f_{\bar{x}}$, where f_x is the function f evaluated at $x = 1$ and $f_{\bar{x}}$ is the function f evaluated at $x = 0$. $S_x f$ represents the minimum set containing f and independent of x. Smoothing a function f with respect to a set of inputs $I = \{x, y, z\}$ is denoted $S_x S_y S_z f$ or simply, $S_I f$.

The consensus operator C (universal quantifier) for a boolean function f is defined as follows: $C_x f = f_x \cdot f_{\bar{x}}$. $C_x f$ represents the maximum set contained by f and independent of x.

An input signal x to a particular gate f is said to be **observable** at f if $\frac{\delta f}{\delta x} = 1$, where $\frac{\delta f}{\delta x} = f_x \oplus f_{\bar{x}}$. That is, the signal x is observable at f if setting x to 1 gives a different result at f than setting x to 0.

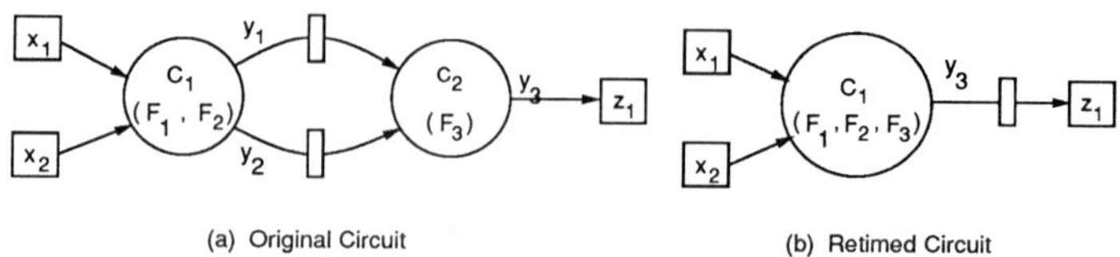

Figure 3: Logic Block View of a Sequential Circuit

3 Preserving Don't Cares after Retiming

A sequential circuit can be viewed as a set of combinational logic blocks separated by registers. The inputs to each block are primary inputs and register outputs, and the outputs of each block are primary outputs and register inputs. The don't care function for each node in a particular block, is expressed in terms of the inputs to the block. The logic block view of the sequential circuit in Figure 1(a) is shown in Figure 3(a). The block C_1 has as inputs the primary inputs x_1 and x_2, and outputs the signals y_1 and y_2 which are register inputs. The don't care functions for the nodes in block C_1 will depend on x_1, x_2.

After the circuit is retimed, the registers acting as boundaries of the block may have been moved forward or backward (but not both; this property of retiming was addressed at the end of Section 2). As a result, the block has different inputs and outputs, and the don't care functions within the block must be updated to reflect this change.

If the registers at the outputs of the block have been moved forward, the block contains new internal nodes and outputs, and the don't care functions in the block should be propagated forward to the new nodes. For example, if the circuit represented in Figure 3(a) is retimed to produce that in Figure 3(b) the node F_3 becomes a new node of block C_1. Don't care information from block C_1 should be propagated forward. If the registers at the outputs have been moved backward, internal nodes within the block become new outputs, and don't care information from the old outputs must be passed to the new outputs. If the circuit represented in Figure 3(b) is retimed to produce that in Figure 3(a), the register is moved back across node F_3 and the block C_1 has new outputs, signals y_1 and y_2. Don't care information from node F_3 should be propagated back to nodes F_1 and F_2 before node F_3 becomes a part of block C_2 and block C_1 loses access to that information.

If the registers at the inputs of a block are moved, the block has a new set of input signals. The don't care functions at each node within the block must be re-expressed in terms of the new input signals.

Thus, there are two operations needed for updating the don't care information in a block: re-expressing the don't care functions in the block in terms of the new inputs, and propagating don't care functions between nodes within a block.

3.1 Re-expressing Don't Care Functions

If the registers at the inputs to a particular block are moved, the don't care functions within that block must be re-expressed in terms of the new inputs so these functions can be used in a meaningful way by the node simplification algorithm. This function re-expression operation can be divided into two cases: one in which the registers are moved backward, and one in which the registers are moved forward.

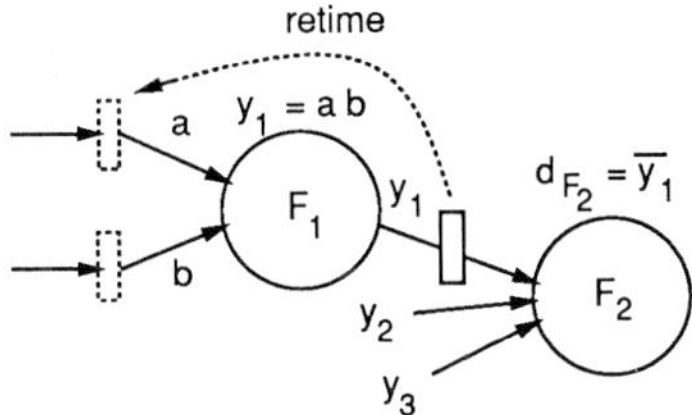

Figure 4: Inputs change as the register moves backward

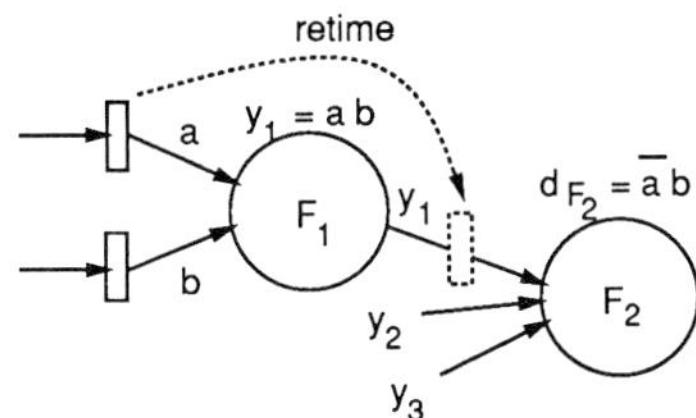

Figure 5: Inputs change as the registers move forward

If the registers are moved backward, the old inputs are functions of the new inputs and the don't care functions can be re-expressed by collapsing the logic between the old and new inputs. This operation is illustrated by the circuit represented in Figure 4. The don't care function for node F_2 is expressed as $d_{F_2} = \bar{y}_1$. If node F_1 is retimed to move the register to its inputs, $r(F_1) = 1$, then the don't care function at node F_2, $d_{F_2} = \bar{y}_1$, can be re-expressed as $d_{F_2} = \overline{ab} = \bar{a} + \bar{b}$.

The case in which the registers at the inputs are moved forward requires more consideration. The don't care functions must be re-expressed in terms of the new inputs, which are functions of the old inputs. For example, in the circuit represented in Figure 5, node F_1 is retimed to move the registers from its inputs to its output, $r(F_1) = -1$. The function d_{F_2} must be expressed in terms of y_1 rather than a and b, since a and b are no longer inputs to the block containing F_2.

This re-expression problem can be formulated in a more general way. Given a function f, with inputs $S = \{s_1, s_2, \ldots, s_m\}$ and $T = \{t_1, t_2, \ldots, t_n\}$, and a set of functions represented by the signals $Y = \{y_1, y_2, \ldots, y_p\}$, which are each functions of S, express f as a function of the signals in Y and T; that is, given $f(S,T)$ and $Y(S)$, determine $f_{new}(Y,T)$.

In general, a function f_{new} which behaves identically to the function f may not exist because the information carried by the signals in S may not be completely captured by the signals in Y, i.e., there is not, in general, a one-to-one mapping from the information carried by the signals S to the information carried by the signals Y. The construction of f_{new} for re-expressing the don't care function f involves retaining the maximal amount of information from f and including information obtained from the SDC set for the signals in Y as a function of S. Let SDC_Y be the SDC for the signals in Y:

$$SDC_Y = y_1 \cdot \bar{f}_{y_1} + \bar{y}_1 \cdot f_{y_1} + y_2 \cdot \bar{f}_{y_2} + \bar{y}_2 \cdot f_{y_2} + \ldots + y_p \cdot \bar{f}_{y_p} + \bar{y}_p \cdot f_{y_p}$$

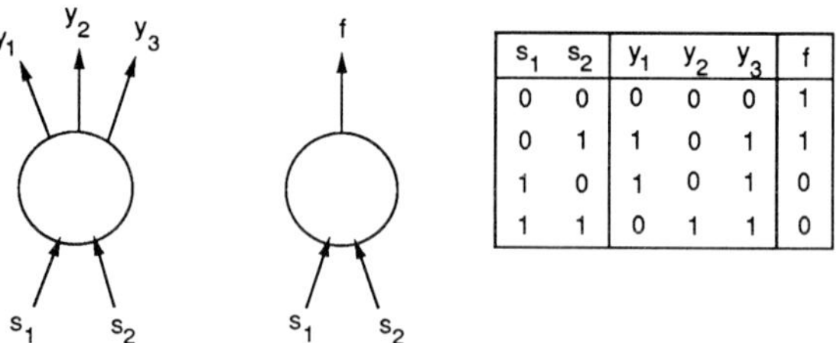

s1	s2	y1	y2	y3	f
0	0	0	0	0	1
0	1	1	0	1	1
1	0	1	0	1	0
1	1	0	1	1	0

Figure 6: Example: Re-expressing f in terms of y_1, y_2, y_3

Then f_{new} is constructed as follows:

$$f_{new} = \mathcal{C}_S \left(f + SDC_Y \right) \tag{1}$$

The consensus produces the **maximal** subset of $f + SDC_Y$ that is independent of S. This conservative approach to constructing f_{new} may result in the loss of some don't care conditions as the next example illustrates. Note that the loss of the don't care conditions is inherent to the problem and is not due to some limitation of this method.

An example of this computation is given in Figure 6, where the functionality of the circuit is given by a truth table. The goal is to represent f in terms of y_1, y_2, and y_3. In this example, $y_1 = s_1 \oplus s_2$, $y_2 = s_1 s_2$, $y_3 = s_1 + s_2$, and $f = \bar{s}_1$. Equation 1 gives

$$f_{new} = y_1 y_2 + \bar{y}_1 \bar{y}_2 + \bar{y}_3.$$

Some of the conditions in f_{new} arise from the SDC alone, and correspond to combinations of the Y values that can never occur: $y_1 y_2$, $\bar{y}_1 \bar{y}_2 y_3$, $\bar{y}_1 y_2 \bar{y}_3$, and $y_1 \bar{y}_2 \bar{y}_3$. One condition arises from the function f: $\bar{y}_1 \bar{y}_2 \bar{y}_3$. Finally, note that the condition $y_1 \bar{y}_2 y_3$ is not included in f_{new} because it is not a don't care under all equivalent assignments to the signals in S.

3.2 Propagating Don't Care Functions

As the registers are moved during a retiming operation, the structure of the logic blocks change, causing some nodes to become part of a new block. The don't care functions for these nodes should be augmented with don't care information from the new blocks. This is done by propagating don't care information either forward or backward within the block to the nodes of interest.

3.2.1 Propagating Don't Cares Backward

Techniques for propagating don't care information backward have been developed in [6]. In the algorithm described in that paper, the computation begins with external don't care information expressed for the outputs, and propagates that information, along with observability don't care information extracted during the computation, backward to the primary inputs. In [6], two formulae were given: one for propagating don't cares from the output of a node to each of its inputs, and another for computing the don't cares of the node based on the don't cares of all of its fanout nodes. Those formulae were proven to generate local **maximal compatible**

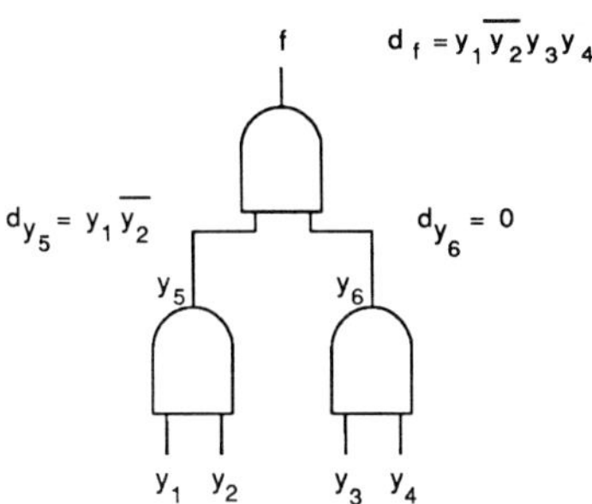

Figure 7: Example: Propagating Don't Care Functions Forward

observability don't cares. It is important that these functions are compatible, so that they can be used simultaneously during node simplification. The same computation as that given in [6] can be used to propagate don't care information backward after retiming. Furthermore, the fact that they are maximal and compatible implies that after retiming, the don't cares obtained can still be simultaneously used during node simplification [1].

3.2.2 Propagating Don't Cares Forward

Given f as a function of $y_1, y_2, \ldots, y_p$, the task is to compute a don't care function for f. A condition can be considered a don't care for f if it can be set to 0 and 1 by using the don't cares on the inputs. An example is given in Figure 7. Suppose the don't care function at y_5 is $d_{y_5} = y_1\bar{y}_2$. In that case, the cube $y_1\bar{y}_2y_3y_4$ is a don't care for f: $y_1\bar{y}_2y_3y_4$ results in $y_5 = 0$, $y_6 = 1$, and $f = 0$, and since $y_1\bar{y}_2$ is a don't care for y_5, the value 1 can be assigned to y_5 in which case $f = 1$.

The don't care function for a node is computed based on the don't care functions at its inputs and the observability of the inputs. Therefore, the following definition of an observable set will aid in developing a formula for propagating don't cares forward.

Definition 1 *Let f be a Boolean function of n input variables in the set $Y = y_1, y_2, \ldots, y_n$, and let $\mathbf{y}^{\mathbf{i}} = [y_1, y_2, \ldots, y_n] \in B^n$ be some assignment of values to the input variables ($f(\mathbf{y}^{\mathbf{i}}) \in B$). Let G be a subset of Y, $G \subseteq Y$, $\mathbf{g}^{\mathbf{i}}$ an assignment to the variables in G, H be the complement of G, $H = Y - G$, and $\mathbf{h}^{\mathbf{k}}$ an assignment to the variables in H. G is an* **observable set** *with respect to f if and only if*

$$\exists\, \mathbf{g}^{\mathbf{i}}, \mathbf{g}^{\mathbf{j}}, \mathbf{h}^{\mathbf{k}} \text{ s.t. } f(\mathbf{g}^{\mathbf{i}}, \mathbf{h}^{\mathbf{k}}) \neq f(\mathbf{g}^{\mathbf{j}}, \mathbf{h}^{\mathbf{k}})$$

$\mathbf{h}^{\mathbf{k}}$ is the condition under which the set is observable. In Section 2 a formula was given for the observability of a single signal. That formula generates a function that represents conditions under which toggling the value of the single input causes the output function to toggle. The definition of an observable set is related in that it conveys observability of a set of input signals;

[1]Recently, Damiani et al [2] proved that this method of propagating locally maximal compatible ODC's may not be maximal compatible when this step is iterated over several gates. Thus a larger maximal compatible set would be possible.

it generates the conditions (values of the signals not in the set) under which there exist two assignments to the signals in the set that cause the output function to toggle.

The following Lemma makes use of Definition 1.

Lemma 1 *Given $Y = y_1, y_2, \ldots, y_n$, the set of input variables for a function f, and G, a subset of Y, G is an observable set at f under the conditions given by $\mathcal{S}_G f \cdot \mathcal{S}_G \bar{f}$.*

Proof. By definition, $\mathcal{S}_G f$ gives the conditions under which there exists an assignment to the variables in G such that $f = 1$, i.e. it gives the conditions, $\mathbf{h^i}$ s.t. $\exists\, \mathbf{g^1}, f(\mathbf{g^1}, \mathbf{h^i}) = 1$. Similarly, $\mathcal{S}_G \bar{f}$ gives the conditions under which there exists an assignment to the variables in G such that $f = 0$, i.e. $\mathbf{h^i}$ s.t. $\exists\, \mathbf{g^2}, f(\mathbf{g^2}, \mathbf{h^i}) = 0$. The product of the two smoothing terms gives the conditions on the variables not in G such that there exists an assignment to the variables in G that sets f to 1, and another assignment to the variables in G that sets f to 0. By definition, each condition in $\mathcal{S}_G f \cdot \mathcal{S}_G \bar{f}$ is a condition under which G is an observable set at f. □

The don't care conditions can be moved forward across a node according to the following theorem.

Theorem 1 *Given a Boolean function f, with n inputs, $Y = y_1, y_2, \ldots, y_n$, each with associated compatible don't care conditions, $D = d_{y_1}, d_{y_2}, \ldots, d_{y_n}$, the* **maximal** *don't care function d_f for f based only on the don't care conditions at the inputs can be expressed as*

$$d_f = \sum_{G \subseteq Y} \left(\prod_{g \in G} d_g \right) \left(\mathcal{S}_G f \cdot \mathcal{S}_G \bar{f} \right) \tag{2}$$

Proof. The proof has two parts: first it must be shown that all the conditions produced by Equation 2 are don't care conditions for f, and second, that this set of don't cares is maximal. For the first part, note that each term in the summation of Equation 2 is comprised of don't care conditions that are common to a subset of input signals intersected with the conditions under which that subset is an observable set with respect to f. Each of the resulting conditions can be considered a don't care for f because 1) the don't care condition at the set of inputs, $\prod_{g \in G} d_g$, allows any assignment of values to those inputs [2], and 2) the observability set condition $\mathcal{S}_G f \cdot \mathcal{S}_G \bar{f}$ guarantees that there exists two assignments to g, one that results in $f = 1$ and the other that results in $f = 0$. To prove that the set of don't cares produced is maximal, suppose there is a don't care condition d on f that is not included in Equation 2. d must arise from one or more of the input don't cares d_{y_i}, since that is the only don't care information given about the circuit for this problem (the function f may be embedded in a larger circuit giving rise to don't cares based on this structure, but these need not be considered for the problem of simply propagating don't cares forward). d also must be observable at the output f in order to be a don't care for f. Finally, regardless of which input(s) the don't care condition arises from, it will be transferred to f provided it is both observable and contained in some subset of the inputs. Therefore, d must be contained by d_f. □

---

[2]By definition, the don't cares d_g are independent or compatible so that all possible assignments within the don't care sets are possible.

function	literals	simplify	retime	retime_dc
$x^2 + y^2$	58	45	45	25
$x^2 > y^2$	28	25	25	18
$a \cdot b + c \cdot d$	56	46	46	40
$x \cdot y$	23	23	20	18

Table 1: Results for Simplification with and without Retiming Don't Cares

We emphasize that Equation 2 can only be applied when the given don't cares at the inputs are CSPF's. The reason for this is that the don't cares at the inputs are used **simultaneously** in the computation of Equation 2, so the don't cares must be compatible.

For the circuit in Figure 7,

$$\begin{aligned} d_f &= d_{y_5}\left(S_{y_5} f \cdot S_{y_5} \bar{f}\right) + d_{y_6}\left(S_{y_6} f \cdot S_{y_6} \bar{f}\right) + d_{y_5} d_{y_6}\left(S_{y_5} S_{y_6} f \cdot S_{y_5} S_{y_6} \bar{f}\right) \\ &= y_1 \bar{y}_2 y_3 y_4 \end{aligned}$$

4 Applications

These techniques have been implemented in SIS, the sequential circuit synthesis program being developed at Berkeley. Whenever a retiming is done, the don't cares are updated accordingly.

There are a number of places in the synthesis procedure where this is beneficial. First, if a circuit has been generated from a high-level description language, don't care conditions inherent in the high-level specification may be passed down to the gate-level implementation. At this level, these represent external don't cares that cannot be computed from the structure of the circuit. If these conditions are not updated during retiming, they are lost and unusable in the remaining synthesis steps. In this case, preserving the don't care conditions can only result in a gain. Another application involves globally retiming a circuit with the don't cares and applying logic optimization to various blocks within the circuit. After retiming, the logic optimization procedure could be applied to the entire circuit and the don't cares completely recomputed, but this may be too time consuming. It may be much more efficient to simply update the don't care conditions for each block in the manner proposed here. Don't cares arising from unreachable states in a state machine can be updated with retiming as well. The unreachable states may be computed once for the entire circuit (which may be a time consuming operation, even with implicit state enumeration techniques) and expressed as don't cares. These functions are then updated incrementally after retiming, avoiding another complete unreachable state computation for the retimed circuit.

To illustrate the need for preserving external don't care conditions, several circuits have been synthesized. These are arithmetic circuits with two-bit operands, comprised of several blocks of sequential logic. Don't care conditions for each block were computed based on information from the other blocks (these are considered external don't cares for the given block). For example, the first circuit in Table 1 computes $x^2 + y^2$. x^2 and y^2 can only take on four values: 0000, 0001, 0100, and 1001. The remaining values are external don't cares for the adder. Each circuit was simplified using the external don't cares, retimed, and simplified again. The number of literals obtained when the external don't cares are not retimed are reported in the **retime** column; the results when the don't cares are retimed are reported in the **retime_dc** column. In both cases, computed don't cares are used in the simplification, but only in the retime_dc case are the external don't cares retained. For all circuits, retiming the don't cares provides an additional area gain. In the last case, an area gain is obtained without retiming the don't cares (due to structural don't cares) but the result is even better when the don't cares are retimed.

5 Conclusions

A method has been proposed for preserving a **maximal** set of don't care conditions across a retiming operation on a sequential circuit. Thus far, the problem of preserving such conditions across retiming has not been addressed in the literature, and hence is a new problem. It is important that the proposed don't care sets are maximal since these don't cares are beneficial in subsequent synthesis algorithms, and since some don't care conditions cannot be recomputed once they are lost.

As a by-product of the technique proposed in this paper, a method was discovered for moving don't care conditions forward in a network (methods for moving them backward are already known, and given in, e.g., [6]). As a result, don't care conditions may be specified for any portion of a sequential circuit, and propagated both forward and backward to other parts of the circuit. The techniques proposed in this paper can be extended slightly to recompute the initial state. In particular, an initial state, or a set of initial states, is expressed in terms of the latch outputs in the circuit (similar to the don't care functions, which are expressed in terms of the latch outputs and the primary inputs). The techniques presented in this paper can be simply extended to generate a new set of initial states given the initial states for the unretimed circuit.

References

[1] K. Bartlett, R.K. Brayton, G. Hachtel, R. Jacoby, C. Morrison, R. Rudell, A. Sangiovanni-Vincentelli, and A. Wang. Multilevel Logic Minimization Using Implicit Don't Cares. *IEEE Transactions on Computer-Aided Design*, 7(6):723–740, June 1988.

[2] M. Damiani and G. De Micheli. Derivation of Don't Care Conditions by Perturbation Analysis of Combination Multiple-Level Logic Circuits. In *Proceedings of the International Workshop on Logic Synthesis*, May 1991.

[3] C.E. Leiserson and J.B. Saxe. Retiming Synchronous Circuitry. In *TM 372, MIT/LCS, 545 Technology Square, Cambridge, Massachusetts 02139*, October 1988.

[4] S. Muroga, Y. Kambayashi, H. C. Lai, and J. N. Culliney. The Transduction Method - Design of Logic Networks Based on Permissible Functions. In *IEEE Transactions on Computers*, October 1989.

[5] A. Saldanha, A. Wang, R.K. Brayton, and A.L. Sangiovanni-Vincentelli. Multi-Level Logic Simplification using Don't Cares and Filters. In *Proceedings of the Design Automation Conference*, pages 277–282, 1989.

[6] H. Savoj and R.K. Brayton. The Use of Observability and External Don't Cares for the Simplification of Multi-Level Networks. In *Proceedings of the Design Automation Conference*, pages 297–301, 1990.

VLSI 91
A. Halaas and P.B. Denyer (Eds.)
Elsevier Science Publishers B.V. (North-Holland)

A fault tolerant and high speed instruction systolic array

Manfred Schimmler[a] and Hartmut Schmeck[b]

[a]Institut für Informatik und Praktische Mathematik, Christian-Albrechts-Universität, W-2300 Kiel 1, Germany

[b]Institut für Angewandte Informatik und Formale Beschreibungsverfahren, Universität Karlsruhe, W-7500 Karlsruhe, Germany

Abstract
In this paper we describe the design and implementation of a high speed and fault tolerant instruction systolic array. It has been designed to operate at a clock frequency of at least 100 MHz. Because of a redundant interconnection structure faulty processing elements can be tolerated by appropriately reconfiguring the array.

1. INTRODUCTION

Instruction systolic arrays (ISA's) have been designed to combine the simplicity of hardwired systolic arrays with the flexible programmability of MIMD-type processor arrays [La86]. In an ISA two orthogonal streams of control information are pumped through a mesh-connected array: a stream of instructions along the columns and a stream of binary selectors ('0' and '1') along the rows. The two streams are combined such that an instruction is executed whenever it meets a '1' and it is turned into a NOOP otherwise. The relevance of the ISA-concept and its applicability to the solution of a large variety of different problems have been demonstrated by extensive studies (see e.g. [La90], [KLSSS88], [Stra89], [Schi87], [Schm86], [DiS88], [ST90]).

Since instruction systolic arrays are supposed to consist of simple cells (or processing elements - PE's) rather large numbers of cells are expected to fit onto a single chip. In order to increase the yield and lifetime of ISA-chips it should be possible to tolerate some faulty cells on a chip without sacrificing too much with respect to performance. Therefore, a redundant interconnection structure has been designed which allows to reconfigure the array such that as many fault-free cells as possible can be utilized in the resulting logical two-dimensional array. Because of the different bandwidths of the two streams of control information, the chosen redundant interconnection structure is asymmetric: there are bypasses in the columns and bypasses plus additional diagonal interconnections in the rows. Strategies for reconfiguring the array based on this interconnection structure are described in [PSS91].

The paper is organized as follows: In the next section the ISA-concept and the processing elements of the array are briefly described. A more detailed description of the full-custom design of a CMOS chip containing a prototype 3×3 array and fabricated with a 1.6μ process is given in Section 3. The redundant interconnection structure is explained in Section 4. The paper ends with some concluding remarks in Section 5.

2. INSTRUCTION SYSTOLIC ARRAYS

In this section we briefly describe the concept of instruction systolic arrays, for more details the reader is referred to [La86], [La90] and [KLSSS88].

Instruction systolic arrays are based on mesh-connected $n \times n$-arrays of processors. Besides some internal registers every processor has a *communication register* K which is accessible to all its four direct neighbors. Every processor can execute instructions from some set of instructions I. At the end of this section we give more details on the specific set of instructions chosen for the design of the array described in this paper. The array is synchronized by a global clock assuming that all the possible instructions take the same time.

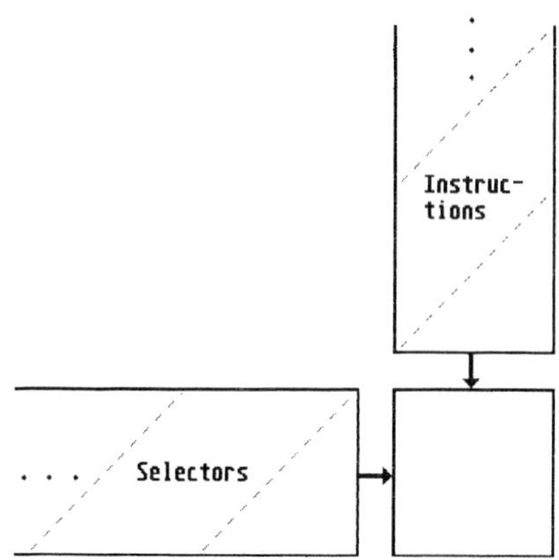

Figure 2.1: The instruction systolic array.

The instruction systolic array is programmed by a sequence $b \in (I^n)^*$ and a sequence $s \in (\mathbb{B}^n)^*$ (where $\mathbb{B} = \{0,1\}$), which move systolically through the array: The stream b of instruction rows moves from the top to the bottom row and the stream s of binary selector columns moves from the left to the right column (see Figure 2.1). Instructions are executed in a processing element whenever they meet a selector "1". Otherwise, a NOOP instruction is executed. Hence, the stream of selector columns is used to select the rows of the array where instructions are to be executed. This leads to a very flexible programmability. Since on their way through the array, the instructions on a diagonal of the instruction stream meet the selectors on a diagonal of the selector stream (cf. Figure 2.1), programs for the ISA are very often "diamond shaped", or defined by a sequence of program (i.e. instruction and selector) diagonals.

Input and output occur at the boundary of the array. A data item is input whenever a processor on the boundary of the array executes an instruction that attempts to read from the communication register of a nonexistent neighbor. For numerous examples of ISA-programs for a large variety of different types of problems the reader is referred to the literature (see e.g. [La90], [Stra89], [Schi87], [Schm86], [DiS88], [ST90]).

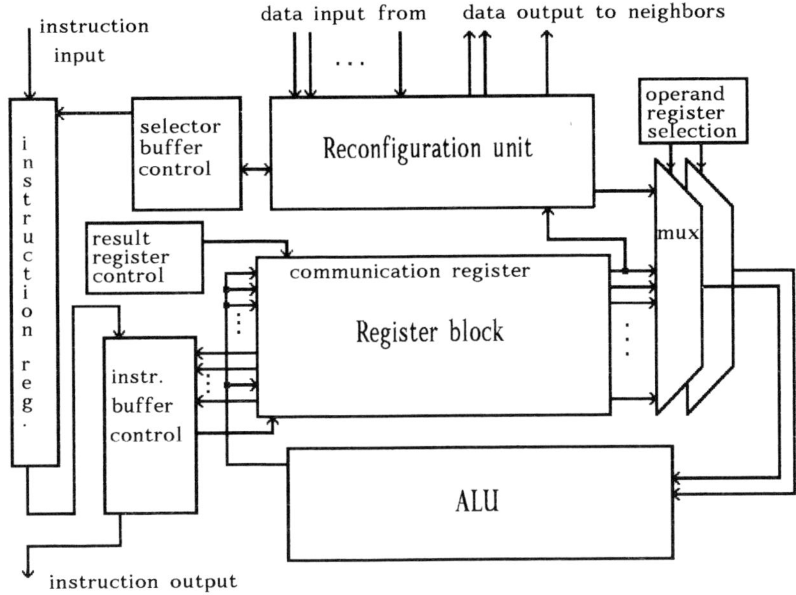

Figure 2.2: Block structure of the processors of the instruction systolic array.

In the instruction systolic array described in this paper, the processing units are designed to understand the instruction set defined in [La90]. Besides a NOOP instruction it contains instructions for moving data between its own registers or from a neighboring communication register into one of its own registers, instructions for performing basic arithmetic/logic operations (but - since this is a prototype design - no multiplication or division), and IF-THEN-ELSE instructions for conditionally moving data between registers depending on the value of binary flags Neg, Zero and Ü which are set, if the result of an addition/subtraction is negative, zero, or if an overflow ("Überlauf") occurs. To allow the use of "macro instructions", i.e. more complex instructions which are expanded into a sequence of simple instructions, the processing units are equipped with an instruction buffer and a selector buffer (cf. [La90]) and there is a special instruction for setting the length of these buffers. All these operations are executed bit-serially on 16-bit operands. The block structure of a processing unit is shown in Figure 2.2. More details on the layout design of these processors are given in the following section.

3. IMPLEMENTATION

A small example of a prototype 3×3 instruction systolic array has been designed as a full custom CMOS chip and fabricated with a 1.6μ double metal n-well process. The processor design was determined mainly by three requirements:

(r1) The array should be clocked with a clock rate of at least 100 MHz.

(r2) The overall power consumption for each processor should be less than 20 mA.

(r3) The area for one processor should be less than 2 mm^2 .

These requirements were motivated by the desired application of the currently processed chip: An 8×8 ISA shall be used as a co-processor for an IBM AT to speed up numerical computations occurring e.g. within finite element methods. The high clock frequency is necessary to get a reasonable speed-up. The restrictions on area and power arise from the maximal available chip size and the heat dissipation, respectively.

To solve the problems of (r1) we use a true single phase clock (TSPC) [YS89]. This clocking technique has been developed for digital circuits with an extremely high degree of pipelining. For our purposes, we use two types of clocked basic cells: a p-latch and an n-latch (Figure 3.1).

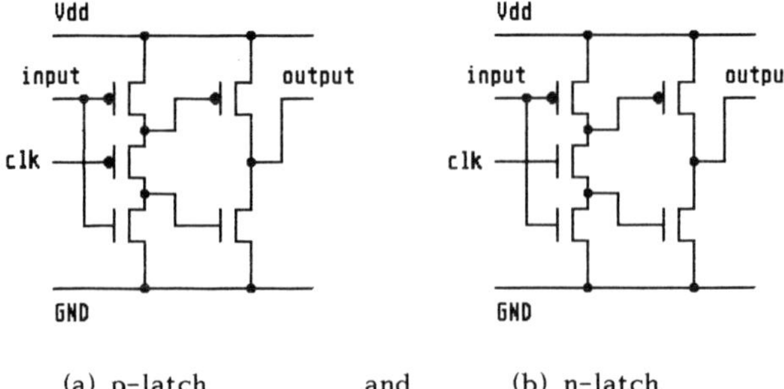

Figure 3.1: (a) p-latch and (b) n-latch

The p-latch transmits its input signal to the output when the clock is low. The n-latch does the same at a high clock signal. Observe that the clock transistor of a p-latch which is a pFET has to transmit a 0 in case of a high input signal. Analogously, the clock transistor of an n-latch is an nFET. When the input signal is low it has to transmit a 1. But since in both cases these signals control only one transistor the circuits perform pretty well in practise. A standard example of the use of these latches is a dynamic shift register which consists of alternating p- and n-latches. Another example is the typical way of synchronising a pipelined circuit with p- and n-latches as depicted in Figure 3.2.

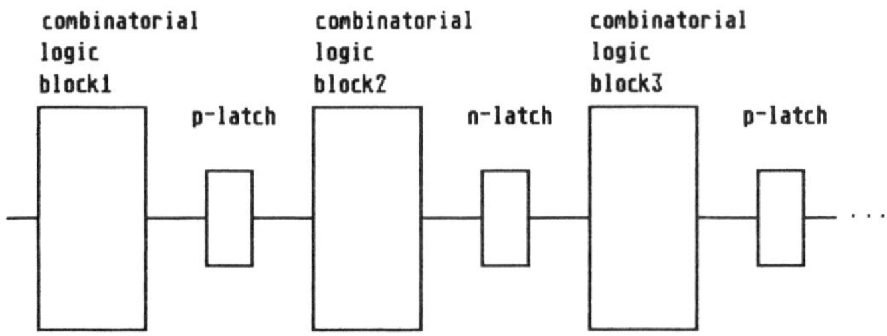

Figure 3.2: Pipeline using p-latches and n-latches for synchronisation

Two main advantages are achieved by using the TSPC. Firstly, the whole clock distribution has to be done only for one signal clk. There are no problems with clock-skew between clk and clkbar or between the four signals as required by a nonoverlapping two phase clock. Therefore, the area for clock drivers and wires is minimized (r3), and an extremely high clock frequency is possible (r1). Secondly, the n- and p-latches are consuming very little power (r2). As usual in C^2MOS there is no direct current from Vdd to GND.

In the following we describe some layout details, in particular those that had to be introduced with respect to the requirements (r1), (r2), and (r3).

As local memory the processor has a block of 11 registers laid out as a 17 bit shift register each. Data items are 16 bits long. There is one additional dummy bit which is used for the selector-bit (the horizontal control). The first of the registers is the communication register, whose output is also available to the four neighbors of the processor. The remaining 10 registers form the local memory of the processor. Up to six registers can optionally be used as an instruction buffer which allows to use some type of microprogramming (cf. [Stra90]). In that case, they are connected to form a long shift register and they are no longer available for storage of data. This is controlled by a fast selection mechanism called the instruction buffer control. The corresponding selector buffer is kept close to the selector buffer control unit, since it has to store at most 7 successive selector bits and so, there is no large memory necessary.

The instructions are read bit-serially from the northern neighbor. They are kept in a shift register from which control information is distributed to where it is used in the processor. Depending on the instruction buffer control the last stage of the instruction shift register is either linked to the current instruction buffer or connected to the instruction input of the southern neighbor.

The instruction cycle consists of 17 bit cycles. Bits 0 to 3 of the instruction word specify the type of instruction, bits 4 to 7 the address of the first operand, bits 8 to 11 the address of the second operand and bits 12 to 15 the address of the target register. Bit 16 is a dummy bit.

The two 16-to-1 operand multiplexers have as inputs the 11 internal registers of the processor, the communication registers of the four neighbors, and one constant (0 or 1, respectively). The high clock frequency required a particular design for these multiplexers:

A conventional layout as a sequence of transmission gates would have been too slow. Instead, for each input line exactly one transmission gate is used. Each of these is controlled by a block of clocked logic, depending on the multiplexer's control signals. All the transmission gates are connected to a common bus connecting the multiplexers with the ALU. This type of multiplexers has turned out to be very fast (in SPICE simulations), but it is rather area- and energy consuming compared with a conventional layout.

Communication with neighboring processors can easily be performed within one clock cycle. As described in more detail in the next section we have implemented a redundant interconnection structure which allows to tolerate faulty processors by bypassing some of the physical neighbors and choosing as logical neighbor a processor which is physically far away. Therefore, we had to take special care in building up the reconfiguration unit and driving these long wires. In the reconfiguration unit this has been solved similarly to the multiplexer design mentioned above. To make the long wires fast, we inserted a driver into each outgoing communication wire directly after the transmission gate of the reconfiguration unit. In this way we achieved a communication speed which allows to bypass up to three processors in sequence without affecting the anticipated clock frequency of 100 MHz.

The whole 3×3 array consists of 43,000 transistors. It covers an area of

25 mm^2 inclusive the pad frame and 17.5 mm^2 without pads. We have run a large number of SPICE simulations of the critical parts of the chip. Particularly, we tested the clock distribution which is performed by a cascade of four driver stages, the power consumption, the shift registers, and the multiplexers. The simulations resulted in an estimated power consumption of less than 200 mA inclusive I-O-devices, running at a clock frequency of 100 MHz.

4. FAULT TOLERANCE

To achieve a large degree of fault tolerance the prototype instruction systolic array has been designed with a redundant interconnection structure providing for a flexible reconfiguration of the array in the presence of faulty processors. The design of the interconnection pattern and of the on-chip logic for reconfiguration was determined by four requirements:

(r1) Each processor of the array should consist of two physically separated units: a processing unit and a reconfiguration unit. The reconfiguration unit should be small in comparison with the processing unit and it should be laid out to be as safe as possible.

(r2) For each processor the processing unit can be bypassed in both horizontal and vertical directions. It should also be possible to choose the diagonal neighbors in the north east or south east and in the north west or south west, respectively, to be logical neighbors of a processor in the east and the west.

(r3) The redundant interconnection structure should not lead to a significant increase in the number of pads on an ISA-chip, even if a large array has to be build from several ISA-chips.

(r4) The interprocessor communication should be fast, such that in spite of long physical distances between logical neighbors the clock frequency is not decreased.

These requirements led to the design of a redundant interconnection structure which is shown in a simplified way in Figure 4.1. In order to make the interconnection structure more visible the reconfiguration unit is not shown as a single block but distributed around the processing unit. Furthermore, the multiplexers responsible for the instruction flow are not shown, since they are analogous to the north-to-south dataflow.

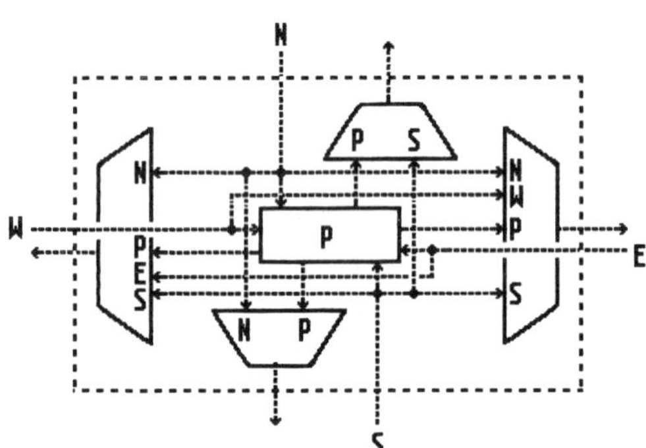

Figure 4.1: Simplified interconnection structure of an inidividual processor.

A specific property of the instruction systolic array has been used in the design of this structure: the output signals generated by a fault free processing unit are identical for the four different directions, since there is only one communication register available to all of the neighbors. As a consequence, a logical neighbor P_3 of a processor P_1 can read the output of P_3 from an arbitrary direction. This is illustrated in Figure 4.2.

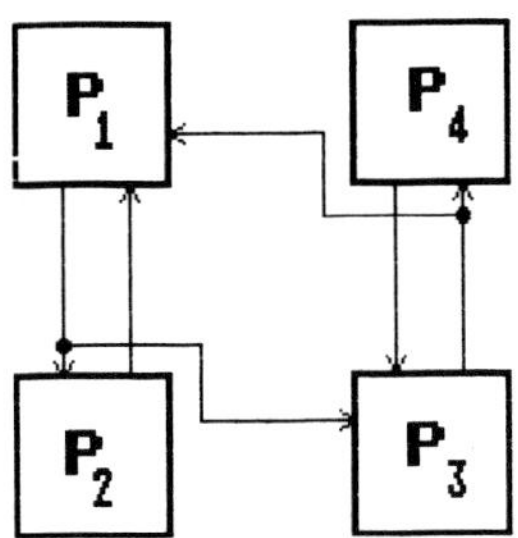

Figure 4.2: Communication between P_1 and P_3

As is indicated in Figure 4.1, the redundancy is hidden within the processor, such that to the outside world such a processor does not look different from a "standard" processor in a mesh-connected array. Therefore, it is possible to build an array consisting of several ISA chips, without suffering from a large number of additional pads for the redundant interconnections.

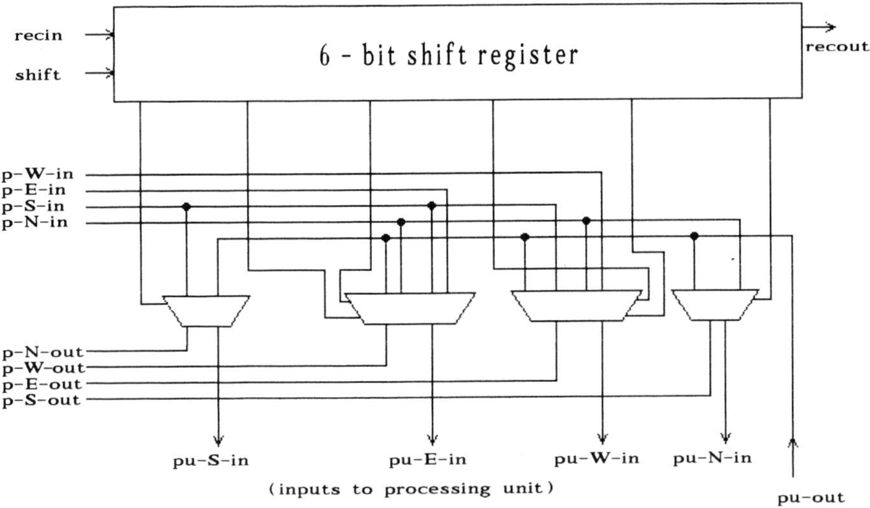

Figure 4.3: The reconfiguration unit of a processor

Nevertheless, in the implementation of the reconfiguration unit of a processor (see Figure 4.3) we need two more lines to feed in and out the signals controlling the interconnection multiplexers and another line carrying a clock signal controlling the "multiplexer control" shift register which turns into a static register after the reconfiguration phase. The 6 bits of this register control the multiplexers in the following way:

The first bit controls a 2:2-multiplexer for the N input of the processing unit of the processor and for the S output of the processor. If this bit is 0 then the N input of the processor is connected to the N input of the processing unit. In addition, the output of the processing unit is switched to the S output of the processor. If the reconfiguration bit is 1, then the N input of the processor is shortcut with its S output, which means that the processing unit is bypassed.

Bits 2 and 3 control the W input of the processing unit and the E output of the processor. If both bits are 1, then the processing unit is bypassed and the E output is switched to the W input. If at least one of the bits is 0, then the W input of the processor is connected to the W input of its processing unit. For 00 in bits 2 and 3 the E output of the processor gets its value from the output of the processing unit. For 10 the signal of the N input is propagated to the E output and for 01 the E output gets its value from the S input.

The multiplexer for the W output of the processor and the E input of the processing unit is analogously controlled by bits 4 and 5 and the one for the N output of the processor and the S input of the processing unit is controlled by bit 6.

Since the instructions of the ISA move systolically from north to south and since this happens synchronously with the data movement in the same direction, the implemented design contains an additional multiplexer for the vertical control flow. In the same way as the north-to-south data movement this is controlled by bit 1. This additional multiplexer is not depicted in Figure 4.3 for simplicity. We do not need any additional multiplexers for the horizontal control flow, since in our design the stream of the binary selectors uses the horizontal data path.

Table 4.1
Possible states of the reconfiguration unit

value of reconfiguration bits in shift register	connections established by reconfiguration multiplexers (p for processor, pu for processing unit, N for north, S for south, E for east, W for west)	
bit 1 = 0	p-N-in = pu-N-in	pu-out = p-S-out
bit 1 = 1	p-N-in = p-S-out	
bit 2 = 0; bit 3 = 0	p-W-in = pu-W-in	pu-out = p-E-out
bit 2 = 0; bit 3 = 1	p-W-in = pu-W-in	p-S-in = p-E-out
bit 2 = 1; bit 3 = 0	p-W-in = pu-W-in	p-N-in = p-E-out
bit 2 = 1; bit 3 = 1	p-W-in = p-E-out	
bit 4 = 0; bit 5 = 0	p-E-in = pu-E-in	pu-out = p-W-out
bit 4 = 0; bit 5 = 1	p-E-in = pu-E-in	p-N-in = p-W-out
bit 4 = 1; bit 5 = 0	p-E-in = pu-E-in	p-S-in = p-W-out
bit 4 = 1; bit 5 = 1	p-E-in = p-W-out	
bit 6 = 0	p-S-in = pu-S-in	pu-out = p-N-out
bit 6 = 1	p-S-in = p-N-out	

Although the reconfiguration information for one processor could be stored in less than 6 bits, we decided to use 6 bits in order to have no decoding effort and thus no additional fault sources.

The effect of the 6 control bits on the multiplexers is summarized in Table 4.1.

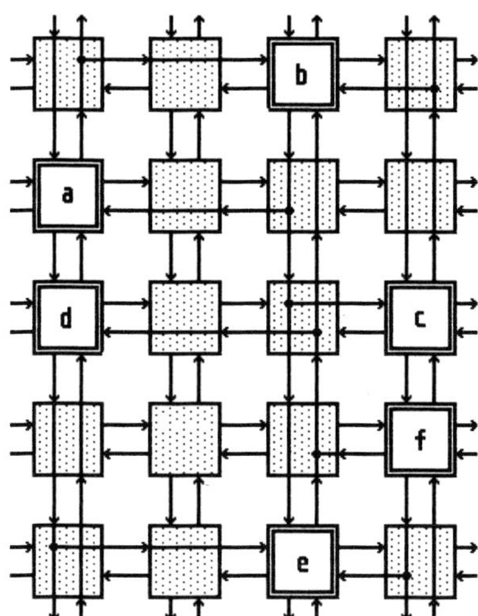

Figure 4.4: Interconnection paths between logically direct neighbors.

A possible reconfiguration based on this redundant interconnection structure is outlined in Figure 4.4 showing a 5×4 array configured into two logical rows. One should observe that whenever a processor is bypassed horizontally we have to bypass all the processors in this column. This is caused by the fact that on the vertical control path there is only a bypass and no pseudo-diagonal interconnection as on the vertical data path. Therefore, the logical neighbors of a processor in the north or the south are always in the same column, whereas any processor in any column to the right (left) of a processor may be its direct logical neighbor in the east (west). This allows for a flexible reconfiguration of the array in the presence of faulty processors.

The degree of fault tolerance achieved by the redundant interconnections depends on the maximal distance allowed between logically direct neighbors. Very often it will not be possible to utilize all the fault-free processors in a reconfigured array. For example, none of the processors shown shaded in Figure 4.4 may be chosen to represent any other active node of the reconfigured array. Strategies for the systematic reconfiguration of an instruction systolic array in the presence of faulty processors are described in [PSS91].

5. CONCLUSION

In this paper we have described the design of a prototype 3×3 instruction systolic array for fabrication in a 1.6 μ CMOS technology. Using a true single phase clock it has been laid out for a maximal clock rate of 100 MHz under strong limits on area and energy consumption. Because of its bit-serial mode of operation the processors could be kept small and simple such that it should be possible to integrate an array of up to 64 processors on a chip of 1 cm^2. Very powerful and flexible fault tolerance capabilities are achieved by adding a physically separated reconfiguration unit to every processor. Although each processor still has no more than four in- and outgoing wires for dataflow, the reconfiguration unit allows to bypass faulty processors in horizontal and vertical directions and to choose processors of different rows as neighbors in horizontal directions. SPICE simulations indicate that even if up to three processors are bypassed the anticipated clock frequency can be sustained.

6. REFERENCES

[DiS88] Dittrich, A., Schmeck, H.: Given's Rotations on an Instruction Systolic Array. In:G. Wolf, T. Legendi, U. Schendel(eds.): Parcella ' 88. Mathematical Research, Bd. 48, Akademie-Verlag, Berlin, (1988), 340-346.

[KLSSS88] Kunde, M., H.-W. Lang, M. Schimmler, H. Schmeck, and H. Schröder: The Instruction Systolic Array and Its Relation to Other Models of Parallel Computers. Parallel Computing (1988), 25-39.

[La86] Lang, H.-W.: The Instruction Systolic Array, a Parallel Architecture for VLSI. Integration, the VLSI Journal 4 (1986), 65-74.

[La90] Lang, H.-W.: Das befehlssystolische Prozessorfeld - Architektur und Programmierung, Dissertation, Universität Kiel, 1990.

[PSS91] Phieler, M., Schimmler, M., Schmeck, H.: A Reconfigurable Instruction Systolic Array. Proc. 5th Int. Conf. Fault-Tolerant Computing Systems, Nürnberg, FRG, Sept. 1991.

[Schi87] Schimmler, M.: Fast Sorting on the Instruction Systolic Array. Bericht 8709, Informatik und Praktische Mathematik, Universität Kiel, 1987.

[Schm86] Schmeck, H.: A Comparison-Based Instruction Systolic Array. In: M. Cosnard, Y. Robert, P. Quinton, M. Tchuente (eds.): Parallel Algorithms and Architectures, North-Holland, Amsterdam (1986), 281-292.

[ST90] Schmeck, H., Thürnau, R.: Skeleton Computations on an Instruction Systolic Array. In: PARCELLA'90, Research in Informatics, Vol. 2, Akademie-Verlag, Berlin (1990), 259-266.

[Stra90] Strazdins, P.E.: Control Structures for Mesh-Connected Networks, Ph.D. Thesis, The Australian National University, Canberra, Australia, 1990.

[YS89] Yuan, J., Svensson, C.: High Speed CMOS Circuit Technique, IEEE Journal of Solid-State Circuits, Vol. 24, (1989), 62-70.

VLSI 91
A. Halaas and P.B. Denyer (Eds.)
Elsevier Science Publishers B.V. (North-Holland)

A reconfigurable fault tolerant module approach to the reliability enhancement for mesh connected processor arrays

Guoning Liao

Department of Informatics, University of Oslo, P.O. Box 1080 Blindern, 0316 Oslo 3, Norway

Abstract

In this paper, we present a new and efficient fault tolerant technique suitable for reliability enhancement for high performance arrays of processing elements(PEs). In the proposed scheme, two spare PEs are located within four PEs, which forms a fault tolerant module. Any spare can functionally replace any one of the primary PEs that is within the fault tolerant module. Because spares are physically close to the PEs that they replace, reconfiguration interconnections are short, thus minimizing the performance degradation. A novel feature of this structure is that the fault tolerant module can detect faulty PEs, locate the faulty PEs, and isolate the faulty PEs by switching the interconnections to the good ones. After presenting the fault tolerant module scheme, we show the analysis results of the reliability enhancement. We also give the design for implementing the switching network needed for the reconfigurations.

1 Introduction

Digital systems that are operated in applications where there is a high cost of failure require high reliability and continuous operation. Since it is impossible to guarantee that parts of a system will never fail, such systems need to be designed to tolerate failures of the system components. The discipline of fault tolerant computing is therefore, one which has attracted a great deal of research interest. Researchers have attempted to derive highly effective and , at the same time, efficient techniques to tolerate failures in complex digital systems.

The high computation needs of many applications can now be met through the use of highly parallel special-purpose systems that can be produced very cost effective through the use of the VLSI technology. Processor arrays, such as ESL systolic arrays[1], the Carnegie Mellon Warp processor[2], and wavefront processor arrays[3] are examples of such systems.

Since regular structures seem well suited for the efficient use of redundancy, the fault tolerant design of regular arrays of processing elements has been an active reasearch area. These include linear arrays[4], rectangular arrays[5], and binary tree architectures[6]. In general, each PE here may consist of a processor and possibly some local memory, with the specific design of the PEs and interconnection topology dependent on the algorithm, or the set of algorithms, to be executed.

There are several methods addressing the fault tolerance problems by employing local redundancy. The interstitial-redundancy scheme[7] is a good illustration of a local redundancy technique. The objective of the reconfiguration strategies is to achieve fair area efficiency by using as many good cells as possible, while also ensuring shorter intercell communication links. Hassan and Agarwal[6] proposed a fault-tolerant architecture for binary trees. Here, a spare element is assigned to every three elements of a sub-binary tree. So, when an element fails, the reconfiguration is taken place locally inside the sub-binary tree. But these techniques have neglected an important problem that has to be considered for realizing fault tolerance — the error detection and fault location. The test of the processing elements often poses a big problem, and introduces quite an amount of hardware overhead. For example, the overhead required by self-checking systems using hardware redundancy and duplication is over 100 percent for error detection, and further hardware and time overhead is necessary for error corrections[8]. In order to reduce the hardware overhead, time redundancy[9] and algorithm-based error detection and fault location[10] may be used. But these test methods are not suitable for fault tolerant schemes with local redundancy, and an error is not found until a partial computation result is obtained at the lower edge of a processor array.

The proposed fault tolerant module approach addresses both local redundancy and fault detection problems. Our design owes its effectiveness to a flexible use of redundancy through new spares and interconnection structures. These greatly reduce the likelyhood that an available spare PE is unable to assume the functions of a failed PE because of interconnection limitations. The test of processing elements is achieved by switching the spare processing elements to duplicate one part of the PEs at one computation cycle, and duplicate the rest part of the PEs at another computation cycle. The rest of the paper is organized as follows: In the next section, we present the fault tolerant module and its application to mesh connected VLSI processor arrays. In section 3, we develop a relaibility analysis model for the mesh connected processor arrays with the fault tolerant modules. The results of the reliability enhancement estimation is obtained. Section 4 dicusses practical implementation issues. We conclude in Section 5.

2 Fault Tolerant Module in Two Dimensional Arrays

A system is said to be failed when it no longer provides the service for which it was designed. The manifestation of a failure will be the errors produced by the system. The cause of an error is denoted as a fault within the system. Faults may be permanent or transient; a permanent fault, of course, will not necessarily produce errors for all system inputs or states. Fault tolerant computing thus deals with techniques designed to prevent erorrs at the output of a system.

A fault can be treated at any level within the system– from a very low level, such as a transistor level, to a higher, processing element level. Such an element-level fault model is ideal for VLSI, where a physical failure can cause some parts of a system to be faulty. In the fault tolerant module approach, the element-level fault model is employed. An element fault is assumed to result in arbitrary erorrs at the output of the element, and it is these errors that must be prevented from appearing at the output of the system computation.

Definition 2.1: A fault tolerant module (FTM) is a regular computational array structure that has two additional spare PEs located within a module of four primary PEs.

Figure 1 depicts a FTM. The spare PEs are connected so as to be able to replace any one of the PEs that has failed. Inside the FTM, any PE may behave like a primary PE, and any PE may be used as a spare PE. So, a combination of four good PEs may form a good computation module.

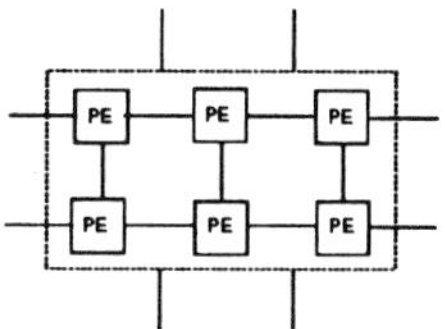

Figure 1: FTM.

Definition 2.2: The redundancy ratio for a FMT is the ratio of the number of spare PEs in the FTM to the number of PEs in the nonredundant processor array.

Observe that in Figure 1, the FTM has 4 primary processing elements and 2 spare elements. If a processor array is composed of such FTM's, then the redundancy ratio will be 0.5, or 50 percent.

Figure 2 shows a 4×4 processor array of FTMs.

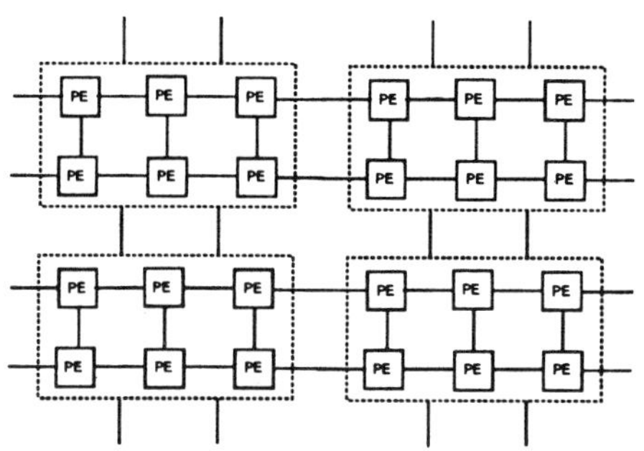

Figure 2: A 4×4 processor array composed of FTMs

2.1 Error Detection and Location

The error detection is performed by comparing the computation results of one column primary PEs (2 PEs) with another column of PEs which is acting as spare PEs.

Figure 3 illustrates the error detection mechanism. When there is no failed PE, the PEs (1.2, 2.2) in the middle column of the FTM undertake the same computation as the PEs (1.1, 2.1) in the left column of the FTM do in one computation cycle, e.g.,

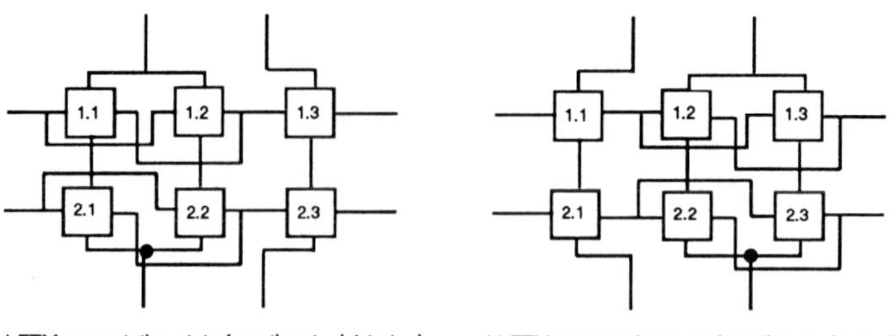

Figure 3: Error detection schemes for FTM.

from time $t = i - 1$ to $t = i$. In the subsequent computation cycle, from time $t = i$ to $t = i + 1$, the PEs in the middle column of the FTM will undertake the computation carried out in the right column of PEs (1.3, 2.3). The small black circles indicate the place where the computation results are compared and output.

Theorem 2.1: In a FTM, a single failed PE can be detected in one computation cycle.

Proof: Suppose PE (1.1) is failed.Then the computation error will propagate to the PE (2.1) and the checking point at the (2.1, 2.2) joint will show that there is a fault inside the PEs (1.1, 1.2, 2.1, 2.2) when the FTM computation state is at state "a" shown in Figure 3a. In another case, if the failed PE is PE (2.3), then the error is detected when the computation state is switched to state "b" shown in Figure 3b.

Theorem 2.2: In a FTM, two computation cycles are needed to locate a single failed PE in a column.

Proof: Let's suppose that PE (1.1) is failed. According to the Theorem 2.1, the error is detected in a computation cycle when the FTM is at state "a". Because the error message only tells that there is a faulty PE inside (1.1,1.2, 2.1, 2.2) PEs group, any one of them may be the faulty PE. So, a second compuatation cycle at the state "b" is performed. Because there is no faulty PE within (1.2, 1.3, 2.2, 2.3) PEs group under our assumption, no error message will be given at the checking point of (2.2, 2.3) joint. By using the two error checking results, we can locate the failed PE in the left column.

The assumption of only one faulty PE exists during one computation cycle is reasonable for processor arrays, because the possibility of more than one PE failing at the same time is low enough to allow us to deal with one faulty PE at a time.

Algorithm 2.1: A failed PE can be detected and located by the following steps based on Theorem 2.1 and Theorem 2.2 under the assumption of one faulty PE inside the FTM: First if there is one failed PE inside the FTM, an error message is obtained from one of the computation states of the FTM. Then another computation state is used as a reference to locate the faulty PE.

1. If the checking point at the (2.1, 2.2) joint has an error message at computation state "a", and an error message is also given at the checking point of (2.2, 2.3) joint in a subsequent computation cycle at state "b", then the faulty PE is located in the middle column.

2. The checking point at the (2.1, 2.2) joint has an error message at computation state "a", and there is no error message given at the checking point of joint (2.2, 2.3) in a subsequent computation cycle at state "b", then the failed PE is located in the left column.

3. The checking point at the (2.2, 2.3) joint has an error message at computation state "b", and there is no error message at the (2.1, 2.2) joint checking point in a subsequent computation cycle at state "a", then the failed PE is located in the right column.

2.2 Fault Isolation

After a faulty PE has been detected and located, some kind of reconfiguration of the FTM will take place in order to bypass the faulty PE. There are three cases for reconfigurations. Figure 4 illustrates the different fault isolation shemes.

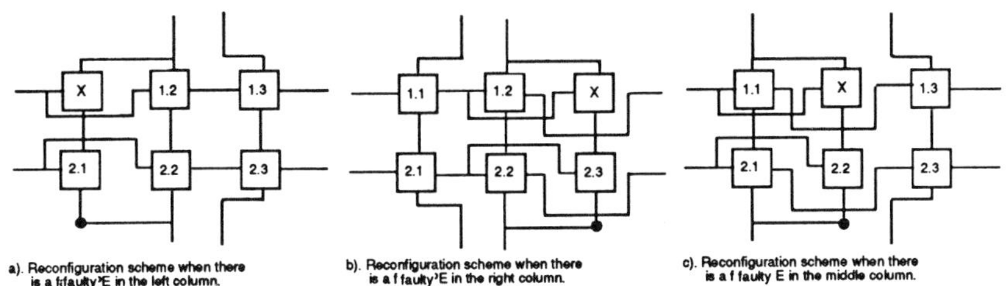

Figure 4: Reconfiguration schemes for FTM.

We note that the faulty column is also fed with data after reconfiguration instead of just bypassed. The computation result of the faulty column is compared with the computation result of the adjacent good column which has the same data input. If the two results are the same for a period of time which is long enough to make us sure that the faulty column can be used as backup again, the FTM will come back to the normal computation condition. This is especially advantageous if transient faults are of primary concern, for example, if the fault tolerant array processors are to be used in space applications. The space radiation effects on the semiconductor devices often cause transient faults rather than permanent faults.

3 Reliability Estimation

The main objective of designing a fault tolerant system is to improve the reliability of the system. In this section, we present an analytic model that enables us to analyse the effectiveness of the fault tolerant module approach in improving the performance of the processor arrays in terms of reliability. The reliability of the FTM is estimated and compared to that of a nonredundant module to examine the enhancement.

Throughout the rest of this section, it is assumed that the failure rate of links is negligible compared to nodes, thus only node failures are considered. Admittedly, this is an optimistic assumption, but it is supported by the fact that some of the link failures

Table I
Reliabilty of a 4x4 processor array system with λ=0.1

t	0.0	0.05	0.1	0.20	0.30	0.40	0.5	0.6	0.70	0.80	0.90
R	1.0000	0.8337	0.6934	0.4715	0.3148	0.2054	0.1303	0.0799	0.0469	0.0261	0.0134
R_{ftm}	1.0000	0.9231	0.8521	0.7261	0.6187	0.5272	0.4493	0.3828	0.3262	0.2780	0.2369

can be treated as failures of nodes connected by the faulty links. A relatively simple reliability analysis of the FTM is possible under this assumption.

Suppose a system consists of N identical primary PEs, and S spare PEs, and the spare PEs can be applied to replace any faulty primary PEs until there is no spare one left. Let $R = e^{-\lambda t}$ represent the reliability of one PE, and R_{nr} represent the reliability of the nonredundant system and R_r represent the reliability of the redundant system with S spare PEs . Accordding to Liao and Tong[11], the reliability of the nonredundant system and the redundant system may be represented by Equation 1 and Equation 2.

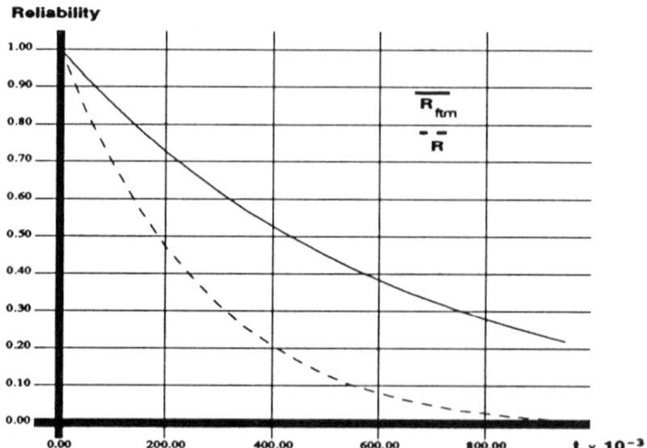

Figure 5: The relaibilty comparation of R and R_{ftm}.

$$R_{nr} = R^N \tag{1}$$

$$R_r = \sum_{j=0}^{S} \binom{N+S}{j} R^{(N+S-j)} (1-R)^j \tag{2}$$

Let R_{ftm} represent the reliability of the FTM. In our case, $S = 2$, $N = 4$, the reliability of the FTM is:

$$R_{ftm} = \sum_{j=0}^{2} \binom{6}{j} R^{(6-j)} (1-R)^j \tag{3}$$

As an example, we consider a 4×4 processor array system for reliability analysis. The reliability of a non-redundant 4×4 processor array system and the redundant 4×4 processor array system with 4 FTM's are shown in Table 1 and plotted in Figure 5.

4 Implementation Issues

The reconfiguration schemes for bypassing the failed PE has been shown in Figure 4. Those drawings only indicate the logical connections. They do not reflect the practical design of the FTM.

A functional diagram proposed for the FTM design is shown in Figure 6, where the following basic units are identified.

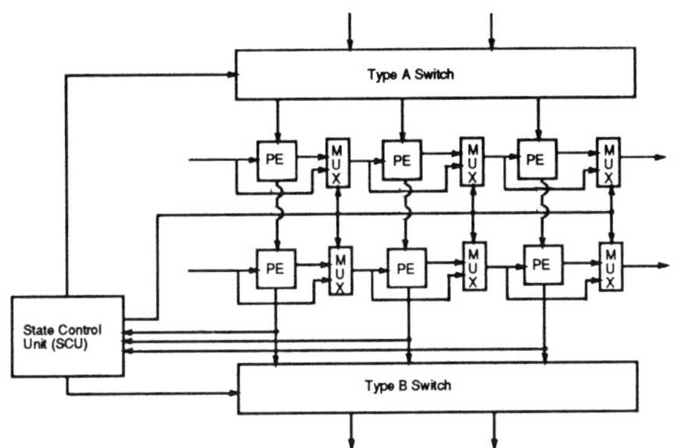

Figure 6: FTM functional diagram.

1. **Type A switch**: This logic unit establishes proper links from two vertical inputs to the three vertical inputs of the left, middle and right column of the FTM. Type A switch assumes two legal states, which is shown in Figure 7a. In_L and In_R are the two inputs and C_A is the switch control signal.

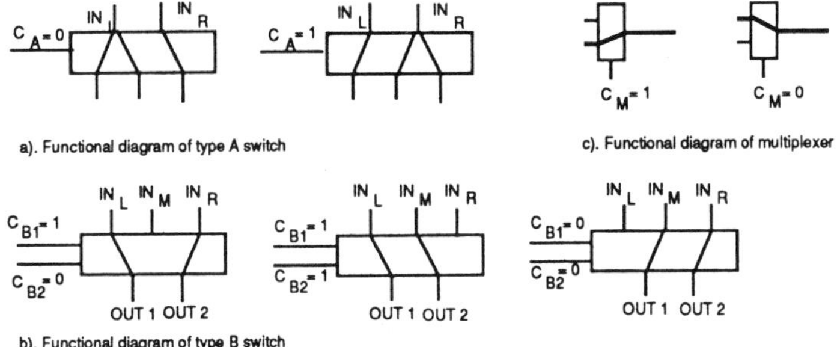

Figure 7: Functional diagram for type A switch, type B switch and multiplexers.

2. **Type B switch**: This logic unit establishes proper links from the vertical outputs of three columns to the FTM's two vertical outputs at the lower edge. Type B switch assumes three legal states which is shown in Figure 7b.

3. **Multiplexer:** This logic unit selects one proper horizontal input from two inputs. There are 6 multiplexers inside a FTM, helping the FTM to make proper horizontal

interconnections inside the FTM. The logic function of the multiplexer is shown in Figure 7c.

4. **State Control Unit (SCU)**: This unit is responsible for detecting faulty PEs inside the FTM, generating control codes for switches and multiplexers. It takes the vertical outputs of three columns as its inputs. Because a faulty PE is detected and located by comparing the computation results of two specified columns, any difference in the two outputs will alter the internal registers representing the FTM states. And the comparision result of the subsequent computation cycle will be used by the SCU to generate the control code which is used to switch off the faulty PEs and establish proper links between the spare PEs and the primary PEs.

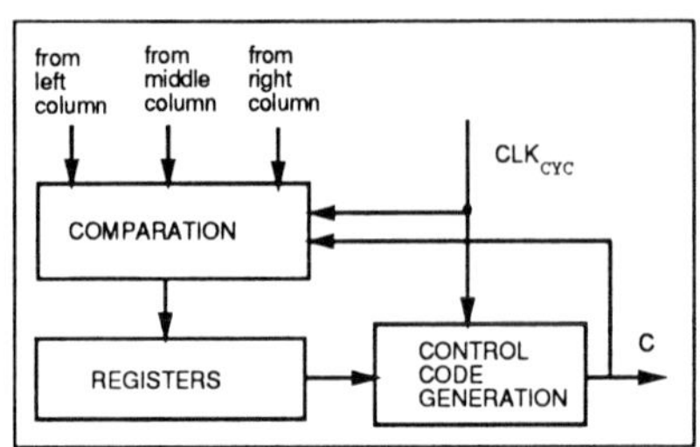

Figure 8: The block diagram of SCU.

Before describing the SCU, we define the notation used in the block diagram of the SCU shown in Figure 8.

CLK_{CYC}: the computation cycle clock. For one computation cycle, the CLK_{CYC} assumes logic one, and for the subsequent computation cycle, the CLK_{CYC} assumes logic zero.

Control code set C: the control code set is defined as the set of control signals to the type A switch, type B switch and the multiplexers. Because two multiplexers in the same column have the same switching patterns, only three control signals, (C_{M1}, C_{M2}, C_{M3}), are needed to control the 6 multiplexers. Here,

$$C = (C_A, C_{B1}, C_{B2}, C_{M1}, C_{M2}, C_{M3}) \tag{4}$$

The COMPARISION circuit:

The comparision circuit takes the outputs of three columns as its inputs. The comparision circuit must know beforehand which of the two columns have the same input data in the upper edge of the FTM. This information is provided by the computation cycle clock, CLK_{CYC} and control signal to the type A switch, C_A, which is part of the control code set signals. The output results of the two columns which have the same input data are compared. If there is a disagreement between the results, the comparision circuit will generate a signal bearing this information to be used by the registers which are described in the following.

The REGISTERS circuit:

The registers are responsible for storing the fault detection information. Any disagreement in the comparision circuit will cause a change in the register. There are four states for the register:

1. No failed PEs;
2. There is a failed PE in the left column;
3. There is a failed PE in the middle column;
4. There is a failed PE in the right column.

The change from one state to another requires two computation cycles.

The CONTROL CODE GENERATION circuit:

According to the state of the registers, appropriate control codes for the type A switch, type B switch, and the multiplexers are generated. There are five cases for the control codes set.

Case 1. No faulty PE, $CLK_{CYC} = 0$, $C = (0,1,0,1,0,0)$, see Figure 9a;
Case 2. No faulty PE, $CLK_{CYC} = 1$, $C = (1,1,0,1,1,0)$, see Figure 9b;
Case 3. There is a faulty PE in the left column, $C = (0,0,0,1,0,0)$, see Figure 9c;
Case 4. There is a faulty PE in the middle column, $C = (0,1,0,1,1,0)$, see Figure 9d;
Case 5. There is a faulty PE in the right column, $C = (1,1,1,0,0,1)$, see Figure 9e.

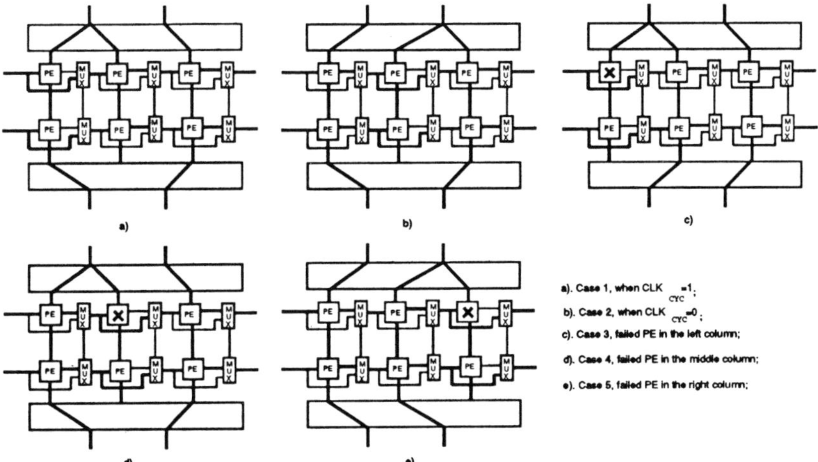

Figure 9: The five cases for control code set. The thick lines indicate the interconnections.

5 Conclusions

The advantages of this fault tolerant module approach are short PE interconnections, relatively high reliability gain and low redundancy ratio. The area overhead associated with the switches and multiplexers is low for a large PE size.

The FTM is designed to tolerate the faults that happen during the processor array's service time. The FTM is best for dealing with the transient faults, because the FTM will return to normal working conditions after checking the previously faulty column to be functional for a sufficient period of time .

The FTM may be treated as a basic fault tolerant component in a large array, and an even better performance in terms of reliability can be achieved by placing redundant FTMs at a higher level thus hierachical redundancy is introduced[12].

The FTM concept can also find use in a multiprocessor array, which has larger individual processing elements.

Acknowlegement

This paper is greatly improved with the help of Professor Oddvar Søråsen. The work is supported in part by The Royal Norwegian Council for Scientific and Industrial Research (NTNF).

References

1 A.V. Kullkarni and D.W.-L. Yen, Systolic Processing and an Implementation for Signal and Image Processing, IEEE Trans. Computers, Oct. 1982, pp.1000-1009.

2 H.T. Kung, Systolic Algorithm for the CMU Warp Processor, Proc. 7th Int'l Conf. Pattern Recognition, July 1984, pp.570-577.

3 S.Y. Kung, S.C. Lo, S.N. Jean, and J.N. Hwang, Wavefront Array Processors–Concept to Inplementation, Computer, July 1987, pp.18-33.

4 D. Fussel and P. Varman, Fault-tolerant wafer-scale Architecture for VLSI, in Proc. 9th Annu. Symp. Comput. Architecture, May 1982.

5 J.W. Greece and A. El Gamal,Configuration of VLSI Arrays in the Presence of defects, J. ACM, vol.31, Oct.,1984,pp.694-717.

6 A.S.M. Hassan and V.K. Agarwal,A Fault Tolerant Modular Architecture for Binary Trees, IEEE Trans. Comput., vol. C-35, April 1986,pp.356-361.

7 Adit D. Singh, Interstitial Redundancy: An Area Effcient Fault Tolerant Scheme for Large Area VLSI Arrays, IEEE Trans, Computers, Nov., 1988, pp1398-1410.

8 J.A. Abraham, et. al., Fault Tolerance Techniques for Systolic Arrays, Computer, July 1987, pp65-74.

9 R.K. Gulati and S.M. reddy, Concurrent Error Detection in VLSI Array Structures, Proc. IEEE Int'l Conf. Computer Design: VLSI in Computers, Oct. 1986, pp488-491.

10 K.-H. Huang and J.A. Abraham, Algorithm-Based Fault Tolerance for Matrix Operation, IEEE Trans. Computers, June 1984, pp518-528.

11 Guoning Liao and Qinyi Tong, On Evaluation of Fault Tolerant Techniques for VLSI Chips, IEEE Asian Electronics Conference, Hong Kong, 1987.

12 Mengly Chean and Jose A.B. Fortes, A Taxonomy of Reconfiguration Techniques for Fault-Tolerant Processor Arrays, Computer, Jan. 1990,pp55-69.

VLSI 91
A. Halaas and P.B. Denyer (Eds.)
Elsevier Science Publishers B.V. (North-Holland)

THE WASP 2 WAFER SCALE INTEGRATION DEMONSTRATOR

Ian.P.Jalowiecki[a], Stephen.J.Hedge[b]

[a]Brunel University, Uxbridge, Middlesex, UB8 3PH, UK

[b]Aspex Microsystems Ltd.Brunel University, Uxbridge, Middlesex, UB8 3PH, UK

Abstract

An Associative String Processor (ASP) is a computer which locates and processes data by content addressing rather than by conventional addressing methods. By its nature, it therefore belongs to the class of SIMD (Single Instruction Multiple Data) processors, since it is able to operate on subsets of data items simultaneously. The ASP is an example of a massively parallel, fine-grain architecture, which has already been shown to be suitable for VLSI [1]. This paper describes the extension of the concept to experimental WSI versions [2].

1 ASP MODULE ARCHITECTURE

ASP (Associative String Processor) modules [3] are building-blocks for the construction of low-MIMD (Multiple Instruction Multiple Data) high-SIMD massively parallel computers.

ASP modules comprise three different component types; these being the ASP blocks (SIMD computing structure), supported by an ASP Data Interface (DI) and an ASP Control Interface (CI), as shown in Figure 1. As shown, the Data Interface (DI) will route data to the ASP substrings via a local ASP Data Buffer (ADB). Similarly the Control Interface will route the instruction stream to the ASP substring.

Typically an ASP substring will have its own autonomous local control unit (or 'ASP Control Unit' (ACU)), operating under the coordination of a single high-level controller. Alternatively, the substring may have no local controller, and instead, use the Control Interface (CI) as a port to a global controller which will provide all the system control.

1.1 ASP substrings

An ASP substring is a programmable, fine-grain SIMD processor, incorporating a homogeneous fault-tolerant string of identical APEs (Associative Processing Elements), a simple interprocessor communications network and a Vector Data

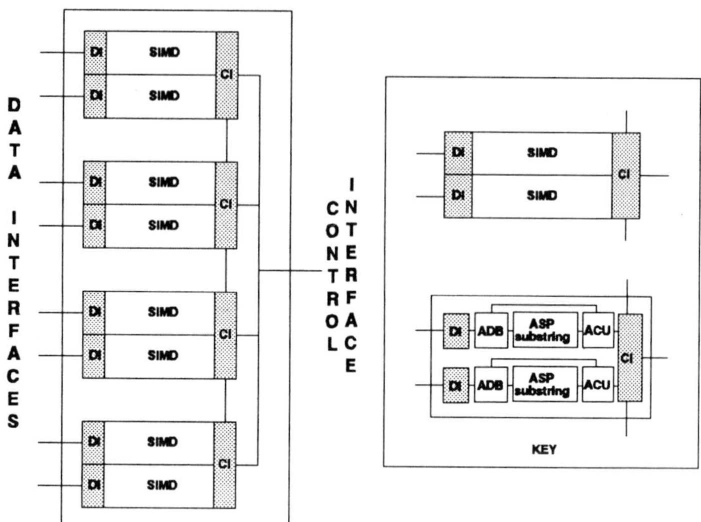

Figure 1 ASP module schematic

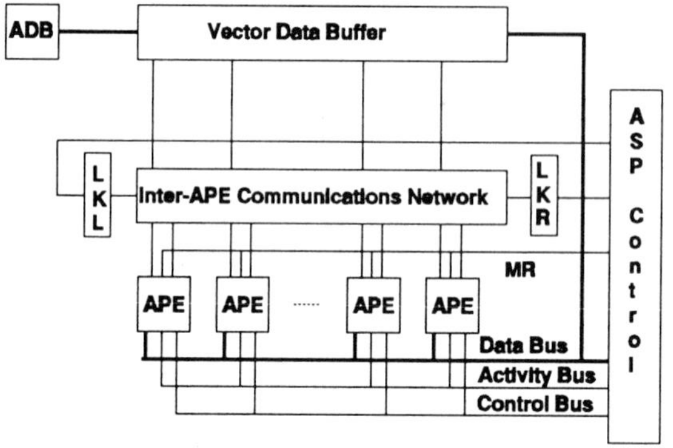

Figure 2 ASP substring schematic

Buffer for overlapped data input-output, as indicated in Figure 2.

Each APE incorporates a 64-bit Data Register and a 6-bit Activity Register, a 70-bit parallel comparator, a single-bit full-adder, 4 single-bit registers (representing carry, Matching and Destination APES and Active APES and logic for local processing and communication.

In operation, data items are distributed over the APE Data Registers. Each ASP substring performs a form of set processing, in which a sub-set of active APEs, selected by a content addressable search operation, may be programmed to perform bit-serial scalar-vector (i.e. between a scalar and items in the Data Registers) and vector-vector (i.e. within the Data Registers) operations.

Active APEs are either directly activated (i.e. selected from those APEs that match data and activity values broadcast from the controller on the relevant buses) or are indirectly activated as a result of an inter-APE communication initiated by some other matching APE(s).

The Match Reply (MR) line provides the control interface with an indication as to whether any APEs have matched.

The inter-APE communications network implements a simply scalable, APE interconnection strategy. The network supports simple extension via the LKL and LKR (Link Left and Right) single-bit ports, shown in Figure 2. The network supports simple circuit-switched asynchronous communication, as well as more conventional synchronous, bi-directional shifting.

At an abstract level, a circularly-linked ASP substring can be considered as a hierarchical chordal-ring structure, with the chords bypassing APE segments (defined as the subset of APEs between two switch blocks), groups of APE segments, and ultimately, ASP blocks. The bypass structures serve to accelerate inter-APE communications signals and provide APE block defect/fault-tolerance. This form of bypassing is simple to implement with minimal logic overhead. Indeed, it is widely recognised that a string topology is highly amenable to defect-tolerant bypassing of this kind [4].

APE blocks failing test in manufacture or service can be simply switched out of the string and bypassed, the content-addressable nature of the string easing the reconfiguration task. It may be possible to tolerate a failure in the bypassing logic, which generally comprises simple transmission gate link bypass switches, provided that a 'higher-level' of hierarchical bypass exists to mask the fault. Should a 'non-maskable' failure occur in these or in any other 'critical' (i.e. non-fault-tolerant) logic block, then the entire ASP block is bypassed.

2 WASP PROTOTYPES

WASP (WSI Associative String Processor) represents a challenging and innovative method of implementing ASP modules, by implementing an ASP substring on a single undiced silicon wafer.

The benefits in moving towards a monolithic WSI systems solution stem from the density advantages over conventional assembly methods, combined with power and speed benefits [5], and, most significantly, very high reliabilities through the elimination of the majority of connections in the system [6].

The proposed architecture for a WSI ASP is indicated in Figure 3. It shows a generic WASP device, composed from 3 different VLSI sized blocks known as Data Routers (DRs), ASP blocks and Control Routers (CRs). These are exposed onto the wafer, such that aligned metal signal traces (e.g. busses) overlap and form electrical connections.

In this manner, the DR, ASP and CR blocks are routed together such that the DR connects rows of ASP blocks to a common Data Interface (DI) and the CR connects the ASP blocks to a common Control Interface (CI). Furthermore, the DR and CR blocks support extension of the communication links from row-to-row.

Each of the DR, ASP and CR blocks integrates a defect/fault tolerant design; such that APE (Associative Processing Element) segments within the ASP blocks, individual ASP blocks and entire ASP substring rows (and, hence, entire WASP

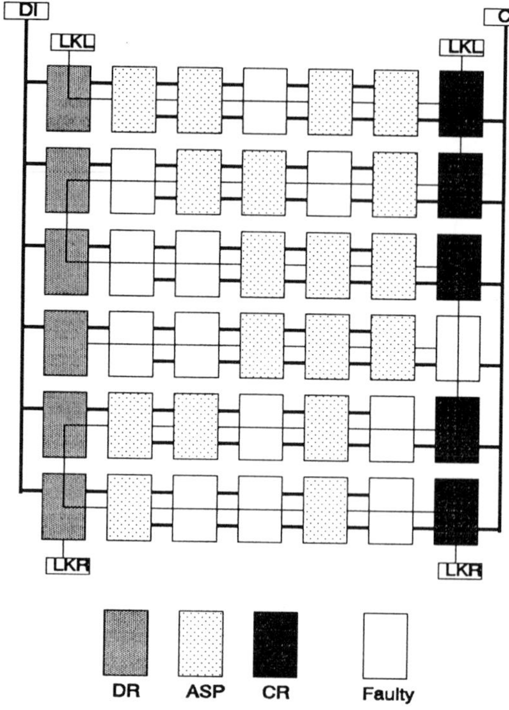

Figure 3 Generic WASP device floorplan

devices) can be bypassed. This is illustrated at the block level in Figure 3.

2.1 WASP technology demonstration

From January 1985 to December 1989 the WASP project was funded as part of a UK Alvey (VLSI) contract, also involving Plessey, GEC, ICL and Middlesex Polytechnic. This included the design (at Brunel University), fabrication (by Plessey) and evaluation (at Brunel University) of defect/fault-tolerant test chips and the WASP 1, WASP 2A and WASP 2B (monolithic) WSI ASP technology demonstrators. The WASP 2A demonstrator was delivered in 2Q90 and tested in 3Q90. The WASP2B demonstrator was delivered in 3Q90.

2.1.1 WASP 1

Designed to 2µm layout rules, the 144-APE WASP 1 devices demonstrated ASP substring rows, each comprising 4 ASP blocks and a simplified Control Router block (see Figure 4). Each ASP substring block integrated 36 APEs, configured in 9 4-APE segments.

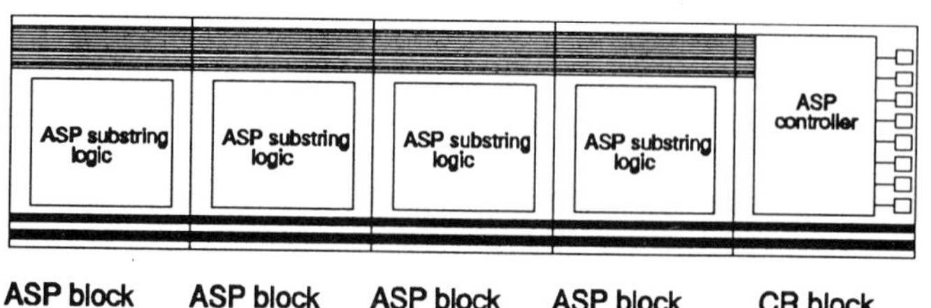

Figure 4 WASP 1 schematic floorplan & layout

Manufacturing methods were based on standard fabrication technology, involving the use of standard steppers and VLSI die masks. Indeed, the WASP 1 &

Table I WASP 1 test results summary

inter-ASP defect-tolerance	block isolation control logic	yes
	block bus isolators	yes
	block signal bypassing	yes
intra-ASP defect-tolerance	APE segment bypassing within the ASP block	yes
Electrical	power, clock & signal distribution	yes
	power isolation	yes
functional	ASP scan paths	yes
	ASP control logic	yes
	ASP processor instructions	yes

2 demonstrators employ only one stepper reticle (i.e. maximum 13mm x 13mm) which is subdivided to achieve cost-effectiveness by manufacturing all blocks from the single mask. Wafer fabrication was based upon the selective exposure of shuttered portions of the reticle by the Canon FPA-1550 stepper used throughout this programme.

The first phase of testing of these assemblies was dedicated to demonstrating basic physical and electrical design issues of this kind of silicon structure. Subsequently, functional testing was carried out. All testing was completed using a wafer-probe interface, which, in combination with the relatively slow test equipment available, meant that only low-speed testing could be attempted, and no ac parametric testing was possible.

A summary of these test results is shown in Table I. They demonstrated the effectiveness of the bypassing strategy, and validated the block isolation. Especially significant was the demonstration of the active power isolation, designed to regulate power and isolate blocks with catastrophic short-circuits.

The outcome of these tests was to identify working processors, which could mapped onto the wafer topology to identify gross loss and long range clustering mechanisms affecting multiple wafers, as illustrated by the sample wafer map shown in Figure 5.

2.1.2 WASP 2

The WASP 2A and 2B technology demonstrators comprise a composition of three VLSI-sized blocks, shown schematically in Figure 6 and described below

1. ASP : Identical to the proven ASP block on WASP 1, the 44,000-transistor design implements an array of 36 APEs, plus local control logic and an ASP substring bus isolator. There are nine 4-APE segments, any or all of which may be bypassed under the control of local test and reconfiguration logic. The size of this block is non-optimal from a yield point-of-view, as illustrated in Table II,

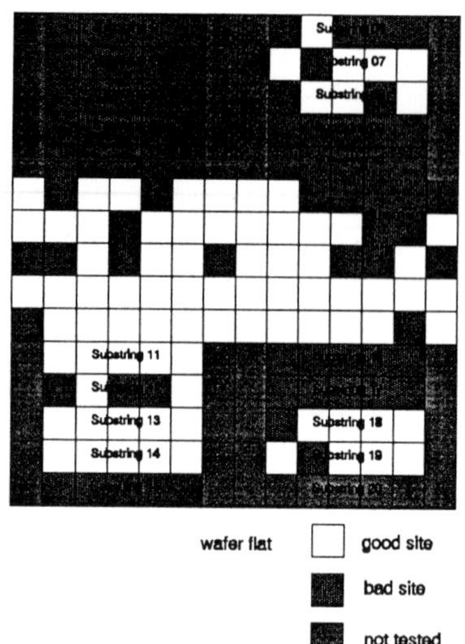

Figure 5 Example distributions of recovered blocks on a WASP 1 wafer

which shows that the overhead of non-fault-tolerant logic is large compared to the fault-tolerant processor array, thus largely negating the benefit of the fault-tolerant bypassing within the processor array.

Whilst not ideal, this was imposed by the maximum size of the shuttered, quarter-reticle mask described in Section 2.1.1 above. Extensive yield studies [8] have shown that a more optimal size for ASP blocks is a 256-APE array, with 16-APE segments, hierarchically bypassed in steps of 16, 64 & 256-APEs.

Each ASP block in an ASP substring (i.e. a WASP 'row') is linked via a 132-bit substring bus, fabricated in planar metal 1 with a line pitch of 8-microns. Each block may be electronically isolated from the ASP substring busses and from the substring power network via the power isolation transistor in the ground rail.

2. DR : An example of a Data Router (DR) block implementing a Data Interface (DI) to each substring, via an ASP Data Buffer (ADB) block, which comprises a double-buffered RAM local memory. ADBs may be simply interconnected via an external data bus. Each ADB module may be isolated from the ASP substring busses and from substring power.

3. CR : Multiple working or partial-working CR blocks are configured by an I/O multiplexer to establish a single global Control Interface (CI). Each CR block handles 8-bits from each of the three external 32-bit system busses, therefore a minimum of four working DR blocks are necessary to multiplex & assemble high-integrity versions of these

Table II ASP block statistics

area	6.5x6.5mm	
device count	44,000 transistors	
clock	16MHz	
area	routing	25%
	non-fault-tolerant logic (overhead)	30%
	fault-tolerant logic	30%
	unused	15%

external system busses onto the internal ASP global ('trunk') bus, which is a 116-bit bus fabricated in planar metal 2, on an 8-micron pitch. This bus spans the CR and links vertically with its neighbouring blocks. This bus is then buffered and repeated onto the horizontal ASP substring busses, via a driver/isolator network, which may isolate and bypass the entire substring in case of a catastrophic failure.

This block also controls the communications links (LKL & LKR) between adjacent rows, and supports link bypassing in case of substring failure.

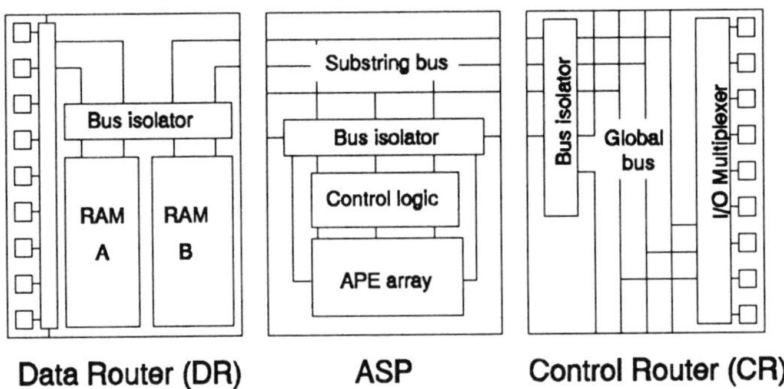

Figure 6 Schematics of the component blocks of WASP2

2.2 WASP 2 STATUS

Two WASP2 variants have been assembled from the module types described above, and these are summarised in Table III and described below.

Table III characteristics of the WASP 2A & 2B devices

	WASP 2A	WASP 2B
#.APEs	864	6480
area	3.9cm x 3.9cm	9.1cm x 9.8cm
transistors	1.26M	8.43M
power	5.2 - 8.0W	29.6-51.2W
external clock	6MHz	6MHz
internal clock	12MHz	12MHz

2.2.1 WASP 2A

The 864-APE WASP 2A integrates 6 ASP substring rows (each comprising a DR block, 4 ASP blocks and a CR block) on a 3.90cm x 3.90cm 1.26M transistor (monolithic) WSI device. The device photograph is shown in Figure 7, where the individual blocks are clearly visible.

Since no suitable packaging was available for testing this experimental module, a prototype ceramic substrate was constructed, on which the wafer was mounted and wire-bonded.

Testing was intended to demonstrate the following objectives

- Achieve a realistic, high-bandwidth wafer interface, with electronic configuration and defect-tolerance
- Demonstrate inter-substring defect-tolerance bypassing (i.e. one hierarchical

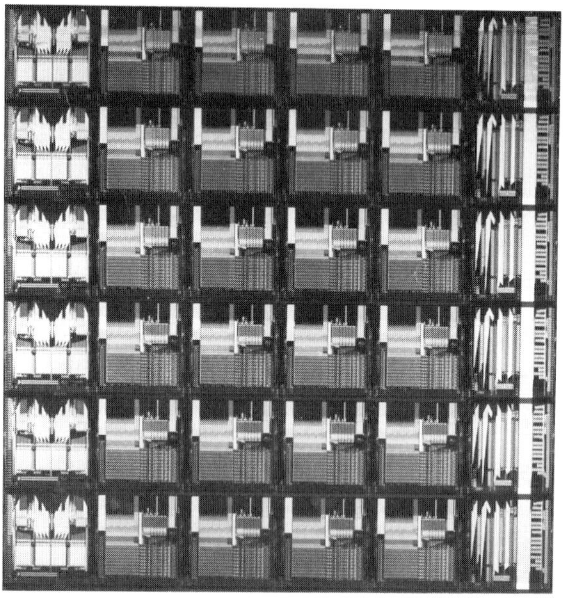

Figure 7 WASP 2A photograph

level above the ASP module bypassing demonstrated in WASP 1).

- Demonstrate on-chip ASP Data Memory.

Initial testing concentrated on the new blocks associated with this composition, since the ASP block was a verified design, whose successful testing has been outlined above.

This testing has successfully demonstrated the correct operation of the realistic, high-bandwidth wafer interface, complete with electronic configuration and defect-tolerance. Similarly, it has proved possible to access the on-wafer DR block host-accessible resources. Unfortunately, a minor design error has precluded successful demonstration of the inter-substring fault-tolerance and communication.

These results are summarised in Table IV.

Of particular significance in this study was the opportunity to undertake tests on the quality of large area bus structures. In particular, the vulnerability of the ASP global ('trunk') bus was critical, since this forms the 'backbone' highway on the assembled device, and a fault here is catastrophic in the existing design.

Of a sample of 34 global interconnect busses ('trunk' busses), 28 proved defect free, representing an upper limit wafer yield of ~80% allowing for no bus defect-tolerance. Since this bus represents the most critical large-area structure present in the device architecture, this is a key parameter. The defects observed

Table IV WASP 2A test results summary

inter-substring defect-tolerance	communications	no
	bypassing	no
wafer-interface	defect-tolerant wafer interface	yes
functional	on-wafer buffer memory	yes
	processor-memory data exchange	no

are summarised in Table V. It is recognised that this represents a small sample from a single batch of devices, however similar problems have been widely reported [7], therefore suggesting that future WASP devices should employ a degree of defect-tolerance in this bus structure, based on the expectation of multiple (i.e. at least line pair) short circuits.

Finally, further microscopic examination of these buses and the substring buses revealed that the bus traces, where they cross block boundaries (i.e. marking the stepper increment and the region of overlapped exposure), suffer very little 'thinning' or line mismatch. Alignment/thinning errors were all judged to be < 0.5-microns, with no special handling or wafer-stepper procedures employed.

2.2.2 WASP 2B

Using the same DSW reticles, the 6,480-APE WASP 2B extends this design to 15 ASP substring rows, each comprising 12 ASP blocks, on a 9.10cm x 9.75cm 8.43M transistor WSI device.

The WASP 2B composition is representative of a whole wafer WASP device, and comprises an array of 180 ASP blocks, on a 6 inch wafer. This device will test some of the fundamental issues of whole wafer monolithic integration, most especially the validation of full-wafer signal and power distribution, as well as providing statistics on full-wafer defect distribution.

Table V WASP 2 global (trunk) bus yield observations

Defect type	Frequency	Comment
2 lines shorted	3	no visible contamination
3 lines shorted	1	contaminant particle
10 lines shorted	1	unetched region
>50 lines shorted	1	large area of random broken and shorted lines

3 CONCLUSIONS

As (monolithic) WSI technology demonstrators, WASP 1 and 2 have been largely successful. However, as functional WASP prototypes, WASP 1 and 2 leave much to be desired, since they provide only a partial implementation of ASP substrings in representative, rather than realistic 'chip' blocks, suitable for proof-of-principle testing rather than development of cost-effective ASP modules.

Although WASP 1 & 2 are not directly representative of the potential Silicon-efficiency of the WASP concept, detailed yield studies [8] have shown that optimum Silicon-efficiency can be achieved with a WASP device comprising ASP blocks with 256-APEs, configured, for fault-tolerance purposes, into 16-APE segments. Hierarchical bypassing at 16 and 64-APE increments maximises the efficiency of the defect-tolerance to withstand cluster defects. Such a component, designed in 1-micron technology, based on an array of 48 ASP blocks, should be capable of yielding in excess of 8K usable APEs in a 58mm x 58mm area.

Meanwhile, this project has served to demonstrate the key manufacturing, electrical and physical design issues, using only standard CMOS process technology. However, the wafer exposure cycle, involving non-standard alignment marks and stepper control, is unusual. Such procedures clearly demand the full knowledge and cooperation of the foundry, and have been shown to require at least three times the operator involvement of a conventional wafer process.

Finally, guidelines have been established for employing appropriate defect-tolerance measures in the design of critical 'top-level' bus structures.

4 REFERENCES

1 I.P.Jalowiecki and R.M.Lea, A 256-element Associative Parallel Processor, IEEE International Solid State Circuits Conference Digest, pp. 196-198, 1987.

2 I.P.Jalowiecki, K.D.Warren and R.M.Lea, WASP: A WSI Associative String Processor, Proc. IEEE Int. Conf. on Wafer Scale Integration (eds. Swartzlander and Brewer), IEEE Computer Society Press, pp. 83-93, 1989.

3 R.M.Lea, The ASP: a cost-effective parallel microcomputer, IEEE Micro, pp.10-29, October 1989

4 S.K.Tewksbury, Wafer-level Integrated Systems: Implementation Issues, Klewer Academic Publishers, 1989.

5 R.McKirdy, An assessment of the benefits of monolithic WSI technology, PhD thesis, Brunel University, 1989.

6 M.J.Davies, A.G.Munns and D.J.Pedder, Silicon Substrate Test Structures for Hybrid Wafer Scale Technology, Proceedings of the 1991 International Conference on Wafer Scale Integration, IEEE Computer Society Press, pp 185-192, 1991.

7 Stephen W.Director, Wojciech Maly and Andrzej J.Strjwas, VLSI Design for Manufacturing: Yield Enhancement, Kluwer Academic Publishers 1990.

8 H.Boulouri and R.M.Lea, Yield Considerations in the Design of WASP 3, Proceedings of the 1991 International Conference on Wafer Scale Integration, IEEE Computer Society Press, pp 170-178, 1991.

IFIP

The INTERNATIONAL FEDERATION FOR INFORMATION PROCESSING is a multinational federation of professional and technical organisations (or national groupings of such organisations) concerned with information processing. From any one country, only one such organisation – which must be representative of the national activities in the field of information processing – can be admitted as a Full Member. In addition a regional group of developing countries can be admitted as a Full Member. On 1 October 1991, 46 organisations were Full Members of the Federation, representing 70 countries.

The aims of IFIP are to promote information science and technology by:

- fostering international co-operation in the field of information processing;
- stimulating research, development and the application of information processing in science and human activity;
- furthering the dissemination and exchange of information about the subject;
- encouraging education in information processing.

IFIP is dedicated to improving worldwide communication and increased understanding among practitioners of all nations about the role information processing can play in all walks of life.

Information technology is a potent instrument in today's world, affecting people in everything from their education and work to their leisure and in their homes. It is a powerful tool in science and engineering, in commerce and industry, in education and adminstration. It is truly international in its scope and offers a significant opportunity for developing countries. IFIP helps to bring together workers at the leading edge of the technology to share their knowledge and experience, and acts as a catalyst to advance the state of the art.

IFIP came into official existence in January, 1960. It was established to meet a need identified at the first International Conference on Information Processing which was held in Paris in June, 1959, under the sponsorship of UNESCO.

Organisational Structure

The Federation is governed by a GENERAL ASSEMBLY, which meets once every year and consists of one representative from each Member organisation. The General Assembly decides on all important matters, such as general policy, the programme of activities, admissions, elections and budget.

IFIP ORGANISATION CHART

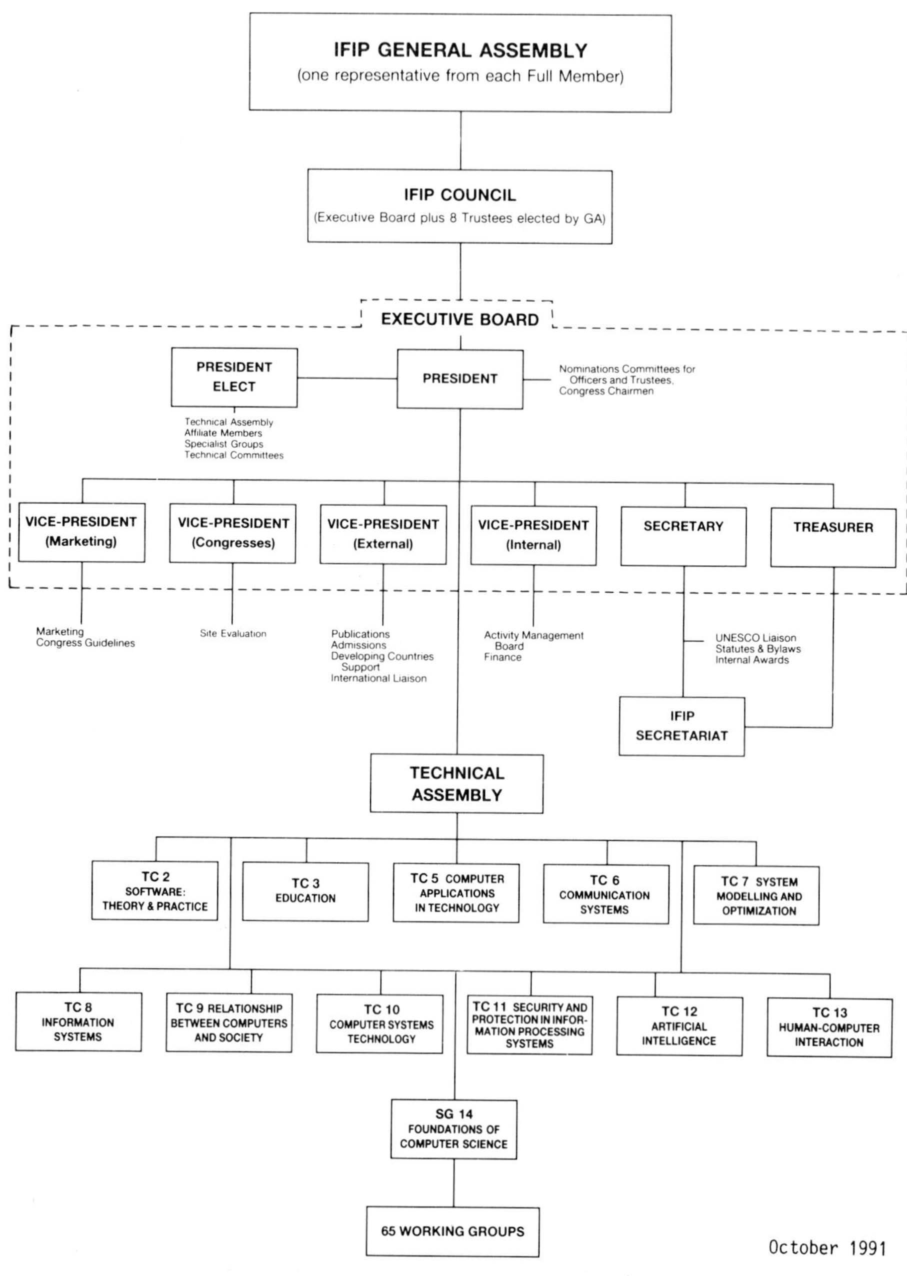

October 1991

The day-to-day work of IFIP is directed by its Officers: the President, Vice-Presidents, Secretary and Treasurer, who are elected by the General Assembly and together constitute the EXECUTIVE BOARD.

The COUNCIL, consisting of the Officers and up to eight Trustees elected from the General Assembly, meets twice a year and takes decisions which become necessary between General Assembly meetings.

The headquarters of the Federation are in Geneva, Switzerland where the IFIP Secretariat administers its affairs.

For further information please contact:

IFIP Secretariat
attn. Mme. GWYNETH ROBERTS
16 Place Longemalle
CH-1204 Geneva, Switzerland
telephone: 41 (22) 28 26 49
facsimile: 41 (22) 781 23 22
Bitnet: ifip@cgeuge51

IFIP's MISSION STATEMENT

IFIP's mission is to be the leading, truly international, apolitical organisation which encourages and assists in the development, exploitation and application of Information Technology for the benefit of all people.

Principal Elements

1. To stimulate, encourage and participate in research, development and application of Information Technology (IT) and to foster international co-operation in these activities.
2. To provide a meeting place where national IT Societies can discuss and plan courses of action on issues in our field which are of international significance and thereby to forge increasingly strong links between them and with IFIP.
3. To promote internation co-operation directly and through national IT Societies in a free environment between individuals, national and international governmental bodies and kindred scientific and professional organisations.
4. To pay special attention to the needs of developing countries and to assist them in appropriate ways to secure the optimum benefit from the application of IT.
5. To promote professionalism, incorporating high standards of ethics and conduct, among all IT practitioners.
6. To provide a forum for assessing the social consequences of IT applications; to campaign for the safe and beneficial development and use of IT and the protection of people from abuse through its improper application.

7. To foster and facilitate co-operation between academics, the IT industry and governmental bodies and to seek to represent the interest of users.
8. To provide a vehicle for work on the international aspects of IT development and application including the necessary preparatory work for the generation of international standards.
9. To contribute to the formulation of the education and training needed by IT practitioners, users and the public at large.

Note to Conference Organizers

Organizers of upcoming IFIP Working Conferences are urged to contact the Publisher. Please send full details of the Conference to:

Mrs. STEPHANIE SMIT
Administrative Editor – IFIP Publications
ELSEVIER SCIENCE PUBLISHERS
P.O. Box 103, 1000 AC Amsterdam
The Netherlands
telephone: 31 (20) 5862481
facsimile: 31 (20) 5862616
email: s.smit@elsevier.nl

IFIP TRANSACTIONS

IFIP TRANSACTIONS is a serial consisting of 15,000 pages of valuable scientific information from leading researchers, published in 35 volumes per year. The serial includes contributed volumes, proceedings of the IFIP World Conferences, and conferences at Technical Committee and Working Group level. Mainstream areas in the IFIP TRANSACTIONS can be found in Computer Science and Technology, Computer Applications in Technology, and Communication Systems.

Please find below a detailed list of topics covered.

IFIP TRANSACTIONS A:
Computer Science and Technology
1992: 19 Volumes, US $ 1181.00/Dfl. 2185.00
ISSN 0926-5473

IFIP Technical Committees
Software: Theory and Practice (TC2)
Education (TC3)
System Modelling and Optimization (TC7)
Information Systems (TC8)
Relationship Between Computers and Society (TC9)

Computer Systems Technology (TC10)
Security and Protection in Information Processing Systems (TC11)
Artificial Intelligence (TC12)
Human-Computer Interaction (TC13)
Foundations of Computer Science (SG14)

IFIP TRANSACTIONS B:
Applications in Technology
1992: 8 Volumes. US $ 497.00/Dfl. 920.00
ISSN 0926-5481

IFIP Technical Committee
Computer Applications in Technology (TC5)

IFIP TRANSACTIONS C:
Communication Systems
1992: 8 Volumes. US 497.00/Dfl. 920.00
ISSN 0926-549X

IFIP Technical Committee
Communication Systems (TC6)

IFIP TRANSACTIONS FULL SET: A, B & C
1992: 35 Volumes. US $ 1892.00/Dfl. 3500.00

The Dutch Guilder prices (Dfl.) are definitive. The US $ prices mentioned above are for your guidance only and are subject to exchange rate fluctuations. Prices include postage and handling charges.

The volumes are also available separately in book form.

Please address all orders and correspondence to:

ELSEVIER SCIENCE PUBLISHERS
attn. PETRA VAN DER MEER
P.O. Box 103, 1000 AC Amsterdam
The Netherlands
telephone: 31 (20) 5862602
facsimile: 31 (20) 5862616
email: m.haccou@elsevier.nl